Brief Contents

Contents

CHAPTER 6   COGNITION   120

CHAPTER 7   **DECISION MAKING**   **156**

CHAPTER 8   **DISPLAYS**   **184**

CHAPTER 9   CONTROL   218

CHAPTER 16   AUTOMATION   418

CHAPTER 17   TRANSPORTATION HUMAN FACTORS   436

Preface

We wrote this book because we saw a need for engineers and system designers and other professionals to understand how knowledge of human strengths and limitations, both mental and physical, can lead to better system design, more effective training of the user, and better assessment of the usability of a system. The knowledge and methods to accomplish these goals are embodied in the study of human factors engineering. As we point out in the early chapters, a *cost-benefit analysis* of human factors applications in system design usually provides a favorable evaluation of those applications.

Our intention in this book is to focus on the clear and intuitive explanation of human factors *principles*. We illustrate these principles with real-world design examples and, where relevent, show how these principles are based on understanding of the human's psychological, biological, and physical characteristics to give the reader an understanding of why the principles are formulated. Because of our focus on principles, we intentionally do not spend a great deal of time addressing psychological theory or research paradigms and experiments. We trust that the reader will know that the principles we describe are indeed based on valid research conclusions, and where relevent we provide citations as to where that research can be examined.

Also, we do not expect that this will be a stand-alone reference manual for applying human factors in design. Many specific numbers, values, and formulae, necessary for fabricating systems with human limitations in mind, were not included in this text in the interest of space. However, we point to ample references where designers can proceed to find these details.

Because of the way we have structured the book, emphasizing design principles and methodologies over theory and research, our primary target audience is the engineering undergraduate, who may well be participating in the design process. Hence we do not assume that the reader will necessarily have had an introductory course in psychology, and so we try to present some of the necessary psychological fundamentals. We also believe, however, that the book will be useful for applied psychology or undergraduate-level engineering psychology courses within a psychology department. This usefulness derives in part, because the book demonstrates how many aspects of psychological science are relevant to the effective design of systems in the workplace and on the highway.

Human factors is a growing field. In many small industries, personnel are assigned to the position of human factors engineer who have no formal training in the discipline. Thus we hope that the book will not only reach the academic classroom in both engineering colleges and psychology departments but will also be available as a reference for personnel and managers in the workplace.

We believe that the strengths of this book lie in its relatively intuitive and readable style, which attempts to illustrate principles clearly, with examples, and without excessive detail and which points to references where more information can be obtained. We have also tried to strike a balance between presenting the human factors associated with different aspects of human performance on the one hand (e.g., physical limitations, display processing, memory failures) and particularly important domains of current applications on the other. For example, there are separate chapters devoted to the human factors of transportation systems and of human computer interaction.

In the second edition, we have not made fundamental changes to content or organization. Professor John Lee of the University of Iowa Industrial Engineering Department has been added as a co-author. He is an expert in automation and highway safety research. In addition to addressing some of the shortcomings of the previous edition, revealed by its users, we have included new sections on a variety of topics such as driver distraction, organizational aspects of human error, human factors applications to law enforcement, meta cognition, and task management. We have also increased the amount of cross referencing between chapters, to highlight the extent to which human factors is an integrated science. A single integrated reference list is compiled at the end of the chapter.

ACKNOWLEDGMENTS

Several people have contributed to this endeavor. The following individuals provided valuable comments on earlier drafts of the work: Terence Andre, United States Air Force Academy, Mark Detweiler, Pennsylvania State University, John Casali, Virginia Polytechnical Institute; Joseph Meloy, Milwaukee School of Engineering; Edward Rinalducci, University of Central Florida; Joseph Goldberg, Pennsylvania State University; Philip Allen, Cleveland State University; and David Carkenord, Longwood College. Peter Hancock, University of Central Florida; Bernard Martin, University of Michigan; and Wendy Rogers, Georgia Institute of Technology provided reviews for which the authors are appreciative. Sallie Gordon wishes to thank Justin Hollands for his helpful suggestions, Nicki Jo Rich for her review/editing work, and Greg, Erin, and Shannon for their patience. The authors also gratefully acknowledge the great assistance of Margaret Dornfeld in the technical editing of the book, as well as the invaluable contributions of Mary Welborn for much of the administrative and technical assistance in preparation of both the first and second editions.

Christopher D. Wickens
John Lee
Yili Liu
Sallie Gordon Becker

Chapter 1

Introduction to Human Factors

In a midwestern factory, an assembly-line worker had to reach to an awkward location and position a heavy component for assembly. Toward the end of a shift, after grabbing the component, he felt a twinge of pain in his lower back. A trip to the doctor revealed that the worker had suffered a ruptured disc, and he missed several days of work. He filed a lawsuit against the company for requiring physical action that endangered the lower back.

Examining a bottle of prescription medicine, an elderly woman was unable to read the tiny print of the dosage instructions or even the red-printed safety warning beneath it. Ironically, a second difficulty prevented her from potentially encountering harm caused by the first difficulty. She was unable to exert the combination of fine motor coordination and strength necessary to remove the "childproof" cap.

In a hurry to get a phone message to a business, an unfortunate customer found herself "talking" to an uncooperative automated voice response system. After impatiently listering to a long menu of options, she accidentally pressed the number of the wrong option and now has no clue as to how to get back to the option she wanted, other than to hang up and repeat the lengthy process.

WHAT IS THE FIELD OF HUMAN FACTORS?

While the three episodes described in the introduction are generic in nature and repeated in many forms across the world, a fourth, which occurred in the Persian Gulf in 1987, was quite specific. The USS *Vincennes*, a U.S. Navy cruiser, was on patrol in the volatile, conflict-ridden Persian Gulf when it received ambiguous information regarding an approaching aircraft. Characteristics of the radar system displays on board made it difficult for the crew to determine whether it was climbing or descending. Incorrectly diagnosing that the aircraft was descending, the crew tentatively identified it as a hostile approaching fighter. A combination of the short time to act in potentially life-threatening circumstances, further breakdowns in communication between people (both onboard the ship and from the aircraft), and crew expectancies that were driven by the hostile environment conspired to produce the captain's decision to fire at the approaching aircraft. Tragically, the aircraft was actually an Iranian passenger airline, which had been climbing rather than descending.

These four episodes illustrate the role of *human factors*. In these cases human factors are graphically illustrated by breakdowns in the interactions between humans and the systems with which they work. It is more often the case that the interaction between the human and the system work well, often exceedingly so. However, it is characteristic of human nature that we notice when things go wrong more readily than when things go right. Furthermore, it is the situation when things go wrong that triggers the call for diagnosis and solution, and understanding these situations represents the key contributions of human factors to system design.

We may define the **goal** of human factors as making the human interaction with systems one that

- Enhances performance.
- Increases safety.
- Increases user satisfaction.

Human factors involves the **study** of factors and **development** of tools that facilitate the achievement of these goals. We will see in chapters 3 and 15 how the goals of productivity and error reduction are translated into the concept of **usability,** which is often applied to the design of computer systems.

In considering these goals, it is useful to realize that there may be tradeoffs between them. For example, *performance* is an all-encompassing term that may involve the reduction of errors or an increase in productivity (i.e., the speed of production). Hence, enhanced productivity may sometimes cause more operator errors, potentially compromising safety. As another example, some companies may decide to cut corners on time-consuming safety procedures in order to meet productivity goals. Fortunately, however, these tradeoffs are not inevitable. Human factors interventions often can satisfy both goals at once (Hendrick, 1996; Alexander, 2002). For example, one company that improved its workstation design reduced worker's compensation losses in the first year after the improvement from $400,000 to $94,000 (Hendrick, 1996). Workers were more able to continue work (increasing productivity), while greatly reducing the risk of injury (increasing safety).

In the most general sense, the three goals of human factors are accomplished through several procedures in the human factors cycle, illustrated in Figure 1.1, which depicts the human operator (brain and body) and the system with which he or she is interacting. At point A, it is necessary to *diagnose* or identify the problems and deficiencies in the human–system interaction of an existing system. To do this effectively, core knowledge of the nature of the physical body (its size, shape, and strength) and of the mind (its information-processing

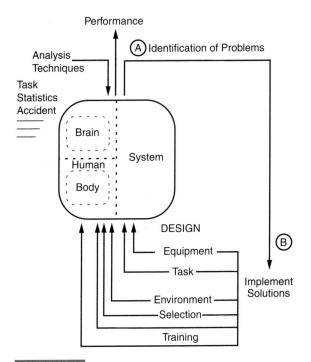

FIGURE 1.1

The cycle of human factors. Point A identifies a cycle when human factors solutions are sought because a problem (e.g., accident or incident) has been observed in the human–system interaction. Point B identifies a point where good human factors are applied at the beginning of a design cycle.

characteristics and limitations) must be coupled with a good understanding of the physical or information systems involved, and the appropriate *analysis* tools must be applied to clearly define the cause of breakdowns. For example, why did the worker in our first story suffer the back injury? Was it the amount of the load or the awkward position required to lift it? Was this worker representative of others who also might suffer injury? Task analysis, statistical analysis, and incident/accident analysis are critical tools for gaining such an understanding, and each is discussed in later chapters. Having identified the problem, the five different approaches shown at point B may be directed toward implementing a solution (Booher, 1990, 2003), as shown at the bottom of the figure.

Equipment design changes the nature of the physical equipment with which humans work. The medicine bottle in our example could be given a more readable label and an easier-to-open top. The radar display on the USS *Vincennes* might be redesigned to provide a more integrated representation of lateral and vertical motion of the aircraft.

Task design focuses more on changing what operators do than on changing the devices they use. The workstation for the assembly-line worker might be redesigned to eliminate manual lifting. Task design may involve assigning part or all of tasks to other workers or to *automated* components. For example, a robot might be designed to accomplish the lift of the component. Of course, automation is not always the answer, as illustrated by the example of the automated voice response system.

Environmental design implements changes, such as improved lighting, temperature control, and reduced noise in the physical environment where the task is carried out. A broader view of the environment could also include the organizational climate within which the work is performed. This might, for example, represent a change in management structure to allow workers more participation in implementing safety programs or other changes in the organization.

Training focuses on better preparing the worker for the conditions that he or she will encounter in the job environment by teaching and practicing the necessary physical or mental skills.

Selection is a technique that recognizes the individual differences across humans in almost every physical and mental dimension that is relevant for good system performance. Such performance can be optimized by selecting operators who possess the best profile of characteristics for the job. For example, the lower-back injury in our leading scenario might have been caused by asking a worker who had neither the necessary physical strength nor the body proportion to lift the component in a safe manner. The accident could have been prevented with a more stringent operator-selection process.

As we see in the figure, any and all of these approaches can be applied to "fix" the problems, and performance can be measured again to ensure that the fix was successful. Our discussion has focused on *fixing* systems that are deficient, that is, intervening at point A in Figure 1.1. In fact, the practice of good human factors is just as relevant to *designing* systems that are effective at the start and thereby anticipating and avoiding the human factors deficiencies before they are inflicted on system design. Thus, the role of human factors in the design loop can just as easily enter at point B as at point A. If consideration for good human factors is given early in the design process, considerable savings in both money and possibly human suffering can be achieved (Booher, 1990; Hendrick, 1996). For example, early attention given to workstation design by the company in our first example could have saved the several thousand dollars in legal costs resulting from the worker's lawsuit. Alexander (2002) has estimated that the percentage cost to an organization of incorporating human factors in design grows from 2 percent of the total product cost when human factors is addressed at the earliest stages (and incidents like workplace accidents are prevented) to between 5 percent and 20 percent when human factors is addressed only in response to those accidents, after a product is fully within the manufacturing stage. In Chapter 3 we talk in greater detail about the role of human factors in the design process.

The Scope of Human Factors

While the field of human factors originally grew out of a fairly narrow concern for human interaction with physical devices (usually military or industrial), its scope has broadened greatly during the last few decades. Membership in the primary North American professional organization of the *Human Factors and Ergonomics Society* has grown to 5,000, while in Europe the *Ergonomics Society* has realized a corresponding growth. A survey indicates that these membership numbers may greatly underestimate the number of people in the workplace who actually consider themselves as doing human factors work (Williges, 1992).

This growth plus the fact that the practice of human factors is *goal-oriented* rather than content-oriented means that the precise boundaries of the discipline of human factors cannot be tightly defined. One way of understanding what human factors professionals do is illustrated in Figure 1.2. Across the top of the matrix is an (incomplete) list of the major categories of *systems* that define the environments or contexts within which the human operates. On the left are those system environments in which the focus is the individual operator. Major categories include the industrial environment (e.g. manufacturing, nuclear power, chemical processes); the computer or information environment; healthcare; consumer products (e.g., watches, cameras, and VCRs); and transportation. On the right are those environments that focus on the interaction between

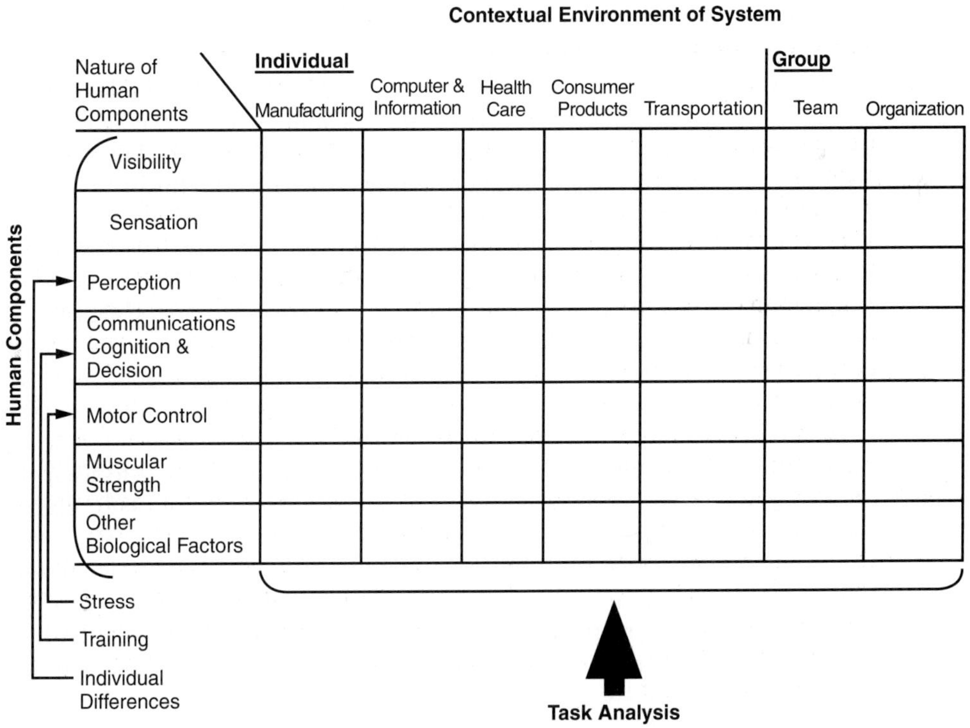

Contextual Environment of System

Nature of Human Components	Individual					Group	
	Manufacturing	Computer & Information	Health Care	Consumer Products	Transportation	Team	Organization
Visibility							
Sensation							
Perception							
Communications Cognition & Decision							
Motor Control							
Muscular Strength							
Other Biological Factors							

Human Components — Stress — Training — Individual Differences

Task Analysis

FIGURE 1.2

This matrix of human factors topics depicts human performance issues against contextual environments within which human factors may be applied. The study of human factors may legitimately belong within any cell or combination of cells in the matrix.

two or more individuals. A distinction can be made between the focus on *teams* involved in a cooperative project and *organizations,* a focus that involves a wider concern with management structure.

Figure 1.2 lists various components of the human user that are called on by the system in question. Is the information necessary to perform the task visible? Can it be sensed and adequately perceived? These components were inadequate for the elderly woman in the second example. What communications and cognitive processes are involved in understanding the information and deciding what to do with it? Decisions on the USS *Vincennes* suffered because personnel did not correctly understand the situation due to ambiguous communications. How are actions to be carried out, and what are the physical and muscular demands of those actions? This, of course, was the cause of the assembly-line worker's back injury. What is the role of other biological factors related to things like illness and fatigue? As shown at the far left of the figure, all of these processes may be influenced by *stresses* imposed on the human operator, by *training,* and by the *individual differences* in component skill and strength.

Thus, any given task environment listed across the top of the matrix may rely upon some subset of human components listed down the side. A critical role of *task analysis* that we discuss in Chapter 3 is to identify the mapping from tasks to human components and thereby to define the scope of human factors for any particular application.

A second way of looking at the scope of human factors is to consider the relationship of the discipline with other related domains of science and engineering. This is shown in Figure 1.3. Items within the figure are placed close to other items to which they are related. The core discipline of human factors is shown at the center of the circle, and immediately surrounding it are various subdomains of study within human factors; these are boldfaced. Surrounding these are disciplines within the study of psychology (on the top) and engineering (toward the bottom) that intersect with human factors. At the bottom of the figure are *domain-specific* engineering disciplines, each of which focuses on a particular kind of system that itself has human factors components. Finally, outside of the circle are other disciplines that also overlap with some aspects of human factors.

Closely related to human factors are ergonomics, engineering psychology, and cognitive engineering. Historically, the study of *ergonomics* has focused on the aspect of human factors related to physical work (Grandjean, 1988): lifting, reaching, stress, and fatigue. This discipline is often closely related to aspects of human physiology, hence its closeness to the study of biological psychology and bioengineering. Ergonomics has also been the preferred label in Europe to describe all aspects of human factors. However, in practice the domains of human factors and ergonomics have been sufficiently blended on both sides of the Atlantic so that the distinction is often not maintained.

Engineering psychology is a discipline within psychology, whereas the study of human factors is a discipline within engineering. The distinction is clear: The ultimate goal of the study of human factors is toward system design, accounting for those factors, psychological and physical, that are properties of the human

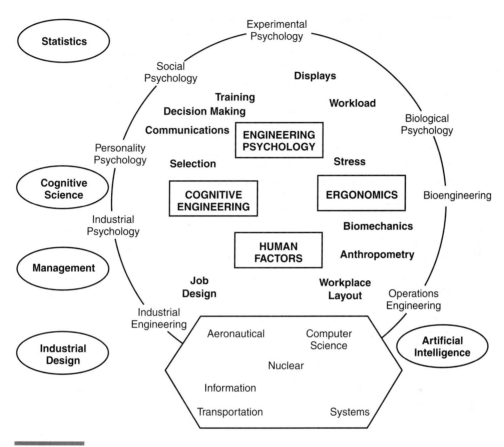

FIGURE 1.3

The relationship between human factors, shown at the center, and other related disciplines of study. Those more closely related to psychology are shown at the top, and those related to engineering are shown toward the bottom.

component. In contrast, the ultimate goal of engineering psychology is to understand the human mind *as is relevant to the design of systems* (Wickens & Hollands, 2000). In that sense, engineering psychology places greater emphasis on discovering generalizable psychological principles and theory, while human factors places greater emphasis on developing usable design principles. But this distinction is certainly not a hard and fast one.

   Cognitive engineering, also closely related to human factors, is slightly more complex in its definition (Rasmussen et al., 1995; Vicente, 1999) and cannot as easily be placed at a single region of Figure 1.3. In essence, it focuses on the complex, *cognitive* thinking and knowledge-related aspects of system performance, whether carried out by human or by machine agents, the latter dealing closely with elements of artificial intelligence and cognitive science.

The Study of Human Factors as a Science

Characteristics of human factors as a science (Meister, 1989) relate to the search for *generalization* and *prediction*. In the problem diagnosis phase (Figure 1.1) investigators wish to *generalize* across classes of problems that may have common elements. As an example, the problems of communications between an air traffic control center and the aircraft may have the same elements as the communications problems between workers on a noisy factory floor or between doctors and nurses in an emergency room, thus enabling similar solutions to be applied to all three cases. Such generalization is more effective when it is based on a deep understanding of the physical and mental components of the human operator. It also is important to be able to *predict* that solutions designed to create good human factors will actually succeed when put into practice.

A critical element to achieving effective generalization and prediction is the nature of the *observation* or study of the human operator that we discuss in Chapter 2. Humans can be studied in a range of environments, which vary in the realism with which the environment simulates the relevant system, from the laboratory for highly controlled observations and experiments, to human behavior (normal behavior, incidents, and accidents) of real users of real systems. Researchers have learned that the most effective understanding, generalization, and prediction depend on the combination of observations along all levels of this continuum. Thus, for example, the human factors engineer may couple an analysis of the events that led up to the USS *Vincennes* tragedy with an understanding, based on laboratory research, of principles of communications (Chapter 6), decision making (Chapter 7), display integration (Chapter 8), and performance degradation under time stress (Chapter 13) to gain a full appreciation of the causes of the *Vincennes'* incident and suggestions for remediation.

OVERVIEW OF THE BOOK

The following chapters are divided into four basic sections. In Chapters 2 and 3 we describe different research techniques and design methodologies. The second section addresses the nature of human information processing: visual and auditory systems (Chapters 4 and 5), perception and cognition (Chapter 6), decision making (Chapter 7), display processing (Chapter 8), and control (Chapter 9). The third section addresses many of the nonpsychological issues of human factors: workspace layout (Chapter 10), strength (Chapter 11), physiology (Chapter 12), stress (Chapter 13), and safety issues (Chapter 14). Chapters on safety and stress address psychological issues as well. The final section focuses on specific domains of application of human factors study: automation (Chapter 15), human–computer interaction (Chapter 16), transportation (Chapter 17), training and selection (Chapter 18), and group and organizational behavior (Chapter 19).

Several fine books cover similar and related material: Sanders and McCormick (1993), Bailey (1996), and Proctor and Van Zandt (1994) offer comprehensive coverage of human factors. Norman (1988) examines human factors manifestations in the kinds of consumer systems that most of us encounter every day, and Meister (1989) addresses the science of human factors. Wickens and Hollands (2000) provide coverage of engineering psychology, foregoing treatment of those human components that are not related to psychology (e.g., visibility, reach, and strength). In complementary fashion, Wilson and Corlett (1991), Chaffin, Andersson, and Martin (1999), and Kroemer and Grandjean (1997) focus more on the physical aspects of human factors (i.e., classical "ergonomics"). Finally, a comprehensive treatment of nearly all aspects of human factors can be found in Salvendy's (1997) *Handbook of Human Factors and Ergonomics,* and issues of system integration can be found in Booher (2003).

Several journals address human factors issues, but probably the most important are *Ergonomics,* published by the International Ergonomics Society, and *Theoretical Issues in Ergonomics Sciences,* both published in the United Kingdom, and three publications offered by the Human Factors and Ergonomics Society in the United States: *Human Factors, Ergonomics in Design,* and the annual publication of the *Proceedings of the Annual Meeting of the Human Factors and Ergonomics Society.*

Chapter

Research Methods

2

A state legislator suffered an automobile accident when another driver ran a stop sign while talking on a cellular phone. The resulting concern about cell phones and driving safety led the legislator to introduce legislation banning the use of cellular phones while a vehicle is in motion. But others challenged whether the one individual's experience could justify a ban on all others' cellular phone use. After all, a single personal experience does not necessarily generalize to all, or even most others. To resolve this debate, a human factors company was contracted to provide the evidence regarding whether or not use of cellular phones compromises driver safety.

Where and how should that evidence be obtained? The company might consult accident statistics and police reports, which could reveal that cell phone use was no more prevalent in accidents than it was found to be prevalent in a survey of drivers asked to report how frequently they talked on their cellular phone. But how reliable and accurate is this evidence? Not every accident report may have a place for the officer to note whether a cellular phone was or was not in use; and those drivers filling out the survey may not have been entirely truthful about how often they use their phone while driving. The company might also perform its own research in an expensive driving simulator, comparing driving performance of people while the cellular phone was and was not in use. But how much do the conditions of the simulator replicate those on the highway? On the highway, people choose when they want to talk on the phone. In the simulator, people are asked to talk at specific times. The company might also rely on more basic laboratory research that characterizes the degree of dual task interference between conversing and carrying out a "tracking task" like vehicle control, while detecting events that may represent pedestrians (Strayer & Johnston, 2001). But isn't a computer-driven tracking task unlike the conditions of real automobile driving?

The approaches to evidence-gathering described above represent a sample of a number of **research methods** that human factors researchers can employ to discover "the truth" (or something close to it) about the behavior of humans interacting with systems in the "real world." Research involves the scientific gathering of observations or data and the interpretation of the meaning of these data regarding the research questions involved. In human factors, such meaning is often expressed in terms like *what works? what is unsafe? which is better?* (in terms of criteria of speed, accuracy, and workload). It may also be expressed in terms of general principles or models of how humans function in the context of a variety of different systems. Because human factors involves the application of science to system design, it is considered by many to be an *applied science.* While the ultimate goal is to establish principles that reflect performance of people in real-world contexts, the underlying scientific principles are gained through research conducted in both laboratory and real-world environments.

Human factors researchers use standard methods for developing and testing scientific principles that have been developed over the years in traditional physical and social sciences. These methods range from the "true scientific experiment" conducted in highly controlled laboratory environments to less controlled but more realistic observational studies in the real world. Given this diversity of methods, a human factors researcher must be familiar with the range of research methods that are available and know which methods are best for specific types of research questions. It is equally important for researchers to understand how practitioners ultimately uses their findings. Ideally, this enables a researcher to direct his or her work in ways that are more likely to be useful to design, thus making the science applicable (Chapanis, 1991).

Knowledge of basic research methods is also necessary for human factors design work. That is, standard design methods are used during the first phases of product or system design. As alternative design solutions emerge, it is sometimes necessary to perform formal or informal studies to determine which design solutions are best for the current problem. At this point, designers must select and use appropriate research methods. Chapter 3 provides an overview of the more common design methods used in human factors and will refer you back to various research methods within the design context.

INTRODUCTION TO RESEARCH METHODS

As we noted in Chapter 1, comprehensive human factors research spans a variety of disciplines, from a good understanding of the mind and how the brain processes information to a good understanding of the physical and physiological limits of the body. But the human factors researcher must also understand how the brain and the body work in conjunction with other systems, whether these systems are physical and mechanical, like the handheld cellular phone, the shovel, or the airplane; or are informational, like the dialogue with a copilot, with a 911 emergency dispatch operator, or with an instructional display. Because of this, much of the scientific research in human factors cannot be as simple or "context-free" as more basic research in psychology, physics, and physiology, although many of the tenets of basic research remain relevant.

Basic and Applied Research

It should be apparent that scientific study relevant to human factors can range from basic to very applied research. *Basic research* can be defined as "the development of theory, principles, and findings that generalize over a wide range of people, tasks, and settings." An example would be a series of studies that tests the theory that as people practice a particular activity hundreds of times, it becomes automatic and no longer takes conscious, effortful cognitive processing. *Applied research* can be defined loosely as "the development of theory, principles, and findings that are relatively specific with respect to particular populations, tasks, products, systems, and/or environments." An example of applied research would be measuring the extent to which the use of a particular cellular phone while driving on an interstate highway takes driver attention away from primary driving tasks.

While some specialists emphasize the dichotomy between basic and applied research, it is more accurate to say that there is a continuum, with all studies falling somewhere along the continuum depending on the degree to which the theory or findings generalize to other tasks, products, or settings. Both basic and applied research have complementary advantages and disadvantages. Basic research tends to develop basic principles that have greater generality across a variety of systems and environments than does applied research. It is conducted in rigorously controlled laboratory environments, an advantage because it prevents intrusions from other confounding variables, and allows us to be more confident in the cause-and-effect relationships we are studying. Conversely, research in a highly controlled laboratory environment is often simplistic and artificial and may bear little resemblance to performance in real-world environments. Caution is required in assuming that theory and findings developed through basic research will be applicable for a particular design problem (Kantowitz, 1990). For this reason, people doing controlled research should strive to conduct controlled studies with a variety of tasks and within a variety of settings, some of which are conducted in the field rather than in the lab. This increases the likelihood that their findings are generalizable to new or different tasks and situations.

We might conclude from this discussion that only applied research is valuable to the human factors designer. After all, applied research yields principles and findings specific to particular tasks and settings. A designer need only locate research findings corresponding to the particular combination of factors in the current design problem and apply the findings. The problem with this view is that many, if not most, design problems are somehow different from those studied in the past. The advantage of applied research is also its downfall. It is more descriptive of real-world behavior, but it also tends to be much more narrow in scope.

In addition, applied research such as field studies is often very expensive. It often uses expensive equipment (for example, driving simulators or real cars in answering the cellular phone question), and may place the human participant at risk for accidents, an issue we address later in this chapter.

Often there are so few funds available for answering human factors research questions, or the time available for such answers is so short, that it is impossible to address the many questions that need asking in applied research designs. As a consequence, there is a need to conduct more basic, less expensive and risky laboratory research, or to draw conclusions from other researchers who have published their findings in journals and books. These research studies may not have exactly duplicated the conditions of interest to the human factors designer. But if the findings are strong and reliable, they may provide useful guidance in addressing that design problem, informing the designer or applied researcher, for example, of the driving conditions that might make cellular phone use more or less distracting; or the extent of benefits that could be gained by a voice-dialed over a hand-dialed phone.

Overview of Research Methods

The goal of scientific research is to describe, understand, and predict relationships between variables. In our example, we are interested in the relationship between the variable of "using a cellular phone while driving" and "driving" performance." More specifically, we might hypothesize that use of a cellular phone will result in poorer driving performance than not using the phone.

As noted earlier, we might collect data from a variety of sources. The data source of more basic research is generally the experiment, although applied research and field studies also often involve experiments. The *experimental method* consists of deliberately producing a change in one or more causal or *independent variables* and measuring the effect of that change on one or more *dependent variables.* The key to good experiments is **control.** That is, only the independent variable should be changed, and all other variables should be held constant or controlled. However, control becomes progressively more difficult in more applied research, where participants perform their tasks in the context of the environment to which the research results are to generalize.

As control is loosened, out of necessity, the researcher depends progressively more on *descriptive methods:* describing relations that exist, even though they could not be actually manipulated or controlled by the researcher. For example, the researcher might describe the greater frequency of cell phone accidents in city than in freeway driving to help draw a conclusion that cell phones are more likely to disrupt the busier driver. A researcher might also simply observe drivers while driving in the real world, objectively recording and later analyzing their behavior.

In human factors, as in any kind of research, collecting data, whether experimental or descriptive, is only half of the process. The other half is inferring the meaning or message conveyed by the data, and this usually involves **generalizing** or predicting from the particular data sampled to the broader population. Do cell phones compromise (or not) driving safety in the broad section of automobile drivers, and not just in the sample of drivers used in the simulator experiment or the sample involved in accident statistics? The ability to generalize involves care in both the *design* of experiments and in the statistical analysis.

EXPERIMENTAL RESEARCH METHODS

An experiment involves looking at the relationship between causal *independent variables* and resulting changes in one or more *dependent variables,* which are typically measures of performance, workload, preference, or other subjective evaluations. The goal is to show that the independent variable, and no other variable, is responsible for causing any quantitative differences that we measure in the dependent variable. When we conduct an experiment, we proceed through a process of five steps or stages.

Steps in Conducting an Experiment

Step 1. Define problem and hypotheses. A researcher first hypothesizes the relationships between a number of variables and then sets up experimental designs to determine whether a cause-and-effect relationship does in fact exist. For example, we might hypothesize that changing peoples' work shifts back and forth between day and night produces more performance errors than having people on a constant shift. Once the independent and dependent variables are defined in an abstract sense (e.g., fatigue or attention) and hypotheses are stated, the researchers must develop more detailed experimental specifications.

Step 2. Specify the experimental plan. Specifying the experimental plan consists of identifying all the detail of the experiment to be conducted. Here we must specify exactly what is meant by the dependent variable. What do we mean by performance? What task will our participants be asked to perform, and what aspects of those tasks do we measure? For example, we could define performance as the number of keystroke errors in data entry. We must also define each independent variable in terms of how it will be manipulated. For example, we would specify exactly what we mean by alternating between day and night shifts. Is this a daily change or a weekly change? Defining the independent variables is an important part of creating the *experimental design.* Which independent variables do we manipulate? How many levels of each? For example, we might decide to examine the performance of three groups of workers: those on a day shift, those on a night shift, and those alternating between shifts.

Step 3. Conduct the study. The researcher obtains participants for the experiment, develops materials, and prepares to conduct the study. If he or she is unsure of any aspects of the study, it is efficient to perform a very small experiment, a *pilot study,* before conducting the entire "real" study. After everything is checked through a pilot study, the experiment is carried out and data collected.

Step 4. Analyze the data. In an experiment, the dependent variable is measured and quantified for each subject (there may be more than one dependent variable). For our example, you would have a set of numbers representing the keystroke errors for the people on changing work shifts, a set for the people on day shift, and a set for the people on night shift. Data are analyzed using both descriptive and inferential statistics to see whether there are significant differences among the three groups.

Step 5. Draw conclusions.　Based on the results of the statistical analysis, the researchers draw conclusions about the cause-and-effect relationships in the experiment. At the simplest level, this means determining whether hypotheses were supported. In applied research, it is often important to go beyond the obvious. For example, our study might conclude that shiftwork schedules affect older workers more than younger workers or that it influences the performance of certain tasks, and not others. Clearly, the conclusions that we draw depend a lot on the experimental design. It is also important for the researcher to go beyond concluding what was found, to ask "why". For example, are older people more disrupted by shiftwork changes because they need more sleep? Or because their natural circadian (day-night rhythms) are more rigid? Identifying underlying reasons, whether psychological or physiological, allows for the development of useful and generalizable principles and guidelines.

Experimental Designs

For any experiment, there are different designs that can be used to collect the data. Which design is best depends on the particular situation. Major features that differ between designs include whether each independent variable has two levels or more, whether one or more independent variable is manipulated, and whether the same or different subjects participate in the different conditions defined by the independent variables (Keppel, 1992; Elmes et al., 1995; Williges, 1995).

The Two-Group Design.　In a *two-group design,* one independent variable or factor is tested with only two conditions or levels of the independent variable. In the classic two-group design, a *control* group gets no treatment (e.g., driving with no cellular phone), and the *experimental* group gets some "amount" of the independent variable (e.g., driving while using a cellular phone). The dependent variable (driving performance) is compared for the two groups. However, in human factors we often compare two different experimental treatment conditions, such as performance using a trackball versus using a mouse. In these cases, a control group is unnecessary: A control group to compare with mouse and trackball users would have no cursor control at all, which does not make sense.

Multiple Group Designs.　Sometimes the two-group design does not adequately test our hypothesis of interest. For example, if we want to assess the effects of VDT brightness on display perception, we might want to evaluate several different levels of brightness. We would be studying one independent variable (brightness) but would want to evaluate many *levels* of the variable. If we used five different brightness levels and therefore five groups, we would still be studying one independent variable but would gain more information than if we used only two levels/groups. With this design, we could develop a *quantitative model* or equation that predicts performance as a function of brightness. In a different multilevel design, we might want to test four different input devices for cursor control, such as trackball, thumbwheel, traditional mouse, and key-mouse. We would have four different experimental conditions but still only one independent variable (type of input device).

Factorial Designs. In addition to increasing the number of levels used for manipulating a single independent variable, we can expand the two-group design by evaluating more than one independent variable or *factor* in a single experiment. In human factors, we are often interested in complex systems and therefore in simultaneous relationships between many variables rather than just two. As noted above, we may wish to determine if shiftwork schedules (Factor A) have the same or different effects on older versus younger workers (Factor B).

A multifactor design that evaluates two or more independent variables by combining the different levels of each independent variable is called a *factorial design*. The term *factorial* indicates that all possible combinations of the independent variable levels are combined and evaluated. Factorial designs allow the researcher to assess the effect of each independent variable by itself and also to assess how the independent variables **interact** with one another. Because much of human performance is complex and human–machine interaction is often complex, factorial designs are the most common research designs used in both basic and applied human factors research.

Factorial designs can be more complex than a 2×2 design in a number of ways. First, there can be more than two levels of each independent variable. For example, we could compare driving performance with two different cellular phone designs (e.g., hand-dialed and voice-dialed), and also with a "no phone" control condition. Then we might combine that first three-level variable with a second variable consisting of two different driving conditions: city and freeway driving. This would result in a 3×2 factorial design. Another way that factorial designs can become more complex is by increasing the number of factors or independent variables. Suppose we repeated the above 2×3 design with both older and younger drivers. This would create a $2 \times 3 \times 2$ design. A design with three independent variables is called a three-way factorial design.

Adding independent variables has three advantages: (1) It allows designers to vary more system features in a single experiment: It is **efficient.** (2) It captures a greater part of the complexity found in the real world, making experimental results more likely to generalize. (3) It allows the experimenter to see if there is an *interaction* between independent variables, in which the effect of one independent variable on performance depends on the level of the other independent variable, as we describe in the box.

Between-Subjects Design. In most of the previous examples, the different levels of the independent variable were assessed using separate groups of subjects. For example, we might have one group of subjects use a cellular car phone in heavy traffic, another group use a cellular phone in light traffic, and so on. We compare the driving performance **between** groups of subjects and hence use the term *between-subjects*. A between-subjects variable is an independent variable whereby different groups of subjects are used for each level or experimental condition.

A *between-subjects design* is a design in which all of the independent variables are between-subjects, and therefore each combination of independent variables is administered to a different group of subjects. Between-subjects

EXAMPLE OF A SIMPLE FACTORIAL DESIGN

To illustrate the logic behind factorial designs, we consider an example of the most simple factorial design. This is where two levels of one independent variable are combined with two levels of a second independent variable. Such a design is called a 2 × 2 factorial design. Imagine that a researcher wants to evaluate the effects of using a cellular phone on driving performance (and hence on safety). The researcher manipulates the first independent variable by comparing driving with and without use of a cellular phone. However, the researcher suspects that the driving impairment may only occur if the driving is taking place in heavy traffic. Thus, he or she may add a second independent variable consisting of light versus heavy traffic driving conditions. The experimental design would look like that illustrated in Figure 2.1: four groups of subjects derived from combining the two independent variables.

Imagine that we conducted the study, and for each of the subjects in the four groups shown in Figure 2.1, we counted the number of times the driver strayed outside of the driving lane as the dependent variable. We can look at the general pattern of data by evaluating the *cell means;* that is, we combine the scores of all subjects within each of the four groups. Thus, we might obtain data such as that shown in Table 2.1.

If we look only at the effect of cellular phone use (combining the light and heavy traffic conditions), we might be led to believe that use of cell phones impairs driving performance. But looking at the entire picture, as shown in Figure 2.2, we see that the use of a cell phone

DRIVING CONDITIONS

	Light traffic	Heavy traffic
No car phone	No cell phone while driving in light traffic	No cell phone while driving in heavy traffic
Car phone	Use cell phone while driving in light traffic	Use cell phone while driving in heavy traffic

FIGURE 2.1
The four experimental conditions for a 2 × 2 factorial design.

impairs driving only in heavy traffic conditions (as defined in this particular study). When the lines connecting the cell means in a factorial
study are not parallel, as in Figure 2.2, we know that there is some
type of interaction between the independent variables: The effect of
phone use depends on driving conditions. Factorial designs are popular
for both basic research and applied questions because they allow researchers to evaluate interactions between variables.

**TABLE 2.1   Hypothetical Data for Driving Study: Average Number
of Lane Deviations**

Cell Phone Use	Light Traffic	Heavy Traffic
No cell phone	2.1	2.1
Cell phone	2.2	5.8

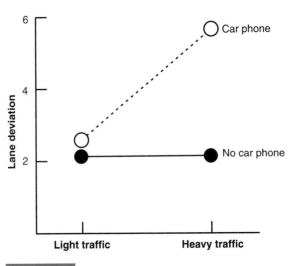

FIGURE 2.2
Interaction between cellular phone use and driving conditions.

designs are most commonly used when having subjects perform in more than
one of the conditions would be problematic. For example, if you have subjects
receive one type of training (e.g., on a simulator), they could not begin over
again for another type of training because they would already know the material. Between-subjects designs also eliminate certain confounds related to *order
effects,* which we discuss shortly.

Within-Subject Designs.   In many experiments, it is feasible to have the same subjects participate in all of the experimental conditions. For example, in the driving study, we could have the same subjects drive for periods of time in each of the four conditions shown in Table 2.1. In this way, we could compare the performance of each person with him- or herself across the different conditions. This within-subject performance comparison illustrates where the methods gets its name. When the same subject experiences all levels of an independent variable, it is termed a within-subjects variable. An experiment where all independent variables are within-subject variables is termed a *within-subjects design*. Using a within-subjects design is advantageous in a number of respects, including that it is more sensitive and easier to find statistically significant differences between experimental conditions. It is also advantageous when the number of people available to participate in the experiment is limited.

Mixed Designs.   In factorial designs, each independent variable can be either between-subjects or within-subjects. If both types are used, the design is termed a *mixed design*. If one group of subjects drove in heavy traffic with and without a cellular phone, and a second group did so in light traffic, this is a mixed design.

Multiple Dependent Variables.   In the previous sections, we described several different types of experimental design that were variations of the same thing—multiple independent variables combined with a single dependent variable or "effect." However, the systems that we study, including the human, are very complex. We often want to measure how causal variables affect several *dependent variables* at once. For example, we might want to measure how use of a cellular phone affects a number of driving variables, including deviations from the lane, reaction time to brake for cars or other objects in front of the vehicle, time to recognize objects in the driver's peripheral vision, speed, acceleration, and so forth.

Selecting the Apparatus and Context

Once the experimental design has been specified with respect to dependent variables, the researcher must decide what tasks the person will be performing and under what context. For applied research, we try to identify tasks and environments that will give us the most generalizable results. This often means conducting the experiments under real-world or high-fidelity conditions.

Selecting Experimental Participants

Participants should represent the population or group in which the researcher is interested. For example, if we are studying pilot behavior, we would pick a sample of pilots who represent the pilot population in general. If we are studying elderly, we define the population of interest (e.g., all people aged 65 and older who are literate); then we obtain a sample that is representative of that population. Notice that it would be difficult to find a sample that has all of the qualities of all elderly people. If lucky, we might get a sample that is representative of all elderly people living in the United States who are healthy, speak English, and so on.

Experimental Control and Confounding Variables

In deciding how the study will be conducted, it is important to consider all variables that might impact the dependent variable. Extraneous variables have the potential to interfere in the causal relationship and must be **controlled** so that they do not interfere. If these extraneous variables do influence the dependent variable, we say that they are *confounding variables.* One group of extraneous variables is the wide range of ways participants differ from one another. These variables must be controlled, so it is important that the different groups of people in a between-subjects experiment differ only with respect to the treatment condition and not on any other variable or category. For example, in the cellular phone study, you would not want elderly drivers using the car phone and young drivers using no phone. Then age would be a confounding variable. One way to make sure all groups are equivalent is to take the entire set of subjects and randomly put them in one of the experimental conditions. That way, on the average, if the sample is large enough, characteristics of the subjects will even out across the groups. This procedure is termed *random assignment.* Another way to avoid having different characteristics of subjects in each group is to use a within-subjects design. However, this design creates a different set of challenges for experimental control.

Other variables in addition to subject variables must be controlled. For example, it would be a poor experimental design to have one condition where cellular phones are used in a Jaguar and another condition where no phone is used in an Oldsmobile. There may be driving characteristics or automobile size differences that cause variations in driving behavior. The phone versus no-phone comparison should be carried out in the same vehicle (or same type of vehicle). We need to remember, however, that in more applied research, it is sometimes impossible to exert perfect control.

For within-subjects designs, there is another variable that must be controlled: the order in which the subject receives his or her experimental conditions, which creates what are called *order effects.* When people participate in several treatment conditions, the dependent measure may show differences from one condition to the next simply because the treatments, or levels of the independent variable, are experienced in a particular order. For example, if participants use five different cursor-control devices in an experiment, they might be fatigued by the time they are tested on the fifth device and therefore exhibit more errors or slower times. This would be due to the order of devices used rather than the device per se. In contrast, if the cursor-control task is new to the participant, he or she might show learning and actually do best on the fifth device tested, not because it was better, but because the cursor-control skill was more practiced. These order effects of fatigue and practice in between-subjects designs are both potential confounding variables; while they work in opposite directions, to penalize or reward the late-tested conditions, they do not necessarily balance each other out.

As a safeguard to keep order from confounding the independent variables, we use a variety of methods. For example, extensive practice can reduce learning effects. Time between conditions can reduce fatigue. Finally, researchers often use a technique termed *counterbalancing*. This simply means that different subjects receive the treatment conditions in different orders. For example, half of the participants in a study would use a trackball and then a mouse. The other half would use a mouse and then a trackball. There are specific techniques for counterbalancing order effects; the most common is a Latin-square design. Research methods books (e.g., Keppel, 1992) provide instruction on using these designs.

In summary, the researcher must control extraneous variables by making sure they do not covary with the independent variable. If they do covary, they become *confounds* and make interpretation of the data impossible. This is because the researcher does not know which variable caused the differences in the dependent variable.

Conducting the Study

After designing the study and identifying a sample of participants, the researcher is ready to conduct the experiment and collect data (sometimes referred to as "running subjects"). Depending on the nature of the study, the experimenter may want to conduct a small pretest, or pilot study, to check that manipulation levels are set right, that participants (subjects) do not experience unexpected problems, and that the experiment will generally go smoothly. When the experiment is being conducted, the experimenter should make sure that data collection methods remain constant. For example, an observer should not become more lenient over time; measuring instruments should remain calibrated. Finally, all participants should be treated ethically, as described later.

Data Analysis

Once the experimental data have been collected, the researcher must determine whether the dependent variable(s) actually did change as a function of experimental condition. For example, was driving performance really "worse" while using a cellular phone? To evaluate the research questions and hypotheses, the experimenter calculates two types of statistics: *descriptive* and *inferential statistics*. Descriptive statistics are a way to summarize the dependent variable for the different treatment conditions, while inferential statistics tell us the likelihood that any differences between our experimental groups are "real" and not just random fluctuations due to chance.

Descriptive Statistics.   Differences between experimental groups are usually described in terms of averages. Thus, the most common descriptive statistic is the *mean*. Research reports typically describe the mean scores on the dependent variable for each group of subjects (e.g., see the data shown in Table 2.1 and Figure 2.2). This is a simple way of conveying the effects of the independent variable(s) on the dependent variable. Standard deviations are also sometimes given to convey the spread of scores.

Inferential Statistics.   While experimental groups may show different means for the various conditions, it is possible that such differences occurred solely on the basis of chance. Humans almost always show random variation in performance, even without manipulating any variables. It is not uncommon to get two groups of subjects who have different means on a variable, without the difference being due to any experimental manipulation, in the same way that you are likely to get a different number of "heads" if you do two series of 10 coin tosses. In fact, it is unusual to obtain means that are exactly the same. So, the question becomes, Is the difference big enough that we can rule out chance and assume the independent variable had an affect? Inferential statistics give us, effectively, the probability that the difference between the groups is due to chance. If we can rule out the "chance" explanation, then we infer that the difference was due to the experimental manipulation.

For a two-group design, the inferential statistical test usually used is a t-test. For more than two groups, we use an analysis of variance (ANOVA). Both tests yield a score; for a t-test, we get a value for a statistical term called t, and for ANOVA, we get a value for F. Most important, we also identify the probability, p, that the t or F value would be found by chance for that particular set of data if there was no effect or difference. The smaller the p probably is, the more significant our result becomes and the more confident we are that our independent variable really did cause the difference. This p value will be smaller as the difference between means is greater, as the variability between our observations within a condition (standard deviation) is less, and, importantly, as the **sample size** of our experiment increases (more subjects, or more measurements per subject). A greater sample size gives our experiment greater *statistical power* to find significant differences.

Drawing Conclusions

Researchers usually assume that if p is less than .05, they can conclude that the results are not due to chance and therefore that there was an effect of the independent variable. Accidentally concluding that independent or causal variables had an effect when it was really just chance is referred to as making a *Type I error.* If scientists use a .05 cutoff, they will make a Type I error only one time in 20. In traditional sciences, a Type I error is considered a "bad thing" (Wickens, 1998). This makes sense if a researcher is trying to develop a cause-and-effect model of the physical or social world. The Type 1 error would lead to the development of false theories.

Researchers in human factors have also accepted this implicit assumption that making a Type I error is bad. Research where the data result in inferential statistics with $p > .05$ is not generally accepted for publication in most journals. Experimenters studying the effects of system design alternatives often conclude that the alternatives made no difference. Program evaluation where introduction of a new program resulted in statistics of $p > .05$ often conclude that the new program did not work, all because there is greater than a 1-in-20 chance that spurious factors *could* have caused the results.

The cost of setting this arbitrary cutoff of $p = .05$ is that researchers are more likely to make *Type II errors,* concluding that the experimental manipulation did not have an effect when in fact it did. (Keppel, 1992). This means, for example, that a safety officer might conclude that a new piece of equipment is no easier to use under adverse environmental conditions, when in fact it is easier. The likelihood of making Type I and Type II errors are inversely related. Thus, if the experimenter showed that the new equipment was not **statistically significantly better** ($p < .05$) than the old, the new equipment might be rejected even though it might actually be better, and if the p level had been set at 0.10 instead of .05, it would have been concluded to be better.

The total dependence of researchers on the $p = .05$ criterion is especially problematic in human factors because we frequently must conduct experiments and evaluations with relatively low numbers of subjects because of expense or the limited availability of certain highly trained professionals (Wickens, 1998). As we saw, using a small number of subjects makes the statistical test less powerful and more likely to show no significance, or $p > .05$, even when there is a difference. In addition, the variability in performance between different subjects or for the same subject but over time and conditions is also likely to be great when we try to do our research in more applied environments, where all confounding extraneous variables are harder to control. Again, these factors make it more likely that the results will show no significance, or $p > .05$. The result is that human factors researchers frequently conclude that there is no difference in experimental conditions simply because there is more than a 1-in-20 chance that it *could* be caused by random variation in the data.

In human factors, researchers should consider the probability of a Type II error when their difference is not significant at the conventional .05 level and consider the consequences if others use their research to conclude that there is no difference (Wickens, 1998). For example, will a safety-enhancing device fail to be adopted? In the cellular phone study, suppose that performance really was worse with cell phones than without, but the difference was not quite big enough to reach .05 significance. Might the legislature conclude, in error, that cell phone use was "safe"? There is no easy answer to the question of how to balance Type I and Type II statistical errors (Keppel, 1992; Nickerson, 2001). The best advice is to realize that the higher the sample size, the less either type of error will occur, and to consider the consequences of *both* types of errors when, out of necessity, the sample size and power of the design of a human factors experiment must be low.

Statistical Significance Versus Practical Significance

Once chance is ruled out, meaning $p < .05$, researchers discuss the differences between groups as though they are a fact. However, it is important to remember that two groups of numbers can be statistically different from one another without the differences being very *large*. Suppose we compare two groups of Army trainees. One group is trained in tank gunnery with a low-fidelity personal computer. Another group is trained with an expensive, high-fidelity simulator. We might find that when we measure performance, the mean percent correct for the personal computer group is 80, while the mean percent correct for the simulator group is 83. If we used a large number of subjects in a very powerful design, there may be a statistically significant difference between the two groups, and we would therefore conclude that the simulator is a better training system.

However, especially for applied research, we must look at the difference between the two groups in terms of *practical* significance. Is it worth spending millions to place simulators on every military base to get an increase from 80 percent to 83 percent? This illustrates the tendency for some researchers to place too much emphasis on statistical significance and not enough emphasis on practical significance.

DESCRIPTIVE METHODS

While experimentation in a well controlled environment is valuable for uncovering basic laws and principles, there are often cases where research is better conducted in the real world. In many respects, the use of complex tasks in a real-world environment results in more generalizable data that capture more of the characteristics of a complex, real-world environment. Unfortunately, conducting research in real-world settings often means that we must give up the "true" experimental design because we cannot directly manipulate and control variables. One example is *descriptive research,* where researchers simply measure a number of variables and evaluate how they are related to one another. Examples of this type of research include evaluating the driving behavior of local residents at various intersections, measuring how people use a particular design of ATM (automatic teller machine), and observing workers in a manufacturing plant to identify the types and frequencies of unsafe behavior.

Observation

In many instances, human factors research consists of recording behavior during tasks performed under a variety of circumstances. For example, we might install video recorders in cars (with the drivers' permission) to film the circumstances in which they place or receive calls on a cellular phone during their daily driving.

In planning observational studies, a researcher identifies the variables to be measured, the methods to be employed for observing and recording each variable, conditions under which observation will occur, the observational timeframe, and so forth. For our cellular phone study, we would develop a series of "vehicle status categories" in which to assign each phone use (e.g., vehicle stopped, during turn, city street, freeway, etc.) These categories define a *taxonomy.* Otherwise, observation will result in a large number of specific pieces of information that cannot be reduced into any meaningful descriptions or conclusions. It is usually most convenient to develop a taxonomy based on pilot data. This way, an observer can use a checklist to record and classify each instance of new information, condensing the information as it is collected.

In situations where a great deal of data is available, it may be more sensible to sample only a part of the behavioral data available or to sample behavior during different sessions rather than all at once. For example, a safety officer is better off sampling the prevalence of improper procedures or risk-taking behavior on the shop floor during several different sessions over a period of time than all at once during one day. The goal is to get representative samples of behavior, and this is more easily accomplished by sampling over different days and during different conditions.

Surveys and Questionnaires

Both basic and applied research frequently rely on surveys or questionnaires to measure variables. The design of questionnaires and surveys is a challenging task if it is to be done in a way that yields reliable and valid data, and the reader is referred to Salvendy and Carayan (1997) and for proper procedures. Questionnaires and surveys sometimes gather qualitative data from open-ended questions (e.g., "what features on the device would you like to see?" or "what were the main problems in operating the device?"). However more rigorous treatment of the survey results can typically be obtained from quantitative data, often obtained from a numerical rating scale, often with endpoints ranging between, say, 1–7 or 1–10. Such quantitative data has the advantage of being addressed by statistical analysis.

A major concern with questionnaires is their **validity.** Aside from assuring that questions are designed to appropriately assess the desired content area, under most circumstances, respondents should be told that their answers will be both confidential and anonymous. It is common practice for researchers to place identifying numbers rather than names on the questionnaires. Employees are more likely to be honest if their names will never be directly associated with their answers.

A problem is that many people do not fill out questionnaires if they are voluntary. If the sample of those who do and who do not return questionnaires is different along some important dimension related to the topic surveyed, the survey results will obviously be biased. For example, in interpreting the results of an anonymous survey of unsafe acts in a factory, those people who are time-stressed in their job are more likely to commit unsafe acts, but also do not have time to complete the survey. Hence, their acts will be underrepresented in the survey results.

Questionnaires and surveys are, by definition, subjective. Their outputs can often be contrasted with objective performance data, such as error rates or response times. The difference between these two classes of measures is important, given that subjective measures are often easier and less expensive to obtain, with a high sample size.

Several good papers have been published on the objective versus subjective measurement issue (e.g., Hennessy, 1990; Muckler, 1992). If we evaluate the literature, it is clear that both objective and subjective measures have their uses. For example, in a study of factors that lead to stress disorders in soldiers, Solomon, Mikulincer, and Hobfoll (1987) found that objective and subjective indicators of event stressfulness and social support were predictive of combat stress reaction and later posttraumatic stress disorder and that "subjective parameters were the stronger predictors of the two" (p. 581). In considering subjective measures, however, it is important to realize that what people subjectively rate as "preferred" is not always the system feature that supports best performance (Andre & Wickens, 1995). For example, people almost always prefer a colored display to a monochrome one, even when the color is used in such a way that it can be detrimental to performance.

Incident and Accident Analysis

Sometimes a human factors analyst must determine the overall functioning of a system, especially with respect to safety. There are a number of methods for evaluating safety, including the use of surveys and questionnaires. Another method is to evaluate the occurrence of incidences, accidents, or both. An *incident* is where a noticeable problem occurs during system operation, but an actual accident does not result from it. Some fields, such as the aerospace community, have formalized databases for recording reported incidents and accidents (Rosenthal & Reynard, 1991). The Aviation Safety Reporting System's (ASRS) database is run by NASA and catalogs approximately 30,000 incidents reported by pilots or air traffic controllers each year.

While this volume of information is potentially invaluable, there are certain difficulties associated with the database (Wickens, 1995). First, the sheer size of the qualitative database makes it difficult to search to develop or verify causal analyses. Second, even though people who submit reports are guaranteed anonymity, not all incidents are reported. A third problem is that the reporting person may not give information that is necessary for identifying the root causes of the incident or accident. The more recent use of follow-up interviews has helped reduce but not completely eliminated the problem.

Accident prevention is a major goal of the human factors profession, especially as humans are increasingly called upon to operate large and complex systems. Accidents can be systematically analyzed to determine the underlying root causes, whether they arose in the human, machine, or some interaction. Accident analysis has pointed to a multitude of cases where poor system design has resulted in human error, including problems such as memory failures in the 1989 Northwest Airlines Detroit crash, training and decision errors in the 1987 Air Florida crash at Washington National Airport, and high mental workload and poor decision making at Three-Mile Island. Accidents are usually the result of several coinciding breakdowns within a system. This means that most of the time, there are multiple unsafe elements such as training, procedures, controls and displays, system components, and so on that would ideally be detected before rather than after an accident. This requires a proactive approach to system safety analysis rather than a reactive one such as accident analysis. This topic is addressed in greater length in Chapter 14.

Data Analysis for Descriptive Measures

Most descriptive research is conducted in order to evaluate the relationships between a number of variables. Whether the research data has been collected through observation or questionnaires, the goal is to see whether relationships exist and to measure their strength. Relationships between variables can be measured in a number of ways.

Relationships Between Continuous Variables. If we were interested in determining if there is a relationship between job experience and safety attitudes within an organization, this could be done by performing a *correlational analysis.* The correlational analysis measures the extent to which two variables covary such that the value of one can be somewhat predicted by knowing the value of the other. For example, in a positive correlation, one variable increases as the value of another variable increases; for example, the amount of illumination needed to read text will be positively correlated with age. In a negative correlation, the value of one variable decreases as the other variable increases; for example, the intensity of a soft tone that can be just heard is negatively correlated with age. By calculating the correlation coefficient, r, we get a measure of the strength of the relationship. Statistical tests can be performed that determine the probability that the relationship is due to chance fluctuation in the variables. Thus, we get information concerning whether a relationship exists (p) and a measure of the strength of the relationship (r). As with other statistical measures, the likelihood of finding a significant correlation increases as the sample size—the number of items measured on both variables—increases.

One caution should be noted. When we find a statistically significant correlation, it is tempting to assume that one of the variables caused the changes seen in the other variable. This causal inference is unfounded for two reasons. First, the direction of causation could actually be in the opposite direction. For example, we might find that years on the job is negatively correlated with risk-taking. While it is possible that staying on the job makes an employee more cautious, it is also possible that being more cautious results in a lower likelihood of injury or death. This may therefore cause people to stay on the job. Second, a third variable might cause changes in both variables. For example, people who try hard to do a good job may be encouraged to stay on and may also behave more cautiously as part of trying hard.

Complex Modeling and Simulation

Researchers sometimes collect a large number of data points for multiple variables and then test the relationships through models or simulations (Pew & Mavor, 1998). According to Bailey (1989), a model is "a mathematical/physical system, obeying specific rules and conditions, whose behavior is used to understand a real (physical, biological, human–technical, etc.) system to which it is analogous in certain respects." Models range from simple mathematical equations, such as the equation that might be used to predict display perception as a function of brightness level, to highly complex computer simulations (runnable models); but in all cases, models are more restricted and less "real" than the system they reflect.

Models are often used to describe relationships in a physical system or the physiological relationships in the human body. Mathematical models of the human body have been used to create simulations that support workstation design. As an example, COMBIMAN is a simulation model that provides graphical displays of the human body in various workstation configurations (McDaniel & Hofmann, 1990). It is used to evaluate the physical accommodation of a pilot to existing or proposed crew station designs.

Mathematical models can be used to develop complex simulations (see Elkind et al., 1990; Pew & Mavor, 1998; Laughery & Corker, 1997). That is, key variables in some particular system and their interrelationships are mathematically modeled and coded into a runnable simulation program. Various scenarios are run, and the model shows what would happen to the system. The predictions of a simulation can be validated against actual human performance (time, errors, workload). This gives future researchers a powerful tool for predicting the effects of design changes without having to do experiments. One important advantage of using models for research is that they can replace evaluation using human subjects to assess the impact of harmful environmental conditions (Kantowitz, 1992; Moroney, 1994).

Literature Surveys

A final research method that should be considered is the careful literature search and survey. While this often proceeds an experimental write-up, a good literature search can often substitute for the experiment itself if other researchers have already answered the experimental question. One particular form of literature survey, known as a *meta-analysis,* can integrate the statistical findings of a lot of other experiments that have examined a common independent variable in order to draw a collective and very reliable conclusion regarding the effect of that variable (Rosenthal & Reynard, 1991).

ETHICAL ISSUES

It is evident that the majority of human factors research involves the use of people as participants in research. Many professional affiliations and government agencies have written specific guidelines for the proper way to involve participants in research. Federal agencies rely strongly on the guidelines found in the Code of Federal Regulations HHS, Title 45, Part 46; Protections of Human Subjects (Department of Health and Human Services, 1991). The National Institute of Health has a Web site where students can be certified in human subjects testing (http://ohsr.od.nih.gov/cbt/). Anyone who conducts research using human participants should become familiar with the federal guidelines as well as APA published guidelines for ethical treatment of human subjects (American Psychological Association, 1992). These guidelines fundamentally advocate the following principles:

- Protection of participants from mental or physical harm
- The right of participants to privacy with respect to their behavior
- The assurance that participation in research is completely voluntary
- The right of participants to be informed beforehand about the nature of the experimental procedures

When people participate in an experiment, or to provide data for research by other methods they are told the general nature of the study. Often, they cannot be told the exact nature of the hypotheses because this will bias their behavior. Participants should be informed that all results will be kept anonymous and confidential. This is especially important in human factors because often participants are employees who fear that their performance will be evaluated by management. Finally, participants are generally asked to sign a document, an *informed consent form,* stating that they understand the nature and risks of the experiment, or data gathering project, that their participation is voluntary, and that they understand they may withdraw at any time. In human factors field research, the experiment is considered to be reasonable in risk if the risks are no greater than those faced in the actual job environment. Research boards in the university or organization where the research is to be conducted certify the adequacy of the consent form and that the potential for any risks to the participant is outweighed by the overall benefits of the research to society.

As one last note, experimenters should always treat participants with respect. Participants are usually self-conscious because they feel their performance is being evaluated (which it is, in some sense) and they fear that they are not doing well enough. It is the responsibility of the investigator to put participants at ease, assuring them that the system components are being evaluated and not the people themselves. This is one reason that the term *user testing* has been changed to *usability testing* (see next chapter) to refer to situations where people are asked to use various system configurations in order to evaluate overall ease of use and other factors.

Chapter 3

Design and Evaluation Methods

Thomas Edison was a great inventor but a poor businessman. Consider the phonograph. Edison invented it, he had better technology than his competitors, but he built a technology-centered device that failed to consider his customers' needs, and his phonograph business failed. One of Edison's important failings was to neglect the practical advantages of the disc over the cylinder in terms of ease of use, storage, and shipping. Edison scoffed at the scratchy sound of the disc compared to the superior sound of his cylinders. Edison thought phonographs could lead to a paperless office in which dictated letters could be recorded and the cylinders mailed to the recipients without the need for transcription. The real use of the phonograph, discovered after much trial and error by a variety of other manufacturers, was to provide prerecorded music. Once again, he failed to understand the real desires of his customers. Edison decided that big-name, expensive artists did not sound that different from the lesser-known professionals. He is probably correct. Edison thought he could save considerable money at no sacrifice to quality by recording those lesser-known artists. He was right; he saved a lot of money. The problem was, the public wanted to hear the well-known artists, not the unknown ones. He thought his customers only cared about the music; he didn't even list the performers' names on his disc records for several years. Edison pitted his taste and his technology-centered analysis on belief that the difference was not important: He lost. The moral of this story is to know your customer. Being first, being best, and even being right do not matter; what matters is understanding what your customers want and need. Many technology-oriented companies are in a similar muddle. They develop technology-driven products, quite often technology for technology's sake, without understanding customer needs and desires. (Adapted from Norman, 1988)

The goal of a human factors specialist is to make systems successful by enhancing performance, satisfaction, and safety. In addition to conducting basic and applied research to broaden our understanding, this is done primarily by applying human factors principles, methods, and data to the design of new products or systems. However, the concept of "design" can be very broad, including activities such as the following:

- Design or help design new products or systems, especially their interface.
- Modify the design of existing products to address human factors problems.
- Design ergonomically sound environments, such as individual workstations, large environments with complex work modules and traffic patterns, home environments for the handicapped, and gravity-free environments.
- Perform safety-related activities, such as conduct hazard analyses, implement industrial safety programs, design warning labels, and give safety-related instructions.
- Develop training programs and other performance support materials such as checklists and instruction manuals.
- Develop methods for training and appraising work groups and teams.
- Apply ergonomic principles to organizational development and restructuring.

In this chapter, we review some of the methods that human factors specialists use to support design, with particular emphasis on the first activity, designing products or systems. Human factors methods and principles are applied in all product design phases: predesign analysis, technical design, and final test and evaluation. Although interface design may be the most visible design element, human factors specialists generally go beyond interface design to design the interaction or job and even redesign work by defining the organization of people and technology. Cooper (1999) argues that focusing solely on interface design is ineffective and calls it "painting the corpse." Making a pretty, 3-D graphical interface cannot save a system that does not consider the job or organization it supports. While the material in this chapter provides an overview of the human factors *process,* later chapters provide some of the basic *content* information necessary to carry out those processes. Later chapters provide both the content and specialized methods needed to address broader design considerations listed above.

OVERVIEW OF DESIGN AND EVALUATION

Many, if not most, products and systems are still designed and manufactured without adequate consideration of human factors. Designers tend to focus primarily on the technology and its features without fully considering the use of the product from the human point of view. In a book that every engineer should read, Norman (1988) writes congently,

> Why do we put up with the frustrations of everyday objects, with objects that we can't figure out how to use, with those neat plastic-wrapped packages that seem impossible to open, with doors that trap people, with

washing machines and dryers that have become too confusing to use, with audio-stereo-television-video-cassette-recorders that claim in their advertisements to do everything, but that make it almost impossible to do anything?

Poor design is common, and as our products become more technologically sophisticated, they frequently become more difficult to use.

Even when designers attempt to consider human factors, they often complete the product design first and only then hand off the blueprint or prototype to a human factors expert. This expert is then placed in the unenviable position of having to come back with criticisms of a design that a person or design team has probably spent months and many thousands of dollars to develop. It is not hard to understand why engineers are less than thrilled to receive the results of a human factors analysis. They have invested in the design, clearly believe in the design, and are often reluctant to accept human factors recommendations. The process of bringing human factors analysis in at the end of the product design phase inherently places everyone involved at odds with one another. Because of the investment in the initial design and the designer's resistance to change, the result is often a product that is not particularly successful in supporting human performance, satisfaction, and safety.

As we noted in Chapter 1, human factors can ultimately save companies time and money. But to maximize the benefits achieved by applying human factors methods, the activities must be introduced early in the system design cycle. The best way to demonstrate the value of human factors to management is to perform a cost/benefit analysis.

Cost/Benefit Analysis of Human Factors Contributions

Human factors analysis is sometimes seen as an extra expense that does not reap a monetary reward equal to or greater than the cost of the analysis. A human factors expert may be asked to somehow justify his or her involvement in a project and explicitly demonstrate a need for the extra expense. In this case, a *cost/benefit analysis* can be performed to demonstrate to management the overall advantages of the effort (Alexander, 2002; Bias & Mayhew, 1994; Hendrick, 1996).

In a cost/benefit analysis, one calculates the expected costs of the human factors effort and estimates the potential benefits in monetary terms. Mayhew (1992) provides a simple example of such an analysis. Table 3.1 shows a hypothetical example of the costs of conducting a usability study for a software prototype.

In most instances, estimating the *costs* for a human factors effort is relatively easy because the designer tends to be familiar with the costs for personnel and materials. Estimating the *benefits* tends to be more difficult and must be based on assumptions (Bias & Mayhew, 1994). It is best if the designer errs on the conservative side in making these assumptions. Some types of benefits are more common for one type of manufacturer or customer than another. For example, customer support costs may be a big consideration for a software developer like

TABLE 3.1 Hypothetical Costs for Conducting a Software Usability Study

Human Factors Task	Hours
Determine Testing Issues	24
Design Test and Materials	24
Test 20 Users	48
Analyze Data	48
Prepare/Present Results	16
TOTAL HP (Human factors professional) HOURS	160

	Cost
160 HP (Human factors professional) hours @ $45	$7,200
48 Assistant hours @ $20	960
48 Cameraman hours @ $30	1,440
Videotapes	120
TOTAL COST	$9,720

Source: D. T. Mayhew, 1992. *Principles and guidelines in software user interface design.* Englewood Cliffs, NJ: Prentice Hall. Adapted by permission.

Microsoft, which spends $800 million each year to help customers overcome difficulties with their products. In contrast, a confusing interface led pilots to enter the wrong information into an onboard computer, which then guided them into the side of a mountain, killing 160 people (Cooper, 1999). Estimating the dollar value of averting such catastrophic failures can be quite difficult. Mayhew (1992) lists nine benefits that might be applicable and that can be estimated quantitatively: increased sales, decreased cost of providing training, decreased customer support costs, decreased development costs, decreased maintenance costs, increased user productivity, decreased user errors, improved quality of service, decreased training time, decreased user turnover.

Other quantifiable benefits are health or safety related (Alexander, 1995), such as increased employee satisfaction (lower turnover) or decreases in sick leave, number of accidents or acute injuries, number of chronic injuries (such as cumulative trauma disorders), medical and rehabilitation expenses, number of citations or fines, or number of lawsuits.

The total benefit of the effort is determined by first estimating values for the relevant variables without human factors intervention. The same variables are then estimated, assuming that even a moderately successful human factors analysis is conducted. The estimated benefit is the total cost savings between the two.

For example, in a software usability testing effort, one might calculate the average time to perform certain tasks using a particular product and/or the average number of errors and the associated time lost. The same values are estimated for performance if a human factors effort is conducted. The difference is then calculated. These numbers are multiplied by the number of times the tasks are performed and by the number of people performing the task (e.g., over a year or five years time). Mayhew (1992) gives an example for a human factors software analysis that would be expected to decrease the throughput time for fill-in screens by three seconds per screen. Table 3.2 shows the estimated benefits. It is easy to see that even small cost savings per task can add up over the course of a year. In this case, the savings of $43,125 in one year easily outweighs the cost of the usability study, which was $9,720. Karat (1990) reports a case where human factors was performed for development of software used by 240,000 employees. She estimated after the fact that the design effort cost $6,800, and the time-on-task monetary savings added up to a total of $6,800,000 for the first year alone. Designers who must estimate performance differences for software screen changes can refer to the large body of literature that provides specific numbers based on actual cases (see Bias & Mayhew, 1994). Manufacturing plants can likewise make gains by reducing costs associated with product assembly and maintenance (e.g., Marcotte et al., 1995), and for injury- and health-related analyses, the benefits can be even greater. Refer to Alexander (1995), Bias and Mayhew (1994), Mantei and Teorey (1988), and Hendrick, 1996 for a more detailed description of cost/benefit analysis. A cost/benefit analysis clearly identifies the value of human factors contributions to design.

Human Factors in the Product Design Lifecycle

One major goal in human factors is to support the design of products in a cost-effective and timely fashion, such that the products support, extend, and transform user work (Wixon et al., 1990). As noted earlier, in order to maximally benefit the final product, human factors must be involved as early as possible in the product (or system) design rather than performed as a final evaluation *after* product design.

There are numerous systematic design models, which specify a sequence of steps for product analysis, design, and production (e.g., see Bailey, 1996; Blanchard & Fabrycky, 1990; Dix et al., 1993; Meister, 1987; Shneiderman, 1992). Product design models are all relatively similar and include stages reflecting pre-design or front-end analysis activities, design of the product, production, and field test and evaluation. Product *lifecycle models* also add product implementation, utilization and maintenance, and dismantling or disposal.

While many people think of human factors as a "product evaluation" step done predominantly towards the end of the design process, as we describe

---

TABLE 3.2   Hypothetical Estimated Benefit for a 3-Second Reduction in Screen Use

250 users
× 60 screens per day
× 230 days per year
× processing time reduced by 3 seconds per screen
× hourly rate of $15
= $43,125 savings per year

Source: D. J. Mayhew, 1992. *Principles and guidelines in software user interface design.* Englewood Cliffs, NJ: Prentice Hall. Adapted by permission.

below, human factors activities occur in many of the stages, and indeed most of the human factors analyses are performed early.

As we will describe in the following pages, six major stages of human factors in the product life cycle include: (1) front end analysis, (2) iterative design and test (3) system production, (4) implementation and evaluation, (5) system operation and maintenance, (6) system disposal. Before describing these six stages in detail, we discuss the sources of data that human factors practitioners use in achieving their goal of user-centered design.

The most effective way to involve human factors in product design is to have multidisciplinary design team members working together from the beginning. This is consistent with industry's emphasis on concurrent engineering (Chao, 1993) in which design teams are made up from members of different functional groups who work on the product from beginning to end. Team members often include personnel from marketing, engineers and designers, human factors specialists, production or manufacturing engineers, service providers, and one or more users or customers. For large-scale projects, multiple teams of experts are assembled.

User-Centered Design

All of the specific human factors methods and techniques that we will review shortly are ways to carry out the overriding methodological principle in the field of human factors: to center the design process around the user, thus making it a *user-centered design* (Norman & Draper, 1986). Other phrases that denote similar meaning are "know the user" and "honor thy user." Obviously, these phrases suggest the same thing. For a human factors specialist, system or product design revolves around the central importance of the user. How do we put this principle into practice? Primarily by adequately determining user needs and by involving the user at all stages of the design process. This means the human factors specialist will study the users' job or task performance, elicit their needs and preferences, ask for their insights and design ideas, and request their response to design solutions. User-centered design does not mean that the user *designs* the product or has control of the design process. The goal of the human factors specialist is to find a system design that supports the user's needs rather than making a system to which users must adapt. User-centered design is also embodied in a subfield known as *usability engineering* (Gould & Lewis, 1985; Nielson, 1993; Rubin, 1994; Wiklund 1994, 1993). Usability engineering has been most rigorously developed for software design (e.g., Nielson, 1993) and involves four general approaches to design:

- *Early focus on the user* and tasks
- *Empirical measurement* using questionnaires, usability studies, and usage studies focusing on quantitative performance data
- *Iterative design* using prototypes, where rapid changes are made to the interface design
- *Participatory design,* where users are directly involved as part of the design team.

Sources for Design Work

Human factors specialists usually rely on several sources of information to guide their involvement in the design process, including previous published research, data compendiums, human factors standards, and more general principles and guidelines.

Data Compendiums. As the field of human factors has matured, many people have emphasized the need for sources of information to support human factors aspects of system design (e.g., Boff et al., 1991; Rogers & Armstrong, 1977; Rogers & Pegden, 1977). Such information is being developed in several forms. One form consists of condensed and categorized databases, with information such as tables and formulas of human capabilities. An example is the four-volume publication by Boff and Lincoln (1988), *Engineering Data Compendium: Human Perception and Performance,* which is also published on CD-ROM under the title "Computer-Aided Systems Human Engineering" (CASHE).

Human Factors Design Standards. Another form of information to support design is engineering or human factors *design standards.* Standards are precise recommendations that relate to very specific areas or topics. One of the commonly used standards in human factors is the military standard MIL-STD-1472D (U.S. Department of Defense, 1989). This standard provides detailed requirements for areas such as controls, visual and audio displays, labeling, anthropometry, workspace design, environmental factors, and designing for maintenance, hazards, and safety. Other standards include the relatively recent ANSI/HFES-100 VDT standard and the ANSI/HFES-200 design standard for software ergonomics (Reed & Billingsley, 1996). Both contain two types of specifications: requirements and recommendations.

Human Factors Principles and Guidelines. Existing standards do not provide solutions for all design problems. For example, there is no current standard to tell a designer where to place the controls on a camera. The designer must look to more abstract principles and guidelines for this information.

Human factors principles and guidelines cover a wide range of topics, some more general than others. On the very general end, Donald Norman gives principles for designing products that are easy to use (Norman, 1992), and Van Cott and Kinkade provide general human factors guidelines for equipment design (Van Cott & Kinkade, 1972). Some guidelines pertain to the design of physical facilities (e.g., McVey, 1990), while others are specific to video display units (e.g., Gilmore, 1985) or software interfaces (e.g., Galitz, 1993; Helander, 1988; Mayhew, 1992; Mosier & Smith, 1986; Shneiderman, 1992). Other guidelines focus on information systems in cars (Campbell et al., 1998; Campbell et al., 1999). Even the Association for the Advancement of Medical Instrumentation has issued human factors guidelines (AAMI, 2001). We present several such guidelines in the following chapters.

It is important to point out that many guidelines are just that: guides rather than hard-and-fast rules. Most guidelines require careful consideration and application by designers, who must think through the implications of their design solutions (Woods et al., 1992).

FRONT-END ANALYSIS

The purpose of front-end analysis is to understand the users, their needs, and the demands of the work situation. Not all of the activities are carried out in detail for every project, but in general, the designer should be able to answer the following questions *before* design solutions are generated in the design stage:

1. Who are the product/system users? (This includes not only users in the traditional sense, but also the people who will dispense, maintain, monitor, repair, and dispose of the system.)
2. What are the major functions to be performed by the system, whether by person or machine? What tasks must be performed?
3. What are the environmental conditions under which the system/product will be used?
4. What are the user's preferences or requirements for the product?

These questions are answered by performing various analyses, the most common of which are described below.

User Analysis

Before any other analysis is conducted, potential system users are identified and characterized for each stage of the system lifecycle. The most important user population are those people who will be regular users or "operators" of the product or system. For example, designers of a more accessible ATM than those currently in use might characterize the primary user population as people ranging from teenagers to senior citizens with an education ranging from junior high to Ph.D. and having at least a third-grade English reading level, or possible physical disabilities (see Chpt 18). After identifying characteristics of the user population, designers should also specify the people who will be installing or maintaining the systems.

It is important to create a complete description of the potential user population. This usually includes characteristics such as age, gender, education level or reading ability, physical size, physical abilities (or disabilities), familiarity with the type of product, and task-relevant skills. For situations where products or systems already exist, one way that designers can determine the characteristics of primary users is to sample the existing population of users. For example, the ATM designer might measure the types of people who currently use ATMs. Notice, however, that this will result in a description of users who are capable of using, and do use, the *existing* ATMs. This is not an appropriate analysis if the goal is to attract, or design for, a wider range of users.

Even if user characteristics are identified, a simple list of characteristics often fails to influence design. Disembodied user characteristics may result in an "elastic user" whose characteristics shift as various features are developed. Designing for an elastic user may create a product that fails to satisfy any real user. Cooper (1999) developed the concept of *personas* to represent the user characteristics in a concrete and understandable manner. A *persona* is a hypothetical person developed through interviews and observations of real people. Personas are not real people, but they represent key characteristics of the user population in the design process. The description of the persona includes not only physical characteristics and abilities, but also the persona's goals, work environment, typical activities, past experience, and precisely what he or she wishes to accomplish. The persona should be specific to the point of having a name. For most applications, three or four personas can represent the characteristics of the user population. Separate personas may be needed to describe people with other roles in the system, such as maintenance personnel. The personas exist to define the goals that the system must support and describe the capabilities and limits of users in concrete terms. Personas enable programmers and other members of the design team to think about specific user characteristics and prevent the natural tendency to assume users are like themselves.

Environment Analysis

In most cases, the user characteristics must be considered in a particular environment. For example, if ATMs are to be placed indoors, environmental analysis would include a somewhat limited set of factors, such as type of access (e.g., will the locations be wheelchair accessible?), weather conditions (e.g., will it exist in a lobby type of area with outdoor temperatures?), and type of clothing people will be wearing (i.e., will they be wearing gloves?). The environment analysis can be performed concurrently with the user and task analysis. Activities or basic tasks that are identified in the task analysis should be described with respect to the specific *environment* in which the activities are performed (Wixon et al., 1990).

Function and Task Analysis

Much of the front-end analysis activity is invested in performing detailed analysis of the functions to be accomplished by the human/machine/environment system and the tasks performed by the human to achieve those functions.

Function Analysis.   Once the population of potential users has been identified, the human factors specialist performs an analysis of the basic functions performed by the "system" (which may be defined as human–machine, human–software, human–equipment–environment, etc.). The functional description lists the general categories of functions served by the system. Functions for an ATM system might simply be *transfer a person's funds into bank account, get funds from bank account to person,* and so forth. Functions represent general transformations of information and system state that help people achieve their goals but do not specify particular tasks.

Task Analysis.   Task analysis is one of the most important tools for understanding the user and can vary substantially in its level of detail. Depending on the nature of the system being designed, the human factors specialist might need to perform a preliminary task analysis (Nielson, 1993), sometimes called an *activity analysis* (Meister, 1971). The preliminary task analysis traditionally specifies the jobs, duties, tasks, and actions that a person will be doing. For example, in designing a chain saw, the designer writes a list of the tasks to be performed with the saw. The tasks should be specific enough to include the types of cuts, type of materials (trees, etc.) to be cut, and so forth. As a simple example, the initial task analysis for design of an ATM might result in a relatively short list of tasks that users would like to perform, such as withdrawing & depositing money from either checking or savings accounts, and determining balances.

In general, the more complex the system, such as air traffic control, the more detailed the function and task analysis. It is not unusual for ergonomists to spend several months performing this analysis for a product or system. The analysis would result in an information base that includes user goals, functions, and major tasks to achieve goals, information required, output, and so on. A task analysis for a digital camera might first specify the different types of photos regularly taken by people—group snapshots, portraits, landscapes, action shots, and so forth. Then, we must add more specific tasks, such as buying film, loading the camera, positioning camera and subject with respect to distance and light, using flash, and so on. Finally, the analysis should also include evaluation of any other activities that may be performed at the same time as the primary tasks being studied. For example, task analysis of a cellular phone for automobile use should include a description of other activities (e.g., driving) that are performed concurrently.

Goals, functions, and tasks are often confused, but they are not the same. A goal is an end condition or reason for performing the tasks. Functions represent the general transformations needed to achieve the goal, and tasks represent the specific activities of the person needed to carry out a function. Goals do not depend on technology, but remain constant; however, technology can change the tasks substantially. Often it is difficult to discriminate the function list from the preliminary task list because the preliminary task list does not provide a detailed description of what the person actually does. For example, a letter opener has the function of opening letters (and perhaps packages), and the task is also to open letters. A more detailed task list would describe the subtasks involved in opening the letter. Similarly, goals and functions are sometimes confused in preliminary analyses of simple systems because the end state (e.g., have the letter open) is quite similar to the function or transformation needed to achieve that state (e.g., open the letter). The short list of a preliminary task analysis is often adequate at the beginning of the design process, but a more extensive task analysis may be needed as the design process progresses.

How to Perform a Task Analysis

Most generally, a task analysis is a way of systematically describing human interaction with a system to understand how to match the demands of the system to human capabilities. The following steps describe the basic elements of a task analysis:

- Define the analysis purpose and identify the type of data required.
- Collect task data.
- Summarize task data.
- Analyze task data.

Kirwan and Ainsworth (1992) provide an exhaustive description of task analysis techniques.

Define Purpose and Required Data.   The first step of task analysis is to define what design considerations the task analysis is to address. Because a task analysis can be quite time consuming, it is critical to focus the analysis on the end use of the data. Typical reasons for performing a task analysis include defining training requirements, identifying software and hardware design requirements, redesigning processes, assessing system reliability, evaluating staffing requirements, and estimating workload.

Both the purpose and the type of the task will influence the information gathered. Tasks can be *physical tasks,* such as setting the shutter speed on a camera, or they can be *cognitive tasks,* such as deciding what the shutter speed should be. Because an increasing number of jobs have a large proportion of cognitive subtasks, the traditional task analysis is being increasingly augmented to describe the cognitive processes, skills, strategies, and use of information required for task performance (Schragen, Chipman, & Shalin, 2000; Gordon & Gill, 1997). While many methods are currently being developed specifically for cognitive task analysis, we will treat these as extensions of standard task analyses, referring to all as *task analysis.* However, if any of the following characteristics are present, designers should pay strong attention to the cognitive components in conducting the analysis (Gordon, 1994).

- Complex decision making, problem solving, diagnosis, or reasoning
- Large amounts of conceptual knowledge needed to perform tasks
- Large and complex rule structures that are highly dependent on situational characteristics

Tasks can be described by several types of information. A particularly important type of information collected in many task analyses is the *hierarchical relationships,* which describe how tasks are composed of subtasks and how groups of tasks combine into functions. With the camera example, a function is *take a picture,* a task that is part of this function is *turn on camera,* and a subtask that is part of this task is *press the on/off switch.* Describing the hierarchical relationships between functions, tasks, and subtasks makes the detail of hundreds of subtasks understandable. Hierarchical grouping of functions, tasks, and subtasks also provides useful information for designing training programs because it identifies natural groupings of tasks to be learned.

A second important type of information in describing tasks is *information flow,* which describes the communication between people and the roles that people and automated systems play in the system. With the camera example, important roles might include the photographer and the recipient of the picture. In this situation, the flow of information would be the image and any annotations or messages that describe the moment captured. For some systems, a complex network of people and automation that must be coordinated. In other systems, it may be only a single person and the technology. However, most systems involve multiple people who must be coordinated, and thinking about the individuals and their roles can identify important design considerations regarding the flow of information and resources that might otherwise go unnoticed, such as how to get the photograph attached to an email message or posted on a Web site.

A third type of information describing tasks is the *task sequence,* which describes the order of tasks and the relationship between tasks over time. In the camera example, important task sequence information would be that the user must first turn on the camera, then frame the picture, and finally depress the shutter button. Performed in a different order, these tasks would not achieve the goal of taking the picture. Task sequence information can be particularly useful in determining how long a set of tasks will take to complete or in estimating the number of people required to complete them. Specific task sequence information includes the goal or intent of task, sequential relationship (what tasks must precede or follow), trigger or event that starts a task sequence, results or outcome of performing the tasks, duration of task, number and type of people required, and the tasks that will be performed concurrently.

A fourth type of information describing tasks is the *location and environmental conditions,* which describe the physical world in which the tasks occur. In the camera example, important location information might be the layout of the user's desk and whether the desk space available makes it difficult to transfer pictures from the camera to the computer. Location of equipment can greatly influence the effectiveness of people in production-line settings. The physical space can also have a surprisingly large effect on computer-based work, as anyone who has had to walk down the hall to a printer knows. Specific location and environmental information include

- Paths that people take to get from one place to another.
- Places where particular tasks occur.
- Physical structures, such as walls, partitions, and desks.
- Tools and their location.
- Conditions under which the tasks are performed.
- Layout of places, paths, and physical structures.

These four categories describe tasks from a different perspective and are all required for a comprehensive task analysis. Other useful information can be included in these four categories, such as the probability of performing the task incorrectly, the frequency with which an activity occurs, and the importance of the task. For example, the frequency of occurrence can describe an information flow between people or the number of times a particular path is taken. Most importantly, a task analysis should record instances where the current system makes it difficult for users to achieve their objectives; such data identify opportunities for redesigning and improving the system.

After the purpose of the task analysis is defined and relevant data identified, task data must be collected, summarized, and analyzed. Many methods exist to support these steps. One of the best resources is Kirwan and Ainsworth (1992), *A Guidebook to Task Analysis,* which describes 41 different methods for task analysis (with detailed examples). Schraagen et al (2000) describe several cognitive task analyses methods. There are a wide range of methods currently in use, organized according to three stages of the task analysis process: methods for collecting task analysis data, methods for representing the task data, and methods for analyzing task data. We review only the most commonly used methods; for a lengthier review of the techniques, see Gordon (1994).

Task analysis tends to be characterized by periods of data collection, analysis, developing new questions, making design changes, and then collecting more data. The following methods can be used in any combination during this iterative process.

Collect Task Data

A task analysis is conducted by interacting extensively with multiple users (Diaper, 1989; Johnson, 1992; Nielson, 1993). The particular data collection approach depends on the information required for the analysis. Ideally, human factors specialists observe and question users as they perform tasks. This is not always possible, and it may be more cost effective to collect some information with other techniques, such as surveys or questionnaires.

Observation. One of the most useful data collection methods is to observe users using existing versions of the product or system if such systems exist (Nielson, 1993; Wixon et al., 1990). For analysis of a camera, we would find users who represent the different types of people who would use the camera, observe how they use their cameras, and identify activities or general tasks performed with the camera. System users are asked to perform the activities under a variety of typical scenarios, and the analyst observes the work, asking questions as needed. It is important to identify different methods for accomplishing a goal rather than identifying only the one typically used by a person. Observation can be performed in the field where the person normally accomplishes the task, or it can be done in a simulated or laboratory situation.

Observations can often be much more valuable than interviews or focus groups because what people say does not always match what they do. In addition, people may omit critical details of their work, they may find it difficult to imagine new technology, and they may distort their description to avoid appearing incompetent or confused. It is often difficult for users to imagine and describe how they would perform a given task or activity. As Wixon and colleagues (1990) note, the structure of users' work is often revealed in their thoughts, goals, and intentions, and so observations alone are not sufficient to understand the tasks. This is particularly true with primarily cognitive tasks that may generate little observable activity.

Think-Aloud Verbal Protocol. Many researchers and designers conduct task analyses by having users think out loud as they perform various tasks. This yields insight into underlying goals, strategies, decisions, and other cognitive components. The verbalizations regarding task performance are termed *verbal protocols,* and analysis or evaluation of the protocols is termed *verbal protocol analysis.* Verbal protocols are usually one of three types: *concurrent* (obtained during task performance), *retrospective* (obtained after task performance via memory or videotape review), and *prospective* (users are given a hypothetical scenario and think aloud as they imagine performing the task). Concurrent protocols are sometimes difficult to obtain. If the task takes place quickly or requires concentration, the user may have difficulty verbalizing thoughts. Retrospective protocols can thus be easier on the user, and a comparative evaluation by Ohnemus and Biers (1993) showed that retrospective protocols actually yield more useable information than do concurrent protocols. Bowers and Snyder (1990) note that concurrent protocols tend to yield procedural information, while retrospective protocols yield more by way of explanations.

Task Performance with Questioning.   A variation on the collection of the verbal protocol is to ask users to perform the tasks while answering questions. The advantage of this method over standard verbal protocols is that it may cue users to verbalize their underlying goals or strategies more frequently. The disadvantage is that it can be disruptive. For this reason, retrospective analysis of videotapes is an effective method for task analysis. Users can be asked to provide think-aloud verbalizations, and when they fail to provide the types of information being requested, the human factors specialist can pause the tape and ask the necessary questions. This functions like a structured interview with the added memory prompt of watching task performance.

Unstructured and Structured Interviews.   Users are often interviewed, with the human factors specialist asking them to describe the general activities they perform with respect to the system. It is common to begin with relatively short unstructured interviews with users. It is necessary for the analyst to ask about not only how the users go about the activities but also their preferences and strategies. Analysts should also note points where users fail to achieve their goals, make errors, show lack of understanding, and seem frustrated or uncomfortable (Nielson, 1993).

In an unstructured interview, the specialist asks the user to describe his or her activities and tasks but does not have any particular method for structuring the conversation. Unstructured interviews tend to revolve around questions or statements such as Tell me about . . . ; What kinds of things do you do . . .? ; and, How do you. . . .? Structured interviews include types of questions or methods that make the interview process more efficient and complete (Creasy, 1980; Graesser et al., 1987). Gordon and Gill (1992) have suggested the use of question probes, relating to when, how and why a particular task is performed, and the consequences of *not* performing the task.

Usually, the specialist conducts several interviews with each user, preparing notes and questions beforehand and tape-recording the questions and answers. Hierarchical network notation (graphs) works especially well because interviews can be structured with questions about the hierarchical relationships between functions, tasks, and subtasks (Gordon & Gill, 1992). Sometimes small groups of users are gathered for the interviewing process, known as conducting a *focus group* (Caplan, 1990; Greenbaum, 1993). Focus groups are groups of between six and ten users led by a facilitator familiar with the task and system (Caplan, 1990; Nielson, 1993). The facilitator should be neutral with respect to the outcome of the discussion. Focus groups are advantageous because they are more cost effective than individual interviews (less time for the analyst), and discussion among users often draws out more information because the conversation reminds them of things they would not otherwise remember.

Surveys and Questionnaires.   Surveys and questionnaires are usually written and distributed after designers have obtained preliminary descriptions of activities or basic tasks. See Chapter 2 for a more complete discussion of surveys and their limits. The questionnaires are used to affirm the accuracy of the information, determine the frequency with which various groups of users perform the tasks, and identify any user preferences or biases. These data help designers prioritize different design functions or features.

Limitations.   For all of these methods to collect task data, designers should remember that there are certain limitations if the task analysis is done in too much detail using existing products or systems. As Roth and Woods (1989) pointed out, overreliance on activity and task analysis using existing systems means that new controls, displays, or other performance aids may be designed to enhance the ability to carry out existing operator strategies that "merely cope with the surface demands created by the impoverished representation of the current work environment." This is why the analysis should focus on the basic user goals and needs, and not exactly on *how they are carried out* using the existing products. It is critical to analyze the task data to identify new design concepts that help people achieve their goals rather than to design to fit the current tasks.

One way to go beyond describing existing tasks is to evaluate the underlying characteristics of the environment and the control requirements of the system (Vicente, 1999). In a nuclear power plant, this would be the underlying physics of the reactor. Often, such an analysis reveals new ways to doing things that might not be discovered by talking with users. Finally, it is important to remember that the task analysis should be completed before product/system design begins. The only exception is the case where a new mock-up or prototype is used for analyzing user activities because they cannot be sufficiently performed on any existing system.

Summarize Task Data

Once task-related information has been gathered, it must be documented and organized in some form. Often, several forms are commonly used in conjunction with one another: (1)lists, outlines, and matrices; (2)hierarchies and networks; and (3)flow charts, timelines, and maps.

Lists, Outlines, and Matrices.   Task analysis usually starts with a set of lists and then breaks the tasks down further into subtasks. An example is shown in Table 3.3. After the hierarchical outlines are relatively complete, the analyst might develop tables or matrices specifying related information for each task or subtask, such as information input, required actions, task duration, and so forth. Such a matrix typically has a row for each task, and the columns describe the tasks.

TABLE 3.3   Part of Task Analysis for Using a Digital Camera, Shown in Outline Form

Step 1. Identify a good view of an interesting subject
    A. Pick subject
    B. Change position to avoid obstacles
    C. Adjust angle relative to the sun
Step 2. Prepare camera
    A. Remove lens cap
    B. Turn on camera
    C. Select proper mode for taking pictures
Step 3. Take picture
    A. Frame picture
        i. Select proper mode (e.g., wide angle, panorama)
        ii. Adjust camera orientation
        iii. Adjust zoom
    B. Focus
    C. Press shutter button

Hierarchies.   The disadvantage of using outlines or tables is that tasks tend to have a complex hierarchical organization, and this is easiest to represent and analyze if the data is graphically depicted. This can be done by using either hierarchical charts or hierarchical networks. An example of a hierarchical chart is the frequently used method known as *hierarchical task analysis* (HTA) (e.g., Kirwan & Ainsworth, 1992). This is a versatile graphical notation method that organizes tasks as sets of actions used to accomplish higher level goals. As an illustration, consider the HTA shown in Figure 3.1 for conducting an accident investigation. The tasks are organized into plans, clusters of tasks that define the preferred order of tasks, and conditions that must be met to perform the tasks.

Another type of hierarchical graph is the representational format known as GOMS, short for goals, operators, methods, and selection rules (Card et al., 1983; Kieras, 1988a). The GOMS model is mostly used to analyze tasks performed when using a particular software interface (e.g., John et al., 1994; Kieras, 1988a) and is described in more detail in Chapter 15. Neither HTA nor GOMS represent detailed levels of cognitive information processing or decision making. For tasks that have a greater proportion of cognitive components, conceptual graphs or computer simulations are frequently used to represent information because they are more capable of depicting abstract concepts, rules, strategies, and other cognitive elements (Gordon & Gill, 1997).

Flow Charts, Timelines, and Maps.   Another graphical notation system frequently used for task analysis is a flow-chart format. Flow charts capture the chronological sequence of subtasks as they are normally performed and depict the decision points for taking alternate pathways. A popular type of flow chart is the *operational sequence diagram* (Kirwan & Ainsworth, 1992). Operational

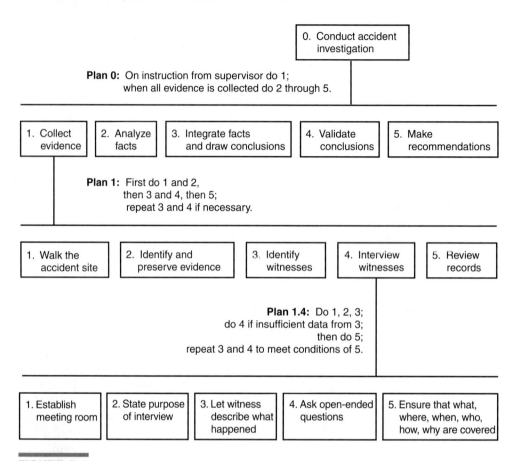

FIGURE 3.1

Hierarchical task analysis for conducting an industrial accident investigation. (*Source:* McCallister, D., unpublished task analysis, University of Idaho. Used with permission.)

sequence diagrams (OSDs), such as that shown in Figure 3.2, show the typical sequence of activity and categorize the operations into various behavioral elements, such as decision, operation, receive, and transmit. They show the interaction among individuals and task equipment. Timelines are useful when the focus is the timing of tasks, and maps are useful when the focus is the physical location of activities.

All of these methods have advantages and disadvantages, and choosing the most appropriate method depends on the type of activity being analyzed. If the tasks are basically linear and usually done in a particular order, as is changing a flat tire, for example, it is appropriate to use an outline or flow chart. If there are more cognitive elements and many conditions for choosing among actions, hierarchical formats are more appropriate. There is one major disadvantage to flow charts that is often not readily apparent. There is evidence that people *mentally* represent goals and tasks in clusters and hierarchies. The design of controls and displays should map onto these clusters and hierarchies. However, when describing or performing a task, the actions will appear as a linear se-

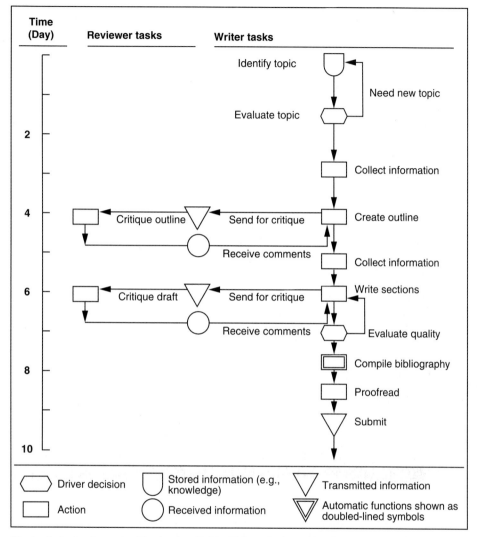

The basic tasks of report writing begin with identifying a topic and involve several iterative steps that result in a polished product. The double square indicates how the bibliography might be compiled using specialized software. This OSD does not include the consequences of procrastination that often dramatically alter the writing process

FIGURE 3.2

Operational sequence diagram for report writing.

quence. If the task analysis is represented in a flow-chart format, the cognitive groupings or "branches" are not evident. This makes it harder for the designer to match the interface with the mental model of the user. To develop efficient interfaces, designers must consider the hierarchal structure and the linear sequence of tasks.

Analyze Task Data

The analysis of these data can include intuitive inspection, such as examining a flow-chart diagram to identify redundant tasks. Frequently, simply inspecting graphics or summary tables cannot make sense of complex systems. More sophisticated analysis approaches are needed. One simple analysis is to use a spreadsheet to calculate the mean and standard deviation of individual task times or sort the tasks to identify tasks that require certain skills or that people find difficult. The spreadsheet can also be used to combine the frequency of occurrence and duration to determine the total time devoted to particular tasks. More sophisticated approaches use computer simulations that combine task data to predict system performance under a variety of conditions (Brown et al., 2001). These quantitative techniques provide a way of going beyond the intuitive analysis of a diagram.

Network analysis. Matrix manipulations can be used to examine information flows in a network. Figure 3.3 shows a matrix representation of information flows between functions. Adding across the rows and down the columns identifies central functions. This simple calculation shows that function 2 is central in providing input to other functions and that function 3 is central in receiving input from other functions. More sophisticated matrix manipulations can identify clusters of related functions (Kusiak, 1999; Wasserman & Faust, 1994). This

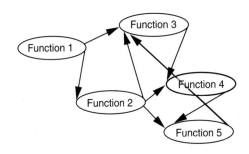

	Function 1	Function 2	Function 3	Function 4	Function 5	
Function 1		1	1			2
Function 2			1	1	1	3
Function 3			1			1
Function 4				1		1
Function 5			1			1
	0	1	3	2	2	

FIGURE 3.3

Graphical and matrix representations of information flows among functions.

approach is most useful when there are many functions and the graphs become too complex to interpret by looking at them. Chapter 10 describes how this approach can be used for determining the appropriate layout for equipment.

Workload Analysis.   The product or system being designed may be complex enough to evaluate whether it is going to place excessive mental workloads on the user, either alone or in conjunction with other tasks. When this is the case, the human factors specialist performs an analysis to predict the workloads that will be placed on the user during various points of task performance, an issue we discuss in Chapters 6 and 13. Sometimes this can be done using the results of the task analysis if the information is sufficiently detailed.

Simulation and Modeling.   Computer simulation and modeling, an evaluation tool discussed in Chapter 2, can also be viewed as a tool for task analysis, whereby software can effectively analyze the output of tasks performed by a human, whether these involve physical operations, like reaching and grasping (See Chapters 10 & 11), or cognitive ones like decision making (Laughery & Corker, 1997: Pew and Mavor, 1998, Elkind et al, 1990).

Safety Analysis.   Any time a product or system has implications for human safety, analyses should be conducted to identify potential hazards or the likelihood of human error. There are several standard methods for performing such analyses. Safety analysis is covered in Chapter 14.

Scenario Specification.   A useful way of making task sequence data concrete is to create scenarios (McGraw & Harbison, 1997). *Scenarios* describe a situation and a specific set of tasks that represent an important use of the system or product. Scenarios are a first step in creating the sequence of screens in software development, and they also define the tasks users might be asked to complete in usability tests. In creating a scenario, tasks are examined, and only those that directly serve users' goals are retained. Those associated with the specific characteristics of the old technology are discarded. Two types of scenarios are useful for focusing scenario specification on the design. The first is *daily use* scenarios, which describe the common sets of tasks that occur daily. In the camera example, this might be the sequence of activities associated with taking a picture indoors using a flashbulb. The second is *necessary use* scenarios, which describe infrequent but critical sets of tasks that must be performed. In the camera example, this might be the sequence of activities associated with taking a picture using a sepia setting to create the feel of an old photograph. Scenarios can be thought of as the script that the personas follow in using the system (Cooper, 1999).

Identify User Preferences and Requirements

Identifying user preferences and requirements is a logical extension of the task analysis. Human factors analysts attempt to determine key needs and preferences that correspond to the major user activities or goals already identified. Sometimes, these preferences include issues related to automation; that is, do users prefer to do a task themselves, or would they rather the system do it automatically?

As an example, for designing a camera, we might ask users (via interview or questionnaire) for information regarding the extent to which water resistance is important, the importance of different features, whether camera size (compactness) is more important than picture quality and so on.

It is easy to see that user preference and requirements analysis can be quite extensive. Much of this type of analysis is closely related to market analysis, and the marketing expert on the design team should be a partner in this phase. Finally, if there are extensive needs or preferences for product characteristics, some attempt should be made to weight or prioritize them.

ITERATIVE DESIGN AND TESTING

Once the front-end analysis has been performed, the designers have an understanding of the user's needs. This understanding must then be consolidated and used to identify initial system specifications and create initial prototypes. As initial prototypes are developed, the designer or design team begins to characterize the product in more detail. The human factors specialist usually works with the designer and one or more users to support the human factors aspects of the design. Much of this work revolves around analyzing the way in which users must perform the functions that have been allocated to the human. More specifically, the human factors specialist evaluates the functions to make sure that they require physical and cognitive actions that fall within the human capability limits. In other words, can humans perform the functions safely and easily?

The initial evaluation is based on the task analysis and is followed by other activities, such as heuristic design evaluation, tradeoff studies, prototyping, and usability testing. The evaluation studies provide feedback for making modifications to the design or prototype. Frequently, early prototypes for software development are created by drawing potential screens to create a paper prototype. Because paper prototypes can be redrawn with little cost, they are very effective at the beginning of the development process because they make it possible to try out many design alternatives (see Chapter 15 for more details on paper prototypes). Paper prototypes are used to verify the understanding of the users' needs identified in the front-end analysis. The purpose of this design stage is to identify and evaluate how technology can fulfill users' needs and address the work demands. This redesign and evaluation continues for *many iterations,* sometimes as many as 10 or 20. The questions answered during this stage of the design process include

1. Do the identified features and functions match user preferences and meet user requirements?
2. Are there any existing constraints with respect to design of the system?
3. What are the human factors criteria for design solutions?
4. Which design alternatives best accommodate human limits?

Providing Input for System Specifications

Once information has been gathered, with respect to user characteristics, basic tasks or activities, the environment(s), and user requirements, the design team writes a set of system specifications and conceptual design solutions. These start out as relatively vague and become progressively more specific. Design solutions are often based on previous products or systems. As the design team generates alternative solutions, the human factors specialist focuses on whether the design will meet system specifications for operator performance, satisfaction, and safety, bringing to bear the expertise gained from the sources of knowledge for design work discussed earlier in the chapter.

System specifications usually include (1) the overall *objectives* the system supports, (2) *performance requirements and features,* and (3) design *constraints.* The challenge is to generate system specifications that select possible features and engineering performance requirements that best satisfy user objectives and goals.

The objectives are global and are written in terms to avoid premature design decisions. They describe what must be done to achieve the user's goals, but not how to do it. The system objectives should reflect the user's goals and not the technology used to build the system. As an example, the objectives for a digital camera targeted at novice to intermediate photographers might include the following (partial) list:

- Capacity to take many pictures
- Take photos outdoors or indoors in a wide range of lighting conditions
- Review pictures without a computer connection
- Take group photographs including the user
- Take close-up pictures of distant objects
- Take pictures without making adjustments

The objectives do not specify any particular product configuration and should not state specifically how the user will accomplish goals or perform tasks.

After the objectives are written, designers determine the means by which the product/system will help the user achieve his or her goals. These are termed *performance requirements and features.* The features state what the system will be able to do and under what conditions. Examples for the camera design might include items such as tripod mount, flash and fill-in flash for distances up to 15 feet, zoom lens, automatic focus and shutter timing capability, at least 16 MB of memory and LCD display.

The performance requirements and system features provide a design space in which the design team develops various solutions. Finally, in addition to the objectives and system features, the specifications document lists various design *constraints,* such as weight, speed, cost, abilities of users, and so forth. More generally, design constraints include cost, manufacturing, development time, and environmental considerations. The constraints limit possible design alternatives.

Translating the user needs and goals into system specifications requires the human factors specialist to take a *systems design* approach, analyzing the entire human–machine system to determine the best configuration of features. The focus should not be on the technology or the person, but on the person–technology system as a unit. The systems design approach draws upon several tools and analyses, discussed as follows.

Quality Function Deployment.   What is the role of the human factors specialist as the system specifications are written? He or she compares the system features and constraints with user characteristics, activities, environmental conditions, and especially the users' preferences or requirements (Bailey, 1996; Dockery & Neuman, 1994). This ensures that the design specifications meet the needs of users and do not add a great number of technical features that people do not necessarily want. Human factors designers often use a simple yet effective method for this process known as the QFD (quality function deployment), which uses the "house of quality" analysis tool (Barnett et al., 1992; Hauser & Clausing, 1988). This tool uses a decision matrix to relate objectives to system features, allowing designers to see the degree to which the proposed features will satisfy customer needs. The matrix also supports analysis of potential conflicts between objectives and the system features.

Figure 3.4 shows a simplified house of quality for the digital camera design. The rows represent the objectives. The columns represent the performance requirements and system features. The task analysis and user preferences identify the importance or weighting of each requirement, which is shown in the column to the right of the objectives. These weightings are often determined by asking people to assign numbers to the importance of the objectives, 9 for very important, 3 for somewhat important, and 1 for marginally important objectives. The rating in each cell in the matrix represents how well each system feature satisfies

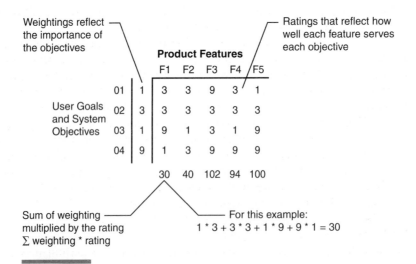

FIGURE 3.4
Simplified house of quality decision matrix for evaluating the importance of features (F) relative to objectives (O).

each objective. These weightings of objectives and ratings of their relationship to features are typically defined using the same 9/3/1 rating scale used to define the weighting, where 9 is most important and 1 is least important. The importance of any feature can then be calculated by multiplying the ratings of each feature by the weighting of each objective and adding the result. This calculation shows the features that matter most for achieving the user's goals. This analysis clearly separates technology-centered features from user-centered features and keeps system development focused on supporting the objectives.

Cost/Benefit Analysis.   The QFD analysis identifies the relative importance of potential system features based on how well they serve users' goals. The importance of the potential features can serve as the input to cost/benefit analysis, which compares different design features according to their costs relative to their benefits. The cost and benefit can be defined monetarily or by a 9/3/1 rating scale. The most common method for doing a quantitative cost/benefit analysis is to create a decision matrix similar to that shown in Figure 3.4. The features, or variables, on which the design alternatives differ are listed as rows on the left side of a matrix, and the different design alternatives are listed as columns across the top. Example features for the camera include the tripod mount and LCD display. Each feature or variable is given a weight representing importance of the feature—the result of the QFD analysis. For the features in Figure 3.4 this would be the total importance shown in the bottom row of the decision matrix. Then, each design alternative is assigned a rating representing how well it addresses the feature. This rating is multiplied by the weighting of each feature and added to determine the total benefit for a design. The cost is divided by this number to determine the cost/benefit ratio. Features with the lowest cost/benefit ratio contribute most strongly to the value of the product.

Tradeoff Analysis.   Sometimes a design feature, such as a particular display, can be implemented in more than one way. The human factors analyst might not have data or guidelines to direct a decision between alternatives. Many times, a small-scale study is conducted to determine which design alternative results in the best performance (e.g., fastest or most accurate). These studies are referred to as trade studies. Sometimes, the analysis can be done by the designer without actually running studies, using methods such as modeling or performance estimates. If multiple factors are considered, the design tradeoffs might revolve around the design with the greatest number of advantages and the smallest number of disadvantages. Alternatively, a decision matrix similar to that used for the QFD and cost/benefit analysis can be constructed. The matrix would assess how well features, represented as rows in the matrix, are served by the different means of implementation, represented as columns.

Although the decision matrix analyses can be very useful, they all share the tendency of considering a product in terms of independent features. Focusing on individual features may fail to consider global issues concerning how they interact as a group. People use a product, not a set of features—a product is more than the sum of its features. Because of this, matrix analyses should be complemented with other approaches, such as scenario specification, so that the product is a coherent whole that supports the user rather than simply a set of highly important but disconnected features.

Human Factors Criteria Identification.   Another role for the human factors specialist is adding human factors criteria to the list of system requirements. This is especially common for software usability engineering (Dix et al., 1993). Human factors criteria, sometimes termed *usability requirements,* specify characteristics that the system should include that pertain directly to human performance and safety. For software usability engineering, human factors requirements might include items such as error recovery, or supporting user interaction pertaining to more than one task at a time.

As another example, for an ergonomic keyboard design, McAlindon (1994) specified that the new keyboard must eliminate excessive wrist deviation, eliminate excessive key forces, and reduce finger movement. The design that resulted from these requirements was a "keybowl" drastically different from the traditional QWERTY keyboard currently in use, but a design that satisfied the ergonomic criteria.

Functional Allocation.   Many functions can be accomplished by either a person or technology, and the human factors specialist must identify an appropriate function for each. To do this, the specialist first evaluates the basic functions that must be performed by the human–machine system in order to support or accomplish the activities identified earlier (Kirwan & Ainsworth, 1992). He or she then determines whether each function is to be performed by the system (automatic), the person (manual), or some combination. This process is termed *functional allocation* and is an important, sometimes critical, step in human factors engineering (Price, 1990).

An example of functional allocation can be given for our camera analysis. We may have determined from the predesign analysis that users prefer a camera that will always automatically determine the best aperture and shutter speed when the camera is held up and focused. Given that the technology exists and that there are no strong reasons against doing so, these functions would then be allocated to the camera. The functional analysis is usually done in conjunction with a cost/benefit analysis to determine whether the allocation is feasible.

However, functional allocation is sometimes not so simple. There are numerous complex reasons for allocating functions to either machine or person. In 1951, Paul Fitts provided a list of those functions performed more capably by humans and those performed more capably by machines (Fitts, 1951). Many such lists have been published since that time, and some researchers have suggested that allocation simply be made by assigning a function to the more "capable" system component. Given this traditional view, where function is simply allocated to the most capable system component (either human or machine), we might ultimately see a world where the functional allocation resembles that depicted in Figure 3.5.

This figure demonstrates the functional allocation strategy now known as the *leftover approach.* As machines have become more capable, human factors

FIGURE 3.5

Ultimate functional allocation when using a "capability" criterion. (*Source:* Cheney, 1989. New Yorker Magazine, Inc.)

specialists have come to realize that functional allocation is more complicated than simply assigning each function to the component (human or machine) that is most capable in some absolute sense. There are other important factors, including whether the human would simply *rather* perform the function. Most importantly, functions should be shared between the person and the automation so that the person is left with a coherent set of tasks that he or she can understand and respond to when the inherent flexibility of the person is needed. Several researchers have written guidelines for performing functional allocation (Kantowitz & Sorkin, 1987; Meister, 1971; Price, 1985, 1990) although it is still more art than science. Functional allocation is closely related to the question of automation and is covered in more depth in Chapter 16.

Support Materials Development.　Finally, as the product specifications become more complete, the human factors specialist is often involved in design of support materials, or what Bailey calls "facilitators" (Bailey, 1996). Frequently, these materials are developed only after the system design is complete. This is unfortunate. The design of the support materials should begin as part of the system specifications that begin with the front-end analyses. Products are often accompanied by manuals, assembly instructions, owner's manuals, training programs, and so forth. A large responsibility for the human factors member of the design team is to make sure that these materials are compatible with the characteristics and limitations of the human user. For example, the owner's manual accompanying a table saw contains very important information on safety and correct procedures. This information is critical and must be presented in a way that maximizes the likelihood that the user will read it, understand it, and comply with it. The development of support materials is discussed in Chapters 14 and 18.

Organization Design

Some of the work performed by ergonomists concerns programmatic design and analysis that address interface, interaction, and organization design. Organization design concerns the training, procedure, and staffing changes. For example, a human factors specialist might conduct an ergonomic analysis for an entire manufacturing plant. This analysis would consider a wide range of factors, including

- Design of individual pieces of equipment from a human factors perspective.
- Hazards associated with equipment, workstations, environments, and so on.
- Safety procedures and policies.
- Design of workstations.
- Efficiency of plant layout.
- Efficiency of jobs and tasks.
- Adequacy of employee training.
- Organizational design and job structures.
- Reward or incentive policies.
- Information exchange and communication.

After evaluating these facets, the human factors specialist develops a list of recommendations for the plant. These recommendations go beyond interface and interaction design for individual pieces of equipment.

An example is given by Eckbreth (1993), who reports an ergonomic evaluation and improvement study for a telecommunications equipment manufacturer. This company had experienced a variety of employee injuries and illness among cable formers in its shops. A team consisting of process engineer, supervisor, plant ergonomist, production associates, and maintenance personnel evaluated the shop. The team assessed injury and accident records and employee complaints, and reviewed task performance videotapes. An ergonomic analysis was carried out, and the team came up with recommendations and associated costs. The recommendations included

Training: Thirty-six employees were taught basic ergonomic principles, including the best working positions, how to use the adjustability of their workstations, and positions to avoid.

Changes to existing equipment: Repairs were made to a piece of equipment, which changed the force required to rotate a component (from 58 pounds down to 16).

Equipment redesign or replacement: Some equipment, such as the board for forming cables, was redesigned and constructed to allow proper posture and task performance in accordance with ergonomic principles. Other equipment, such as scissors, was replaced with more ergonomically sound equipment.

Purchase of step stools: The purchase of step stools eliminated overhead reaching that had occurred with certain tasks.

Antifatigue mats: Floor mats to reduce fatigue and cumulative trauma disorder were purchased.

Job rotation: Job rotation was recommended but could not be implemented because it was the only level-2 union job in the company.

This example shows that a workstation or plant analysis frequently results in a wide variety of ergonomic recommendations. After the recommended changes are instituted, the human factors specialist should evaluate the effects of the changes. Obviously, the most common research design for program evaluation is the pretest-posttest comparison. Because the design is not a true experiment, there are certain factors that can make the results uninterpretable. Ergonomists should design program evaluation studies carefully in order to avoid drawing conclusions that are unfounded (see Cook et al., 1991, for detailed information on the limitations and cautions in making such comparisons).

It is clear that human factors concerns more than just the characteristics or interface of a single product or piece of equipment. An increasing number of human factors specialists are realizing that often an entire reengineering of the organization, including the beliefs and attitudes of employees, must be addressed for long-term changes to occur. This global approach to system redesign, termed *macroergonomics,* is a new and growing subfield in human factors. We briefly review the basic concepts of macroergonomics in chapter 19, which deals with social factors. New technology often changes roles of the users considerably, and ignoring the social and organization implications of these changes undermine system success.

Prototypes

To support interface and interaction design, usability testing, and other human factors activities, product *mock-ups* and *prototypes* are built very early in the design process. Mock-ups are very crude approximations of the final product, often made of foam or cardboard. Prototypes frequently have more of the look and feel of the final product but do not yet have full functionality. Paper prototypes of software systems are useful because screen designs can be sketched on paper, then quickly created and modified with little investment. For this reason, they can be useful early in the design process. The use of prototypes during the design process has a number of advantages:

- Confirming insights gathered during the front-end analysis.
- Support of the design team in making ideas concrete.
- Support of the design team by providing a communication medium.
- Support for heuristic evaluation.
- Support for usability testing by giving users something to react to and use.

In designing computer interfaces, specialists often use *rapid prototyping* tools that allow extremely quick changes in the interface so that many design iterations can be performed in a short time. Bailey (1993) studied the effectiveness of prototyping and iterative usability testing. He demonstrated that user performance improved 12 percent with each design iteration and that the average time to perform software-based tasks decreased 35 percent from the first to the final design iteration. Prototypes may potentially be used for any of the evaluations listed next.

Heuristic Evaluation

A heuristic evaluation of the design(s) means analytically considering the characteristics of a product or system design to determine whether they meet human factors criteria (Desurvire & Thomas, 1993). For usability engineering, heuristic evaluation means examining every aspect of the interface to make sure that it meets usability standards (Nielson, 1993; Nielson & Molich, 1990). However, there are important aspects of a system that are not directly related to usability, such as safety and comfort. Thus, in this section heuristic evaluation will refer to a systematic evaluation of the product design to judge compliance with human factors guidelines and criteria (see O'Hara, 1994, for a detailed description of one method). Heuristic evaluations are usually performed by comparing the system interface with the human factors criteria listed in the requirements specification and also with other human factors standards and guidelines. This evaluation is done by usability experts and does not include the users of the system. For simple products/systems, checklists may be used for this purpose. Heuristic evaluation can also be performed to determine which of several system characteristics, or design alternatives, would be preferable from a human factors perspective. While an individual analyst can perform the heuristic evaluation, the odds are great that this person will miss most of the usability or other human factors problems. Nielson (1993) reports that, averaged over six projects, only 35 percent of the interface usability problems were found by single evaluators. Since different evaluators find different problems, the difficulty can be overcome by having multiple evaluators perform the heuristic evaluation. Nielson recommends using at least three evaluators, preferably five. Each evaluator should inspect the product design or prototype in isolation from the others. After each has finished the evaluation, they should be encouraged to communicate and aggregate their findings.

Once the heuristic evaluations have been completed, the results should be conveyed to the design team. Often, this can be done in a group meeting, where the evaluators and design team members discuss the problems identified and brainstorm to generate possible design solutions (Nielson, 1994a). Heuristic evaluation has been shown to be very cost effective. For example, Nielson (1994b) reports a case study where the cost was $10,500 for the heuristic evaluation, and the expected benefits were estimated at $500,000 (a 48:1 ratio).

Usability Testing

Designers conduct heuristic evaluations and other studies to narrow the possible design solutions for the product/system. They can determine whether it will cause excessive physical or psychological loads, and they analyze associated hazards. However, if the system involves controls and displays with which the user must interact, there is one task left. The system must be evaluated with respect to usability. *Usability* is primarily the degree to which the system is easy to use, or "user friendly." This translates into a cluster of factors, including the following five variables (from Nielson, 1993):

- *Learnability:* The system should be easy to learn so that the user can rapidly start getting some work done.
- *Efficiency:* The system should be efficient to use so that once the user has learned the system, a high level of productivity is possible.
- *Memorability:* The system should be easy to remember so that the casual user is able to return to the system after some period of not having used it, without having to learn everything all over again.
- *Errors:* The system should have a low error rate so that users make few errors during the use of the system and so that if they do make errors, they can easily recover from them. Further, catastrophic errors must not occur.
- *Satisfaction:* The system should be pleasant to use so that users are subjectively satisfied when using it; they like it.

Designers determine whether a system is usable by submitting it to *usability testing*. Usability testing is the process of having users interact with the system to identify human factors design flaws overlooked by designers. Usability testing conducted early in the design cycle can consist of having a small number of users evaluate rough mock-ups. As the design evolves, a larger number of users are asked to use a more developed prototype to perform various tasks. If users exhibit long task times or a large number of errors, designers revise the design and continue with additional usability testing.

Comprehensive human factors test and evaluation has a long history and provides a more inclusive assessment of the system than does a usability evaluation (Chapanis, 1970; Fitts, 1951). Usability is particularly limited when considering complex systems and organization design. Because usability testing has evolved primarily in the field of human–computer interaction, are (Chapter 15). However, usability methods generalize to essentially any interaction when a system has control and display components, but are more limited than comprehensive test and evaluation methods.

FINAL TEST AND EVALUATION

We have seen that the human factors specialist performs a great deal of evaluation during the system design phases. Once the product has been fully developed, it should undergo final test and evaluation. In traditional engineering, system evaluation would determine whether the physical system is functioning correctly. For our example of a camera, testing would determine whether the product meets design specifications and operates as it should (evaluating factors such as mechanical functions, water resistance, impact resistance, etc.). For human factors test and evaluation, designers are concerned with any aspects of the system that affect human performance, safety, or the performance of the entire human–machine system. For this reason, evaluation inherently means involving users. Data are collected for variables such as acceptability, usability, performance of the user or human–machine system, safety, and so on. Most of the methods used for evaluation are the same experimental methods used for research. Therefore, the material presented in Chapter 2 is applicable. However, evaluation is a complex topic, and readers who will conduct evaluation studies should seek more detailed information from publications such as Weimer (1995) or Meister (1986) and an extensive treatment of testing and evaluation procedures by Carlow International (1990).

CONCLUSION

In this chapter we have seen some of the techniques human factors specialists use to understand user needs and to design systems to meet those needs. Designers who skip the front-end analysis techniques that identify the users, their needs, and their tasks risk creating technology-centered designs that tend to fail. The techniques described in this chapter provide the basic outline for creating user-centered systems. A critical step in designing user-centered systems is to provide human factors criteria for design. Many of these criteria depend on human perceptual, cognitive and control characteristics. The following chapters describe these characteristics in detail.

Chapter

Visual Sensory Systems

The 50-year-old traveler, arriving in an unfamiliar city on a dark, rainy night, is picking up a rental car. The rental agency bus driver points to "the red sedan over there" and drives off, but in the dim light of the parking lot, our traveler cannot easily tell which car is red and which is brown. He climbs into the wrong car, realizes his mistake, and settles at last in the correct vehicle. He pulls out a city map to figure out the way to his destination, but in the dim illumination of the dome light, the printed street names on the map are just a haze of black. Giving up on the map, he remains confident that he will see the appropriate signage to Route 60 that will direct him toward his destination, so he starts the motor to pull out of the lot. The streaming rain forces him to search for the wiper switch, but the switch is hard to find because the dark printed labels cannot be read against the gray color of the interior. A little fumbling, however, and the wipers are on, and he emerges from the lot onto the highway. The rapid traffic closing behind him and bright glare of headlights in his rearview mirror force him to accelerate to an uncomfortably rapid speed. He cannot read the first sign to his right as he speeds by. Did that sign say Route 60 or Route 66? He drives on, assuming that the turnoff will be announced again; he peers ahead, watching for the sign. Suddenly, there it is on the left side of the highway, not the right where he had expected it, and he passes it before he can change lanes. Frustrated, he turns on the dome light to glance at the map again, but in the fraction of a second his head is down, the sound of gravel on the undercarriage signals that his car has slid off the highway. As he drives along the berm, waiting to pull back on the road, he fails to see the huge pothole that unkindly brings his car to an abrupt halt.

Our unfortunate traveler is in a situation that is far from unique. Night driving in unfamiliar locations is one of the more hazardous endeavors that humans undertake (Evans, 1991), especially as they become older (see Chapter 17). The reasons the dangers are so great relate to the pronounced limits of the visual sensory system. Many of these limits reside within the peripheral features of the eyeball itself and the neural pathways that send messages of visual information to the brain. Others relate more directly to brain processing and to many of the perceptual processes we discuss in Chapter 6. In this chapter we discuss the nature of light stimulus and the eyeball anatomy as it processes this light. We then discuss several of the important characteristics of human visual performance as it is affected by this interaction between characteristics of the stimulus and the human perceiver.

THE STIMULUS: LIGHT

Essentially all visual stimuli that the human can perceive may be described as a wave of electromagnetic energy. The wave can be represented as a point along the visual *spectrum*. As shown in Figure 4.1a, this point has a *wavelength*, typically expressed in nanometers along the horizontal axis, and an amplitude on the vertical axis. The wavelength determines the *hue* of the stimulus that is perceived, and the amplitude determines its *brightness*. As the figure shows, the range of wavelengths typically visible to the eye runs from short wavelengths of around 400 nm (typically observed as blue-violet) to long wavelengths of around 700 nm (typically observed as red). In fact, the eye rarely encounters "pure" wavelengths. On the one hand, mixtures of different wavelengths often

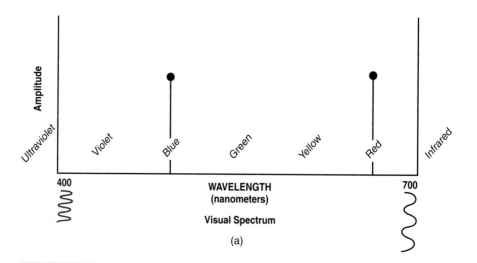

FIGURE 4.1a

(a) The visible spectrum of electromagnetic energy (light). Very short (ultraviolet) and very long (infrared) wavelengths falling just outside of this spectrum are shown. Monochromatic (black, gray, white) hues are not shown because these are generated by the combinations of wavelengths. (b) The CIE color space, showing some typical colors created by levels of *x* and *y* specifications. (*Source:* Helander, M., 1987. The design of visual displays. In *Handbook of Human Factors*. G. Salvendy, ed., New York: Wiley, Fig. 5.1.35, p. 535; Fig. 5.1.36, p. 539. Reprinted by permission of John Wiley and Sons, Inc.).

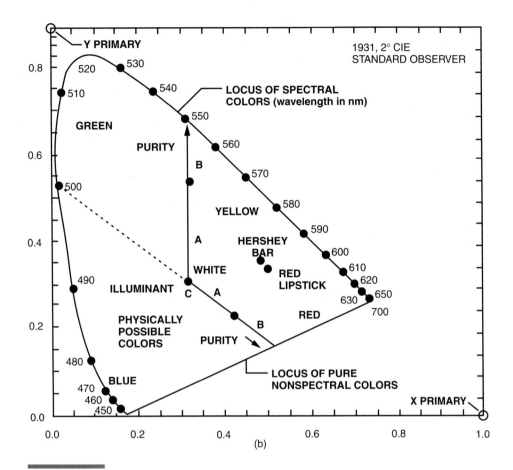

FIGURE 4.1b

act as stimuli. For example, Figure 4.1a depicts a spectrum that is a mixture of red and blue, which would be perceived as purple. On the other hand, the pure wavelengths, characterizing a hue, like blue or yellow, may be "diluted" by mixture with varying amounts of gray or white (called *achromatic* light). This is light with no dominant hue and therefore not represented on the spectrum). When wavelengths are not diluted by gray, like pure red, they are said to be *saturated*. Diluted wavelengths, like pink, are of course unsaturated. Hence, a given light stimulus can be characterized by its hue (spectral values), saturation, and brightness.

The actual hue of a light is typically specified by the combination of the three primary colors—red, green, and blue—necessary to match it (Helander, 1987). This specification follows a procedure developed by the Commission Internationel de L'Elairage and hence is called the CIE color system.

As shown in Figure 4.1b, the CIE color space represents all colors in terms of two primary colors of long and medium wavelengths specified by the *x* and *y* axes respectively (Wyszecki, 1986). Those colors on the rim of the curved lines defining the space are pure, saturated colors. A monochrome light is represented at point C in the middle of the space. The figure does not represent brightness, but this could be shown as a third dimension running above and below the color space of 4.1b. Use of this standard coordinate system allows common specification of colors across different users. For example a "lipstick red" color would be established as having .5 units of long wavelength and .33 units of medium wavelength (see Post, 1992, for a more detailed discussion of color standardization issues).

While we can measure or specify the hue of a stimulus reaching the eyeball by its wavelength, the measurement of brightness is more complex because there are several different meanings of light *intensity*. (Boyce, 1997) This is shown in Figure 4.2, where we see a source of light, like the sun or, in this case, the headlight of our driver's car. This source may be characterized by its *luminous intensity*, or luminous flux, which is the actual light energy of the source. It is measured in units of *candela*. But the amount of this energy that actually strikes the surface of an object to be seen—the road sign, for example—is a very different measure, described as the illuminance and measured in units of lux or foot candles. Hence, the term *illumination* characterizes the lighting quality of a given working environment. How much illuminance an object receives depends

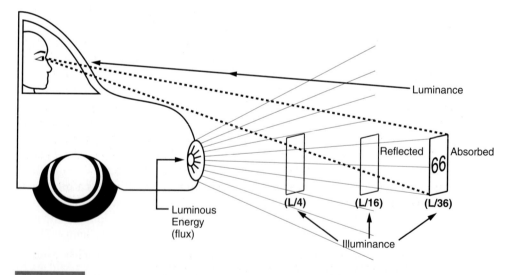

FIGURE 4.2

Concepts behind the perception of visual brightness. Luminance energy (flux) is present at the source (the headlight), but for a given illuminated area (illuminance), this energy declines with the square of the distance from the source. This is illustrated by the values under the three signs at increasing intervals of two units, four units, and six units away from the headlight. Some of the illuminance (solid rays) is absorbed by the sign, and the remainder is reflected back to the observer, characterizing the luminance of the viewed sign. Brightness is the subjective experience of the perceiver.

on the distance of the object from the light source. As the figure shows, the illuminance declines with the square of the distance from the source.

Although we may sometimes be concerned about the illumination of light sources in direct viewing, the amount of glare produced by headlights shining from the oncoming vehicles for example (Theeuwes et al., 2002), and human factors is also concerned about the illumination of work place (see Chapter 14), human factors is also concerned with the amount of light reflected off of objects to be detected, discriminated, and recognized by the observer when these objects are not themselves the source of light. This may characterize, for example, the road sign in Figure 4.2. We refer to this measure as the *luminance* of a particular stimulus typically measured in foot lamberts (FL). Luminance is different from illuminance because of differences in the amount of light that surfaces either reflect or absorb. Black surfaces absorb most of the illuminance striking the surface, leaving little luminance to be seen by the observer. White surfaces reflect most of the illuminance. In fact, we can define the *reflectance* of a surface as the following ratio:

$$\text{Reflectance (\%)} = \frac{\text{luminance (FL)}}{\text{illuminance (FC)}} \tag{4.1}$$

(A useful hint is to think of the illuminance light, leaving some of itself [the "il"] on the surface and sending back to the eye only the luminance.)

The *brightness* of a stimulus, then, is the actual experience of visual intensity, an intensity that often determines its visibility. From this discussion, we can see how the visibility or brightness of a given stimulus may be the same if it is a dark (poorly reflective) sign that is well illuminated or a white (highly reflective) sign that is poorly illuminated. In addition to brightness, the ability to see an object—its visibility—is also affected by the *contrast* between the stimulus and its surround, but that is another story that we shall describe in a few pages.

Table 4.1 summarizes these various measures of light and shows the units by which they are typically measured. A photometer is an electronic device that measures luminous intensity in terms of foot lamberts. An illumination meter is a device that measures illuminance.

TABLE 4.1   Physical Quantities of Light and Their Units

Quantity	Units
Luminous flux	1 candela or 12.57 lumins
Illuminance	Foot candle or 10.76 LUX
Luminance	Candela/M^2 or foot lambert
Reflectance	A ratio
Brightness	

THE RECEPTOR SYSTEM: THE EYEBALL AND THE OPTIC NERVE

Light, or electromagnetic energy, must be transformed to electrochemical neural energy, a process that is accomplished by the eye. Figure 4.3 presents a schematic view of the wonderful receptor system for vision, the eyeball. As we describe certain key features of its anatomy and how this anatomy affects characteristics of the light energy that passes through it, we identify some of the distortions that disrupt our ability to see in many working environments and therefore should be the focus of concern for the human factors engineer.

The Lens

As we see in the figure, the light rays first pass through the cornea, which is a protective surface that absorbs some of the light energy (and does so progressively more as we age). Light rays then pass through the pupil, which opens or dilates (in darkness) and closes or constricts (in brightness) to admit adaptively more light when illumination is low and less when illumination is high. The lens of the eye is responsible for adjusting its shape, or *accommodating*, to bring the image to a precise focus on the back surface of the eyeball, the *retina*. This accommodation is accomplished by a set of ciliary muscles surrounding the lens. Sensory receptors located within the ciliary muscles send information regarding accommodation to the higher perceptual centers of the brain. When we view images up close, the light rays emanating from the images converge as they approach the eye, and the muscles must accommodate by changing the lens to a rounder shape, as reflected in Figure 4.3. When the image is far away and the

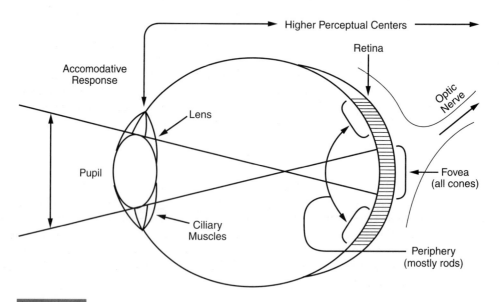

FIGURE 4.3
Key aspects of the anatomy of the eyeball.

light rays reach the eye in essentially parallel fashion, the muscles accommodate by creating a flatter lens. Somewhere in between is a point where the lens comes to a natural "resting" point, at which the muscles are doing little work at all. This is referred to as the *resting state* of accommodation.

The amount of accommodation can be described in terms of the distance of a focused object from the eye. Formally, the amount of accommodation required is measured in *diopters*, which equal 1/viewing distance (meters). Thus, 1 diopter is the accommodation required to view an object at 1 meter.

As our driver discovered when he struggled to read the fine print of the map, our eyeball does not always accommodate easily. It takes time to change its shape, and sometimes there are factors that limit the amount of shape change that is possible. *Myopia*, or nearsightedness, results when the lens cannot flatten and hence distant objects cannot be brought into focus. *Presbyopia*, or farsightedness, results when the lens cannot accommodate to very near stimuli. As we grow older, the lens becomes less flexible in general, but farsightedness in particular becomes more evident. Hence, we see that the older reader, when not using corrective lenses, must hold the map farther away from the eyes to try to gain focus, and it takes longer for that focus to be achieved.

While accommodation may be hindered by limits on flexibility of the lens and compensated by corrective lenses, it is also greatly influenced by the amount of visibility of the image to be fixated, which is determined by both its brightness and its contrast.

The Visual Receptor System

An image, whether focused or not, eventually reaches the retina at the back of the eyeball. The image may be characterized by its intensity (luminance), its wavelengths, and its size. The image size is typically expressed by its *visual angle*, which is depicted by the two-headed arrows in front of the eyes in Figure 4.3. The visual angle of an object of height H, viewed at distance D, is approximately equal to arctan (H/D) (the angle whose tangent = H/D). Knowing the distance of an object from a viewer and its size, one can compute this ratio. For visual angles less than around 10 degrees, the angle may be expressed in minutes of arc rather than degrees (60 minutes = 1 degree) and approximated by the formula

$$VA = 5.7 \times 60 \times (H/D) \qquad (4.2)$$

Importantly, the image can also be characterized by *where* it falls on the back of the retina because this location determines the types of visual receptor cells that are responsible for transforming electromagnetic light energy into the electrical impulses of neural energy to be relayed up the optic nerve to the brain. There are two types of receptor cells, *rods* and *cones*, each with six distinctly different properties. Collectively, these different properties have numerous implications for our visual sensory processing.

1. *Location.* The middle region of the retina, the *fovea*, consisting of an area of around 2 degrees of visual angle, is inhabited exclusively by cones (Figure 4.3).

Outside of the fovea, the *periphery* is inhabited by rods as well as cones, but the concentration of cones declines rapidly moving farther away from the fovea (i.e., with greater *eccentricity*.)

2.   *Acuity*. The amount of fine detail that can be resolved is far greater when the image falls on the closely spaced cones than on the more sparsely spaced rods. We refer to this ability to resolve detail as the *acuity*, often expressed as the inverse of the smallest visual angle (in minutes of arc) that can just be detected. Thus, an acuity of 1.0 means that the operator can resolve a visual angle of 1 minute of arc (1/60 of 1 degree). Table 4.2 provides various ways of measuring visual acuity. Since acuity is higher with cones than rods, it is not surprising that our best ability to resolve detail is in the fovea, where the cone density is greatest. Hence, we "look at" objects that require high acuity, meaning that we orient the eyeball to bring the image into focus on the fovea. While visual acuity drops rapidly toward the periphery, the sensitivity to *motion* declines at a far less rapid rate. We often use the relatively high sensitivity to motion in the periphery as a cue for something important on which we later fixate. That is, we notice motion in the periphery and move our eyes to focus on the moving object.

3.   *Sensitivity*. Although the cones have an advantage over the rods in acuity, the rods have an advantage in terms of sensitivity, characterizing the minimum amount of light that can be detected, or the *threshold*. Sensitivity and threshold are reciprocally related: As one increases, the other decreases. Since there are no rods in the fovea, it is not surprising that our fovea is very poor at picking up dim illumination (i.e., it has a high threshold). To illustrate this, note that if you try to look directly at a faint star, it will appear to vanish. *Scotopic vision* refers to vision at night when only rods are operating. *Photopic vision* refers to vision when the illumination is sufficient to activate both rods and cones (but when most of our visual experience is due to actions of cones).

4.   *Color sensitivity*. Rods cannot discriminate different wavelengths of light (unless they also differ in intensity). Rods are "color blind," and so the extent to which hues can be resolved declines both in peripheral vision (where fewer cones are present) and at night (when only rods are operating). Hence, we can understand how our driver, trying to locate his car at night, was unable to discriminate the poorly illuminated red car from its surrounding neighbors.

5.   *Adaptation*. When stimulated by light, rods rapidly lose their sensitivity, and it takes a long time for them to regain it (up to a half hour) once they are returned to the darkness that is characteristic of the rods' "optimal viewing envi-

TABLE 4.2   Some Measures of Acuity

Minimum separable acuity	General measurement of smallest detail detectable
Vernier acuity	Are two parallel lines aligned?
Landolt ring	Is the gap in a ring detectable?
Snellen acuity	Measurement of detail resolved at 20 feet, relative to the distance at which a normal observer can resolve the same detail (e.g., 20/30)

ronment." This phenomenon describes the temporary "blindness" we experience when we enter a darkened movie theater on a bright afternoon. Environments in which operators are periodically exposed to bright light but often need to use their scotopic vision are particularly disruptive. In contrast to rods, the low sensitivity of the cones is little affected by light stimulation. However, cones may become *hypersensitive* when they have received little stimulation. This is the source of *glare* from bright lights, particularly at night. We discuss glare further in Chapter 13.

6. *Differential wavelength sensitivity.* Whereas cones are generally sensitive to all wavelengths, rods are particularly insensitive to long (i.e., red) lengths. Hence, red objects and surfaces look very black at night. More important, illuminating objects in red light in an otherwise dark environment will not destroy the rods' dark adaptation. For example, on the bridge of a ship, the navigator may use a red lamp to stimulate cones in order to read the fine detail of a chart, but this stimulation will not destroy the rods' dark adaptation and hence will not disrupt the ability of personnel to scan the horizon for faint lights or dark forms.

Collectively, these pronounced differences between rods and cones are responsible for a wide range of visual phenomena. We consider some of the more complex implications of these phenomena to human factors issues related to three important aspects of our sensory processing: contrast sensitivity (CS), night vision, and color vision.

SENSORY PROCESSING LIMITATIONS

Contrast Sensitivity

Our unfortunate driver could not discern the wiper control label, the map detail, or the pothole for a variety of reasons, all related to the vitally important human factors concept of *contrast sensitivity*. Contrast sensitivity may be defined as the reciprocal of the minimum contrast between a lighter and darker spatial area that can just be detected; that is, with a level of contrast below this minimum, the two areas appear homogeneous. Hence, the ability to detect contrast is necessary in order to detect and recognize shapes, whether the discriminating shape of a letter or the blob of a pothole. The contrast of a given visual pattern is typically expressed as the ratio of the *difference* between the luminance of light, L, and dark, D, areas to the *sum* of these two luminance values:

$$c = (L - D)/(L + D) \qquad (4.3)$$

The higher the contrast sensitivity that an observer possesses, the smaller the minimum amount of contrast that can just be detected, C_M, a quantity that describes the *contrast threshold*. Hence,

$$CS = 1/C_M \qquad (4.4)$$

The minimum separable acuity (the width of light separating two dark lines) represents one measure of contrast sensitivity, because a gap that is smaller than this minimum will be perceived as a uniform line of constant brightness.

Contrast sensitivity may often be measured by a *grating*, such as that shown along the *x* axis of Figure 4.4. If the grating appears to be a smooth bar like the grating on the far right of the figure (if it is viewed from a distance), the viewer is unable to discern the alternating patterns of dark and light, and the contrast is below the viewer's CS threshold.

Expressed in this way, we can consider the first of several influences on contrast sensitivity: the spatial frequency of the grating. As shown in Figure 4.4, *spatial* frequency may be expressed as the number of dark-light pairs that occupy 1 degree of visual angle (cycles/degrees or c/d). If you hold this book approximately 1 foot away, then the spatial frequency of the left grating is 0.6 c/d, of the next grating is 1.25 c/d, and of the third grating is 2.0 c/d. We can also see that the spatial frequency is inversely related to the width of the light or dark bar. The human eye is most sensitive to spatial frequencies of around 3 c/d, as shown by the two CS functions drawn as curved lines across the axis of Figure 4.4.

When the contrast (between light and dark bars) is greater, sensitivity is greater across all spatial frequencies.

The high spatial frequencies on the right side of Figure 4.4 characterize our sensitivity to small visual angles and fine detail (and hence reflect the standard measurement of visual acuity), such as that involved in reading fine print or making fine adjustments on a vernier *scale*. Much lower frequencies characterize the recognition of *shapes* in *blurred* or degraded conditions, like the road sign sought by our lost driver or the unseen pothole that terminated his trip. Low contrasts at low spatial frequencies often characterize the viewing of images that

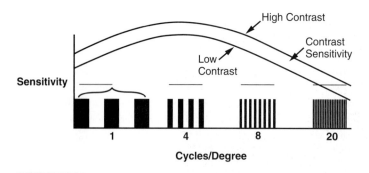

FIGURE 4.4

Spatial frequency gratings, used to measure contrast sensitivity. The particular values on the *x* axis will vary as a function of visual angle and therefore the distances at which the figure is held from the eyes. The line above each grating will occupy 1 degree of visual angle when the book is viewed at a distance of 52 cm. The two curves represent contrast sensitivity as a function of spatial frequency for two different contrast levels.

are degraded by poor "sensor resolution," like those from infrared radar (Uttal et al., 1994).

A second important influence on contrast as seen in Figure 4.4 is that *lower* contrasts are *less easily* discerned. Hence, we can understand the difficulty our driver had in trying to read the label against the gray dashboard. Had the label been printed against a white background, it would have been far easier to read. Many users of products like VCRs are frustrated by the black on black raised printing instructions (Figure 4.5). Color contrast does not necessarily produce good luminance–contrast ratios. Thus, for example, slides that produce black text against a blue background may be very hard for the viewing audience to read.

A third influence on contrast sensitivity is the level of *illumination* of the stimulus ($L + D$, the denominator of formula 4.3). Not surprisingly, *lower* illumination reduces the sensitivity and does so more severely for sensing high spatial frequencies (which depend on cones) than for low frequencies. This explains the obvious difficulty we have reading fine print under low illumination. However, low illumination can also disrupt vision at low spatial frequencies: Note the loss of visibility that our driver suffered for the low spatial frequency pothole.

Two final influences on contrast sensitivity are the resolution of the eye itself and the *dynamic* characteristics of the viewing conditions. Increasing age reduces the amount of light passing through the cornea and greatly reduces the sensitivity. This factor, coupled with the loss of visual accommodation ability at close viewing, produces a severe deficit for older readers in poor illumination. Constant sensitivity declines also when the stimulus is moving relative to the viewer, as our driver found when trying to read the highway sign.

All of these factors, summarized in Table 4.3, are critical for predicting whether or not detail will be perceived and shapes will be recognized in a variety of degraded viewing conditions, and hence these factors are critical for

FIGURE 4.5
Difficult visibility of low-contrast, raised-plastic printing. With small letters and black plastic, such information is often nearly illegible in poor illumination. (*Source:* Courtesy of Anthony D. Andre, Interface Analysis Associates, San Jose, CA.)

TABLE 4.3   **Some Variables That Affect Contrast and Visibility**

Variable	Effect	Example
↓ Contrast	↓ Visibility	Black print on gray
↓ Illumination	↓ Contrast sensitivity	Reading map in poor light
Polarity	Black on white better than white on black	Designing viewgraphs
Spatial frequency	Optimum *CS* at 3 *C/D*	Ideal size of text font given viewing distance
Visual accommodation	*CS*	Map reading during night driving
Motion	↓ CS	Reading a road sign while moving

indirectly informing the designer of certain standards that should be adhered to in order to guarantee viewability of critical symbols. Many of these standards may be found in handbooks like Boff and Lincoln (1988) or textbooks such as Salvendy (1997).

Human factors researchers are also trying to develop models to show how all the influences in Table 4.3 *interact* in a way that would, for example, allow one to specify the minimum text size for presenting instructions to be viewed by someone with 20/40 vision in certain illumination or to determine the probability of recognizing targets at night at a particular distance (Owens et al., 1994). However, the accuracy of such models has not yet reached a point where they are readily applicable when several variables are involved. What can be done instead is to clearly identify how these factors influence the best design whenever print or symbols must be read under less than optimal circumstances. We describe some of these guidelines as they pertain to the readability of the printed word. We discuss other human factors issues of language-based instructions in Chapter 8.

Reading Print.   Most obviously, print should not be too fine in order to guarantee its readability. When space is not at a premium and viewing conditions may be less than optimal, one should seek to come as close to the 3 cycles/degrees value as possible (i.e., stroke width of 1/6 degree of visual angle) to guarantee maximum readability. Fine print and very narrow stroke widths are dangerous choices. Similarly, one should maximize contrast by employing black letters on white background rather than, for example, using the "sexier" but less readable hued backgrounds (e.g., black on blue). Black on red is particularly dangerous with low illumination, since red is not seen by rods. Because of certain asymmetries in the visual processing system, dark text on lighter background ("negative contrast") also offers higher contrast sensitivity than light on dark ("positive contrast"). The disruptive tendency for white letters to spread out or "bleed" over a black background is called *irradiation*.

The actual character font matters too. Fonts that adhere to "typical" letter shapes like the text of this book are easier to read because of their greater familiarity than those that create block letters or other ***nonstandardized shapes***. Another effect on readability is the *case* of the print. For single, isolated words, UP-PERCASE appears to be as good as if not better than lowercase print, as, for example, the label of an "on" switch. This advantage results in part because of the wider visual angle and lower spatial frequency presented. However, for multiword text, UPPERCASE PRINT IS MORE DIFFICULT TO READ than lowercase or mixed-case text. This is because lowercase text typically offers a greater variety of *word shapes*. This variety conveys sensory information at lower spatial frequencies that can be used to discern some aspects of word meaning in parallel with the high spatial frequency analysis of the individual letters (Broadbent & Broadbent, 1980; Allen et al., 1995). BLOCKED WORDS IN ALL CAPITALS will eliminate the contributions of this lower spatial frequency channel. Other guidelines for text size and font type may be found in Sanders and McCormick (1993).

Color Sensation

Color vision is a facility employed in the well-illuminated environment. Our driver had trouble judging the color of his red sedan because of the poor illumination in the parking lot. A second characteristic that limits the effectiveness of color is that approximately 7 percent of the male population is *color deficient*; that is, they are unable to discriminate certain hues from each other. Most prevalent is red-green "color blindness" (*protanopia*) in which the wavelengths of these two hues create identical sensations if they are of the same luminance intensity. Many computer graphics packages use color to discriminate lines. If this is the only discriminating feature between lines, the graph may be useless for the color-blind reader or the reader of the paper passed through a monochrome photocopier.

Because of these two important sensory limitations on color processing, a most important human factors guideline is to *design for monochrome first* (Shneiderman, 1987) and use color only as a redundant backup to signal important information. Thus, for example, a traffic signal uses the location of the illuminated lamp (top, middle, bottom) redundantly with its color to signal the important traffic command information.

Two additional characteristics of the sensory processing of color have some effect on its use. *Simultaneous contrast* is the tendency of some hues to appear different when viewed adjacent to other hues (e.g., green will look deeper when viewed next to red than when viewed next to a neutral gray). This may affect the usability of multicolor-coded displays, like maps, as the number of colors grows large, an issue we treat further in our discussion of absolute judgment in Chapter 8. The *negative afterimage* is a similar phenomenon to simultaneous contrast but describes the greater intensity of certain colors when viewed after prolonged viewing of other colors.

Night Vision

The loss of contrast sensitivity at all spatial frequencies can inhibit the perception of print as well as the detection and recognition of objects by their shape or color in poorly illuminated viewing conditions. Coupled with the loss of contrast sensitivity due to age, it is apparent that night driving for the older population is a hazardous undertaking, particularly in unfamiliar territory (Waller, 1991; Shinar & Schieber, 1991).

Added to these hazards of night vision are those associated with *glare*, which may be defined as irrelevant light of high intensity. Beyond its annoyance and distraction properties, *glare* has the effect of temporarily destroying the rod's sensitivity to low spatial frequencies. Hence, the glare-subjected driver is less able to spot the dimly illuminated road hazard (the pothole or the darkly dressed pedestrian; Theeuwes et al., 2002).

BOTTOM-UP VERSUS TOP-DOWN PROCESSING

Up to now, we have discussed primarily the factors of the human visual system that effect the *quality* of the sensory information that arrives at the brain in order to be perceived. As shown in Figure 4.6, we may represent these influences as those that affect processing from the *bottom* (lower levels of stimulus processing) *upward* (toward the higher centers of the brain involved with perception and understanding). As examples, we may describe loss of acuity as a degradation in bottom-up processing or the high-contrast sensitivity as an enhancement of bottom-up processing. In contrast, an equally important influence on processing operates from the *top* downward. This is perception based on our knowledge (and desire) of what *should be* there. Thus, if I read the instructions, "After the procedure is completed, turn the system off," I need not worry as much if the last word happens to be printed in very small letters or is visible with low con-

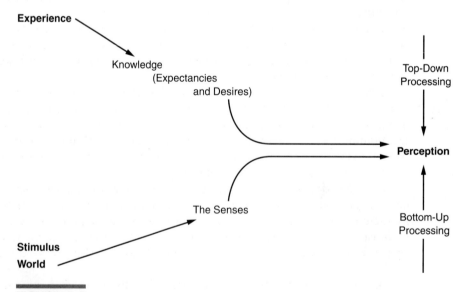

FIGURE 4.6

The relation between bottom-up and top-down processing.

trast because I can pretty much guess what it will say. Much of our processing of perceptual information depends on the delicate interplay between top-down processing, signaling what *should be* there, and bottom-up processing, signaling what is there. Deficiencies in one (e.g., small, barely legible text) can often be compensated by the operation of the other (e.g., expectations of what the text should say). Our initial introduction to the interplay between these two modes of processing is in a discussion of depth perception, and the distinction between the two modes is amplified further in our treatment of signal detection.

DEPTH PERCEPTION

Humans navigate and manipulate in a three-dimensional (3-D) world, and we usually do so quite accurately and automatically (Gibson, 1979). Yet there are times when our ability to perceive where we and other things are in 3-D space breaks down. Airplane pilots flying without using their instruments are also very susceptible to dangerous illusions of where they are in 3-D space and how fast they are moving (O'Hare & Roscoe, 1990; Hawkins & Orlady, 1993; Leibowitz, 1988).

In order to judge our distance from objects (and the distance between objects) in 3-D space, we rely on a host of *depth cues* to inform us of how far away things are. The first three cues we discuss—accommodation, binocular convergence, and binocular disparity—are all inherent in the physiological structure and wiring of the visual sensory system. Hence, they may be said to operate on *bottom-up* processing.

Accommodation, as we have seen, is when an out-of-focus image triggers a change in lens shape to accommodate, or bring the image into focus on the retina. As shown in Figure 4.3, sensory receptors, within the ciliary muscles that accomplish this change, send signals to the higher perceptual centers of the brain that inform those centers how much accommodation was accomplished and hence the extent to which objects are close or far (within a range of about 3 m). (As we discuss in Chapter 5, these signals from the muscles to the brain are called *proprioceptive input*.)

Convergence is a corresponding cue based on the amount of inward rotation ("cross-eyedness") that the muscles in the eyeball must accomplish to bring an image to rest on corresponding parts of the retina on the two eyes. The closer the distance at which the image is viewed, the greater the amount of proprioceptive "convergence signal" sent to the higher brain centers by the sensory receptors within the muscles that control convergence.

Binocular disparity, sometimes called *stereopsis*, is a depth cue that results because the closer an object is to the observer, the greater the amount of disparity there is between the view of the object received by each eyeball. Hence, the brain can use this disparity measure, computed at a location where the visual signals from the two eyes combine in the brain, to estimate how far away the object is.

All three of these bottom-up cues are only effective for judging distance, slant, and speed for objects that are within a few meters from the viewer (Cutting & Vishton, 1995). (However, stereopsis can be created in stereoscopic displays to simulate depth information at much greater distances, as we discuss in Chapter 8.) Judgment of depth and distance for more distant objects and surfaces depends on a host of what are sometimes called "pictorial" cues because they are the kinds of cues that artists put into pictures to convey a sense of depth. Because the effectiveness of most pictorial cues is based on past experience, they are subject to top-down influences. As shown in Figure 4.7, some of the important pictorial cues to depth are

Linear perspective: The converging of parallel lines (i.e., the road) toward the more distant points.

Relative size: A cue based on the knowledge that if two objects are the same true size (e.g., the two trucks in the figure), then the object that occupies a smaller visual angle (the more distant vehicle in the figure) is farther away.

Interposition: Nearer objects tend to obscure the contours of objects that are farther away (see the two buildings).

Light and shading: Three-dimensional objects tend to cast shadows and reveal reflections and shadows on themselves from illuminating light. These shadows provide evidence of their location and their 3-D form (Ramachandran, 1988).

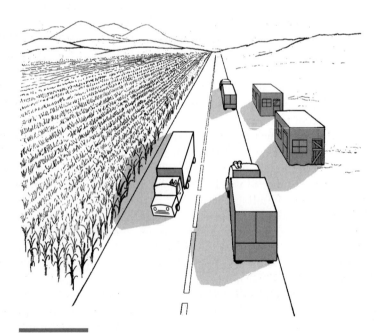

FIGURE 4.7

Some pictorial depth cues. (*Source:* Wickens, C. D., 1992. *Engineering Psychology and human performance.* New York: HarperCollins. Reprinted by permission of Addison-Wesley Educational Publishers, Inc.)

Textural gradients: Any textured surface, viewed from an oblique angle, will show a gradient or change in texture density (spatial frequency) across the visual field (see the Illinois cornfield in the figure). The finer texture signals the more distant region, and the amount of texture change per unit of visual angle signals the angle of slant relative to the line of sight.

Relative motion, or *motion parallax,* describes the fact that more distant objects show relatively smaller movement across the visual field as the observer moves. Thus, we often move our head back and forth to judge the relative distance of objects. Relative motion also accounts for the accelerating growth in the retinal image size of things as we approach them in space, a cue sometimes called *looming* (Regan et al., 1986). We would perceive the vehicle in the left lane of the road in Figure 4.7 to be approaching, because of its growing image size on the retina.

Collectively, these cues provide us with a very rich sense of our position and motion in 3-D space as long as the world through which we move is well illuminated and contains rich visual texture. Gibson (1979) clearly described how the richness of these cues in our natural environment support very accurate space and motion perception. However, when cues are degraded, impoverished, or eliminated by darkness or other unusual viewing circumstances, depth perception can be distorted. This sometimes leads to dangerous circumstances. For example, a pilot flying at night or over an untextured snow cover has very poor visual cues to help determine where he or she is relative to the ground (O'Hare & Roscoe, 1990), so pilots must rely on precision flight instruments (see Chapter 8 & 17). Correspondingly, the implementation of both edge markers and high-angle lighting on highways greatly enriches the cues available for speed (changing position in depth) for judging distance hazards and allows for safer driving (see Chapter 17). In Chapter 8 we discuss how this information is useful for the design of 3-D displays.

Just as we may predict poorer performance in tasks that demand depth judgments when the quality of depth cues is impoverished, we can also predict that certain *distortions* of perception will occur when features of the world violate our expectations, and top-down processing takes over to give us an inappropriate perception. For example, Eberts and MacMillan (1985) established that the higher-than-average rate at which small cars are hit from behind results because of the cue of relative size. A small car is perceived as more distant than it really is from the observer approaching it from the rear. Hence, a small car is approached faster (and braking begins later) than is appropriate, sometimes leading to the unfortunate collision.

Of course, clever application of human factors can sometimes turn these distortions to advantage, as in the case of the redesign of a dangerous traffic circle in Scotland (Denton, 1980). Drivers tended to overspeed when coming into the traffic circle with a high accident rate as a consequence. In suggesting a solution, Denton decided to trick the driver's perceptual system by drawing lines across the roadway of diminishing separation, as the circle was approached. Approaching the circle at a constant (and excessive) speed, the driver experiences the "flow" of texture past the vehicle as signaling increasing in speed (i.e., accelerating). Because of the nearly automatic way in which many aspects of perception are carried out, the driver should instinctively brake in response to the perceived acceleration, bringing the speed closer to the desired safe value. This is exactly the effect that was observed in relation to driving behavior after the marked pavement was introduced, resulting in a substantial reduction in fatal accidents at the traffic circle, a result that has been sustained for several years (Godley, 1997).

VISUAL SEARCH AND DETECTION

A critical aspect of human performance in many systems concerns the closely linked processes of visual search and object or event detection. Our driver at the beginning of the chapter was searching for several things: the appropriate control for the wipers, the needed road sign, and of course any number of possible hazards or obstacles that could appear on the road (the pothole was one that was missed). The goal of these searches was to *detect* the object or event in question. These tasks are analogous to the kind of processes we go through when we search the phone book for the pizza delivery listing, search the index of this book for a needed topic, search a cluttered graph for a data point, or when the quality control inspector searches the product (say, a circuit board) for a flaw. In all cases, the search may or may not successfully end in a detection.

Despite the close link between visual search and detection, it is important to separate our treatment of these topics, both because different factors affect each and because human factors personnel are sometimes interested in detection when there is no search (e.g., the detection of a fire alarm). We consider the process of search itself, but to understand visual search, we must first consider the nature of eye movements, which are heavily involved in searching large areas of space. Then we consider the process of detection.

Eye Movements

Eye movements are necessary to search the visual field (Monty & Senders, 1976; Hallett, 1986). Eye movements can generally be divided into two major classes. *Pursuit* movements are those of constant velocity that are designed to follow moving targets, for example, following the rapid flight of an aircraft across the sky. More related to visual search are *saccadic* eye movements, which are abrupt, discrete movements from one location to the next. Each saccadic movement can be characterized by a set of three critical features: an *initiation latency*, a *movement time* (or speed), and a *destination*. Each destination, or *dwell*, can be characterized by both its *dwell duration* and a *useful field of view* (UFOV). In continuous search, the initiation latency and the dwell duration cannot be distinguished.

The actual movement time is generally quite fast (typically less than 50 msec) and is not much greater for longer than for shorter movements. The greatest time is spent during dwells and initiations. These time limits are such that even in rapid search there are no more than about 3 to 4 dwells per second (Moray, 1986), and this frequency is usually lower because of variables that prolong the dwell. The destination of a scan is usually driven by top-down processes (i.e., expectancy; Senders, 1964), although on occasion a saccade may be drawn by salient bottom-up processes (e.g., a flashing light). The dwell duration is governed jointly by two factors: (1) the *information content* of the item fixated (e.g., when reading, long words require longer dwells than short ones), and (2) the ease of *information extraction*, which is often influenced by stimulus quality (e.g., in target search, longer dwells on a degraded target). Finally, once the eyes have landed a saccade on a particular location, the useful field of view defines how large an area, surrounding the center of fixation, is available for information extraction (Sanders, 1970; Ball et al., 1988). The useful field of view defines the diameter of the region within which a target might be detected if it is present.

The useful field of view should be carefully distinguished from the area of *foveal vision*, defined earlier in the chapter. Foveal vision defines a specific area of approximately 2 degrees of visual angle surrounding the center of fixation, which provides high visual acuity and low sensitivity. The diameter of the useful field of view, in contrast, is task-dependent. It may be quite small if the operator is searching for very subtle targets demanding high visual acuity but may be much larger than the fovea if the targets are conspicuous and can be easily detected in peripheral vision.

Recent developments in technology have produced more efficient means of measuring eye movements with *oculometers*, which measure the orientation of the eyeball relative to an image plane and can therefore be used to infer the precise destination of a saccade.

Visual Search

The Serial Search Model. In describing a person searching any visual field for something, we distinguish between *targets* and *nontargets* (nontargets are sometimes called *distractors*). The latter may be thought of as "visual noise" that must be inspected in order to determine that it is not in fact the desired target. Many searches are *serial* in that each item is inspected in turn to determine whether it is or is not a target. If each inspection takes a relatively constant time, *I*, and the expected location of the target is unknown beforehand, then it is possible to predict the average time it will take to find the target as

$$T = (N \times I)/2 \qquad\qquad (4.5)$$

where I is the average inspection time for each item, and N is the total number of items in the search field (Neisser et al., 1964). Because, *on the average*, the target will be encountered after *half* of the targets have been inspected (sometimes earlier, sometimes later), the product $(N \times I)$ is divided by two. This serial search model has been applied to predicting performance in numerous environments in which people search through maps or lists, such as phone books or computer menus (Lee & MacGregor, 1985; Yeh & Wickens 2001).

If the visual search space is organized coherently, people tend to search from top to bottom and left to right. However, if the space does not benefit from such organization (e.g., searching a map for a target or searching the ground below the aircraft for a downed airplane [Stager & Angus, 1978]), then people's searches tend to be considerably more random in structure and do not "exhaustively" examine all locations (Wickens, 1992; Stager & Angus, 1978). If targets are not readily visible, this nonexhaustive characteristic leads to a search-time function that looks like that shown in Figure 4.8 (Drury, 1975). The figure suggests that there are diminishing returns associated with giving people too long to search a given area if time is at a premium. Drury has used such a model to defined the optimum inspection time that people should be allowed to examine each image in a quality-control inspection task.

Search models can be extremely important in human factors (Brogan, 1993) for predicting search time in time-critical environments; for example, how long will a driver keep eyes off the highway to search for a road sign? Unfortunately, however, there are two important circumstances that can render the strict serial model inappropriate, one related to bottom-up processing and the other to top-down processing. Both factors force models of visual search to become more complex and less precise.

Conspicuity.   The bottom-up influence is the *conspicuity* of the target. Certain targets are so conspicuous that they may "pop out" no matter where they are in the visual field, and so nontarget items need not be inspected (Yantis, 1993; Treisman, 1986). Psychologists describe the search for such targets as *parallel* because, in essence, all items are examined at once (i.e., in parallel), and in contrast to the equation 4.5, search time does not increase with the total number of items. Such is normally the case with "attention grabbers," such as a flashing warning signal, a moving target, or a uniquely colored, highlighted item on a checklist, a computer screen, or in a phone book.

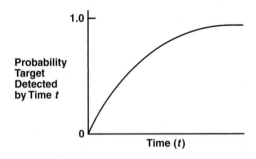

FIGURE 4.8
Predicted search success probability as a function of the time spent searching. (*Source:* Adapted from Drury, C., 1975. "Inspection of sheet metal: Models and data." Reprinted with permission from *Human Factors, 17.* Copyright 1975 by the Human Factors and Ergonomics Society.

Conspicuity is a desirable property if the task requires the target to be processed, but an undesirable one if the conspicuous item is irrelevant to the task at hand. Thus, if I am designing a checklist that highlights emergency items in red, this may help the operator in responding to emergencies but will be a distraction if the operator is using the list to guide normal operating instructions; that is, it will be more difficult to focus attention on the normal instructions. As a result of these dual consequences of conspicuity, the choice of highlighting (and the effectiveness of its implementation) must be guided by a careful analysis of the likelihood that the user will *need* the highlighted item as a target (Fisher & Tan, 1989). Table 4.4 lists some key variables that can influence the conspicuity of targets and, therefore, the likelihood that the field in which they are embedded will be searched in parallel.

Expectancies. The second influence on visual search that leads to departures from the serial model has to do with the top-down implications of *searcher expectancies* of where the target might be likely to lie. Expectancies, like all top-down processes, are based upon prior knowledge. Our driver did not expect to see the road sign on the left of the highway and, as a result, only found it after it was too late. As another example, when searching a phone book we do not usually blanket the entire page with fixations, but our *knowledge* of the alphabet allows us to start the search near or around the spelling of the target name. Similarly, when searching an index, we often have an idea what the topic is likely to be called, which guides our starting point.

It is important to realize that these expectancies, like all knowledge, come only with experience. Hence, we might predict that the skilled operator will have more top-down processes driving visual search than the unskilled one and as a result will be more in the efficient, a conclusion born out by research (Parasuraman, 1986). These top-down influences also provide guidance for designers who develop search fields, such as indexes and menu pages, to understand the subjective orderings and groupings the items that users have. This topic is addressed again in Chapter 15.

Conclusion. In conclusion, research on visual search has four general implications, all of which are important in system design.

TABLE 4.4 Target Properties Inducing Parallel Search

1. Discriminability from background elements.
    a. In color (particularly if nontarget items are uniformly colored)
    b. In size (particularly if the target is larger)
    c. In brightness (particularly if the target is brighter)
    d. In motion (particularly if background is stationary)
2. Simplicity: Can the target be defined only by one dimension (i.e., "red") and not several (i.e., "red and small")
3. Automaticity: a target that is highly familiar (e.g., one's own name)

Note that unique *shapes* (e.g., letters, numbers) do not generally support parallel search (Treisman, 1986).

1. Knowledge of conspicuity effects can lead the designer to try to enhance the visibility of target items (consider, for example, reflective jogging suits [Owens et al., 1994] or highlighting critical menu items). In dynamic displays, automation can highlight critical targets to be attended by the operator (Yeh & Wickens 2001b; Dzindolet et al., 2002. See Chapter 16).

2. Knowledge of the serial aspects of many visual search processes should forewarn the designer about the costs of *cluttered* displays (or search environments). When too much information is present, many maps present an extraordinary amount of clutter. For electronic displays, this fact should lead to consideration of *decluttering* options in which certain categories of information can be electronically turned off or deintensified (Mykityshyn et al., 1994; Stokes et al., 1990; Yeh & Wickens 2001a). However, careful use of color and intensity as discriminating cues between different classes of information can make decluttering unnecessary (Yeh & Wickens, 2001a).

3. Knowledge of the role of top-down processing in visual search should lead the designer to make the *structure* of the search field as apparent to the user as possible and consistent with the user's knowledge (i.e., past experience). For verbal information, this may involve an alphabetical organization or one based on the semantic similarity of items. In positioning road signs, this involves the use of *consistent* placement (see Chapter 17).

4. Knowledge of all of these influences can lead to the development of *models* of visual search that will predict how long it will take to find particular targets, such as the flaw in a piece of sheet metal (Drury, 1975), an item on a computer menu (Lee & MacGregor, 1985; Fisher & Tan, 1989), or a traffic sign by a highway (Theeuwes, 1994). For visual search, however, the major challenge of such models resides in the fact that search appears to be guided much more by top-down than by bottom-up processes (Theeuwes, 1994), and developing precise mathematical terms to characterize the level of expertise necessary to support top-down processing is a major challenge.

Detection

Once a possible target is located in visual search, it becomes necessary to *confirm* that it really is the item of interest (i.e., *detect* it). This process may be trivial if the target is well known and reasonably visible (e.g., the name on a list), but it is far from trivial if the target is degraded, like a faint flaw in a piece of sheet metal, a small crack in an x-rayed bone, or the faint glimmer of the lighthouse on the horizon at sea. In these cases, we must describe the operator's ability to *detect signals*. Signal detection is often critical even when there is no visual search at all. For example, the quality-control inspector may have only one place to look to examine the product for a defect. Similarly, human factors is concerned with detection of auditory signals, like the warning sound in a noisy industrial plant, when search is not at all relevant.

Signal Detection Theory.   In any of a variety of tasks, the process of signal detection can be modeled by *signal detection theory* (SDT) (Green & Swets, 1988; Swets, 1996; T. D. Wickens, 2002), which is represented schematically in Figure 4.9. SDT assumes that "the world" (as it is relevant to the operator's task) can be modeled as either one in which the "signal" to be detected is present or absent, as shown across the top of the matrix in Figure 4.9. Whether the signal is present or absent, the world is assumed to contain noise: Thus, the luggage inspected by the airport security guard may contain a weapon (signal) in addition to a number of things that might look like weapons (i.e., the noise of hair blowers, calculators, carabiners, etc.), or it may contain the noise alone, with no signal.

The goal of the operator in detecting signals is to *discriminate* signals from noise. Thus, we may describe the relevant behavior of the observer as that represented by the two rows of Figure 4.9—saying, "Yes (I see a signal)" or "No (there is only noise)." This combination of two states of the world and two responses yields four joint events shown as the four cells of the figure labeled *hits, false alarms, misses*, and *correct rejections*. Two of these cells (hits and correct rejections) clearly represent "good" outcomes and ideally *should* characterize much of

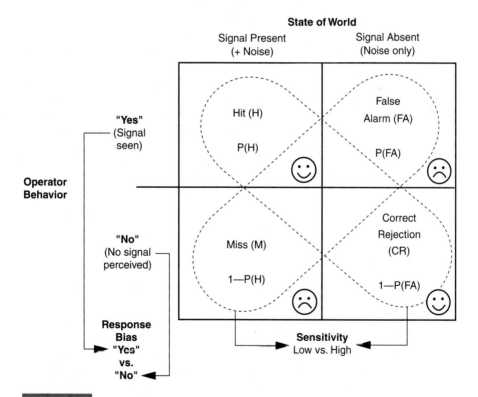

FIGURE 4.9

Representation of the outcomes in signal detection theory. The figure shows how changes in the four joint events within the matrix influence the primary performance measures of response bias and sensitivity, shown at the bottom.

the performance, while two are "bad" (misses and false alarms) and ideally should never occur. If several encounters with the state of the world (signal detection trials) are aggregated, some involving signals and some involving noise alone, we may then express the numbers within each cell as the *probability* of a hit [#hits/#signals = $p(hit)$]; the *probability* of a miss [$1 - p(hit)$]; the probability of a false alarm [#FA/#no-signal encounters] and the probability of a correct rejection [$1 - p(FA)$]. As you can see from these equations, if the values of $p(hit)$ and $p(FA)$ are measured, then the other two cells contain entirely redundant information.

Thus, the data from a signal detection environment (e.g., the performance of an airport security inspector) may easily be represented in the form of the matrix shown in Figure 4.9, if a large number of trials are observed so that the probabilities can be reliably estimated. However, SDT considers these same numbers in terms of two fundamentally different influences on human detection performance: *sensitivity* and *response bias*. We can think of these two as reflecting bottom-up and top-down processes respectively.

Sensitivity and Response Bias.   As Figure 4.9 shows at the bottom, the measure of sensitivity, often expressed by the measure d' (d prime) expresses how good an operator is at discriminating the signal from the noise, reflecting essentially the number of good outcomes (hits and correct rejections) relative to the total number of both good and bad outcomes. Sensitivity is higher if there are more correct responses and fewer errors. It is influenced both by the keenness of the senses and by the strength of the signal relative to the noise (i.e., the *signal-to-noise ratio*). For example, sensitivity usually improves with experience on the job up to a point; it is degraded by poor viewing conditions (including poor eyesight). An alert inspector has a higher sensitivity than a drowsy one. The formal calculation of sensitivity is not discussed in this book, and there are other related measures that are sometimes used to capture sensitivity (T. D. Wickens, 2002). However, Table 4.5 presents some values of d' that might be observed from signal detection analysis.

TABLE 4.5   Some Values of d'

P(hit)	P (false alarm)					
	0.01	0.02	0.05	0.10	0.20	0.30
0.51	2.34	2.08	1.66	1.30	0.86	0.55
0.60	2.58	2.30	1.90	1.54	1.10	0.78
0.70	2.84	2.58	2.16	1.80	1.36	1.05
0.80	3.16	2.89	2.48	2.12	1.68	1.36
0.90	3.60	3.33	2.92	2.56	2.12	1.80
0.95	3.96	3.69	3.28	2.92	2.48	2.16
0.99	4.64	4.37	3.96	3.60	3.16	2.84

Source: Selected values from *Signal Detection and Recognition by Human Observers* (Appendix 1, Table 1) by J. A. Swets, 1969, New York: Wiley. Copyright 1969 by John Wiley and Sons, Inc. Reproduced by permission.

The measure of response bias, or *response criterion*, shown in the left of Figure 4.9, reflects the *bias* of the operator to respond "yes, signal" versus "no, noise." Although formal signal detection theory characterizes response bias by the term *beta*, which has a technical measurement (Green & Swets, 1988; Wickens & Hollands, 2000), one can more simply express response bias as the probability that the operator will respond yes [(#*yes*)/(Total *responses*)]. Response bias is typically affected by two variables, both characteristic of top-down processing. First, increases in the operator's *expectancy* that a signal will be seen leads to corresponding increases in the probability of saying yes. For example, if a quality-control inspector has knowledge that a batch of products may have been manufactured on a defective machine and therefore may contain a lot of defects, this knowledge should lead to a shift in response criterion to say "signal" (defective product) more often. The consequences of this shift are to generate both more hits *and* more false alarms.

Second, changes in the *values*, or costs and benefits, of the four different kinds of events can also shift the criterion. The air traffic controller cannot afford to miss detecting a signal (a conflict between two aircraft) because of the potentially disastrous consequences of a midair collision (Bisseret, 1981). As a result, the controller will set the response criterion at such a level that misses are very rare, but the consequences are that the less costly false alarms are more frequent. In representing the air traffic controller as a signal detector, these false alarms are circumstances when the controller detects a potentially conflicting path and redirects one of the aircraft to change its flight course even if this was not necessary.

In many cases, the outcome of a signal detection analysis may be plotted in what is called a *receiver operating characteristic* (*ROC*) space, as shown in Figure 4.10 (Green & Swets, 1988). Here $p(FA)$ is plotted on the *x* axis, $P(FA)$ is plotted on the *y* axis, and a single point in the space (consider point **A**) thereby represents all of the data from one set of detection conditions. In different conditions,

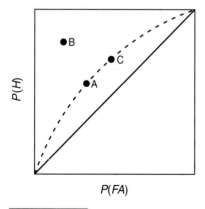

FIGURE 4.10

A receiver operating characteristic, or ROC curve. Each point represents the signal detection data from a single matrix, such as that shown in Figure 4.9.

detection performance at **B** would represent improved sensitivity (higher d'). Detection performance at **C** would represent only a shift in the response criterion relative to **A** (here a tendency to say yes more often, perhaps because signals occurred more frequently). More details about the ROC space can be found in Green and Swets (1988) T. D. Wickens (2002) and Wickens and Hollands (2000).

Interventions. The distinction between sensitivity and response criterion made by SDT is important because it allows the human factors practitioner to understand the consequences of different kinds of job interventions that may be intended to improve detection performance in a variety of circumstances. For example, any instructions that "exhort" operators to "be more vigilant" and not miss signals will probably increase the hit rate but will also increase the false-alarm rate. This is because the instruction is a motivational one reflecting costs and values, which typically affects the setting of the response criterion, as the shift from point A to point C in the ROC of Figure 4.10. (Financially rewarding hits will have the same effect.) Correspondingly, it has been found that directing the radiologist's attention to a particular area of an x-ray plate where an abnormality is likely to be found will tend to shift the response criterion for detecting abnormalities at that location but will not increase the sensitivity (Swennsen et al., 1977). Hence, the value of such interventions must consider the relative costs of misses and false alarms.

However, there are certain things that *can* be done that do have a more desirable direct influence on increasing sensitivity (that is, moving from point A to point B in Figure 4.10). As we have noted, training the operator for what a signal looks like can improve sensitivity. So also can providing the inspector with a "visual template" of the potential signal that can be compared with each case that is examined (Kelly, 1955). Several other forms of interventions to influence signal detection and their effects on sensitivity or response bias are shown in Table 4.6. These are described in more detail in Wickens and Hollands (2000). We describe in Chapter 5 how signal detection theory is also important in the design of auditory alarms.

TABLE 4.6   Influences on Signal Detection Performance

Payoffs (typically influence response bias)

Introducing "false signals" to raise signal rate artificially [response bias: P (yes) increase]

Providing incentives and exhortations (response bias)

Providing knowledge of results (usually increases sensitivity, but may calibrate response bias if it provides observer with more accurate perception of probability of signal)

Slowing down the rate of signal presentation (slowing the assembly line; increases sensitivity)

Differentially amplifying the signal (more than the noise; increases sensitivity)

Making the signal dynamic (increases sensitivity)

Giving frequent rest breaks (increases sensitivity)

Providing a visual (or audible) template of the signal (increases sensitivity)

Providing experience seeing the signal (increases sensitivity)

Providing redundant representations of the signal (increases sensitivity)

In Chapter 13 we describe its role in characterizing the loss of vigilance of operators in low arousal monitoring tasks, like the security guard at night. For inspectors on an assembly line, the long-term decrement in performance may be substantial, sometimes leading to miss rates as high as 30 to 40 percent. The guidance offered in Table 4.6 suggests some of the ways in which these deficiencies might be addressed. To emphasize the point made above, however, it is important for the human factors practitioner to realize that any intervention that shifts the response criterion to increase hits will have a consequent increase in false alarms. Hence, it should be accepted that the costs of these false alarms are less severe than the costs of misses (i.e., are outweighed by the benefits of more hits). The air traffic control situation is a good example. When it comes to detecting possible collisions, a false alarm is less costly than a miss (a potential collision is not detected), so interventions that increase false alarm rate can be tolerated if they also decrease miss rate. Formal development of SDT shows how it is possible to set the optimal level of the response criterion, given that costs, benefits, and signal probabilities can be established (Wickens & Hollands, 2000).

DISCRIMINATION

Very often, issues in human visual sensory performance are based on the ability to *discriminate* between one of two signals rather than to *detect* the existence of a signal. Our driver was able to *see* the road sign (detect it) but, in the brief view with dim illumination, failed to discriminate whether the road number was 60 or 66 (or in another case, perhaps, whether the exit arrow pointed left or right). He was also clearly confused over whether the car color was red or brown. Confusion, the failure to discriminate, results whenever stimuli are similar. Even fairly different stimuli, when viewed under degraded conditions, can produce confusion. As one example, it is believed that one cause of the crash of a commercial jet liner in Europe was that the automated setting that controlled its flight path angle with the ground (3.3 degrees) looked so similar to the automated setting that controlled its vertical speed (3,300 feet/minute; Billings, 1996; see Figure 4.11). As a result, pilots could easily have confused the two, thinking that they had "dialed in" the 3.3-degree angle when in fact they had set the 3,300 ft/min vertical speed (which is a much more rapid decent rate than that given by the 3.3-degree angle). Gopher and colleagues (1989) have pointed out the dangers in medicine that result from the extreme visual similarity of very different drug names. Consider such names as capastat and cepastat, mesantoin and metinon, and Norflox and Norflex; each has different health implications, yet the names are quite similar in terms of visual appearance. Such possible confusions are likely to be amplified when the prescription is filtered through the physician's (often illegible) handwriting.

Thus, it is important for the designer of controls that must be reached and manipulated or of displays that must be interpreted to consider the alternative controls (or displays) that *could* be activated (or perceived). Can they be adequately discriminated? Are they far enough apart in space or distinguished by other features like color, shape, or other labels so that confusion will not occur?

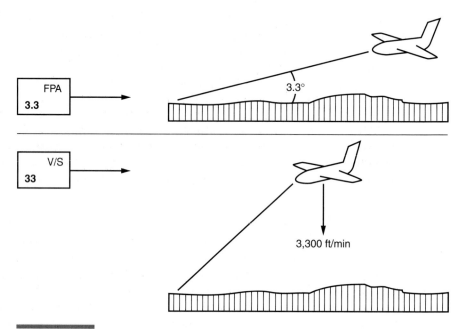

FIGURE 4.11

Confusion in the automation setting feedback believed to have contributed to the cause of a commercial airline crash. The pilots believed the top condition to exist, when in fact the bottom existed. The single display illustrating the two conditions was very similar, and hence the two were quite confusable.

It is important to remember, however, that if only verbal labels are used to discriminate the displays or controls from each other, then attention *must* be given to the visibility and readability issues discussed earlier. We discuss the important and often overlooked issues of discrimination and confusion further as we address the issues of working memory in Chapter 6 and displays in Chapter 8.

An even simpler form of discrimination limits characterizes the ability of people to notice the change or difference in simple dimensional values, for example, a small change in the height of a bar graph or the brightness of an indicator. In the classic study of *psychophysics* (the relation between the psychological sensations and physical stimulation), such difference thresholds are called *just noticeable difference*, or JND. Designers should not assume that users will make judgments of displayed quantities that are less than a JND. For example, if a user monitoring a power meter should be aware of fluctuations greater than a certain amount, the meter should be scaled so that those fluctuations are greater than a JND.

Along many sensory continua, the JND for judging intensity **differences** increases in proportion to the absolute amount of intensity, a simple relationship described by **Weber's law;**

$$JND = K(DI)/I \qquad (4.6)$$

where *DI* is the change in intensity, *I* is the absolute level of intensity, and *K* is a constant, defined separately for different sensory continua (such as the brightness of lights, the loudness of sounds, or the length of lines). Importantly, Weber's law also describes the psychological reaction to changes in other non-sensory quantities. For example, how much a change in the cost of an item means to you (i.e., whether the cost difference is above or below a JND) depends on the cost of the item. You may stop riding the bus if the bus fare is increased by $1.00, from $0.50 to $1.50; the increase was clearly greater than a JND of cost. However, if an air fare increased by the same $1.00 amount (from $432 to $433), this would probably have little influence on your choice of whether or not to buy the ticket. The $1.00 increase is less than a JND compared to the $432 cost.

ABSOLUTE JUDGMENT

Discrimination refers to judgment of differences between two sources of information that are actually (or potentially) present, and generally people are good at this task as long as the differences are not small and the viewing conditions are favorable. In contrast, *absolute judgment* refers to the limited human capability to judge the absolute value of a variable signaled by a coded stimulus. For example, estimating the height of a bar graph to the nearest digit is an absolute judgment task with 10 levels. Judging the color of a traffic signal (ignoring its spatial position) is an absolute judgment task with only three levels of stimulus value. People are not generally very good at these absolute value judgments of attaching "labels to levels" (Wickens & Hollands, 2000). It appears that they can be guaranteed to do so accurately only if fewer than around five levels of any sensory continuum are used (Miller, 1956) and that people are even less accurate when making absolute value judgments in some sensory continua like pitch or sound loudness; that is, even with five levels they may be likely to make a mistake, such as confusing level three with level four.

The lessons of these absolute judgment limitations for the designer are that the number of levels that should be judged on the basis of some absolute coding scheme, like position on a line or color of a light, should be chosen conservatively. It is recommended, for example, that no more than seven colors be used if precise accuracy in judgment is required (and an adjacent color scale for comparison is not available. The availability of such a scale would turn the absolute judgment task into a relative judgment task). Furthermore, even this guideline should be made more stringent under potentially adverse viewing conditions (e.g., a map that is read in poor illumination).

CONCLUSION

We have seen in this chapter how limits of the visual system influence the nature of the visual information that arrives at the brain for more elaborate perceptual interpretation. We have also begun to consider some aspects of this interpretation, as we considered top-down influences like expectancy, learning, and values. In Chapter 5, we consider similar issues regarding the processing of auditory and other sensory information. Together, these chapters describe the sensory processing of the "raw" ingredients for the more elaborative perceptual and cognitive aspects of understanding the world. Once we have addressed these issues of higher processing in Chapter 6, we can consider how all of this knowledge— of bottom-up sensory processing, perception, and understanding—can guide the design of *displays* that support tasks confronting the human user. This is the focus of Chapter 8.

Chapter 5

Auditory, Tactile, and Vestibular System

The worker at the small manufacturing company was becoming increasingly frustrated by the noise level at her workplace. It was unpleasant and stressful, and she came home each day with a ringing in her ears and a headache. What concerned her in particular was an incident the day before when she could not hear the emergency alarm go off on her own equipment, a failure of hearing that nearly led to an injury. Asked by her husband why she did not wear earplugs to muffle the noise, she said, "They're uncomfortable. I'd be even less likely to hear the alarm, and besides, it would be harder to talk with the worker on the next machine, and that's one of the few pleasures I have on the job." She was relieved that an inspector from Occupational Safety and Health Administration (OSHA) would be visiting the plant in the next few days to evaluate the complaints that she had raised.

The worker's concerns illustrate the effects of three different types of sound: the undesirable *noise* of the workplace, the critical *tone* of the alarm, and the important communications through *speech*. Our ability to process these three sources of acoustic information, whether we want to (alarms and speech) or not (noise), and the influence of this processing on performance, health, and comfort are the focus of the first part of this chapter. We conclude by discussing three other sensory channels: tactile, proprioceptive-kinesthetic, and vestibular. These senses have played a smaller but nevertheless significant role in the design of human–machine systems.

SOUND: THE AUDITORY STIMULUS

As shown in Figure 5.1a, the stimulus for hearing is sound, a vibration (actually compression and rarefaction) of the air molecules. The acoustic stimulus can therefore be represented as a sine wave, with amplitude and frequency. This is

SPEECH PERCEPTION

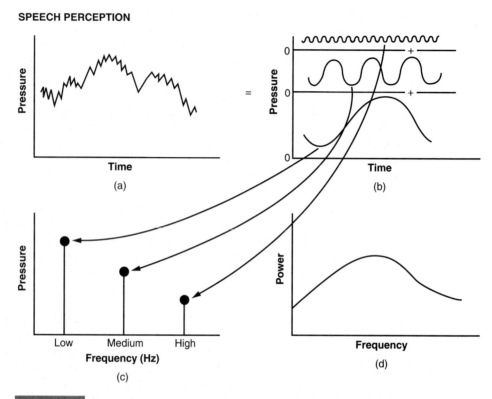

FIGURE 5.1

Different schematic representations of speech signal: (a) time domain; (b) three frequency components of (a); (c) the power spectrum of (b); (d) a continuous power spectrum of speech.

analogous to the representation of spatial frequency discussed in Chapter 3; however, the frequency in sound is played out over time rather than space. Figure 5.1b shows three frequencies, each of different values and amplitudes. These are typically plotted on a spectrum, as shown in Figure 5.1c. The position of each bar along the spectrum represents the actual frequency, expressed in cycles/second or *Hertz* (*Hz*). The height of the bar reflects the amplitude of the wave and is typically plotted as the square of the amplitude, or the *power*.

Any given sound stimulus can be presented as a single frequency, a small set of frequencies, as shown in Figure 5.1c, or a continuous *band* of frequencies, as shown in Figure 5.1d. The frequency of the stimulus more or less corresponds to its *pitch*, and the amplitude corresponds to its *loudness*. When describing the effects on hearing, the amplitude is typically expressed as a *ratio* of sound pressure, *P*, measured in *decibels* (dB). That is,

$$\text{Sound intensity (dB)} = 20 \log (P1/P2).$$

As a ratio, the decibel scale can be used in either of two ways: First, as a measure of *absolute* intensity, the measure P2 is fixed at a value near the threshold of hearing (i.e., the faintest sound that can be heard under optimal conditions). This is a pure tone of 1,000 Hz at 20 micro Newtons/square meter. In this context, decibels represent the ratio of a given sound to the threshold of hearing. Table 5.1 provides some examples of the absolute intensity of everyday sounds along the decibel scale. Second, because it is a ratio measure, the decibel scale can also be employed to characterize the *ratio* of two hearable sounds; for example, the OSHA inspector at the plant may wish to determine how much *louder* the alarm is than the ambient background noise. Thus, we might say it is 15 dB more intense. As another example, we might characterize a set of earplugs as reducing the noise level by 20 dB.

Sound intensity may be measured by the sound intensity meter. This meter has a series of scales that can be selected, which enable sound to be measured more specifically within particular frequency ranges. In particular, the A scale differentially weights sounds to reflect the characteristics of human hearing, providing greatest weighting at those frequencies where we are most sensitive. The C scale weights all frequencies nearly equally and therefore is less closely correlated with the characteristics of human hearing.

In addition to amplitude (intensity) and frequency (pitch), two other critical dimensions of the sound stimulus are its temporal characteristics, sometimes referred to as the *envelope* in which a sound occurs, and its location. The temporal characteristics are what may distinguish the wailing of the siren from the steady blast of the car horn, and the location (relative to the hearer) is, of course, what might distinguish the siren of the firetruck pulling up from behind from that of the firetruck about to cross the intersection in front (Casali & Porter, 1980).

TABLE 5.1   The Decibel Scale

Sound Pressure Level (db)		
140	———	Ear damage possible; jet at take-off
130	———	Painful sound
120	———	Propeller plane at take-off
110	———	Loud thunder
100	———	Subway train
90	———	Truck or bus
80	———	
70	———	Average auto; loud radio
60	———	Normal conversation
50	———	Quiet restaurant
40	———	Quiet office, household sounds
30	———	
20	———	Whisper
10	———	Normal breathing
0	———	Threshold of hearing

THE EAR: THE SENSORY TRANSDUCER

The ear has three primary components responsible for differences in our hearing experience. As shown in Figure 5.2, the *pinnea* both collects sound and, because of its asymmetrical shape, provides some information regarding where the sound is coming from (i.e., behind or in front). Mechanisms of the *outer* and *middle ear* (the ear drum or tympanic membrane, and the hammer, anvil, and stirrup bones) conduct and amplify the sound waves into the inner ear and are potential sources of breakdown or deafness (e.g., from a rupture of the eardrum or buildup of wax). The muscles of the middle ear are responsive to loud noises and reflexively contract to attenuate the amplitude of vibration before it is conveyed to the inner ear. This *aural reflex* thus offers some protection to the inner ear.

The *inner ear*, consisting of the *cochlea*, within which lies the basilar membrane, is that portion where the physical movement of sound energy is transduced to electrical nerve energy that is then passed up the auditory nerve to the brain. This transduction is accomplished by displacement of tiny hair cells along the basilar membrane as the membrane moves differently to sounds of different

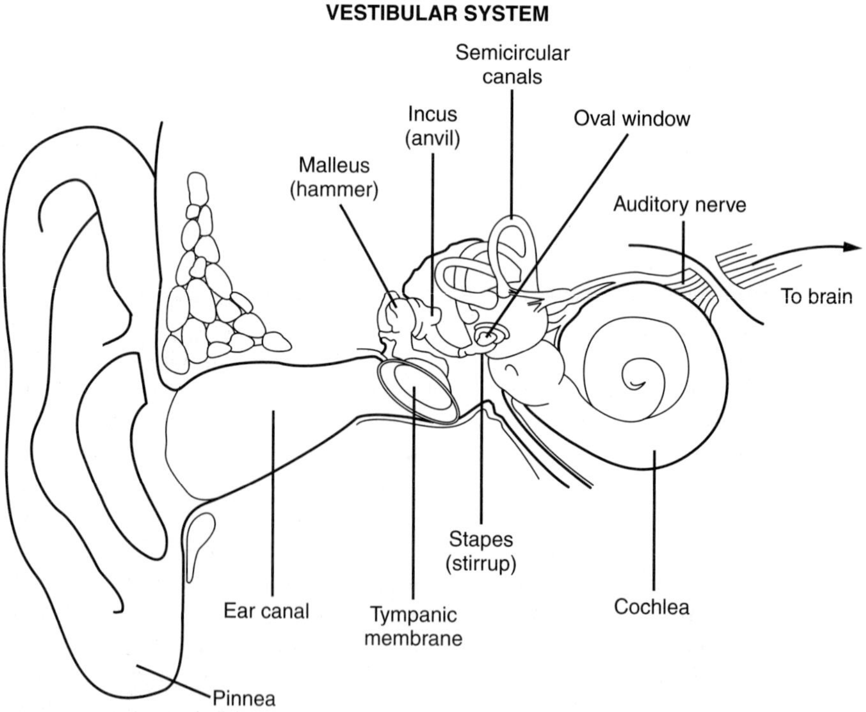

FIGURE 5.2

Anatomy of the ear. (*Source:* Bernstein, D., Clark-Stewart, A., Roy, E., & Wickens, C. D. 1997. *Psychology,* 4th ed. Copyright 1997 by Houghton-Mifflin. Reprinted with permission).

frequency. Intense sound experience can lead to selective hearing loss at particular frequencies as a result of damage to the hair cells at particular locations along the basilar membrane. Finally, the neural signals are compared between the two ears to determine the delay and amplitude differences between them. These differences provide another cue for sound localization, because these features are identical only if a sound is presented directly along the midplane of the listener.

THE AUDITORY EXPERIENCE

To amplify our previous discussion of the sound stimulus, the four dimensions of the raw stimulus all map onto psychological experience of sound: Loudness maps to intensity, pitch maps to frequency, and perceived location maps to location. The quality of the sound is determined both by the set of frequencies in the stimulus and by the envelope. In particular, the *timbre* of a sound stimulus— what makes the trumpet sound different from the flute—is determined by the set of higher *harmonic* frequencies that lie above the *fundamental* frequency (which determines the pitch of the note). Various temporal characteristics, including the envelope and the rhythm of successive sounds, also determine the sound quality. As we shall see, differences in the envelope are critically important in distinguishing speech sounds.

Loudness and Pitch

Loudness is a psychological experience that correlates with, but is not identical to, the physical measurement of sound intensity. Two important reasons why loudness and intensity do not directly correspond are reflected in the psychophysical scale of loudness and the modifying effect of pitch. We discuss each of these in turn.

Psychophysical Scaling. Equal increases in sound intensity (on the decibel scale) do not create equal increases in loudness; for example, an 80-dB sound does not sound twice as loud as a 40-dB sound, and the increase from 40 to 50 dB is not judged as the same loudness increase as that from 70 to 80 dB. Instead, the *scale* that relates physical intensity to the psychological experience of loudness, expressed in units called *sones,* is that shown in Figure 5.3.

One sone is established arbitrarily as the loudness of a 40-dB tone of 1,000 Hz. A tone twice as loud will be two sones. As an approximation, we can say that loudness doubles with each 10-dB increase in sound intensity. It is important to distinguish two critical levels along the loudness scale shown in Figure 5.3. As noted, the *threshold* is the minimum intensity at which a sound can be detected. At some higher intensity, around 85 to 90 dB, is the second critical level at which potential danger to the ear occurs. Both of these levels, however, as well as the loudness of the intensity levels in between, are influenced by the frequency (pitch) of the sound, and so we must now consider that influence.

Frequency Influence. Figure 5.4 plots a series of equal-loudness curves shown by the various wavy lines. That is, every point along a line sounds just as loud as

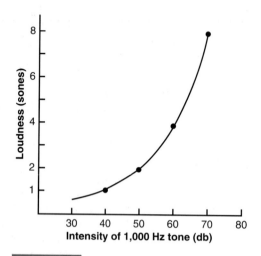

FIGURE 5.3

Relation between sound intensity and loudness.

any other point along the same line. For example, a 100-Hz tone of around 70 dB has the same perceived loudness as a 500-Hz tone of around 57 dB. The equal loudness contours follow more or less parallel tracks. As shown in the figure, the frequency of a sound stimulus, plotted on the *x* axis, influences all of the critical levels of the sound experience: threshold, loudness, and danger levels. The range of human hearing is limited between around 20 Hz and 20,000 Hz. Within this range, we are most sensitive (lowest threshold) to sounds of around 4,000 Hz. (In the figure, all equal loudness curves are described in units of *phons.* One phon = 1 dB of loudness of a 1,000-Hz tone, the standard for calibration. Thus, all tones lying along the 40-phon line have the same loudness—1 sone—as a 1,000-Hz tone of 40 dB.)

Masking.   As our worker at the beginning of the chapter discovered, sounds can be *masked* by other sounds. The nature of masking is actually quite complex (Yost, 1992), but a few of the most important principles for design are the following:

1. The minimum intensity difference necessary to ensure that a sound can be heard is around 15 dB (above the mask), although this value may be larger if the pitch of the sound to be heard is unknown.
2. Sounds tend to be masked most by sounds in a critical frequency band surrounding the sound that is masked.
3. Low-pitch sounds mask high-pitch sounds more than the converse. Thus, a woman's voice is more likely to be masked by other male voices than a man's voice would be masked by other female voices even if both voices are speaking at the same intensity level.

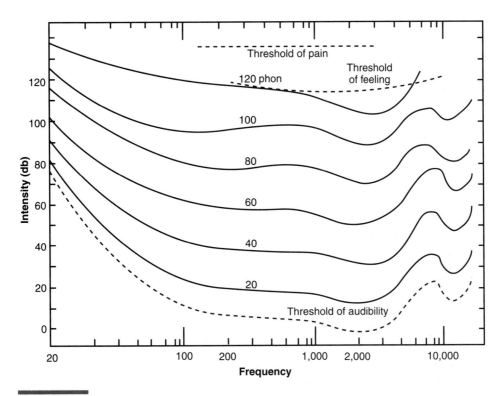

FIGURE 5.4

Equal loudness contours showing the intensity of different variables as a function of frequency. All points lying on a single curve are perceived as equally loud. Thus, a 1,000-Hz tone of 40 dB sounds about the same loudness (40 phons) as an 8,000-Hz tone of around 60 dB. (*Source:* Kryter, K. D. Speech Communications, in Van Cott, H. P., & R. G. Kinkade, eds., 1972. *Human Engineering Guide to System Design.* Figures 4–6. Washington, DC: U.S. Government Printing Office.).

ALARMS

The design of effective alarms, the critical signal that was nearly missed by the worker in our opening story, depends very much on a good understanding of human auditory processing (Stanton, 1994; Bliss & Gilson, 1998; Pritchett, 2001; Woods, 1995). Alarms tend to be a uniquely auditory design for one good reason: The auditory system is *omnidirectional;* that is, unlike visual signals, we can sense auditory signals no matter how we are oriented. Furthermore, it is much more difficult to "close our ears" than it is to close our eyes (Banbury et al., 2001). For these and other reasons, auditory alarms induce a greater level of compliance than do visual alarms (Wolgalter et al., 1993). Task analysis thus dictates that if there is an alarm signal that *must be sensed,* like a fire alarm, it should be given an auditory form (although redundancy in the visual or tactile channel may be worthwhile in certain circumstances).

While the choice of modality is straightforward, the issue of how auditory alarms should be designed is far from trivial. Consider the following quotation from a British pilot, taken from an incident report, which illustrates many of the problems with auditory alarms.

> I was flying in a jetstream at night when my peaceful revelry was shattered by the stall audio warning, the stick shaker, and several warning lights. The effect was exactly what was *not* intended; I was frightened numb for several seconds and drawn off instruments trying to work out how to cancel the audio/visual assault, rather than taking what should be instinctive actions. The combined assault is so loud and bright that it is impossible to talk to the other crew member and action is invariably taken to cancel the cacophony before getting on with the actual problem. (Patterson, 1990)

Criteria for Alarms.   Patterson (1990) has discussed several properties of a good alarm system, as shown in Figure 5.5, that can prevent the two opposing problems of detection, experienced by our factory worker at the beginning of the chapter, and "overkill" experienced by the pilot.

1. Most critically, the alarm must be *heard* above the background ambient noise. This means that the noise spectrum must be carefully measured at the hearing location of all users who must respond to the alarm. Then, the alarm should be tailored to be at least 15 dB *above* the thresh-

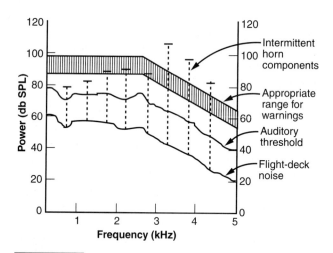

FIGURE 5.5

The range of appropriate levels for warning sound components on the flight deck of the Boeing 737 (vertical line shading). The minimum of the appropriate-level range is approximately 15 dB above auditory threshold (broken line), which is calculated from the spectrum of the flight deck noise (solid line). The vertical dashed lines show the components of the intermittent warning horn, some of which are well above the maximum of the appropriate-level range. (*Source:* Patterson, R. D., 1990. Auditory warning sounds in the work environment. *Phil. Trans. R. Soc. London B.,* 327, p. 487, Figure 1).

old of hearing above the noise level. This typically requires about a 30-dB difference above the noise level in order to *guarantee* detection, as shown in Figure 5.5. It is also wise to include components of the alarm at several different frequencies, well distributed across the spectrum, in case the particular malfunction that triggered the alarm creates its own noise (e.g., the whine of a malfunctioning engine), which exceeds the ambient level.

2. The alarm should not be above the danger level for hearing, whenever this condition can be avoided. (Obviously, if the ambient noise level is close to the danger level, one has no choice but to make the alarm louder by criterion 1, which is most important.) This danger level is around 85 to 90 dB. Careful selection of frequencies of the alarm can often be used to meet both of the above criteria. For example, if ambient noise is very intense (90 dB), but only in the high frequency range, it would be counterproductive to try to impose a 120-dB alarm in that same frequency range when several less intense components in a lower frequency range could adequately be heard.

3. Ideally, the alarm should not be overly startling or abrupt. This can be addressed by tuning the *rise time* of the alarm pulse.

4. In contrast to the experience of the British pilot, the alarm should not disrupt the perceptual understanding of other signals (e.g., other simultaneous alarms) or any background speech communications that may be essential to deal with the alarm. This criterion in particular implies that a careful *task analysis* should be performed of the conditions under which the alarm might sound and of the necessary communications tasks to be undertaken as a consequence of that alarm.

5. The alarm should be *informative,* signaling to the listener the nature of the emergency and, ideally, some indication of the appropriate action to take. The criticality of this informativeness criterion can be seen in one alarm system that was found in an intensive care unit of a hospital (an environment often in need of alarm remediation [Patterson, 1990]). The unit contained six patients, each monitored by a device with 10 different possible alarms: 60 potential signals that the staff may have had to rapidly identify. Some aircraft have been known to contain at least 16 different auditory alerts, each of which, when heard, is supposed to automatically trigger in the pilot's mind the precise identification of the alarming condition. Such alarms are often found to be wanting in this regard. Hence, in addition to being informative, the alarm must not be *confusable* with other alarms that may be heard in the same context. As you will recall from our discussion of vision in Chapter 4, this means that the alarm should not impose on the human's restrictive limits of *absolute judgment.* Just four different alarms may be the maximum allowable to meet this criterion if these alarms differ from each other on only a single physical dimension, such as pitch.

Designing Alarms.   How should an alarm system be designed to avoid, or at least minimize, the potential costs described above?

First, as we have noted, *environmental and task analysis* must be undertaken to understand the quality and intensity of the other sounds (noise or communications) that might characterize the environment in which the alarm is presented to guarantee detectability and minimize disruption of other essential tasks.

Second, to guarantee informativeness *and* to minimize confusability, designers should try to stay within the limits of absolute judgments. However, within these limits, one can strive to make the parameters of the different alarm sounds as different from each other as possible by capitalizing on the various dimensions along which sounds differ. For example, a set of possible alarms may contain three different dimensions: their pitch (fundamental pitch or frequency band), their envelope (e.g., rising, *woop woop;* constant *beep beep*), and their rhythm (e.g., synchronous *da da da* versus asynchronous da ***da*** da ***da***). A fourth dimension that could be considered (but not easily represented graphically in the figure) is the timbre of the sound that may contrast, for example, a horn versus a flute. Two alarms will be most discriminable (and least confusable) if they are constructed at points on opposite ends of all three (or four) dimensions. Correspondingly, three alarms can be placed far apart in the multidimensional space, although the design problem becomes more complex with more possible alarms. However, the philosophy of maintaining wide separation (discriminability) along each of several dimensions can still be preserved.

A third step involves designing the specifics of the individual sound. Patterson (1990) recommends the procedure outlined in Figure 5.6, a procedure that has several embedded rationales. At the top of the figure, each individual pulse in the alarm is configured with a rise envelope that is not too abrupt (i.e., at least 20 msec) so that it will avoid the "startle" created by more abrupt rises. The *set* of pulses in the alarm sequence, shown in the middle of the figure, are configured with two goals in mind: (1) The unique set of pauses between each pulse can be used to create a unique rhythm that can be used to help avoid confusions; and (2) the increase then decrease in intensity gives the perception of an approaching then receding sound, which creates a psychological sense of urgency. Edworthy, Loxley, and Dennis (1991), and Hellier and colleagues (2002) provide more elaborate guidelines for creating the psychological perception of urgency from alarms.

Finally, the bottom row of Figure 5.6 shows the philosophy by which repeated presentations of the alarm sequence can be implemented. The first two presentations may be at high intensity to guarantee their initial detection (first sequence) and identification (first or second sequence). Under the assumption that the operator has probably been alerted, the third and fourth sequences may be diminished in intensity to avoid overkill and possible masking of other sounds by the alarm (e.g., the voice communications that may be initiated by the alarming condition). However, an intelligent alarm system may infer, after a few sequences, that no action has been taken and hence repeat the sequence a couple of times at an even higher intensity.

Voice Alarms and Meaningful Sounds. Alarms composed of synthetic voice provide one answer to the problems of discriminability and confusion. Unlike

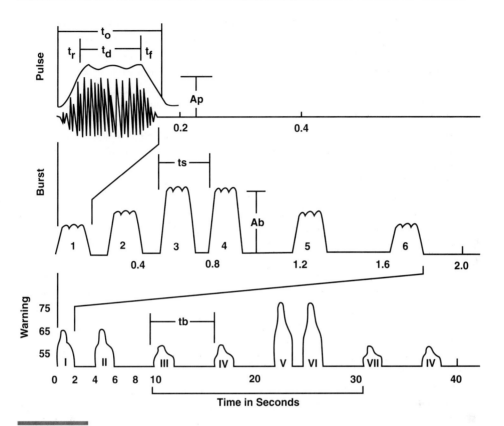

FIGURE 5.6

The modules of a prototype warning sound: The sound pulse at the top is an acoustic wave with rounded onsets and offsets and a distinctive spectrum; the burst shown in the middle row is a set of pulses with a distinctive rhythm and pitch contour; the complete warning sound sequence, shown at the bottom, is a set of bursts with varying intensity and urgency. (*Source:* Patterson, R. D., 1990. Auditory warning sounds in the environment. *Phil. Trans. R Soc. London B.,* 327, p. 490, Figure 3.)

"symbolic" sounds, the hearer does not need to depend on an arbitrary learned connection to associate sound with meaning. The loud sounds *Engine fire!* or *Stall!* in the cockpit mean exactly what they seem to mean. Voice alarms are employed in several circumstances (the two aircraft warnings are an example). But voice alarms themselves have limitations that must be considered. First, they are likely to be more confusable with (and less discriminable from) a background of other voice communications, whether this is the ambient speech background at the time the alarm sounds, the task-related communications of dealing with the emergency, or concurrent voice alarms. Second, unless care is taken, they may be more susceptible to frequency-specific masking noise. Third, care must be taken if the meaning of such alarms is to be interpreted by listeners in a multilingual environment who are less familiar with the language of the voice.

The preceding concerns with voice alarm suggest the advisability of using a *redundant* system that combines the alerting, distinctive features of the (nonspeech) alarm sound with the more informative features of synthetic voice (Simpson & Williams, 1980). Echoing a theme that we introduced at the end of the last chapter, *redundancy gain* is a fundamental principle of human performance that can be usefully employed in alarm system design.

Another possible design that can address some of the problems associated with comprehension and masking is to synthesize alarm sounds that sound like the condition they represent, called auditory icons or *earcons* (Gaver, 1986). Belz, Robinson, and Casali (1999), for example, found that representing hazard alarms to automobile drivers in the form of earcons (e.g., the sound of squealing tires representing a potential forward collision) significantly shortened driver response time relative to conventional auditory tones.

False Alarms. An alarm is of course one form of *automation* in that it typically monitors some process for the human operator and alerts the operator whenever it infers that the process is getting out of hand and requires some form of human intervention (see Chapter 16). Alarms are little different from the human signal detector described in Chapter 4. When sensing low-intensity signals from the environment (a small increase in temperature, a wisp of smoke), the system sometimes makes mistakes, inferring that nothing has happened when it has (the miss) or inferring that *something* has happened when it has not (the false alarm).

Most alarm designers and users set the alarm's criterion as low as possible to minimize the miss rate for obvious safety reasons. But as we learned, when the low-intensity signals on which the alarm decision is made, are themselves noisy, the consequence of setting a miss-free criterion is a higher than desirable false alarm rate: To paraphrase from the old fable, the system "cries wolf" too often Bliss & Gilson, 1998). Such was the experience with the initial introduction of the ground proximity warning system in aircraft, designed to alert pilots that they might be flying dangerously close to the ground. Unfortunately, when the conditions that trigger the alarm occur very rarely, an alarm system that guarantees detection will, almost of necessity, produce a fair number of false alarms, or "nuisance alarms" (Parasuraman et al., 1997).

From a human performance perspective, the obvious concern is that users may come to distrust the alarm system and perhaps ignore it even when it provides valid information (Pritchett, 2001; Parasuraman & Riley, 1997). More serious yet, users may attempt to disable the annoying alarms (Sorkin, 1989). Many of these concerns are related to the issue of *trust* in automation, discussed in Chapter 16 (Muir, 1988; Lee & Moray, 1992).

Five logical steps may be taken to avoid the circumstances of "alarm false-alarms." First, it is possible that the alarm criterion itself has been set to such an extremely sensitive value that readjustment to allow fewer false alarms will still not appreciably increase the miss rate. Second, more sophisticated decision algorithms within the system may be developed to improve the *sensitivity* of the alarm system, a step that was taken to address the problems with the ground proximity warning system. Third, users can be trained about the inevitable tradeoff between misses and false alarms and therefore can be taught to accept the false alarm rates as an inevitable consequence of automated protection in an uncertain probabilistic world rather than as a system failure. (This acceptance will be more likely if care is taken to make the alarms noticeable by means other than shear loudness; Edworthy et al., 1991.) Fourth, designers should try to provide the user with the "raw data" or conditions that triggered the alarm, at least by making available the tools that can verify the alarm's accuracy.

Finally, a logical approach suggested by Sorkin, Kantowitz, and Kantowitz (1988) is to consider the use of *graded* or *likelihood alarm systems* in which more than a single level of alert is provided. Hence, two (or more) levels can signal to the human the system's *own* confidence that the alarming conditions are present. That evidence in the fuzzy middle ground (e.g., the odor from a slightly burnt piece of toast), which previously might have signaled the full fire alarm, now triggers a signal of noticeable but reduced intensity.

The concept of the likelihood alarm is closely related to the application of **fuzzy signal detection theory** (Parasuraman, et al., 2000). Crisp signal detection theory, described in Chapter 4, characterizes circumstances in which a "signal" either was or was not present (and a response is either yes or no). In fuzzy signal detection theory, one speaks instead of the **degree** of signal present, or the degree of danger or threat—a variable that can take on a continuous range of values. This might represent the degree of future threat of a storm, fire, disease outbreak, or terrorist attack. All of these events can happen with various degrees of seriousness. As a consequence, they may be addressed with various degrees of "signal present" responses. The consequences of applying fuzzy boundaries to both the states of the world and the classes of detection responses are that the concepts of joint outcomes (hits, false alarms, correct rejections, and misses) are themselves fuzzy, as are the behavioral measures of sensitivity and response bias.

An important facet of alarms is that experienced users often employ them for a wide range of uses beyond those that may have been originally intended by the designer (i.e., to alert to a dangerous condition of which the user is not aware; Woods, 1995). For example, in one study of alarm use in hospitals, anesthesiologists Seagull and Sanderson (2001) noted how anesthesiologists use alarms as a means of verifying the results of their decisions or as simple reminders of the time at which a certain procedure must be performed.

SOUND LOCALIZATION

In Chapter 3 we described the role of the visual system in searching spatial worlds as guided by eye movements. The auditory system is somewhat less well suited for precise spatial localization but nevertheless has some very useful capabilities in this regard, given the differences in the acoustic patterns of a single sound, processed by the two ears (McKinley et al., 1994; Begault & Pittman, 1996). The ability to process the location of sounds is better in azimuth (e.g., left-right) than it is in elevation, and front-back confusions are also prominent. Overall, precision is less than the precision of visual localization. However, in some environments, where the eyes are heavily involved with other tasks or where signals could occur in a 360-degree range around the head (whereas the eyes can cover only about a 130-degree range with a given head fixation), sound localization can provide considerable value. An example might be providing the pilot with guidance as to the possible location of a midair conflict Begault & Pittman, 1996). In particular, a redundant display of visual and auditory location can be extremely useful in searching for targets in a 3-D 360-degree volume. The sound can guide the head and eyes very efficiently to the general direction of the target, allowing the eyes then to provide the precise localization (Bolia et al., 1999).

THE SOUND TRANSMISSION PROBLEM

Our example at the beginning of the chapter illustrated the worker's concern with her ability to communicate with her neighbor at the workplace. A more tragic illustration of communications breakdown is provided by the 1979 collision between two jumbo jets on the runway at Tenerife airport in the Canary Islands, in which over 500 lives were lost (Hawkins & Orlady, 1993). One of the jets, a KLM 747, was poised at the end of the runway, engines primed, and the pilot was in a hurry to take off while it was still possible before the already poor visibility got worse and the airport closed operations. Meanwhile, the other jet, a Pan American airplane that had just landed, was still on the same runway, trying to find its way off. The air traffic controller instructed the pilot of the KLM: "Okay, stand by for takeoff and I will call." Unfortunately, because of a less than perfect radio channel and because of the KLM pilot's extreme desire to proceed with the takeoff, he apparently *heard* just the words "Okay . . . take off." The take-off proceeded until the aircraft collided with the Pan Am 747, which had still not steered itself clear from the runway.

In Chapter 4, we discussed the influences of both bottom-up (sensory quality) and top-down (expectations and desires) processing on perception. The Canary Island accident tragically illustrates the breakdown of both processes. The communications signal from ATC was degraded (loss of bottom-up quality), and the KLM pilot used his own expectations and desires to "hear what he wanted to hear" (inappropriate top-down processing) and to interpret the message as authorization to take off. In this section we consider in more detail the role of both of these processes in what is arguably the most important kind of auditory communications, the processing of human speech. (We have already discussed the communications of warning information. We first describe the nature of the speech stimulus and then discuss how it may be distorted in its transmission by changes in signal quality and by noise. Finally, we consider possible ways of remediating breakdowns in the speech transmission process.

The Speech Signal

The Speech Spectrograph.   The sound waves of a typical speech signal look something like the pattern shown in Figure 5.7a. As we have seen, such signals are more coherently presented by a spectral representation, as shown in Figure 5.7b. However, for speech, unlike noise or tones, many of the key properties are captured in the *time-dependent changes* in the spectrum; that is, in the *envelope*

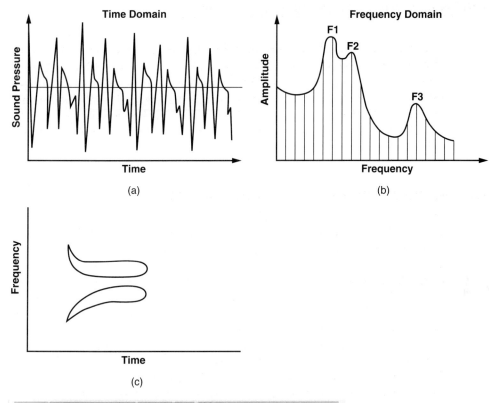

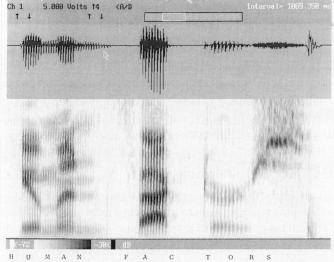

FIGURE 5.7

(a) Voice time signal;

(b) Voice spectrum (*Source:* Yost, W. A., 1994. *Fundamentals of Hearing,* 3rd ed. San Diego: Academic Press);

(c) Schematic speech spectrograph (the sound *dee*);

(d) A real speech spectrograph of the words "human factors." (*Source:* Courtesy of Speech and Hearing Department, University of Illinois.)

of the sound. To represent this information graphically, speech is typically described by the *speech spectrograph,* as shown in Figure 5.7c. One can think of each vertical slice of the spectrograph as the momentary spectrum, existing at the time labeled on the *x* axis. Where there is darkness (or thickness), there is power (and greater darkness represents more power). However, the spectral content of the signal changes as the time axis moves from left to right. Thus, the particular speech signal shown at the bottom of Figure 5.7c represents a very faint initial pitch that increases in its frequency value and intensity over the first few msec to reach a steady state at a higher frequency. Collectively, the two bars shown in the figure characterize the sound of the human voice saying the letter *d* (dee). Figure 5.7d shows the spectrum of more continuous speech.

Masking Effects of Noise.   The potential of any auditory signal to be masked by other sounds depends on both the intensity (power) and frequency of that signal (Crocker, 1997). These two variables are influenced by the speaker's gender and by the nature of the speech sound. First, since the female voice typically has a higher base frequency than the male, it is not surprising that the female voice is more vulnerable to masking of noise. Second, as Figure 5.7c illustrates, the power or intensity of speech signals (represented by the thickness of the lines) is much greater in the vowel range *eee* than in the initial consonant part *d.* This difference in salience is further magnified because, as also seen in Figure 5.7c, the vowel sounds often stretch out over a longer period of time than do the consonants. Finally, certain consonant sounds, like *s* and *ch,* have distinguishing features at very high frequencies, and high frequencies are more vulnerable to masking by low frequencies than the converse. Hence, it is not surprising that consonants are much more susceptible to masking and other disruptions than are vowels. This characteristic is particularly disconcerting because consonants typically transmit more information in speech than do vowels (i.e., there are more of them). One need only think of the likely possibility of confusing "fly to" with "fly through" in an aviation setting to realize the danger of such consonant confusion (Hawkins & Orlady 1993). Miller and Nicely (1955) provide a good analysis of the confusability between different consonant sounds. We return to the issue of sound confusion in Chapter 6.

Measuring Speech Communications

Human factors engineers know that noise degrades communications, but they must often assess (or predict) precisely how much communications will be lost in certain degraded conditions. For this, we must consider the measurement of speech communications effectiveness.

There are two different approaches to measuring speech communications, based on bottom-up and top-down processing respectively. The bottom-up approach derives some objective measure of speech quality. It is most appropriate in measuring the potential degrading effects of noise. Thus, the *articulation index* (AI) computes the signal-to-noise ratio (db of speech sound minus db of background noise) across a range of the spectrum in which useful speech information is imparted. Figure 5.8 presents a simple example of how the AI might be computed with four different frequency bands. This measure can be weighted

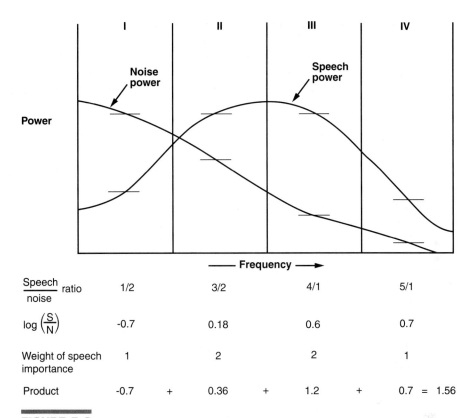

	I		II		III		IV
$\dfrac{\text{Speech}}{\text{noise}}$ ratio	1/2		3/2		4/1		5/1
$\log\left(\dfrac{S}{N}\right)$	-0.7		0.18		0.6		0.7
Weight of speech importance	1		2		2		1
Product	-0.7	+	0.36	+	1.2	+	0.7 = 1.56

FIGURE 5.8

Schematic representation of the calculation of an AI. The speech spectrum has been divided into four bands, weighted in importance by the relative power that each contributes to the speech signal. The calculations are shown in the rows below the figure. (*Source:* Wickens, C. D. *Engineering Psychology and Human Performance,* 2nd ed., New York: HarperCollins, 1992. Reprinted by permission of Addison-Wesley Educational Publishers, Inc.)

by the different frequency bands, providing greater weight to the ratios within bands that contribute relatively more heavily to the speech signal.

While the objective merits of the bottom-up approach are clear, its limits in predicting the understandability of speech should become apparent when one considers the contributions of top-down processing to speech perception. For example, two letter strings, *abcdefghij* and *wcignspexl,* might both be heard at intensities with the same AI. But it is clear that more letters of the first string would be correctly understood (Miller et al., 1951). Why? Because the listener's knowledge of the predictable *sequence* of letters in the alphabet allows perception to "fill in the gaps" and essentially guess the contents of a letter whose sensory clarity may be missing. This, of course, is the role of top-down processing.

A measure that takes top-down processing into account is the *speech intelligibility level* (SIL). This index measures the percentage items correctly heard. Naturally, at any given bottom-up AI level, this percentage will vary as a function of the listener's expectation of and knowledge about the message communicated, a variable that influences the effectiveness of top-down processing. This complementarity relationship between bottom-up and top-down processing is illustrated in Figure 5.9, which shows, for example, that sentences that are known to listeners can be recognized with just as much accuracy as random isolated words, even though the latter are presented with nearly twice the bottom-up sensory quality.

Speech Distortions. While the AI can objectively characterize the damaging effect of noise on bottom-up processing of speech, it cannot do the same thing with regard to *distortions*. Distortions may result from a variety of causes, for example, clipping of the beginning and ends of words, reduced bandwidth of high-demand communications channels, echoes and reverberations, and even the low quality of some digitized synthetic speech signals (Pisoni, 1982).

While the bottom-up influences of these effects cannot be as accurately quantified as the effects of noise, there are nevertheless important human factors guidelines that can be employed to minimize their negative impact on voice

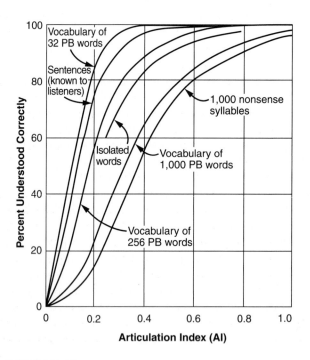

FIGURE 5.9

Relationship between the AI and the intelligibility of various types of speech test materials. Note that at any given AI, a greater percentage of items can be understood if the vocabulary is smaller or if the word strings form coherent sentences. (*Source:* Adapted from Kryter, K., 1972. Speech Communications. In *Human Engineering Guide to System Design*, H. P. Van Cott and R. G. Kinkade, eds., Washington, DC: U.S. Government Printing Office.)

recognition. One issue that has received particular attention from acoustic engineers is how to minimize the distortions resulting when the high-information speech signal must be somehow "filtered" to be conveyed over a channel of lower bandwidth (e.g., through digitized speech).

For example, a raw speech waveform such as that shown in Figure 5.7b may contain over 59,000 bits of information per second (Kryter, 1972). Transmitting the raw waveform over a single communications channel might overly restrict that channel, which perhaps must also be shared with several other signals at the same time. There are, however, a variety of ways to reduce the information content of a speech signal. One may filter out the high frequencies, digitize the signal to discrete levels, clip out bits of the signal, or reduce the range of amplitudes by clipping out the middle range. Human factors studies have been able to inform the engineer which way works best by preserving the maximum amount of speech intelligibility for a given resolution in information content. For example, amplitude reduction seems to preserve more speech quality and intelligibility than does frequency filtering, and frequency filtering is much better if only very low and high frequencies are eliminated (Kryter, 1972).

Of course, with the increasing availability of digital communications and voice synthesizers, the issue of transmitting voice quality with minimum bandwidth is lessened in its importance. Instead, one may simply transmit the symbolic contents of the message (e.g., the letters of the words) and then allow a speech synthesizer at the other end to reproduce the necessary sounds. (This eliminates the uniquely human, nonverbal aspects of communications—a result that may not be desirable when talking on the telephone.) Then, the issue of importance becomes the level of fidelity of the voice synthesizer necessary to (1) produce recognizable speech, (2) produce recognizable speech that can be heard in noise, and (3) support "easy listening." The third issue is particularly important, as Pisoni (1982) has found that listening to synthetic speech takes more mental resources than does listening to natural speech. Thus, listening to synthetic speech can produce greater interference with other ongoing tasks that must be accomplished concurrently with the listening task (see Chapter 6) or will be more disrupted by the mental demands of those concurrent tasks.

The voice, unlike the printed word, is transient. Once a word is spoken, it is gone and cannot be referred back to. The human information-processing system is designed to prolong the duration of the spoken word for a few seconds through what is called *echoic memory*. However, beyond this time, spoken information must be actively rehearsed, a demand that competes for resources with other tasks. Hence, when displayed messages are more than a few words, they should be delivered visually or at least backed up with a redundant visual signal.

Hearing Loss

In addition to noise and distortions, a final factor responsible for loss in voice transmission is the potential loss of hearing of the listener (Crocker, 1997; Kryter, 1995) As shown in Figure 5.10, simple age is responsible for a large portion of hearing loss, particularly in the high-frequency regions, a factor that should be considered in the design of alarm systems, particularly in nursing homes. On

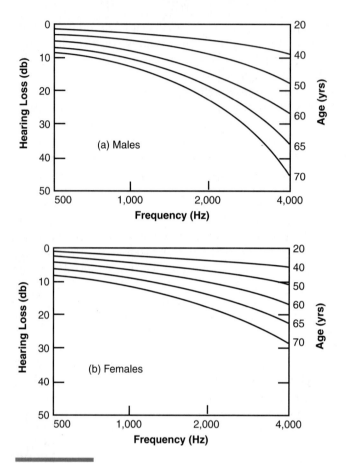

FIGURE 5.10

Idealized median (50th percentile) hearing loss at different frequencies for males and females as a function of age. (*Source:* Kryter, K., 1983. Addendum: Presbycusis, Sociocusis and Nococusis. *Journal of Acoustic Society of America, 74*, pp. 1907–1909. Reprinted with permission. Copyright Acoustic Society of America.)

top of the age-related declines may be added certain occupation-specific losses related to the hazards of a noisy workplace (Crocker, 1997; Taylor et al., 1965). These are the sorts of hazards that organizations (OSHA) try to eliminate.

NOISE REVISITED

We discussed noise as a factor disrupting the transmission of information. In this section we consider two other important human factors concerns with noise: its potential as a health hazard in the workplace and its potential as an irritant in the environment. In Chapter 13, we also consider noise as a stressor that has degrading effects on performance other than the communications masking effect discussed here. In chapter 14, we consider broader issues of workplace safety. We conclude by offering various possible remediations to the degrading effects of noise in all three areas: communications, health, and environment.

The worker in our story was concerned about the impact of noise at her workplace on her ability to hear. When we examine the effects of noise, we consider three components of the potential hearing loss. The first, masking, has already been discussed; this is a loss of sensitivity to a signal *while the noise is present.*

The second form of noise-induced hearing loss is the *temporary threshold shift* (Crocker, 1997). If our worker steps away from the machine to a quieter place to answer the telephone, she may still have some difficulty hearing because of the "carryover" effect of the previous noise exposure. This temporary threshold shift (TTS) is large immediately after the noise is terminated but declines over the following minutes as hearing is "recovered" (Figure 5.11). The TTS is typically expressed as the amount of loss in hearing (shift in threshold in dB) that is present two minutes after the source of noise has terminated. The TTS is increased by a longer prior noise exposure and a greater prior level of that exposure. The TTS can be quite large. For example, the TTS after being exposed to 100 dB noise for 100 minutes is 60 dB.

The third form of noise-induced hearing loss, which has the most serious implications for worker health, is the *permanent threshold shift* (PTS). This measure describes the "occupational deafness" that may set in after workers have been exposed to months or years of high-intensity noise at the workplace. Like the TTS, the PTS is greater with both louder and longer prior exposure to noise.

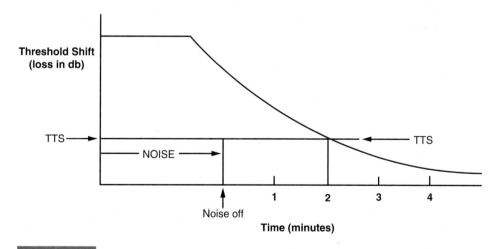

FIGURE 5.11

TTS following the termination of noise. Note that sensitivity is recovered (the threshold shift is reduced over time). Its level at two minutes is arbitrarily defined as the TTS.

Also, like age-related hearing loss, the PTS tends to be more pronounced at higher frequencies, usually greatest at around 4,000 Hz (Crocker, 1997).

During the last few decades in the United States, OSHA has taken steps to try to ensure worker safety from the hazardous effects of prolonged noise in the workplace by establishing standards that can be used to trigger remediating action (OSHA 1983). These standards are based on a *time weighted average* (TWA) of noise experienced in the workplace, which trades off the intensity of noise exposure against the duration of the exposure. If the TWA is above 85 dB, the *action level,* employers are required to implement a hearing protection plan in which ear protection devices are made available, instruction is given to workers regarding potential damage to hearing and steps that can be taken to avoid that damage, and regular hearing testing is implemented. If the TWA is above 90 dB, the *permissible exposure level,* then the employer is *required* to takes steps toward noise reduction through procedures that we discuss below.

Of course, many workers do not experience continuous noise of these levels but may be exposed to bursts of intense noise followed by periods of greater quiet. By addressing the tradeoff between time and intensity, the OSHA standards provide means of converting the varied time histories of noise exposures into the single equivalent standard of the TWA (Sanders & McCormick, 1993). The noise level at a facility cannot be expressed by a single value but may vary from worker to worker, depending on his or her location relative to the source of noise. For this reason, TWAs must be computed on the basis of noise *dosemeters,* which are worn by individual workers and collect the data necessary to compute the TWA over the course of the day.

NOISE REMEDIATION

The steps that should be taken to remediate the effects of noise might be very different, depending on the particular nature of the noise-related problem and the level of noise that exists before remediation. On the one hand, if noise problems relate to communications difficulties in situations when the noise level is below 85 dB (e.g., a noisy phone line), then *signal enhancement* procedures may be appropriate. On the other hand, if noise is above the action levels (a characteristic of many industrial workplaces), then *noise reduction* procedures must be adopted because enhancing the signal intensity (e.g., louder alarms) will do little to alleviate the possible health and safety problems. Finally, if noise is a source of irritation and stress in the environment (e.g., residential noise from an airport or nearby freeway), then many of the sorts of solutions that might be appropriate in the workplace, like wearing earplugs, are obviously not applicable.

Signal Enhancement

Besides obvious solutions of "turning up the volume" (which may not work if this amplifies the noise level as well and so does not change the signal-to-noise ratio) or talking louder, there may be other more effective solutions for enhancing the amplitude of speech or warning sound signals relative to the background noise. First, careful consideration of the *spectral content* of the masking noise may allow one to use signal spectra that have less overlap with the noise content. For example, the spectral content of synthetic voice messages or alarms can be carefully chosen to lie in regions where noise levels are lower. Since lower frequency noise masks higher frequency signals, more than the other way around, this relation can also be exploited by trying to use lower frequency signals. Also, synthetic speech devices or earphones can often be used to bring the source of signal closer to the operator's ear than if the source is at a more centralized location where it must compete more with ambient noise.

There are also signal-enhancement techniques that emphasize more the *redundancy* associated with top-down processing. As one example, it has been shown that voice communications is far more effective in a face-to-face mode than it is when the listener cannot see the speaker, (Sumby & Pollack, 1954). This is because of the contributions made by many of the redundant cues provided by the lips (Massaro & Cohen, 1995), cues of which we are normally unaware unless they are gone or distorted. (To illustrate the important and automatic way we typically integrate sound and lip reading, recall, if you can, the difficulty you may have in understanding the speech of poorly dubbed foreign films when speech and lip movement do not coincide in a natural way.)

Another form of redundancy is involved in the use of the phonetic alphabet ("alpha, bravo, charlie, . . . charlie, . . ."). In this case, more than a single sound is used to convey the content of each letter, so if one sound is destroyed (e.g., the consonant *b*), other sounds can unambiguously "fill in the gap" (*ravo*).

In the context of communications measurement, improved top-down processing can also be achieved through the choice of vocabulary. Restricted vocabulary, common words, and standardization of communications procedures, such as that adopted in air traffic control (and further emphasized following the Tenerife disaster), will greatly restrict the number of *possible* utterances that could be heard at any given moment and hence will better allow perception to "make an educated guess" as to the meaning of a sound if the noise level is high, as illustrated in Figure 5.9.

Noise Reduction in the Workplace

We may choose to reduce noise in the workplace by focusing on the source, the path or environment, or the listener. The first is the most preferred method; the last is the least.

The Source: Equipment and Tool Selection.   Many times, effective reduction can be attained by the appropriate and careful choice of tools or sound-producing equipment. Crocker (1997) provides some good case studies where this has been done. Ventilation or fans, or handtools, for example, vary in the sounds they produce, and appropriate choices in purchasing such items can be made. The noise of vibrating metal, the source of loud sounds in many industrial settings, can be attenuated by using damping material, such as rubber. One should consider also that the irritation of noise is considerably greater in the high-frequency region (the shrill pierced whine) than in the mid- or low-frequency region (the low rumble). Hence, to some extent the choice of tool can reduce the irritating quality of its noise.

The Environment. The *environment* or path from the sound source to the human can also be altered in several ways. Changing the environment near the source, for example, is illustrated in Figure 5.12, which shows the attenuation in noise achieved by surrounding a piece of equipment with a plexiglass shield. Sound absorbing walls, ceilings, and floors can also be very effective in reducing the noise coming from reverberations. Finally, there are many circumstances when repositioning workers relative to the source of noise can be effective. The effectiveness of such relocation is considerably enhanced when the noise emanates from only a single source. This is more likely to be the case if the source is present in a more sound-absorbent environment (less reverberating).

The Listener: Ear Protection. If noise cannot be reduced to acceptable levels at the source or path, then solutions can be applied to the listener. Ear protection devices that must be made available when noise levels exceed the action level are of two generic types: earplugs, which fit inside the ear, and ear muffs, which fit over the top of the ear. As commercially available products, each is provided with a certified *noise reduction ratio* (NRR), expressed in decibels, and each may also have very different spectral characteristics (i.e., different decibel reduction across the spectrum). For both kinds of devices, it appears that the manufacturer's specified NRR is typically greater (more optimistic) than is the actual

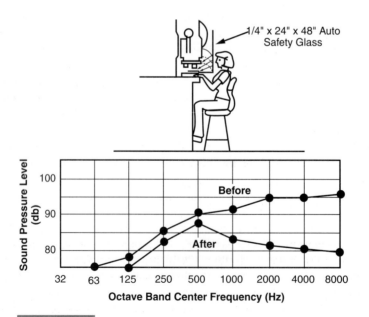

FIGURE 5.12

Use of a ¼-in. (6-mm)-thick safety glass barrier to reduce high-frequency noise from a punch press. (*Source:* American Industrial Hygiene Association, 1975, Figure 11.73. Reprinted with permission by the American Industrial Hygiene Association.)

noise reduction experienced by users in the workplace (Casali et al., 1987). This is because the manufacturer's NRR value is typically computed under ideal laboratory conditions, whereas users in the workplace may not always wear the device properly.

Of the two devices, earplugs can offer a greater overall protection *if properly worn* (Sanders & McCormick, 1993). However, this qualification is extremely important because earplugs are more likely than ear muffs to be worn improperly. Hence, without proper training (and adherence to that training), certain muffs may be more effective than plugs. A second advantage of muffs is that they can readily double as headphones through which critical signals can be delivered, simultaneously achieving signal enhancement and noise reduction.

Comfort is another feature that cannot be neglected in considering protector effectiveness in the workplace. It is likely that devices that are annoying and uncomfortable may be disregarded in spite of their safety effectiveness (see Chapter 14). Interestingly, however, concerns such as that voiced by the worker at the beginning of the chapter that hearing protection may not allow her to hear conversations are not always well grounded. After all, the ability to hear conversation is based on the signal-to-noise ratio. Depending on the precise spectral characteristics and amplitude of the noise and the signal *and* the noise-reduction function, wearing such devices may actually *enhance* rather than reduce the signal-to-noise ratio, even as both signal and noise intensity are reduced. The benefit of earplugs to increasing the signal-to-noise ratio is greatest with louder noises, above about 80 to 85 dB (Kryter, 1972).

Finally, it is important to note that the adaptive characteristics of the human speaker may themselves produce some unexpected consequences on speech comprehension. We automatically adjust our voice level, in part, on the basis of the intensity of sound that we hear, talking louder when we are in a noisy environment (Crocker, 1997) or when we are listening to loud stereo music through headphones. Hence, it is not surprising that speakers in a noisy environment talk about 2 to 4 dB *softer* (and also somewhat faster) when they are wearing ear protectors than when they are not. This means that listening to such speech may be disruptive in environments in which all participants wear protective devices, unless speakers are trained to avoid this automatic reduction in the loudness of their voice.

Environmental Noise

Noise in residential or city environments, while presenting less of a health hazard than at the workplace, is still an important human factors concern, and even the health hazard is not entirely absent. Meecham (1983), for example, reported that the death rate from heart attacks of elderly residents near the Los Angeles Airport was significantly higher than the rate recorded in a demographically equivalent nearby area that did not receive the excessive noise of aircraft landings and takeoffs.

Measurement of the irritating qualities of environmental noise levels follows somewhat different procedures from the measurement of workplace dangers. In particular, in addition to the key component of intensity level, there are a number of other "irritant" factors that can drive the annoyance level upward. For example, high frequencies are more irritating than low frequencies. Airplane noise is more irritating than traffic noise of the same dB level. Nighttime noise is more irritating than daytime noise. Noise in the summer is more irritating than in the winter (when windows are likely to be closed). While these and other considerations cannot be precisely factored into an equation to predict "irritability," it is nevertheless possible to estimate their contributions in predicting the effects of environmental noise on resident complaints (Environmental Protection Agency, 1974). One study (Finegold et al., 1994) found that the percentage of people "highly annoyed" by noise follows a roughly linear function for noise levels above 70dB, of the form

$$\% \text{ highly annoyed} = 20 + 3.2 \text{ dB}$$

Is All Noise Bad?

Before we leave our discussion of noise, it is important to identify certain circumstances in which softer noise may actually be helpful. For example, low levels of continuous noise (the hum of a fan) can mask the more disruptive and startling effects of discontinuous or distracting noise (the loud ticking of the clock at night or the conversation in the next room). Soft background music may accomplish the same objective. Under certain circumstances, noise can perform an alerting function that can maintain a higher level of vigilance (Parasuraman et al., 1987; Broadbent, 1972; see Chapter 4). Furthermore, one person's noise may be another person's "signal" (as is often the case with conversation).

This last point brings us back to reemphasize one final issue that we have touched on repeatedly: the importance of *task analysis*. The full impact of adjusting sound frequency and intensity levels on human performance can never be adequately predicted without a clear understanding of what sounds will be present when, who will listen to them, who *must* listen to them, and what the costs will be to task performance, listener health, and listener comfort if hearing is degraded.

THE OTHER SENSES

Vision and hearing have held the stage during this chapter and the previous one for the important reason that the visual and auditory senses are of greatest implications for the design of human–machine systems. The "other" senses, critically important in human experience, have played considerably less of a role in system design. Hence, we do not discuss the senses of smell and taste, important as both of these are to the pleasures of eating (although smell can provide an important safety function as an advanced warning of fires and overheating engines). We discuss briefly, however, three other categories of sensory experience that have some direct relevance to design: touch and feel (the tactile and haptic sense), limb position and motion (proprioception and kinesthesis), and whole-body orientation and motion (the vestibular senses). All of these offer important channels of information that help coordinate human interaction with many physical systems.

Touch: Tactile and Haptic Senses

Lying just under the skin are sensory receptors that respond to pressure on the skin and relay their information to the brain regarding the subtle changes in force applied by the hands and fingers (or other parts of the body) as they interact with physical things in the environment. Along with the sensation of pressure, these senses, tightly coupled with the proprioceptive sense of finger position, also provide *haptic* information regarding the *shape* of manipulated objects and things (Loomis & Lederman, 1986; Kaczmarer & Bach-T-Rita, 1995). We see the importance of these sensory channels in the following examples:

1. A problem with the membrane keyboards sometimes found on calculators is that they do not offer the same "feel" (tactile feedback) when the fingers are positioned on the button as do mechanical keys (see Chapter 9).
2. Gloves, to be worn in cold weather (or in hazardous operations) must be designed with sensitivity to maintaining some tactile feedback if manipulation is required (Karis, 1987).
3. Early concern about the confusion that pilots experienced between two very different controls—the landing gear and the flaps—was addressed by redesigning the control handles to feel quite distinct. The landing gear felt like a wheel—the plane's tire—while the flap control felt like a rectangular flap. Incidentally this design also made the controls feel and look somewhat like the system that they activate; see Chapter 9.
4. The tactile sense is well structured as an alternative channel to convey both spatial and symbolic information for the blind through the braille alphabet.
5. Designers of *virtual environments* (Durlach & Mavor, 1995; Kaczmarer & Bach-T-Rita, 1995), which we discuss in Chapter 15, attempt to provide artificial sensations of touch and feel via electrical stimulation to the fingers, as the hand manipulates "virtual objects" (Bullinger et al., 1997).
6. In situations of high visual load, tactile displays can be used to call attention to important discrete events (Sklar & Sarter, 1999) and can sometimes provide important information for tracking, or continuous control (Kortelling & van Emmerick, 1998), as we discuss further in Chapter 9.

Proprioception and Kinesthesis

We briefly introduced the *proprioceptive* channel in the previous section in the context of the brain's knowledge of finger position. In fact, a rich set of receptor systems, located within all of the muscles and joints of the body, convey to the brain an accurate representation of muscle contraction and joint angles everywhere and, by extension, a representation of limb position in space. The proprioceptive channel is tightly coupled with the *kinesthetic* channel, receptors within the joints and muscles, which convey a sense of the *motion* of the limbs as exercised by the muscles. Collectively, the two senses of kinesthesis and proprioception provide rich feedback that is critical for our everyday interactions with things in the environment. One particular area of relevance for these senses is in the design of manipulator controls, such as the joystick or mouse with a computer system, the steering wheel on a car, the clutch on a machine tool, and the control on an aircraft (see Chapter 9). As a particular example, an *isometric* control is one that does not move but responds only to pressure applied upon it. Hence, the isometric control cannot benefit from any proprioceptive feedback regarding *how far* a control has been displaced, since the control does not move at all. Early efforts to introduce isometric side-stick controllers in aircraft were, in fact, resisted by pilots because of this elimination of the "feel" of control deflection.

The Vestibular Senses

Located deep within the inner ear (visible in Figure 5.2) are two sets of receptors, located in the *semicircular canals* and in the *vestibular sacs.* These receptors convey information to the brain regarding the angular and linear *accelerations* of the body respectively. Thus, when I turn my head with my eyes shut, I "know" that I am turning, not only because kinesthetic feedback from my neck tells me so but also because there is an angular acceleration experienced by the semicircular canals. Associated with the three axes along which the head can rotate, there are three semicircular canals aligned to each axis. Correspondingly, the vestibular sacs (along with the tactile sense from the "seat of the pants") inform the passenger or driver of linear acceleration or braking in a car. These organs also provide the constant information about the accelerative force of gravity downward, and hence they are continuously used to maintain our sense of balance (knowing which way is up and correcting for departures).

Not surprisingly, the vestibular senses are most important for human–system interaction when the systems either move directly (as vehicles) or *simulate* motion (as vehicle simulators or virtual environments). The vestibular senses play two important (and potentially negative) roles here, related to *illusions* and to *motion sickness.*

Vestibular illusions of motion, occur because certain vehicles, particularly aircraft, place the passenger in situations of sustained acceleration and nonvertical orientation for which the human body is not naturally adapted. Hence, for example, when the pilot is flying in the clouds without sight of the ground or horizon, the vestibular senses may sometimes be "tricked" into thinking that up is in a different direction from where it really is. This illusion presents some real dangers of *spatial disorientation* and the possible loss of control of the aircraft that may result (O'Hare & Roscoe, 1990; Young, 2003).

The vestibular senses also play a key role in motion sickness. Normally, our visual and vestibular senses convey compatible and redundant information to the brain regarding how we are oriented and how we are moving. However, there are certain circumstances in which these two channels become *decoupled* so that one sense tells the brain one thing and the other tells it something else.

These are conditions that invite *motion sickness* (Oman, 1993; Reason & Brand, 1975; Young, 2003; Jackson, 1994). One example of this decoupling results when the vestibular cues signal motion and the visual world does not. When riding in a vehicle with no view of the outside world (e.g., a toddler sitting low in the backseat of the car, a ship passenger below decks with the portholes closed, or an aircraft passenger flying in the clouds), the visual view forward, which is typically "framed" by a man-made rectangular structure, provides no visual evidence of movement (or evidence of where the "true" horizon is). In contrast, the continuous rocking, rolling, or swaying of the vehicle provides very direct stimulation of movement to the vestibular senses to all three of these passengers. When the two senses are in conflict, motion sickness often results (a phenomenon that was embarrassingly experienced by the first author while in the Navy at his first turn to "general quarters" with the portholes closed below decks).

Conflict between the two senses can also result from the opposite pattern. The visual system can often experience a very compelling sense of motion in video games, driving or flight simulators, and virtual environments even when there is no motion of the platform (Bullinger et al., 1997). Again, there is conflict and the danger of a loss of function (or wasted training experience) when the brain is distracted by the unpleasant sensations of motion sickness. We return to this topic in Chapter 13.

CONCLUSION

Audition, when coupled with vision and the other senses, can offer the brain an overwhelming array of information. Each sensory modality appears to have particular strengths and weaknesses, and collectively the ensemble nicely compensates for the collective weaknesses of each sensory channel alone. Clever designers can capitalize on the strengths and avoid the weaknesses in rendering the sensory information available to the higher brain centers for perception, interpretation, decision making, and further processing. In the following two chapters, we consider the characteristics of these higher level information-processing or *cognitive* operations before addressing, in Chapter 8, how sensory processing and information processing may be gracefully connected in human factors by the careful engineering design of displays.

Chapter

Cognition

6

Laura was running late for an appointment in a large, unfamiliar city and relied on her new navigation device to guide her. She had read the somewhat confusing instructions and realized the importance of the voice display mode so that she could hear the directions to her destination without taking her eyes off the road. She had reminded herself to activate it before she got into heavy traffic, but the traffic suddenly increased, and she realized that she had forgotten to do so. Being late, however, she did not pull over but tried to remember the sequence of mode switches necessary to activate the voice mode. She couldn't get it right, but she managed to activate the electronic map. However, transposing its north-up representation to accommodate her south-bound direction of travel was too confusing. Finally lost, she pulled out her cellular phone to call her destination, glanced at the number she had written down, 303-462-8553, and dialed 303-462-8533. Getting no response, she became frustrated. She looked down to check the number and dial it carefully. Unfortunately, she did not see the car rapidly converging along the entrance ramp to her right, and only at the last moment the sound of the horn alerted her that the car was not yielding. Slamming on the brakes, heart beating fast, she pulled off to the side to carefully check her location, read the instructions, and place the phone call in the relative safety of the roadside.

Each day, we process large amounts of information from our environment to accomplish various goals and make our way successfully through the world. The previous illustration represents a typical problem that one might experience because of a poor match between man-made equipment (or the environment) and the human information-processing system. Sometimes these mismatches cause misperceptions, and sometimes people just experience memory failures. While the scenario described above may seem rather mundane, there are dozens of other cases where difficulties result in injury or death (Casey, 1993; Wickens & Hollands, 2000). Some of these cases are discussed in Chapter 14 on safety. In this chapter, we consider the basic mechanisms by which people perceive, think, and remember, processes generally grouped under the label of *cognition,* and we provide a framework for understanding how such information is processed. As we learn about the various limitations of the human cognitive system, we consider the implications of, and some solutions for, design problems.

The human information-processing system is conveniently represented by diferent **stages** at which information gets transformed: (1) perception of information about the environment, (2) central processing or transforming that information, and (3) responding to that information. We highlight the first and second stages as the processes involved in **cognition** and most typically represented in the study of applied cognitive psychology (Durso, 1999), although we present a more elaborate picture than the simple three-stage model. This chapter then picks up where our discussions of the more sensory aspects of auditory and visual processing left off in the previous two chapters. In Chapter 7, we describe more complex cognitive processes that form the basis of decision making, in Chapter 8 we discuss the implications of perception and cognition for display design, and in Chapter 9 we discuss the implications for control. Finally, our discussions of memory have many direct implications for learning, as discussed in Chapter 18.

INFORMATION PROCESSING MODELS

Shown in Figure 6.1 is a model of information processing that highlights those aspects that typically influence cognition: perceiving, thinking about, and understanding the world. The senses, shown to the left of the figure, gather information, which is then **perceived,** providing a meaningful interpretation of what is sensed as aided by prior knowledge, through a mechanism that we described in Figure 4.6 as **top-down processing.** This prior knowledge is stored in long-term memory.

Sometimes, perception leads directly to the selection and execution of a response, as when the driver swerved to avoid the converging car in the opening story. Quite often, however, an action is delayed, or not executed at all, as we "think about" or manipulate perceived information in **working memory.** This stage of information processing plays host to a wide variety of mental activities that are in our consciousness, such as rehearsing, planning, understanding, visualizing, decision making, and problem solving. Working memory is a temporary, effort-demanding store. One of the activities for which working memory is used is to create a more permanent representation of the information in **long-term memory,** where it may be retrieved minutes, hours, days, or years later. These are the processes of learning (putting information into long-term memory) and retrieval. As we see in the figure, information from long-term memory is retrieved every time we perceive familiar information.

At the top of the figure we note that many of the stages of information processing depend upon mental or cognitive **resources,** a sort of pool of attention or *mental effort* that is of limited availability and can be allocated to processes as required. In particular, the figure highlights an important distinction that has

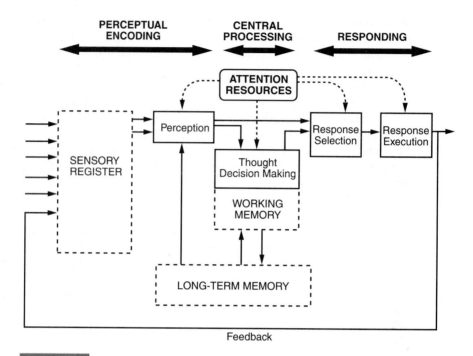

FIGURE 6.1

A model of human information processing.

been quite visible in the research on attention. On the left, we see the role of attention, the limited resources in **selecting** sensory channels for further information processing, as when our eyes focus on one part of the world, and not another. In contrast, the other dashed arrows suggest the role of attention in supporting all aspects of performance as well as in dividing attention between tasks. These two aspects of attention, selection and division, are treated separately in this chapter.

Finally, we note the feedback loop. Our actions often generate new information to be sensed and perceived. The sequence of information processing may start anywhere. For example, sometimes we initiate an action from a decision with no perception driving it. We then may evaluate the consequence of that decision later, through sensation and perception. We consider the importance of this closed feedback loop in Chapter 9.

SELECTIVE ATTENTION

Laura was not attending to the roadway at the time she was looking at her cell phone, and this failure of selective attention nearly caused an accident. We shall see in Chapter 17 that failures of attention, many of them selective, are the major cause of automobile accidents (Malaterre, 1990). Correspondingly, the cause of the greatest number of fatal accidents in commercial aviation, **controlled flight**

into terrain, when a pilot flies a perfectly good airplane into the ground, represents a failure of selective attention to all those sources of information regarding the plane's altitude above the ground (Phillips, 2001; Wiener, 1977).

Selective attention does not guarantee perception, but it is usually considered necessary to achieve it. Stated in other terms, we normally look at the things we perceive and perceive the things we are looking at. We considered the role of visual scanning in selective attention in Chapter 4. While we do not have "earballs" that can index selective auditory attention as we have eyeballs in the visual modality, there is nevertheless a corresponding phenomenon in auditory selection. For example, we may tune our attention to concentrate on one conversation in a noisy workplace while filtering out the distraction of other conversations and noises.

The selection of channels to attend (and filtering of channels to ignore) is typically driven by four factors: **salience, effort, expectancy,** and **value** (Wickens et al., 2003, in press). They can be represented in the same contrasting framework of stimulus-driven bottom-up processes versus knowledge-driven top-down processes that we applied to perception in chapters 4 and 5. Salience is a bottom-up process, characterizing what is described as **attentional capture** (Yantis, 1993). The car horn, for example, clearly captured Laura's attention. Salient stimulus dimensions are chosen by designers to signal important events via alarms and alerts. Abrupt onsets (Yantis, 1993), distinct stimuli and auditory stimuli (Spence & Driver, 2000; Banbury et al., 2001), and tactile stimuli (Sklar & Sarter, 1999) are particularly salient. In contrast, many events that do not have these characteristics may not be noticed, even if they are significant, a phenomenon known as *change blindness* or *attentional blindness* (Rensink 2002).

Expectancy and value together define what are characteristically called as top-down or knowledge-driven factors in allocating attention. That is, we tend to look at, or "sample," the world where we *expect* to find information. Laura looked downward because she expected to see the phone number there. As an example in visual search, discussed in Chapter 4, a radiologist looks most closely at those areas of an x-ray plate most likely to contain an abnormality. Correspondingly, a pilot looks most frequently at the instrument that changes most rapidly because here is where the pilot expects to see change (Senders, 1964; Bellenkes et al., 1997). Conversely, the frequency of looking at or attending to channels is also modified by how *valuable* it is to look at (or how costly it may be to miss an event on a channel; Moray, 1986). This is why a trained airplane pilot will continue to scan the world outside the cockpit for traffic even if that traffic is rare (Wickens et al., 2002); the costs of not seeing the traffic (and colliding with it) are large.

Finally, selective attention may be inhibited if it is effortful. We prefer to scan short distances rather than long ones, and we often prefer to avoid head movements to select information sources. It is for this reason that drivers, particularly fatigued ones (who have not much "effort to give"), fail to look behind them to check their blind spot when changing lanes.

In addition to understanding the high frequency with which failures to notice may contribute to accidents (Jones & Endsley, 1996) and considering ways of training selective attention (which we discuss at the end of this chapter), understanding bottom-up processes of attentional capture are important for the design of alarms (Woods, 1995; Chapters 5 & 8) and automated cueing (Chapter 16). Understanding the role of effort in inhibiting attention movement is important in both designing integrated displays (Chapter 8) and configuring the layout of workspaces (Chapter 10).

PERCEPTION

The most direct consequence of selective attention selection is perception, which involves the extraction of meaning from an array (visual) or sequence (auditory) of information processed by the senses. Our driver, Laura, eventually looked to the roadside (selection) and perceived the hazard of the approaching vehicle. Sometimes, meaning may be extracted (perception) without attention. In this way, our attention at a party can be "captured" in a bottom-up fashion when a nearby speaker utters our name even though we were not initially selecting that speaker. This classic phenomenon is sometimes labeled the "cocktail party effect" (Moray 1969; Wood & Cowan, 1995). Correspondingly, the driver may not be consciously focusing attention on the roadway, even though he is adequately perceiving roadway information enough to steer the car.

Three Perceptual Processes

To elaborate on our discussion in chapters 4 and 5, perception proceeds by three often simultaneous and concurrent processes: (1) bottom-up feature analysis, (2) unitization, and (3) top-down processing. The latter two are based on long-term memory, and all of them have different implications for design.

Perception proceeds by **analyzing** the raw features of a stimulus or event, whether it is a word (the features may be letters), a symbol on a map (the features may be the color, shape, size, and location), or a sound (the features may be the phonemes of the word or the loudness and pitch of an alarm). Every event could potentially consist of a huge combination of features. However, to the extent that past experience has exposed the perceiver to sets of features that occur together and their co-occurrence is familiar (i.e., represented in long-term memory), these sets are said to become *unitized*. The consequence of unitization is more rapid and *automatic* than perceptual processing. Thus, the difference between perceiving the printed words of a familiar and an unfamiliar language is that the former can be perceived as whole units, and their meaning is directly accessed (retrieved from long-term memory), whereas the latter may need to be analyzed letter by letter, and the meaning is more slowly and effortfully retrieved from long-term memory. This distinction between the effortful processing of feature analysis and the more automatic processing of familiar unitized feature combinations (whose combined representation is stored in long-term memory), can be applied to almost any perceptual experience, such as perceiving symbols and icons (see Chapter 8), depth cues (Chapter 4), or alarm sounds (Chapter 5).

Whether unitized or not, stimulus elements and events may be perceived in clear visual or auditory form (reading large text in a well-lighted room or hearing a clearly articulated speech) or may be perceived in a degraded form. For a visual stimulus, short glances, tiny text, and poor illumination or contrast represent such degradation. For an auditory event, masking noise and low intensity or unfamiliar accents produce degradation. We describe these degrading characteristics as producing poor bottom-up processing. The successful perception of such stimuli or events are beneficial to the extent that they are unitized and familiar. However, the third aspect of perceptual processing, top-down processing, provides another support to offset the degradation of bottom-up processing.

Top-down processing may be conceived as the ability to correctly guess what a stimulus or event is, even in the absence of clear physical (bottom-up) features necessary to precisely identify it. Such guesses are based upon expectations, and these expectations are based upon past experience, which is, by definition, stored in long-term memory. That is, we see or hear what we expect to see or hear (see Fig 4.6). High expectations are based on events that we have encountered frequently in the past. They are also based on **associations** between the perceived stimulus or event, and other stimuli or events that are present in the same **context** and have been joined in past experience.

The concepts of frequency and context in supporting top-down processing can be illustrated by the following example. A status indicator for a piece of very reliable equipment can be either green, indicating normal operations, or red, indicating failure.

- Given our past experience of red and green in human-designed systems, the association of these two colors to their meaning is made fairly automatically.
- A brief glance at the light, in the high glare of the sun, makes it hard to see which color it is (poor bottom-up processing).
- The past high reliability of the system allows us to "guess" that it is green (top-down processing based upon frequency) even if the actual color is hard to make out. Hence, the quick glance perceives the light to be green.
- The sound of smooth running and good system output provides a context to amplify the "perception of greenness" (top-down processing based upon context).
- An abnormal sound gradually becomes evident. The context has now changed, and red becomes somewhat more expected. The same ambiguous stimulus (hard to tell the color) is now perceived to be red (changing context).
- Now a very close look at the light, with a hand held up to shield it from the sun, reveals that it in fact *is* red, (improved bottom-up processing), and it turns out that it was red all along. Perception had previously been deceived by expectancy.

We now consider two other examples of the interplay between, and complementarily of, bottom-up and top-down processing. As one example, in reading, bottom-up processing is degraded by speed (brief glances) as well as by legibility, factors discussed in Chapter 4. When such degradation is imposed, we can read words more easily than random digit strings (phone numbers, basketball scores, or stock prices), because each word provides an expectancy-based context for the letters within, and when text is presented, the sentence words provide context for reading degraded words within. For example, if we read the sentence "Turn the machine off when the red light indicates failure" and find the fourth word to be nearly illegible (poor bottom-up), the context of the surrounding words allows us to guess that the word is probably "off."

As a second example, scrolling messages viewed in a small window can present a tradeoff between bottom-up and top-down processing. If words within sentences are presented so that context is available,

Then it may
be better to
use small text

However, if random digits are to be displayed within the same space, top-down processing cannot assist perception, so we must maximize bottom-up processing by making the digits larger (and presenting fewer of them)

72184
64992.

If a line in a phone book can be thought of as a space-limited "message window" in the above example, then the same analysis can be applied, and it makes better sense to display the phone number in a larger font than the name, because the name is more likely to provide contextual cues for its spelling. Furthermore, there are usually less serious consequences for failing to perceive the name correctly than for failing to perceive the phone number correctly. The latter will always lead to a dialing error. Like the digits in the phone number, the letters in an email address should also be larger, since the lack of standardization of email addresses (and the fact that many people don't know the middle initial of an addressee) removes context that could otherwise help support top-down processing. In short,

Adam Humfac Adamjhumfa@xxx.yyy 444-455-2995.

is a better design than is

Adam Humfac adamjhumfa@xxx.yyy 444-455-2995.

Human Factors Guidelines in Perception

The proceeding examples and others lead us to a few simple guidelines for supporting perception.

1. Maximize bottom-up processing (Chapters 4 and 5). This involves not only increasing visible legibility (or audibility of sounds), but also pay-

ing careful attention to confusion caused by similarity of message sets that could be perceived in the same context.

2. Maximize automaticity and unitization by using familiar perceptual representations (those encountered frequently in long-term memory). Examples include the use of familiar fonts and lowercase text (Chapter 4), meaningful icons (Chapter 8), and words rather than abbreviations.

3. Maximize top-down processing when bottom-up processing may be poor (as revealed by analysis of the environment and the conditions under which perception may take place), and when unitization may be missing (unfamiliar symbology or language). This can be done by providing the best opportunities for guessing. For example,

   - Avoid confusions: Maximize discriminating features.
   - Use a smaller vocabulary. This has a double benefit of improving guess rate and allowing the creation of a vocabulary with more discriminating features.
   - Create context. For example, the meaning of "your fuel is low" is better perceived than that of the shorter phrase "fuel low," particularly under noisy conditions (Simpson, 1976).
   - Exploit redundancy. This is quite similar to creating context, but redundancy often involves direct repetition of content in a different format. For example, simultaneous display of a visual and auditory message is more likely to guarantee correct perception in a perceptually degraded environment. The phonetic alphabet exploits redundancy by having each syllable convey a message concerning the identity of a letter (al-pha = a).
   - When doing usability testing of symbols or icons, make sure that the context in which these will eventually be used is instated for the testing conditions (Wolff & Wogalter, 1998). This provides a more valid test of the effective perception of the icons.
   - Be wary of the "conspiracy" to invite perceptual errors when the unexpected may be encountered and when bottom-up processing is low (as revealed by task and environmental analysis). Examples of such conditions are; flying at night and encountering unusual aircraft attitudes, which can lead to illusions; or driving at night and encountering unexpected roadway construction. In all cases, as top-down processing attempts to compensate for the bottom-up degradation, it encourages the perception of the expected, which will not be appropriate. Under such conditions, perception of the unusual must be supported by providing particularly salient cues. A special case here is the poor perception of **negation** in sentences. For example, "do not turn off the equipment" could perceived as "turn off the equipment" if the message is badly degraded, because our perceptual system appears to treat the positive meaning of the sentence as the "default" state of a message (Clark & Chase, 1972). We return to this issue in our discussion of comprehension and working memory. If negation is to be used, it should be highlighted.

One downside of the redundancy and context, as these are employed to support top-down processing, is that the length of perceptual messages is increased, thereby reducing the **efficiency** of information transfer. For example, "alpha" and "your fuel is low" both take longer to articulate than "A" and "fuel low" (although they do not necessarily take longer to understand). The printed message "failure" occupies more space than the letter "F" or a small red light. Thus, redundancy and context can help to gain perceptual accuracy, but at the expense of efficiency. This is a tradeoff that designers must explicitly consider by carefully analyzing the consequences of perceptual errors and the extent of environmental factors and stress factors that may degrade bottom-up processing. We consider these stress factors, such as a stress-induced speed-accuracy tradeoff, in more detail in Chapter 13.

Conclusion

Perception is assumed to be relatively automatic (but becomes less so as bottom-up processing is degraded and top-down and unitization processes become less effective). However, as the duration of the perceptual process increases, we speak less of perception and more of *comprehension*, which is less automatic. The border between perception and comprehension is a fuzzy one, although we usually think of *perceiving* and word, but *comprehending* a series of words that make up a sentence. As we shall see, comprehension, like perception, is very much driven by top-down processing, from past experience and long-term memory. However, comprehension tends to also rely heavily upon the capabilities of *working memory* in a way that perception does not.

WORKING MEMORY

Failures of memory occur for everyone—and relatively frequently (Schacter, 2001). Sometimes, the failures are trivial, such as forgetting a new password that you just created. Other times, memory failures are more critical. For example, in 1915 a railroad switchman at a station in Scotland forgot that he had moved a train to an active track. As a result, two oncoming trains used the same track and the ensuing crash killed over 200 people (Rolt, 1978).

The next few sections focus on the part of cognition that involves human memory systems. Substantial evidence shows that there are two very different types of memory storage. The first, *working memory* (sometimes termed *short-term memory*), is relatively transient and limited to holding a small amount of information that may be rehearsed or "worked on" by other cognitive transformations (Cowan, 2001; Baddeley, 1986, 1990). It is the temporary store that keeps information active while we are using it or until we use it. Some examples are looking up a phone number and then holding it in working memory until we have completed dialing, remembering the information in the first part of a sentence as we hear the later words and integrate them to understand the sentence meaning, "holding" subsums while we multiply two two-digit numbers, and constructing an image of the way an intersection will look from a view on a map. Working memory holds two different types of information: verbal and spatial.

The other memory store, *long-term memory,* involves the storage of information after it is no longer active in working memory and the retrieval of the information at a later point in time. When retrieval fails from either working or long-term memory, it is termed forgetting. Conceptually, working memory is the temporary holding of information that is active, either perceived from the environment or retrieved from long-term memory, while long-term memory involves the relatively passive store of information, which is activated only when it needs to be retrieved. The limitations of working memory hold major implications for system design.

A Model of Working Memory

Working memory can be understood in the context of a model proposed by Baddeley (1986, 1990), consisting of three components. In this model, a **central executive component** acts as an attentional control system that coordinates information from the two "storage" systems.

The **visuospatial sketch pad** holds information in an analog spatial form (e.g., visual imagery) while it is being used (Logie, 1995). These images consist of encoded information that has been brought from the senses or retrieved from long-term memory. Thus, the air traffic controller uses the visual-spatial sketch-pad to retain information regarding where planes are located in the airspace. This representation is essential for the controller if the display is momentarily lost from view. This spatial working-memory component is also used when a driver tries to construct a mental map of necessary turns from a set of spoken navigational instructions. Part of the problem that Laura had in using her north-up map to drive south into the city was related to the mental rotation in spatial working memory that was necessary to bring the map into alignment with the world out her windshield.

The **phonological loop** represents verbal information in an acoustical form (Baddeley, 1990). It is kept active, or "rehearsed," by articulating words or sounds, either vocally or subvocally. Thus, when we are trying to remember a phone number, we subvocally sound out the numbers until we no longer need them.

Whether material is verbal (in the phonetic loop) or spatial (in the visuospatial sketchpad), our ability to maintain information in working memory is limited in four interrelated respects: how *much* information can be kept active (its capacity), how *long* it can be kept active, how *similar* material is to other elements of working memory and ongoing information processing, and how much *attention* is required to keep the material active. We describe each of these influences in turn.

Limits of Working Memory

Capacity. Researchers have defined the upper limit or the *capacity* of working memory to be around 7 ± 2 *chunks* of information (Miller, 1956), although even this limit may be somewhat optimistic (Cowan, 2001). A chunk is the unit of working memory space, defined jointly by the physical and cognitive properties that *bind* items within the chunk together. Thus, the sequence of four unrelated letters, X F D U, consists of four chunks, as does the sequence of four digits, 8 4 7 9. However, the four letters DOOR or the four digits 2004 consist of only one chunk, because these can be coded into a single meaningful unit. As a result, each occupies only one "slot" in working memory, and so our working memory could hold 7 ($\pm$ 2) words or familiar dates as well as 7 $\pm$ 2 unrelated letters or digits.

What then binds the units of an item together to make a single chunk? As the examples suggest, it is familiarity with the links or associations between the units, a familiarity based upon past experience and therefore related to long-term memory. The operation is analogous to the role of unitization in perception, discussed earlier. As a child learns to read, the separate letters in a word gradually become unified to form a single chunk. Correspondingly, as the skilled expert gains familiarity with a domain, an acronym or abbreviation that was once several chunks (individual letters) now becomes a single chunk.

Chunking benefits the operations in working memory in several ways. First, and most directly, it reduces the number of items in working memory and therefore increases the capacity of working memory storage. Second, chunking makes use of meaningful associations in long-term memory, and this aids in retention of the information. Third, because of the reduced number of items in working memory, material can be more easily rehearsed and is more likely to be transferred to long-term memory (which then reduces load on working memory).

Chunks in working memory can be thought of as "memory units," but they also have physical counterparts in that **perceptual chunks** may be formed by providing spatial separation between them. For example, the social security number 123 45 6789 contains three physical chunks. Such physical chunking is helpful to memory, but physical chunking works best when it is combined with cognitive chunking. In order to demonstrate this, ask yourself which of the following would be the easiest to perceive and remember: FBI CIA USA, or FB ICIAU.

Time. The capacity limits of working memory are closely related to the second limitation of working memory, the limit of *how long* information may remain. The strength of information in working memory decays over time unless it is periodically *reactivated,* or "pulsed" (Cowan, 2001), a process called *maintenance rehearsal.* (Craik & Lockhart, 1972). Maintenance rehearsal for acoustic items in verbal working memory is essentially a serial process of subvocally articulating each item. Thus, for a string of items like a phone number or a personal identity number (PIN), the interval for reactivating any particular item depends on the length of time to proceed through the whole string. For a seven-digit phone number, we can serially reactivate all items in a relatively short time, short enough to keep all items active (i.e., so that the first digit in the phone number will still be active by the time we have cycled through the last item). The more chunks contained in working memory (like a seven-digit phone number plus a three-digit area code), the longer it will take to cycle through the items in maintenance rehearsal, and the more likely it will be that items have decayed beyond the point where they can be reactivated.

Two specific features should be noted in the proceeding example, relevant to both time and capacity. First, seven digits is right about at the working memory limit, but 10 digits clearly exceeds it. Hence, requiring area codes to be retained in working memory, particularly unfamiliar ones, is a bad human factors design (and a costly one when wrong numbers are dialed in long-distance calls). Second, familiar area codes create one chunk, not three, and a familiar prefix also reduces three chunks to one. Thus, a familiar combination, such as one's own phone number, will occupy six, not 10, slots of working memory capacity.

To help predict working memory decay for differing numbers of chunks, Card, Moran, and Newell (1986) combined data from several studies to determine the "half-life" of items in working memory (the delay after which recall is reduced by half). The half-life was estimated to be approximately 7 seconds for a memory store of three chunks and 70 seconds for one chunk.

Confusability and Similarity. Just as perceptual confusability was seen as a source of error in Chapter 4, so also in working memory high confusability—similarity—between the features of different items means that as their representation decays before reactivation, it is more likely that the discriminating details will be gone. For example, the ordered list of letters E G B D V C is less likely to be correctly retrieved from working memory than is the list E N W R U J because of the greater confusability of the acoustic features of the first list. (This fact, by the way, demonstrates the dominant auditory aspect of the phonetic loop, since such a difference in working memory confusion is observed no matter whether the lists are heard or seen). Thus, decay and time are more disruptive on material that is more similar, particularly when such material needs to be recalled in a particular order (Cowan, 2001). The repetition of items also leads to confusability. A particularly lethal source of errors concerns the confusability of *which* items are repeated. For example, as Laura discovered in the driving example, the digit string 8553 is particularly likely to be erroneously recalled as 8533. Smith (1981) provides good data on the most likely sources of confusion in digit and letter sequences.

Attention and Similarity. Working memory, whether verbal or spatial, is **resource-limited.** In the context of Figure 6.1 working memory depends very much upon the limited supply of attentional resources. If such resources are fully diverted to a concurrent task, rehearsal will stop, and decay will be more rapid. In addition, if the activity toward which resources are diverted uses similar material, like diverting attention to listening to basketball scores while trying to retain a phone number, the added confusion may be particularly lethal to the contents of working memory. The diversion of attention need not be conscious and intentional in order to disrupt working memory. For example, Banbury and colleagues (2001) describe the particular way that sounds nearly automatically intrude on the working memory for serial order. We return to this issue of auditory disruption at the end of the chapter, just as we highlighted its attention-capturing properties in our discussion of selective attention. In terms of

Baddeley's model of working memory, the visual spatial scratchpad is more disrupted by other spatial tasks, like pointing or tracking, and the phonetic loop is more disrupted by other verbal or language-based tasks, like listening or speaking (Wickens et al., 1983; Wickens, 2002).

Human Factors Implications of Working Memory Limits

1. *Minimize working memory load.* An overall rule of thumb is that both the time and the number of alphanumeric items that human operators have to retain in working memory during task performance should be kept to a minimum (Loftus et al., 1979). In general, designers should try to avoid human use of long codes of arbitrary digit or numerical strings (Peacock & Peacock-Goebel, 2002). Hence, any technique that can offload more information in working memory sooner is of value. Windows in computer systems allow comparisons between side-by-side information sources without requiring the larger demands on working memory imposed by sequencing between screens. Electronic "notepads" can accomplish the same general purpose (Wright et al., 2000).

2. *Provide visual echoes.* Wherever synthetic voice is used to convey verbal messages, these messages can, and ideally should, be coupled with a redundant visual (print) readout of the information so that the human's use of the material is not vulnerable to working memory failures. For example, since automated telephone assistance can now "speak" phone numbers with a synthetic voice, a small visual panel attached to the phone could display the same number in the form of a "visual echo." The visual material can be easily rescanned. In contrast, auditory material whose memory may be uncertain cannot be reviewed without an explicit request to "repeat."

3. *Provide placeholders for sequential tasks.* Tasks that require multiple steps, whose actions may be similar in appearance or feedback, benefit from some visual reminder of what steps have been completed, so that the momentarily distracted operator will not return to the task, forgetting what was done, and needing to start from scratch (Gray, 2000).

4. *Exploit chunking.* We have seen how chunking can increase the amount of material held in working memory and increase its transfer to long-term memory. Thus, any way in which we can take advantage of chunking is beneficial. There are several ways in which this can be done:

   - *Physical chunk size.* For presenting arbitrary strings of letters, numbers, or both, the optimal chunk size is three to four numbers or letters per chunk (Bailey, 1989; Peacock & Peacock-Goebel, 2002; Wickelgren, 1964).
   - *Meaningful sequences.* The best procedure for creating cognitive chunks out of random strings is to find or create meaningful sequences within the total string of characters. A meaningful sequence should already have an integral representation in long-term memory. This means that the sequence is retained as a single item rather than a set of the individual characters. Meaningful sequences include things such as 555, 4321, or a friend's initials.

- *Superiority of letters over numbers.* In general, letters induce better chunking than numbers because of their greater potential for meaningfulness. Advertisers have capitalized on this principle by moving from numbers such as 1-800-663-5900, which has eight chunks, to letter-based chunking such as 1-800-GET HELP, which has three chunks ("1-800" is a sufficiently familiar string that it is just one chunk). Grouping letters into one word, and thus one chunk, can greatly increase working memory capabilities.
- *Keeping numbers separate from letters.* If displays must contain a mixture of numbers and letters, it is better to keep them separated (Preczewski & Fisher, 1990). For example, a license plate containing one numeric and one alphabetic chunk, such as 458 GST, will be more easily kept in working memory than a combination such as 4G58ST or 4G58 ST.

5. *Minimize confusability.* Confusability in working memory can be reduced by building physical distinctions into material to be retained. We have already noted that making words and letters sound more different reduces the likelihood that they will be confused during rehearsal. This can sometimes be accommodated by deleting common elements between items that might otherwise be confused. For example, confusion between 3 and 2 is less likely than between A5433 and A5423. Spatial separation also reduces confusability (Hess, Detweiler, and Ellis, 1999). A display that has four different windows for each of four different quantities will be easier to keep track of than a single window display in which the four quantities are cycled. Spatial location represents a salient, discriminating cue to reduce item confusability in such cases.

6. *Avoid unnecessary zeros in codes to be remembered.* The zeros in codes like 002385, which may be created because of an anticipated hundredfold increase in code number, will occupy excessive slots of working memory.

7. *Consider working memory limits in instructions.* The sentences presented in instructions must be accurately comprehended. There may be no tolerance for error in such instructions when they are designed to support emergency procedures. To understand how we comprehend sentences, it is useful to assume that most words in a sentence may need to be retained in working memory until the sentence meaning is interpreted (Wickens & Carswell, 1997; Kintsch & Van Dijk, 1978; Carlson et al., 1989). Thus, long sentences obviously create vulnerabilities. So too do those with unfamiliar words or codes. Particularly vulnerable are those instructions in which information presented early must be retained (rather than "dumped") until the meaning of the whole string is understood. Such an example might be procedural instructions that read:

Before doing X and Y, do A.

Here, X and Y must be remembered until A is encountered. Better would be the order

Do A. Then do X and Y.

Wickens and Hollands (2000) refer to this improved design as one that maintains *congruence* between the order of text and the order of action. Congruence is a good design principle that reduces working memory load.

Finally, reiterating a point made in the context of perception, designers of comprehension material should remember that **negation** imposes an added chunk in working memory. Even if the negation may be perceived in reading or hearing an instruction, it may be forgotten from working memory as that instruction is retained before being carried out. In such circumstances, the default memory of the positive is likely to be retained, and the user may do the opposite of what was instructed. This is another reason to advocate using positive assertions in instructions where possible (Wickens & Hollands, 2000). More details on text and instructional design are given in Chapter 18.

LONG-TERM MEMORY

We constantly maintain information in working memory for its immediate use, but we also need a mechanism for storing information and retrieving it at later times. This mechanism is termed *long-term memory. Learning* is the processing of storing information in long-term memory, and when specific procedures are designed to facilitate learning, we refer to this as instruction or *training,* an issue treated in depth in Chapter 18. Our emphasis in the current chapter is on retrieval and forgetting and the factors that influence them.

Long-term memory can be distinguished by whether it involves memory for general knowledge, called *semantic memory* (memory for facts or procedures), or memory for specific events, called *event memory.*

The ability to retrieve key information from long-term memory is important for many tasks in daily life. We saw at the beginning of this chapter that Laura's failure to recall instructions was a major source of her subsequent problems. In many jobs, forgetting to perform even one part of a job sequence can have catastrophic consequences. In this section, we review the basic mechanisms that underlie storage and retrieval of information from long-term memory and how to design around the limitations of the long-term memory system.

Basic Mechanisms

Material in long-term memory has two important features that determine the ease of later retrieval: its strength and its associations.

Strength.   The strength of an item in long-term memory is determined by the frequency and recency of its use. Regarding frequency, if a password is used every day (i.e., frequently) to log onto a computer, it will probably be well represented in long-term memory and rarely forgotten. Regarding recency, if a pilot spends a day practicing a particular emergency procedure, that procedure will be better recalled (and executed) if the emergency is encountered in flight the very next day than if it is encountered a month later. In this regard, the fact that emergency procedures are generally *not* used frequently in everyday practice suggests that their use should be supported by external visual checklists rather than reliance upon memory.

Associations.   Each item retrieved in long-term memory may be linked or associated with other items. For example, the sound of a foreign word is associated with its meaning or with its sound in the native language of the speaker. As a different example, a particular symptom observed in an abnormal system failure will, in the mind of the skilled troubleshooter, be associated with other symptoms caused by the same failure as well as with memory of the appropriate procedures to follow given the failure. Associations between items have a strength of their own, just as individual items do. As time passes, if associations are not repeated, they become weaker. For example, at some later point a worker might recognize a piece of equipment but be unable to remember its name.

Working Memory and Long-term Memory.   Information in long-term memory becomes more available as a function of the richness or *number* of associations that can be made with other items. Like strings tied to an underwater object, the more strings there are, the greater likelihood that any one (or several) can be found and pulled to retrieve the object. Thus, thinking about the material you learn in class in many different contexts, with different illustrative examples, improves your ability to later remember that material. Doing the mental work to form meaningful associations between items describes the active role of working memory in learning (Carlson et al., 1989; see Chapter 18). As we noted in the discussion of working memory, storing such relations in long-term memory results in the formation of chunks, which are valuable in reducing the load on working memory. Sometimes, however, when rehearsing items through simple repetition (i.e., the pure phonetic loop) rather than actively seeking meaning through associations, our memories may be based solely on frequency and recency, which is essentially *rote memory.* Rote memory is more rapidly forgotten. This is a second reason that advertisers have moved from solely digit-based phone numbers to items such as 1-800-GET-RICH. Such phone numbers have both fewer items (chunks) and more associative meaning.

Forgetting.   The decay of item strength and association strength occurs in the form of an exponential curve, where people experience a very rapid decline in memory within the first few days. This is why evaluating the effects of training immediately after an instructional unit is finished does not accurately indicate the degree of one's eventual memory. Even when material is rehearsed to avoid forgetting, if there are many associations that must be acquired within a short period of time, they can interfere with each other or become confused, particularly if the associations pertain to similar material. New trainees may well recall the equipment they have seen and the names they have learned, but they confuse which piece of equipment is called which name as the newer associations interfere with the older ones.

   Thus, memory retrieval often fails because of (1) weak strength due to low frequency or recency, (2) weak or few associations with other information, and (3) interfering associations. To increase the likelihood that information will be remembered at a later time, it should be processed in working memory frequently and in conjunction with other information in a meaningful way.

Different forms of long-term memory retrieval degrade at different rates. In particular, *recall*, in which one must retrieve the required item (fact, name, or appropriate action), is lost faster than *recognition*, in which a perceptual cue is provided in the environment, which triggers an association with the required item to be retrieved. For example, a multiple-choice test visually presents the correct item, which must be recognized and discriminated from a set of "foils." In contrast, short-answer questions require recall. In human–computer interaction, discussed in Chapter 15, command languages require recall of the appropriate commands to make something happen. In contrast, menus allow visual recognition of the appropriate command to be clicked.

Organization of Information in Long-Term Memory

It is apparent from the description of working memory that we do not put isolated pieces of information in long-term memory the way we would put papers in a filing cabinet. Instead, we store items in connection with related information. The information in long-term memory is stored in *associative networks* where each piece of information (or image or sound) is associated with other related information. Much of our knowledge that we use for daily activities is *semantic* knowledge, that is, the basic meaning of things. Cognitive psychologists have performed research showing that our knowledge seems to be organized into *semantic networks* where sections of the network contain related pieces of information. Thus, you probably have a section of your semantic network that relates all of your knowledge about college professors, both general information and specific instances, based on previous experience. These semantic networks are then linked to other associated information, such as images, sounds, and so on.

A semantic network has many features in common with the network structure that may underlie a database or file structure, such as that used in an index, maintenance manual, or computer menu structure. It is important that the designer create the structure of the database to be compatible or **congruent** with the organization of the user's semantic network (Roske-Hofstrand & Paap, 1986; Seidler & Wickens, 1992). In this way, items that are close together, sharing the same node in the semantic network, will be close together in the database representation of the information. For example, if the user of a human factors database represents perception and displays as closely associated, the database should also contain links between these two concepts. We see in Chapter 8 how this process can be aided by good displays.

In addition to networks, there are three other ways that psychologists have described the organization of information: **schemas, mental models,** and **cognitive maps.**

Schemas and Scripts. The information we have in long-term memory is sometimes organized around central concepts or topics. The entire knowledge structure about a particular topic is often termed a *schema*. People have schemas about all aspects of their world, including equipment and systems that they use. Examples of common schemas are semantic networks associated with college courses, cups, or vacations. Schemas that describe a typical *sequence* of activities, like getting online in a computer system, shutting down a piece of industrial equipment, or dealing with a crisis at work, are called *scripts* (Schank & Abelson, 1977).

Mental Models.   People also have schemas about equipment or systems. The schemas of dynamic systems are often called *mental models* (Gentner & Stevens, 1983; Norman, 1988; Rouse & Morris, 1986; Wilson & Rutherford, 1989). Mental models typically include our understanding of system components, how the system works, and how to use it. In particular, mental models generate a set of **expectancies** about how the equipment or system will behave. Mental models may vary on their degree of **completeness** and **correctness.** For example, a correct mental model of aerodynamics posits that an aircraft stays aloft because of the vacuum created over the wings. An incorrect model assumes that it stays aloft because of the speed through which it travels through airspace. Mental models may also differ in terms of whether they are personal (possessed by a single individual) or are similar across large groups of people. In the latter case the mental model defines a *population stereotype* (Smith, 1981). Designs that are consistent with the population stereotype are said to be *compatible* with the stereotype (such as turning a knob clockwise should move a radio dial to the right). Later chapters on displays (Chapter 8), controls (Chapter 9), and computer design (Chapter 15) illustrate the importance of knowing the user's mental model.

Cognitive Maps.   Mental representations of spatial information, like the layout of a city, a room, or a workplace, are referred to as cognitive maps. They represent the long-term memory analogy to the visual-spatial scratchpad in working memory. Such maps may not necessarily be accurate renderings of the space they represent (Wickens & Hollands, 2000). For example, cognitive maps of a geographical area often simplify by "mentally straightening" corners that are not at right angles (Chase & Chi, 1979). People also have a preferred or "canonical" orientation by which they typically represent an environment (Sholl, 1987). This may often represent the direction in which you most frequently view the environment. For example, your cognitive map of a classroom may have the orientation of the direction you face when you sit in it. Reorienting one's perspective of a cognitive map through "mental rotation" requires mental effort (Tversky & Franklin, 1981). As we discuss in our treatment of map displays in Chapter 8, this has some implications for how maps are configured.

Long-Term Memory Implications for Design

Designers frequently fail to realize or predict the difficulty people will experience in using their system. One reason is that they are extremely familiar with the system and have a very detailed and complete mental model (Norman, 1988). They know how the system works, when it will do various things, and how to control the system to do what the user wishes. They fail to realize that the average user does not have this mental model and may never interact with the system enough to develop one. When people have to do even simple tasks on an infrequent basis, they forget things. Manufacturers write owners' manuals as if they will be read thoroughly and all of the information will be remembered for the life of the equipment. Neither is necessarily the case. Even if we have very clear and explicit instructions for operating our programmable VCR (which is unlikely), what average owner wants to get the instructions out every time he or she must perform a task?

The following are some ways that we can design the environment and systems within it so that people do not have problems, errors, accidents, and inconveniences due to poor retrieval from long-term memory.

1. *Encourage regular use of information to increase frequency and recency.*

2. *Encourage active verbalization or reproduction of information that is to be recalled.* For example, taking notes in class or requiring active recitation or readback of heard instructions increases the likelihood that the information will be remembered.

3. *Standardize.* One way that we can decrease the load on long-term memory is to standardize environments and equipment, including controls, displays, symbols, and operating procedures. An example from the automotive industry where a control is being standardized is the shift pattern, and where a control has still not been standardized is the location and operation of electronic windows and lighting. Standardization results in development of strong yet simple schemas and mental models that are applicable to a wide variety of circumstances. Of course, the conflict between standardizing across industries and still preserving uniqueness of product style remains a difficult design challenge.

4. *Use memory aids.* When a task will be performed infrequently or when correct task performance is critical, designers should provide computer-based or hardcopy memory aids or job aids as discussed in Chapter 18. These consist of information critical for task performance and can be as simple as a list of procedures.

Norman (1988) characterizes memory aids as putting "knowledge in the world" (i.e., perception) so that the operator does not have to rely on "knowledge in the head" (i.e., long-term memory). In the context of command languages and menus, such aids often replace recall requirements with recognition opportunities. This important human factors topic is reconsidered in Chapters 15 and 18.

5. *Carefully design information to be remembered.* Information that must be remembered and later retrieved unaided should have characteristics such as the following:

- Meaningful to the individual and semantically associated with other information.
- Concrete rather than abstract words when possible.
- Distinctive concepts and information (to reduce interference).
- Well-organized sets of information (grouped or otherwise associated).
- Able to be guessed based on other information (top-down processing).
- Little technical jargon.

6. *Design to support development of correct mental models.* One way to develop correct mental models is to apply the concept of *visibility,* as suggested by Norman (1988). This guideline suggests that a device has visibility if the user can immediately and easily determine the state of the device and the alternatives for action. For example, switches that have different positions when activated have visibility, whereas push/toggle switches do not. The concept of visibility also relates to the ability of a system to show variables intervening between an operator's action and the ultimate system response. An example is an oven display showing that an input has been read, the heat system is warming up, and the temperature has not reached the target temperature. Mental model development can also be encouraged by the appropriate wording of instructional manuals that describe *why* a particular action is required as well as what the action is.

Episodic Memory for Events

In contrast to both procedural and declarative knowledge, which is often embodied in schemas, scripts, and skills and acquired from multiple experiences, the personal knowledge or memory of a specific event or *episode* is, almost by definition, acquired from a single experience. This may be the first encounter with an employer or coworker, a particular incident or accident at home or the workplace, or the eyewitness view of a crime or accident. Such memories are very much based on visual imagery, but the memories themselves are not always faithful "video replays" of the events, having a number of biases.

Episodic memory is of tremendous importance to the psychology of eyewitness testimony. While this is certainly of great importance to legal criminal proceedings (Wright & Davies, 1999; Devenport et al., 1999), it also has considerable relevance to the field of accident investigation, which we discuss in Chapter 14. That is, what does the witness to an accident recall about its circumstances when later interviewed by the investigator?

Through a simple cognitive task analysis, we can represent the processes involved in the formation, storage, and retrieval of episodic memories as shown in Figure 6.2. Here an "event" occurs, which defines some ground truth of what actually happened. The witness observes the event and encodes information about it, which reflects the allocation of selective attention and may reflect some of the top-down biases of expectancy on perception that we described earlier. As time passes, the memory of the episode is maintained in long-term memory, where it will show some degradation (forgetting), and the memory may be distorted by influences related to both schema memory and specific intervening events (Bartlett, 1932). Finally, the memory may be retrieved in a variety of circumstances: For example, a witness picks out a suspect from a police lineup, the witness is interviewed by police as the prosecution develops its case, or the witness responds to queries during actual courtroom testimony. These retrieval tests may have characteristics of both recall and recognition.

Extensive research on eyewitness testimony has revealed that the episodic memory process is far from perfect (e.g., Wright & Davies, 1999; Schacter, 2001; Wells & Seelau, 1995). In one study of police lineup recognition, for example, Wright and McDaid (1996) estimated that an innocent person was chosen (as a guilty perpetrator) approximately 20 percent of the time. The sources of such

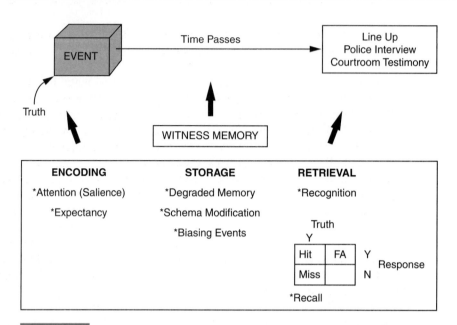

FIGURE 6.2

Episodic memory. The processes involved in episodic memory characteristics and influences on these processes are shown in the box at the bottom. It will be noted that retrieval in a courtroom (testimony) often starts another memory cycle: that of the jury who encodes the witness testimony for later retrieval during the jury deliberations and judgment.

biases can occur at all three stages. For example, at encoding, a well-established bias is the strong focus of witness attention on a weapon when one is used at the scene of the crime. In light of what we know about the limits of attention, it should come as no surprise that this focus degrades the encoding of other information in the scene, particularly the physical appearance of the suspect's face relative to crimes where no weapon is employed (Loftus et al., 1987). In a different application of attention research, Lassiter and his colleagues (e.g., Lassiter, 2002) show how the focus of a video camera during interrogation, on the suspect alone rather than on the suspect *and* the interviewer, can bias the judgment of a jury who views such a video. Focus on the suspect alone leads jurors to substantially increase their later judgment that the suspect is guilty, independent of the contents of the interrogation.

Episodic memory also has an auditory component, but this too may be flawed. John Dean, the former council to President Richard M. Nixon, had a reputation for having a particularly precise memory. Some even called him a "human tape recorder" (Neisser, 1982). His confident recall of dozens of conversations helped bring down the Nixon administration. After his testimony, tapes of the actual conversations were released and Neisser compared the recorded and recalled conversations. He found Dean's memory to be seriously flawed regarding the details of specific conversations. Dean was not a "human tape recorder"; however, he was quite accurate in capturing the general theme or gist of the conversations. Instead of a verbatim recording of conversations, memory relies on extracting the gist and reconstructing the details. The reconstruction of the details may be distorted by the cultural background and self-interests of the individual (Bartlett, 1932).

As Figure 6.2 suggests, two qualitatively different forms of bias may influence the memory during storage (Wright & Davies, 1999). First, a degraded visual recollection may be partially replaced by a long-term memory schema of what the crime might "typically" look like. For example, it may be replaced by the witness's memory of the appearance of the "typical criminal" or by the assumption that the typical automobile accident will occur at a high rate of speed, thereby leading to an overestimation of vehicle speed in a crash. Second, certain events during the storage interview can also bias memory. For example, a chance encounter with a suspect in handcuffs in the hallway prior to a police lineup might increase the likelihood that the suspect will be selected in the lineup. Sometimes, the way questions are phrased in a witness interview can also "suggest" that a particular suspect is guilty or that events occurred in a different way than they actually did, and as a consequence, distort the accuracy of episodic recall, which may be used in trial.

Finally, biases at retrieval can sometimes be represented in the form of a signal detection task discussed in Chapter 4 when recognition tests are used, as they are in a police lineup (Wells, 1993). As shown in the lower right corner of Figure 6.2, a "signal" can be represented as the witness's accurate episodic memory of the suspect's appearance. The witness's response is represented as either selecting the suspect from the lineup (yes) or failing to do so (no). This defines the four classes of events in which the "hit" is the most important for accurately developing the police's case. In contrast, a false alarm, in which an innocent person is positively identified, is clearly an undesirable event and one that has dangerous implications for society.

Within this context, it is in the interest of all parties to maximize the sensitivity (keeping misses and false alarms to a minimum). However, it is also important to avoid a "guilty bias" where witnesses are likely to see the suspect as being guilty. Even one who had not encoded the crime at all would still have a 20 percent chance of picking the suspect from a police lineup if the witness felt certain that the actual perpetrator was in the lineup. Wells and Seelau (1995) describe ways of conducting eyewitness line-up procedures to avoid the unfortunate consequences of a guilty bias (see also Wright & Davies, 1999). For example, they suggest that witnesses be clearly informed that the perpetrator might *not* be in the lineup; furthermore, witnesses can initially be shown a blank lineup in which the suspect is not included. Witnesses who "recognize" the suspect in such a lineup can be assumed to have a strong guilty bias and their testimony can therefore be discounted. If they do not respond yes to the blank lineup, then they can be shown the actual lineup with the suspect included.

Since those who judge a witness's testimony often have no independent means of assessing the accuracy of that testimony (e.g., they do not know the "sensitivity" of a recognition memory test), we might think that asking witnesses to express the confidence in their memory should provide a means of assessing this accuracy. Unfortunately, however, extensive research has shown that the self-rated confidence of a witness's judgment is only weakly correlated with the accuracy of that judgment (Wells & Seelau, 1995; Wright & Davies, 1999; Wells et al., 1979). People aren't very well calibrated in estimating the strength of their own episodic memory.

In one important application of memory research to episodic retrieval, Fisher and Geiselman (1992; Fisher, 1999) developed what is called the **cognitive interview** (CI) technique for assisting police in interviewing witnesses in order to maximize the retrieval of information. Their approach is to avoid recognition tasks because, they argue persuasively, classic recognition tests, approximated by asking witnesses a series of yes-or-no questions ("Did the suspect have red hair?") can be quite biasing and leave vast quantities of encoded information untapped. Instead, they apply a series of principles from cognitive psychology to develop effective **recall** procedures. For example, the CI technique

- Encourages the witness to reinstate the context of the original episode, thereby possibly exploiting a rich **network of associations** that might be connected with the episodic memory.
- Avoids **time-sharing requirements** where the witness must divide cognitive resources between searching episodic memory for details of the crime and listening to the interrogator ask additional questions. We learn about the consequences of such time-sharing later in the chapter.
- Avoids time stress, allowing the witness plenty of time to retrieve information about the crime and ideally allowing the witness multiple opportunities to recall. These multiple opportunities will take advantage of the rich network of associations.

The CI technique has been shown to allow witnesses to generate between 35 and 100 percent more information than standard police interview procedures and to do so without any substantial loss of accuracy; it has been adopted by a number of police forces (Fisher, 1999; Wright & Davies, 1999).

A final important issue regarding cognitive psychological principles in legal proceedings pertains to the admissibility of testimony from expert psychologists regarding the sorts of eyewitness biases described above (Devenport et al., 1999; Levett & Kovera, 2002). Judges may disagree on whether such scientific recommendations are themselves "true" and hence admissible evidence. Research shows that to the extent that such expert testimony from psychologists *is* admissible, it has two effects on jury belief. First it leads to some general down-weighting of the impact of that testimony, as jurors themselves become more skeptical of the "video tape" analogy to episodic memory. Second, it allows the jurors to become more sensitive in discriminating accurate from inaccurate testimony (Devenport et al., 1999).

In conclusion, it is evident that the cognitive psychology of memory and attention has tremendous importance for the quality of criminal and other legal proceedings. We will also see the relevance of the study of decision making in the next chapter. One final implication for every reader is that when you witness a serious episode about which you might be later queried, it is good advice to write down everything about it as soon as the episode has occurred and at that time think clearly about and indicate your degree of certainty or uncertainty about the events within the incident. Your written record will now be "knowledge in the world," not susceptible to forgetting.

Prospective Memory for Future Events

Whereas failures of episodic memory are inaccurate recollection of things that happened in the past, failures of *prospective memory* are forgetting to do something in the future (Harris & Wilkins, 1982). Laura, in the story at the beginning of the chapter, forgot to activate the voice mode while the traffic was still light. In 1991, an air traffic controller positioned a commuter aircraft at the end of a runway and later forgot to move the aircraft to a different location. The unfortunate aircraft was still positioned there as a large transport aircraft was cleared to land on the same runway. Several lives were lost in the resulting collision (NTSB, 1992).

Failures of prospective memory are sometimes called absentmindedness. Several system and task design procedures are incorporated in systems to support prospective memory. Strategies can be adopted to implement *reminders* (Herrmann et al., 1999). These may be things like tying a string around your finger, setting a clock or programming a personal data assistant (PDA) to sound an alarm at a future time, taping a note to the steering wheel of your car, or putting a package you need to mail in front of the door so that you will be sure to notice it (if not trip on it!) on your way out. In systems with multiple operators, sharing the knowledge of what one or the other is to do decreases the likelihood that both will forget that it is to be done. Also, loss of prospective memory is reduced by verbally stating or physically taking some action (e.g., writing down or typing in) regarding the required future activity the moment it is scheduled. Checklists are aids for prospective memory (Degani & Wiener, 1991). Herrmann and colleagues (1999) describe characteristics of ideal reminding devices like the PDA.

SITUATION AWARENESS

In the dynamic sequence of events leading up to Laura's near accident, she was unaware of the converging vehicle until her attention was captured by its horn. Designers, researchers, and users of complex dynamic systems often employ the cognitive concept of *situation awareness,* or SA, to characterize users' awareness of the meaning of dynamic changes in their environment (Durso & Gronlund, 1999; Adams et al., 1995). A pilot loses SA whenever he or she suffers a catastrophic controlled-flight into terrain (Strauch, 1997; Wiener, 1977), and as we shall see in Chapter 16, control room operators at the Three Mile Island nuclear power plant lost SA when they believed the water level in the plant to be too high rather than too low, a misdiagnosis that led to a catastrophic release of radioactive material into the atmosphere (Rubinstein & Mason, 1979).

Endsley (1995) defines SA as "the perception of the elements in the environment within a volume of time and space, the comprehension of their meaning, and the projection of their status in the near future" (p 36). These three stages, perception (and selective attention), understanding, and prediction, must be applied to a specific situation. Thus, a user cannot be simply said to have SA without specifying what that awareness is (or should be) about. A vehicle driver may have good awareness of time and navigational information (where I am and how much time it will take me to drive to where I need to be), but little awareness of the local traffic tailgating behind.

Many elements that are necessary to support SA have been covered elsewhere in this chapter. Selective attention is necessary for the first stage, while the second stage of understanding depends very much upon both working memory and long-term memory. The third stage, projection and prediction, however, is an important construct in cognitive psychology that we have not yet discussed but consider in more detail later when we discuss planning and scheduling.

It is important to note that SA is distinct from **performance.** SA can be maintained even when there is no performance to be observed. For example, a passenger in a vehicle may have very good awareness of the traffic and the navigational situation, even as he or she carries out no actions (other than visual scanning). Great differences in the ability of pilots to deal with unexpected occurrences within their automation system are observed as a function of how well they are aware of changes in an automated state during periods when the pilots are totally passive observers (Sarter & Woods, 2000; Sarter et al., 1997). We can also identify instances in which very good performance is observed with low SA, as when you are so absorbed in doing a task that you lose awareness of the time. A key issue here is that the importance of SA is not so much for understanding and describing the quality of routine performance (e.g., the accuracy in staying in a lane or maintaining speed while driving) as it is for understanding the appropriate and timely response to unexpected events (Wickens, 2000).

Measuring SA

The importance of SA can often be realized after an accident by inferring that the loss of SA was partially responsible. In controlled-flight-into-terrain accidents it is almost always assumed that the pilot lost awareness of the aircraft's altitude over or trajectory toward the terrain (Strauch, 1997). However, "measuring" SA after the fact by assuming its absence (SA = 0) is not the same as measuring how well a particular system or operator preserves SA in the absence of an unexpected event (Endsley & Garland, 2000). A popular technique for SA measurement is the SA global assessment technique (SAGAT; Endsley, 1995) in which the operator is briefly interrupted in the performance of a dynamic task and asked questions about it; for example, identify the location of other road traffic (Gugerty, 1997) or identify the direction of the nearest hazardous terrain (Wickens & Prevett, 1995). While SA can sometimes be measured by a subjective evaluation ("rate your SA on a scale of 1 to 10; Selcon et al., 1991), a concern about the validity of such self-rating techniques is that people are not always aware of what they are not aware. This issue of *metacognition* is addressed later in this chapter.

Importance of SA to Human Factors

Probably the first and most direct application of the SA concept to human factors is its implications for designing easy-to-interpret displays of dynamic systems that can help people notice what is going on (stage 1), interpret and understand the meaning—a challenge when there are several coupled display elements in a complex system (stage 2), and predict the future implications—a challenge when the system is slow or lagged, like a supertanker, industrial oven, or air traffic system. Human factors practitioners have noted how easy it is to lose SA when automation carries out much of the processing for complex systems and hence how critical it is to have SA-supporting displays in the unexpected event that automation does not perform as intended (Parasuraman et al., 2000; Sarter et al., 1997; Sarter & Woods, 2000; see Chapter 16). In this regard, the design of products to maximize the effectiveness of routine performance may not be the same as the design of those to support SA (Wickens, 2000). The support for SA typically imposes the need for added information display to support appropriate behavior when unexpected things go wrong. This information must be carefully integrated in order to avoid issues of information overload.

Second, SA can be an important tool for accident analysis, understanding when its loss was a contributing factor (Strauch, 1997). To the extent that accidents may be caused by SA loss, an added implication is that systems should be designed and, when appropriate, certified to support SA (Wickens, 2000). This becomes important when federal regulators are responsible for certification, such as the case with new aircraft or nuclear power plants.

Third, the SA concept has important implications for training. Training for routine performance may conflict with training to maintain SA. One particularly relevant aspect concerns the training of attentional skills (stage 1 SA) to scan the environment with enough breadth to assure that important and relevant dynamic events are noticed when they occur (Gopher, 1993).

PROBLEM SOLVING AND TROUBLESHOOTING

The cognitive phenomena of problem solving and troubleshooting are often closely linked because they have so many overlapping elements. Both start with a difference between an initial "state" and a final "goal state" and typically require a number of cognitive operations to reach the latter. The identity of those operations is often not immediately apparent to the human engaged in problem-solving behavior. Troubleshooting is often embedded within problem solving in that it is sometimes necessary to understand the identity of a problem before solving it. Thus, we may need to understand why our car engine does not start (troubleshoot) before trying to implement a solution (problem solving). Although troubleshooting may often be a step within a problem-solving sequence, problem solving may occur without troubleshooting if the problem is solved through "trial and error" or if a solution is accidentally encountered through serendipity.

While both problem solving and troubleshooting involve attaining a state of knowledge, both also typically involve performance of specific actions. Thus, troubleshooting usually requires a series of tests whose outcomes are used to diagnose the problem, whereas problem solving usually involves actions to implement the solution. Both are considered to be iterative processes of perceptual, cognitive, and response-related activities involving the full cycle of processing shown in Figure 6.1.

Challenges

Both problem solving and troubleshooting impose heavy cognitive activity, and human performance is therefore often limited (Wickens & Hollands, 2000; Casner, 1994; Teague & Allen, 1997). In troubleshooting, for example, people usually maintain no more than two or three active hypotheses in working memory as to the possible source of a problem (Rasmussen, 1981; Wickens, 1992). More than this number overloads the limited capacity of working memory, since each hypothesis is complex enough to form more than a single chunk. Furthermore, when testing hypotheses, there is a tendency to focus on only one hypothesis at a time in order to confirm it or reject it. Thus, the engine troubleshooter will probably assume one form of the problem and perform tests specifically defined to confirm that it is the problem.

Naturally, troubleshooting success depends closely upon attending to the appropriate cues and test outcomes. This dependency makes troubleshooting susceptible to attention and perceptual biases. The operator may attend selectively to very salient outcomes (bottom-up processing) or to outcomes that are anticipated (top-down processing). As we consider the first of these potential biases, it is important to realize that the least salient stimulus or event is the "nonevent." People do not easily notice the absence of something (Wickens & Hollands, 2000; Hunt & Rouse, 1981). Yet the absence of a symptom can often be a very valuable and diagnostic tool in troubleshooting to eliminate faulty hypotheses of what might be wrong. For example, the fact that a particular warning light might *not* be on could eliminate from consideration a number of competing hypotheses.

An important bias in troubleshooting, resulting from top-down or expectancy-driven processing, is often referred to as **cognitive tunneling,** or **confirmation bias** (Woods & Cook, 1999; Woods et al., 1994; see Chapter 7). In troubleshooting, this is the tendency to stay fixated on a particular hypothesis (that chosen for testing), look for cues to confirm it (top-down expectancy guiding attention allocation), and interpret ambiguous evidence as supportive (top-down expectancy guiding perception). In problem solving, the corresponding phenomenon is to become fixated on a particular solution and stay with it even when it appears not to be working.

These cognitive biases are more likely to manifest when two features characterize the system under investigation. First, high system complexity (the number of system components and their degree of coupling or links) makes troubleshooting more difficult (Meister, 2002; Wohl, 1983). Complex systems are more likely to produce incorrect or "buggy" mental models (Sanderson & Murtash, 1990), which can hinder the selection of appropriate tests or correct interpretation of test outcomes. Second, **intermittent** failures of a given system component turn out to be particularly difficult to troubleshoot (Teague & Allen, 1997).

PLANNING AND SCHEDULING

The cognitive processes of planning and scheduling are closely related to those discussed in the previous section, because informed problem solving and troubleshooting often involve careful planning of future tests and activities. However, troubleshooting and diagnosis generally suggest that something is "wrong" and needs to be fixed. Planning and scheduling do not have this implication. That is, planning may be invoked in the absence of problem solving, as when a routine schedule of activities is generated. We saw earlier in the chapter that prospective memory could be considered a form of planning.

In many dynamic systems, the future may be broken down into two separate components: the **predicted state** of the system that is being controlled and the ideal or **command state** that should be obtained. Thus, a factory manager may have predicted output that can be obtained over the next few hours (given workers and equipment available) and a target output that is requested by external demands (i.e., the factory's client). When systems cannot change their state or productive output easily, we say they are sluggish, or have "high inertia." In these circumstances of sluggish systems, longer range planning becomes extremely important to guarantee that future production matches future demands. This is because sudden changes in demand cannot be met by rapid changes in system output. Examples of such sluggish systems—in need of planning—are the factory whose equipment takes time to be brought online, the airspace in which aircraft cannot be instantly moved to new locations, or any physical system with high inertia, like a supertanker or a train.

You will recognize the importance to planning of two cognitive constructs discussed earlier in the chapter. First, stage 3, SA is another way of expressing an accurate estimate of future state and future demands. Second, skilled operators often employ a mental model of the dynamic system to be run through a *mental simulation* in order to infer the future state from the current state (Klein & Crandall, 1995). The role of mental simulation is discussed in Chapter 7, and the great importance of mental models in controlling complex and sluggish industrial processes is visited in Chapter 16. Here, however, we note the heavy cognitive demands on working memory to run an accurate mental model. Such a task requires a heavy investment of cognitive resources (the "tank" at the top of Figure 6.1). Where these resources are lacking, diverted to other tasks, then prediction and planning may be poor, or not done at all, leaving the operator unprepared for the future.

In general, people tend to avoid complex, optimizing, planning schedules over long time horizons (Tulga & Sheridan, 1980), a decision driven both by a desire to conserve the resources imposed by high working memory load and by the fact that in an uncertain world accurate planning is impossible, and plans may need to be revised or abandoned altogether as the world evolves in a way that is different from what was predicted. Here, unfortunately, people sometimes fail to do so, creating what is known as a *plan continuation error* (Orasanu et al., 2001; Muthard & Wickens, 2003; Goh & Wiegmann, 2001), a form of behavior that has much in common with cognitive tunneling.

As with problem solving and troubleshooting, a variety of automation tools are proposed to reduce these cognitive demands in planning (Gronland et al., 2002). Most effective are predictive **displays** that offer visual representations of the likely future, reducing the need for working memory (Wickens et al., 2000). We discuss these in the next chapter. Also potentially useful are computer-based planning aids that can either recommend plans (Layton et al., 1994; Muthard &

Wickens, 2003) or allow fast-time simulation of the consequence of such plans to allow the operator to try them out and choose the successful one (Sheridan, 2002). Air traffic controllers can benefit from such a planning aid known as the User Request Evaluation Tool (URET) to try out different routes to avoid aircraft conflicts (Wickens et al., 1998).

METACOGNITION AND EFFORT

Performance of nearly all tasks is supported by some combination of perceptual information and long-term-memory knowledge about the task. Norman (1988) refers to these as "knowledge in the world" and "knowledge in the head" respectively. Psychologists have also identified a qualitatively different source of knowledge that is important in many aspects of performance, *metaknowledge* or *metacognition* (Reder, 1996; Bjork, 1999), which refers to people's knowledge about their *own* knowledge and abilities. Consider, for example, a troubleshooter who is trying to diagnose and fix an engine problem before restarting. Conditions are such that if the diagnosis is incorrect and a restart is tried (and fails), it could lead to serious damage. She asks herself whether she knows enough about the nature of the problem and the projected effectiveness of her "fix" to be confident that the start will proceed without damage. In short, she assesses her knowledge about her own knowledge. In a corresponding situation, a student may assess whether he knows enough to stop studying for the test and turn to another activity.

Another example of metacognition might be the eyewitness who is about to testify and applies her awareness of the general tendency toward overconfidence in recognition memory in such a way as to consciously "downgrade" her estimates of self-confidence on the witness stand. Thus, metacognition sometimes modulates people's choices of what they do, assertions of what they know (knowledge in the head), and choices of whether additional information should be sought (knowledge in the world).

Seeking additional information related to selective attention is also related to another construct of metacognition, the **anticipated effort** required to gain that information (Wright et al., 2000; Fennema & Kleinmuntz, 1995). This construct of anticipated effort is closely linked to the strategies people use with information systems, not just seeking information but also performing a wider range of tasks (Gray, 2000). People often ask themselves, implicitly or explicitly, whether the anticipated effort necessary to access information is worth the potential gains in knowledge from acquiring that information. For example, is it worthwhile traveling across campus to the library to check out a particular book that contains the information I need, or to continue an apparently unproductive search for new information (MacGregor et al., 1987)? In a more general sense, people ask themselves similar tradeoff questions regarding whether the effort required to use a particular system feature balances the gain in productivity from using that feature.

One important metacognitive tradeoff often made is between knowledge in the head and knowledge in the world. Sometimes gaining knowledge in the world (accessing perceptual information) is more accurate but requires more effort than using knowledge in the head (relying upon potentially faulty memory; Gray & Fu, 2001). It is important, however, for designers to realize that people are often overconfident in the accuracy of their own knowledge (Bjork, 1999), as was the case with Laura's knowledge of how to activate the voice mode in her vehicle and with the overconfidence of eyewitnesses. Thus, the decision of users to avoid effort-imposing access of perceptual information may not always be a wise one.

Balancing the costs and benefits of attributes like anticipated effort and accuracy is an issue discussed more formally in the context of decision making in Chapter 7. There too, we discuss the manner in which people are effort-conserving in the kinds of decision strategies they use (Bettman et al., 1990), choosing low effort heuristics over high-effort, optimizing decision techniques. With regard to scheduling and planning, people tend to choose simpler schedules rather than complex but optimal ones (Raby & Wickens, 1994). Designers must understand the effort costs generated by potentially powerful features in interfaces. Such costs may be expressed in terms of the cognitive effort required to learn the feature or the mental and physical effort and time cost required to load or program the feature. Many people are disinclined to invest such effort even if the anticipated gains in productivity are high. The feature will go unused as a result. Correspondingly, requiring people to engage in manual activity to retrieve information is more effort-consuming than simply requiring them to scan to a different part of the visual field (Yeh & Wickens, 2001; Gray & Fu, 2001), a characteristic that penalizes the concepts of hidden databases, multilevel menus, and decluttering tools. Solutions to this problem are offered by pop-up messages and other automation features that can infer a user's information needs and provide them without imposing the effort cost of access (Hammer, 1999).

We now turn to a direct examination of this important concept of effort as it is joined with other information processing features to determine people's success or failure in carrying out two tasks at the same time: divided attention.

ATTENTION AND TIME-SHARING

Earlier in this chapter we spoke of attention as acquiring information about the environment. This was **selective attention,** a process that sometimes requires effort. In this section we discuss attention as supporting the ability to do two (or more) things at one time—to **divide attention** between two tasks or mental activities (Wickens, 2002). The two aspects of attention are related, but not identical (Wickens et al., 2003, in press). For example, selecting two sources of information to process—the roadway view and the electronic map for our driver Laura—may or may not allow the successful division of attention between the tasks supported by those sources. In Laura's case, it did not. Researchers of human time-sharing have identified four major factors that contribute to the success or failure of divided attention (Damos, 1991): resource demand, structure, similarity, and resource allocation or task management.

Mental Effort and Resource Demand

In the prior section, we described the effort required to carry out a task or cognitive activity. People, being effort-conserving, tend to avoid high-effort activities or to do them poorly, such as rehearsing an eight-chunk phone number, engaging in mental rotation, or doing prediction. Furthermore, the high mental effort, difficulty, or resource demand of one activity degrades the ability to carry out a second activity at the same time, as if the resources necessary to support one, shown in the "tank" at the top of Figure 6.1, are limited and are therefore less available to the other. For example, one can converse and drive at the same time if the conversation is simple and the driving task is easy. But when the conversation becomes difficult, perhaps solving a tough problem, resources may be diverted from driving at the cost of safety. Alternatively, if the driving suddenly becomes demanding, conversation may cease. This relationship between single-task difficulty and dual-task divided attention decrements is the fundamental feature of **resource theory** (Kahneman et al., 1973). Scarce mental resources are shared by tasks, and more difficult tasks leave fewer resources for concurrent tasks, whose performance declines as a result (line 1 of Figure 6.3).

The concept of mental effort is closely (and inversely) linked to that of **automaticity** (Schneider, 1985; Logan, 1985). A task that is said to be automated, like signing your name or following a familiar computer log-on procedure, has several properties. It is typically highly practiced, carried out rapidly with little conscious thought, and, most importantly, demands few mental resources for its execution, thereby improving the ability to perform other tasks at the same time. Automaticity is a matter of degree, not an all-or-none "thing." So, the degree of

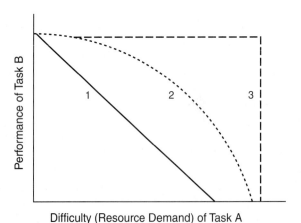

Difficulty (Resource Demand) of Task A

FIGURE 6.3

Relation between performance of one task (B) and the difficulty of a second task (A) carried out concurrently, as resources are shared between them. Lines 1, 2, and 3 represent versions of task B that are progressively more automatized. At high levels of automaticity (line 3), perfect performance of task B can still be attained even when task A is quite difficult.

automaticity dictates the level of performance that can be obtained for a given investment of mental resources (Norman & Bobrow, 1975; Navon & Gopher, 1979; Wickens & Hollands, 2000). Lines 2 and 3 of Figure 6.3 represent dual-task performance of versions of task (B) that are progressively more automatized. As noted earlier, automaticity is one feature that results when perceptual elements are unitized.

Structural Similarity

Automaticity or resource demand is a property of a single task (or mental activity) that directly predicts its success in time-sharing with another. In contrast, **structural similarity** is the similarity between key processing structures of both tasks in a concurrently performed pair. Laura failed to monitor the converging traffic, in part because she could not see the road while looking at her cell phone. As she herself realized before starting the trip, she would have been more successful if she could have heard the navigational instructions via the voice mode, dividing attention between the eye and the ear. Researchers have argued that different structures in human information processing behave as if they were supported by separate or *multiple resources,* so that instead of the single "pool," shown in Figure 6.1, there are multiple pools (Navon & Gopher, 1979; Wickens, 1984, 2002). To the extent that two tasks demand separate resources, time-sharing is improved.

Table 6.1 lists four dichotomous dimensions of multiple resources for which there is generally better time-sharing between than within each end of the dimension. The table provides examples of activities that would "load" each end of a dimension. These four dimensions are partially independent, or "orthogonal," from each other so that, for example, a spatial or a verbal task (code dimension) can involve either perceptual–working memory activity or response activity (stage dimension). However, some of the dichotomies are nested within others. For example, the distinction between the focal and ambient visual channels is one that is only defined within processing that is visual and perceptual-cognitive.

The most important design feature to be derived from the table is that to the extent that two tasks demand common levels on one or more dimension, time-sharing is likely to be worse, and one or the other task will decrease farther from its single task-performance level. For example, a wide variety of research has shown that two tasks involving verbal material on the "code" dimension—like speaking while rehearsing a phone number—interfere more than a verbal and spatial task (Wickens & Liu, 1988). Regarding the modality dimension, research in driving and aviation generally supports the benefits of auditory display of information in the heavy visual demands of vehicle control (Wickens & Seppelt, 2002). Does this mean, for example, that it is always wise to present information auditorally rather than visually to the driver or pilot who has ongoing visual demands of vehicle control? Not necessarily, because sometimes altering the structure of information display may change the resource demand, our first contributor to dual-task interference. As an example, the auditory delivery of

TABLE 6.1 **Four Dimensions of Multiple Resources**

Dimension	Two Levels	Examples
Modalities	Auditory vs. Visual	Synthesized voice display, spatially localized tones
		Print, electronic map
Codes	Spatial vs. Verbal	Tracking, hand pointing, mental rotation, imaging (visuospatial scratchpad)
		Listening to speech, rehearsing (phonetic loop), speaking
Stages	Perceptual– Working Memory vs. Response	Searching, imaging, reading, rehearsing, listening
		Pushing, speaking, pointing, manipulating
Visual Channels	Focal vs. Ambient	Reading, interpreting symbols
		Processing flow fields, visual perception to maintain balance

(From Wickens, 2000).

long messages of five to nine chunks imposes a high resource demand that was not present in the long visual message, since the latter does not need to be rehearsed. Thus, only by considering both resource demand and structural similarity together can the degree of dual-task interference be predicted.

Confusion

We noted that the similarity between items in working memory leads to confusion. We also presented a corresponding argument regarding similarity-based confusion in our discussion of visual sensation in Chapter 4. Here also we find that concurrent performance of two tasks that both have similar material increases task interference (Fracker & Wickens, 1989; Gillie & Broadbent 1989 Wickens & Hollands, 2000). For example, monitoring basketball scores while doing mental arithmetic will probably lead to disruption as digits from one task become confused with digits relevant to the other. Correspondingly, listening to a voice navigational display of turn directions instructing a left turn, while the automobile passenger says, "*right . . .* that's what I thought," could lead to the unfortunate wrong turn. Auditory background information, because of its intrusiveness, may be particularly likely to cause confusion even if it is not part of an ongoing task (Banbury et al., 2001).

Task Management and Interruptions

In the previous section, we described the concept of **total interference** between two ongoing tasks, determined as a joint function of their combined resource demand, structural overlap, and possibly similarity. If these factors produce interference, then one task or the other will suffer a decrement. But will they both suffer? Or will one or the other be "protected"? In analyzing dual-task performance we typically speak of the *primary task* as that which should receive the highest priority and will be buffered from the negative effects of high demand or structural similarity. The task that is degraded is referred to as the *secondary task*. The dual-task performer's decision to treat one task as primary and another as secondary is an example of **task management.** Thus, there would be no problem with cell phone use in cars if drivers consistently treated safe driving as the primary task and cell phone use as the secondary task. Unfortunately, not all drivers adhere to such optimum task management strategies, and cell phone-induced accidents are the result (Violanti, 1998).

At a most basic level, task management is simply the allocation of resources to one task or the other. However, this allocation can become considerably more complex than a simple two-state decision. For example, given that most people know (metacognition) that cell phone use (or other in-vehicle tasks) can divert resources from driving and road monitoring, why do drivers still engage in concurrent tasks? One reason is that successful time-sharing strategies can allow an optimal **switching** of attention between tasks. For example, the driver can sample a competing source of secondary-task visual information at a moment when he or she knows that there is little chance of something happening on the road ahead. When the car is on a straight stretch of freeway, with little traffic on a calm day, the vehicle inertia and absence of hazards can allow the eyes to scan downward for some time. As we described in the context of selective attention, there is little expectancy of important events on the "roadway channel." How long can the eye safely stay "head down"? This depends on a number of factors, such as the speed of the vehicle, the degree of traffic on the highway, and the degree of trust that a driver has that he or she will be warned of an impending event. Thus, the well-skilled driver can develop an accurate mental model of event expectancies and costs to support accurate scheduling of scanning (Moray, 1986; Wickens et al., 2003).

The previous discussion suggests that switching between tasks can be good, and in fact necessary, when parallel processing is impossible, as it is when information to support two tasks is displayed in widely separated locations. Indeed, if attention is switched or alternated fast enough between tasks, the result is indistinguishable from parallel processing. Consistent with this interpretation is the finding that people who more rapidly alternate between tasks may be more effective in their concurrent performance (Raby & Wickens, 1994). At the other end of the spectrum, very slow switching in a multitask environment can lead to cognitive tunneling (Moray & Rotenberg, 1989; Kerstholt et al., 1996); this is the process of keeping attention fixated on one task or channel of information long after a second task or channel should have been attended. In the context of memory failures, one can attribute such errors to forgetting the need to check the neglected task; a breakdown in prospective memory.

Human factors designs to avoid cognitive tunneling are imposed by reminders, as described earlier in the chapter (Herrmann et al., 1999). However, an even more basic human factors solution lies in the design of alarms, as discussed in Chapter 5. Alarms, particularly auditory ones, are specifically designed to interrupt whatever task is ongoing in order to redirect the user's attention to a problem that the system deems worthy of observation (Woods, 1995). It appears important to train people how to handle interruptions in complex multitask environments like the cockpit (Dismukes 2001); (McFarlane & Latorella, 2002.)

Addressing Time-Sharing Overload

As our discussion suggests, there are a number of ways of addressing the multitask environment of the overloaded office secretary, vehicle driver, airline pilot, or supervisor of an organization in crisis. Briefly, we may subdivide these into four general categories:

  1. *Task redesign.* On the one hand, we should avoid asking operators to perform too many tasks that may impose time-sharing requirements. In some environments, the military combat aircraft, for example, there is a temptation to load progressively more "mission tasks" on the pilot (e.g., weapons and surveillance systems). These must inevitably impose challenging time-sharing requirements, inviting overload. We noted earlier in the chapter how the CI interview technique for eyewitnesses explicitly avoids time-sharing of effortful memory retrieval and question comprehension (Fisher, 1999). On the other hand, we can sometimes redesign tasks to make them less resource-demanding. Reducing working memory demands is often successful—for example, users should not be required to remember a 10-digit phone number or even a seven-digit number in multitask situations.

  2. *Interface redesign.* Sometimes interfaces can be changed to offload heavily demanded resources. As noted, there are many circumstances in which synthesized voice display can replace visual text when the eyes are needed for continuous vehicle control or monitoring (Dixon & Wickens, 2003; Wickens, Sandry, & Vidulich 1983).

  3. *Training.* Explicit or implicit training of the operator, as we discuss in Chapter 18, has two different components in multitask environments. First, repeated and consistent practice at component tasks can develop automaticity (Schneider, 1985), thereby reducing resource demands (see Figure 6.3). Second, training in attention management skills can improve the appropriate allocation of resources (Gopher, 1993; Gopher et al., 1994; Wickens, 1989) and the handling of task switching and interruptions (Dismukes, 2001).

  4. *Automation.* Automation also has two aspects relevant to dual-task performance. First, as we discuss in Chapter 16, many aspects of automation can either replace or greatly simplify resource-demanding aspects of performance—cruise control, the computer spell check, and the warning signal are typical examples. Second, designers have recently considered intelligent automation that can serve as a task manager, which can direct users' selective attention dynamically to neglected tasks or assume performance responsibility for those tasks when required (Hammer, 1999).

CONCLUSION

In this chapter we discussed a number of mental processes that define the contents of cognitive psychology and lie at the core of much information processing in complex environments. These components find their relevance in many other chapters of this book. Our discussion of perception links to chapters 4 and 5 on visual and auditory sensation, as well as to Chapter 8 on displays, where we consider designer artifacts that can support perception. Our discussions of attention relate to topics in both chapters 4 and 8 as well as to those of workload overload in Chapter 13. Issues of metacognition underlie people's tendency to engage in unsafe behavior, as discussed in Chapter 14. Cognition of all sorts is involved in computer usage (Chapter 15) and in dealing with automation and complex systems (Chapter 16) and transportation systems (Chapter 17). Cognition is knowledge, and knowledge is acquired through learning and training (Chapter 18). Finally, many aspects of cognition of perception and working memory are involved in the all-important task of decision making, the topic to which we turn in the next chapter.

Chapter 7

Decision Making

An anesthesiology team in a large hospital consisted of four physicians, three of whom were residents in training. The group was asked to assist with four procedures in one building (an in vitro fertilization, a perforated viscus, reconstruction of a leg artery, and an appendectomy) and an exploratory laparotomy in another building. All procedures were urgent and could not be delayed for the regular operating-room scheduling. There were several delays in preoperative preparation, and several surgeons and nurses were pressuring the team to get the procedures finished. The situation was complicated by the fact that the staff was only able to run two operating rooms simultaneously, and the best use of resources was to overlap procedures so that one case was started as another was finishing. The anesthesiologist in charge had to decide how to allocate the four members of the anesthesiology team to the five needed procedures. Also, there was always the possibility that an emergency case would come into the hospital's trauma center, in which case the anesthesiologist in charge was expected to be immediately available. Should she allocate only the other three anesthesiologists to the five procedures, or should she help out also, leaving no one available should a major emergency come in unexpectedly? (Adapted from Cook & Woods, 1994)

Although this scenario happens to occur in the medical domain, everyone makes hundreds of decisions each day—much time is spent considering multiple pieces of information, determining what the information represents or really "means," and selecting the best course of action. The information we process may be simple or complex, clear or distorted, and complete or filled with gaps. Because of this variability in information complexity and completeness, *we adopt different decision processes depending on the situation.* Sometimes, we carefully calculate and evaluate alternatives, but we often just interpret it to the best of our ability and make educated guesses about what to do. Some decisions are so routine that we might not even consider them to be decisions.

In many cases, the increasing complexity of the systems with which we interact makes decision making and problem solving difficult and prone to error. This makes decision making a central concern to the human factors specialist.

In the following, we consider three major classes of decision making models: optimal model based on expected value (and the departures therefrom), an information processing model that highlights heuristics and biases, and a model that addresses the context in which decisions are being made in natural environments. A final section addresses remediations to human decision making challenges, in terms of automation, decision supports, displays and training.

DEFINITION OF DECISION MAKING

What is a decision-making task? Generally, it is a task in which (a) a person must select one option from a **number of alternatives,** (b) there is some amount of **information available** with respect to the option, (c) the **timeframe is relatively long** (longer than a second), and (d) the choice is associated with **uncertainty;** that is, it is not necessarily clear which is the best option. By definition, decision making involves risk, and a good decision maker effectively assesses risks associated with each option (Medin & Ross, 1992). The decisions we discuss in this chapter run the range between those involving a slow deliberative process, involving how to allocate recourses (as in the story above), or diagnostic problem solving, to those which are quite rapid, with few alternatives, like the decision to speed up, or apply the brakes, when seeing a yellow traffic light.

Decision making can generally be represented by three phases, each of which itself can be elaborated into subphases: (1) acquiring and perceiving information or *cues* relevant for the decision (2) generating and selecting *hypotheses* or *situation assessments* about what the cues mean, regarding the current and future state relevant to the decision, (3) planning and selecting *choices* to take, on the basis of the inferred state, and the costs and values of different outcomes. The three stages often cycle and iterate in a single decision.

DECISION-MAKING MODELS

Most of the initial research on decision making focused on the study of optimal, *rational* decision making (Fischhoff, 1982; Luce & Raiffa, 1957). The assumption was that if researchers could specify the values (costs or benefits) associated with different choices, mathematical models could be applied to those values, yielding the optimal choice that would maximize these values (or minimize their costs). Early decision theory was thus a set of formal models that prescribed what people should do when faced with a set of decision choices, and it was also a yardstick by which to judge people's deviations from the optimal decision (Coombs et al., 1970; Edwards, 1954, 1961; Pitz & Sachs, 1984; Slovic et al., 1977). Rational models of decision making are also sometimes called *normative models,* because they specify what people ideally *should* do; they do not necessarily describe how people actually perform decision-making tasks. Normative models are important to understand because they form the basis for many computer-based decision aids (Edwards, 1987). Later researchers became interested in describing the cognitive processes associated with human decision-making behavior and developed a number of *descriptive models.* These models are often based on laboratory studies, which do not reflect the full range of decision-making situations.

Normative Decision Models

Normative decision models revolve around the central concept of *utility,* the overall value of a choice, or how much each outcome or product is "worth" to the decision maker. This model has application in engineering decisions as well as decisions in personal life. Choosing between different corporate investments, materials for product, jobs, or even cars are all examples of choices that can be modeled using *multiattribute utility theory.* The decision matrix described in Chapter 3 (Figure 3.3) is an example of how multiattribute utility theory can be used to guide engineering design decisions. Similarly, it has been used to resolve conflicting objectives, to guide environmental cleanup of contaminated sites (Accorsi et al., 1999), and to support operators of flexible manufacturing systems (Aly & Subramaniam, 1993).

The number of potential options, the number of attributes or features that describe each option, and the difficulty in comparing alternatives on very different dimensions make decisions complicated. Multiattribute utility theory addresses this complexity, using a utility function to translate the multidimensional space of attributes into a single dimension that reflects the overall utility or value of each option. In theory, this makes it possible to compare apples and oranges and pick the best one.

Multiattribute utility theory assumes that the overall value of a decision option is the sum of the magnitude of each attribute multiplied by the utility of each attribute, where $U(v)$ is the overall utility of an option, $a(i)$ is the magnitude of the option on the *ith* attribute, $u(i)$ is the utility (goodness or importance) of the *ith* attribute, and n is the number of attributes.

$$U(v) = \sum_{i=1}^{n} a(i)u(i)$$

Figure 7.1 shows the analysis of four different options, where the options are different cars that a student might purchase. Each car is described by five attributes. These attributes might include the initial purchase price, the fuel economy, insurance costs, sound quality of the stereo, and maintenance costs. The utility of each attribute reflects its importance to the student. For example, the student cannot afford frequent and expensive repairs, so the utility or importance of the fifth attribute (maintenance costs) is quite high (8), whereas the student does not care about the sound quality of the stereo and so the fourth attribute (stereo system quality) is quite low (1). The cells in the decision table show the magni-

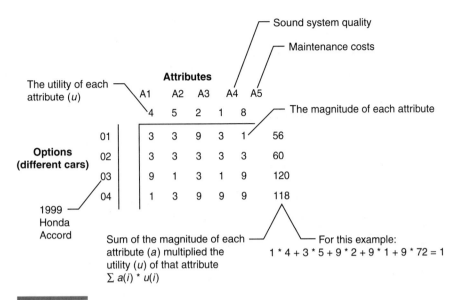

FIGURE 7.1

Multiattribute utility analysis combines information from multiple attributes of each of several options to identify the optimal decision.

tude of each attribute for each option. For this example, higher values reflect a more desirable situation. For example, the third car has a poor stereo but low maintenance costs. In contrast, the first car has a slightly better stereo but high maintenance costs. Combining the magnitude of all the attributes shows that third car (option 3) is most appealing or "optimal" choice and that the first car (option 1) is least appealing.

Multiattributed utility theory, shown in Figure 7.1, assumes that all outcomes are certain. However, life is uncertain, and probabilities often define the likelihood of various outcomes (e.g., you cannot predict maintenance costs precisely). Another example of a normative model is *expected value theory,* which addresses uncertainty. This theory replaces the concept of utility in the previous context with that of *expected value* and applies to any decision that involves a "gamble" type of decision, where each choice has one or more outcomes with an associated worth and probability. For example, a person might be offered a choice between

1. Winning $50 with a probability of .20, or
2. Winning $20 with a probability of .60.

Expected value theory assumes that the overall value of a choice is the sum of the worth of each outcome multiplied by its probability where $E(v)$ is the expected value of the *choice,* $p(i)$ is the probability of the *ith* outcome, and $v(i)$ is the value of the ith outcome.

$$E(v) = \sum_{i=1}^{n} p(i)v(i)$$

The expected value of the first choice for the example is 50×20, or $10, meaning that if the choice were selected many times, one would expect an average gain of $10. The expected value of the second choice is 20×60, or $12, which is a higher overall value. Therefore, the optimal or normative decision maker should always choose the second gamble. In a variety of decision tasks, researchers have compared results of the normative model to actual human decision making and found that people often vary from the optimal choice. This model does not predict the decisions people actually make.

Expected value theory is relatively limited in scope because it quickly becomes clear that many choices in life have different values to different people. For example, one person might value fuel efficiency in an automobile, whereas another might not. This facet of human decision making led to the development of *subjective expected utility (SEU) theory*. SEU theory still relies on the concepts of subjective probability times worth or value for each possible outcome. However, *the worth component is subjective, determined for each person;* that is, instead of an objective (e.g., monetary) worth, an outcome has some value or utility to each individual. Thus, each choice a person can make is associated with one or more outcomes, and each outcome has an associated probability and some subjective utility.

Descriptive Decision Models

Numerous researchers have evaluated the extent to which humans follow normative decision models, especially SEU theory. The conclusion, based on several years of experimentation, is that human decision making frequently violates key assumptions of the normative models.

Because actual decision making commonly showed violations of normative model assumptions, researchers began to search for more descriptive models that would capture how humans actually make decisions. These researchers believed that rational consideration of all factors associated with all possible choices, as well as their outcomes, is frequently just too time consuming and ef-

TABLE 7.1 Hypothetical Values in a Subjective Expected Utility Model for Two Possible States

Option	No emergency probability = 0.80	Emergency probability = 0.20	Total expected utility
Use three anesthesiologists	$-4 \ (-4 \times .80 = -3.2)$	$10 \ (10 \times .20 = 2.0)$	$-3.2 + 2.0 = -1.2$
Use four anesthesiologists	$6 \ (6 \times .80 = +4.8)$	$-10 \ (-10 \times .20 = -2.0)$	$4.8 - 2.0 = 2.8$

Each cell shows the utility of a particular outcome and the calculation ($p \times u$) for that outcome. The column on the right suggests that the option of four anesthesiologists yields the highest expected utility and therefore is optimal.

fort demanding. They suggested descriptive models of decision making where people rely on simpler and less-complete means of selecting among choices. People often rely on simplified shortcuts or rules-of-thumb that are sometimes referred to as *heuristics*. One well-known example of an early descriptive model is Simon's concept of *satisficing*.

Simon (1957) argued that people do not usually follow a goal of making the absolutely best or optimal decision. Instead, they opt for a choice that is "good enough" for their purposes, something satisfactory. This shortcut method of decision making is termed *satisficing*. In satisficing, the decision maker generates and considers choices only until one is found that is acceptable. Going beyond this choice to identify something that is better simply has too little advantage to make it worth the effort.

Satisficing is a very reasonable approach given that people have limited cognitive capacities and limited time. Indeed, if minimizing the time (or effort) to make a decision is itself considered to be an attribute of the decision process, then satisficing or other shortcutting *heuristics* can sometimes be said to be optimal—for example, when a decision must be made before a deadline, or all is lost. Heuristics such as satisficing are often quite effective (Gigerenzer & Todd, 1999) but they can also lead to *biases* and poor decisions, a topic discussed in detail later in this chapter.

Many real-world decisions take place in dynamic, changing environments. These environments have features that are far more complex, like those confronting the anesthesiologist described at the outset of the chapter (Orasanu & Connolly, 1993). The study of *naturalistic decision making* (Zsambok & Klein, 1997, Lipshitz et al. 2001) attempts to identify these features, highlighted in Table 7.2, and describe the decision-making processes of skilled practitioners engaged in real-world choices. The decision making involved in fighting a forest fire is an example of naturalistic decision making (Orasanu & Connolly, 1993).

To see how some of these characteristics combine, consider the anesthesiologist at the beginning of the chapter. There was incomplete, complex, and dynamically changing information; time stress; high risk; and a large set of outcomes, costs, and benefits. Another problem in making this decision is that she had multiple and conflicting goals imposed from the outside: making the surgeons happy, helping the patients needing immediate surgery, keeping hospital costs low, avoiding lawsuits, maintaining good relationships with staff, and keeping resources available for a possible major emergency.

In summary, if the amount of information is relatively small and time is unconstrained, careful analysis of the choices and their utilities is desirable and possible. To the extent that the amount of information exceeds cognitive-processing limitations, time is limited, or both, people shift to using simplifying heuristics. The following section describes some common heuristics and associated biases. Following the discussion of heuristics and biases, we describe the range of decision-making processes that people adopt and how the decision-making process depends on the decision-making context.

TABLE 7.2   Features of Naturalistic Decision-Making Situations

Characteristic	Example
Ill-structured problems	There is no single "best" way of fighting a forest fire.
Uncertain, dynamic environments	The fire is continually changing, presenting new decisions and considerations.
Information-rich environments where situational cues may change rapidly	Smoke and flames can be seen, heard, and felt, and they are constantly changing.
Cognitive processing that proceeds in iterative action/feedback loops	The application of fire retardants is monitored to decide what to do next.
Multiple shifting and/or competing individual and organizational goals	As the forest fire evolves, the goals may shift from protecting property to saving lives.
Time constraints or time stress	Decisions often need to be made quickly because the fire continues to spread as the decision is being made.
High risk	Substantial property damage or loss of life can result from a poor decision.
Multiple persons somehow involved in the decision	Many people contribute information and perspectives to the decisions concerning actions to take in fighting the fire.

HEURISTICS AND BIASES

Cognitive heuristics represent rules-of-thumb that are easy ways of making decisions. Heuristics are usually very powerful and efficient (Gigerenzer & Todd, 1999), but they do not always guarantee the best solution, (Kahneman et al., 1982). Unfortunately, because they represent simplifications, heuristics occasionally lead to systematic flaws and errors. The systematic flaws represent deviations from a rational or normative model and are sometimes referred to as *biases,* and can be represented in terms of a basic information-processing model.

Information Processing Limits in Decision Making

Figure 7.2 shows a relatively simple information-processing framework that highlights some of the cognitive limits critical to conscious, effortful decision making. Just as they were related to troubleshooting and problem solving, selective attention, activities performed within working memory, and information retrieval from long-term memory all have an important influence on decision making. These processes impose important limits on human decision making and are one reason why people use heuristics to make decisions. According to this model, the following occur in working memory:

1. *Cue reception and integration.* A number of cues, or pieces of information, are received from the environment and go into working memory. For example, an engineer trying to identify the problem in a manufacturing process

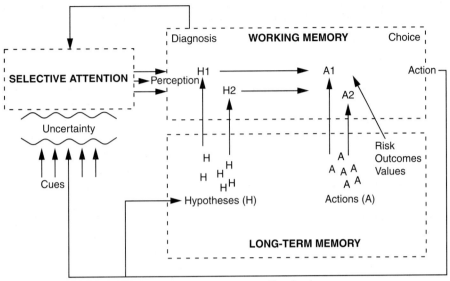

FIGURE 7.2

Information-processing model of decision making. Cues are selectively sampled (on the left); hypotheses are generated through retrieval from long-term memory. Possible actions are retrieved from long-term memory, and an action is selected on the basis of risks and the values of their outcomes. (Adapted from Wickens, C. D., 1992. *Engineering Psychology and Human Performance* (2nd ed.). New York: HarperCollins Publishers.)

might receive a number of cues, including unusual vibrations, particularly rapid tool wear, and strange noises. The cues must be selectively attended, interpreted and somehow integrated with respect to one another. The cues may also be incomplete, fuzzy, or erroneous; that is, they may be associated with some amount of uncertainty. As discussed in Chapter 6, stage-1 situation awareness, perceiving system status, depends on this element of the decision process.

    2.   *Hypothesis generation and selection.* A person may then use these cues to generate one or more hypotheses, "educated" guesses, diagnoses, or inferences as to what the cues mean. This is accomplished by retrieving information from long-term memory. For example, an engineer might hypothesize that the set of cues described above is caused by a worn bearing. Many of the decision tasks studied in human factors require such inferential diagnosis, which is the process of inferring the underlying or "true" state of a system. Examples of inferential diagnosis include medical diagnosis, fault diagnosis of a mechanical or electrical system, inference of weather conditions based on measurement values or displays, and so on. Sometimes this diagnosis is of the current state, and sometimes it is of the predicted or forecast state, such as in weather forecasting or economic projections. As described in Chapter 6, stage-2 and stage-3 situation awareness, comprehending and projecting system state, depend on this element of the decision process. In decision making, SA is sometimes described as situation *assessment*.

The hypotheses brought into working memory are evaluated with respect to how likely they are to be correct. This is accomplished by gathering additional cues from the environment to either confirm or disconfirm each hypothesis. In addition, hypotheses may need to be revised, or a new one may need to be generated. When a hypothesis is found to be adequately supported by the information, that hypothesis is chosen as the basis for a course of action. The process can be seen in a scenario in which a surgeon sees a female patient in the emergency room, who complains of nausea and severe abdominal pain which has lasted several hours. After some tests, the surgeon considers alternative diagnoses of appendicitis or stomach flu. Following additional tests, which provided additional cues, he diagnosed appendicitis as the more likely situation.

3. *Plan generation and action choice.* One or more alternative actions are generated by retrieving possibilities from memory. For example, after diagnosing acute appendicitis, the surgeon in our scenario generated several alternative actions, including waiting, conducting additional tests, and performing surgery. Depending on the decision time available, one or more of the alternatives are generated and considered.

To choose an action, the decision maker might evaluate information such as possible outcomes of each action (where there may be multiple possible outcomes for each action), the likelihood of each outcome, and the negative and positive factors associated with each outcome, following the sorts of procedures laid out in Figure 7.1 and Table 7.1. Each action is associated with multiple possible outcomes, some of which are more likely than others. In addition, these outcomes may vary from mildly to extremely positive (i.e., one outcome from surgery is that the appendix is removed without complication) or from mildly to extremely negative (i.e., he could wait, she could die from a burst appendix, and he could be sued for malpractice). Table 7.3 shows the possible outcomes associated with the actions of operating or testing. The "optimal" choice would be the sum of the utilities of each outcome multiplied by the estimated probability of each state.

If the working hypothesis, plan, or action proves unsatisfactory, the decision maker may generate a new hypothesis, plan, or action. When a plan is finally selected, it is executed, and the person monitors the environment to update his or her situation assessment and to determine whether changes in procedures must be made.

Figure 7.2 shows that the decision process depends on limited cognitive resources, such as working memory. In the following sections, we consider a variety of heuristics and biases that result from limited cognitive resources. Familiarity

TABLE 7.3   Decision Matrix for the Decision to Operate or Order Tests

Options\States	Appendicitis	Stomach flu
Operate	Successful removal	Unneeded operation
Test	Burst appendix, death	Correct diagnosis

with the heuristics can help develop information displays and cognitive support systems that counteract the biases inherent in human information processing (examples are described later in this chapter and also in Chapters 8 & 16).

Heuristics and Biases in Receiving and Using Cues

1. *Attention to a limited number of cues.* Due to working memory limitations, people can use only a relatively small number of cues to develop a picture of the world or system. This is one reason why configural displays that visually integrate several variables or factors into one display are useful (see Chapter 8 for a description).

2. *Cue primacy and anchoring.* In decisions where people receive cues over a period of time, there are certain trends or biases in the use of that information. The first few cues receive greater than average weight or importance. This is a *primacy* effect, found in many information-processing tasks, where preliminary information tends to carry more weight than subsequent information (e.g., Adelman et al., 1996). It often leads people to "anchor" on hypotheses supported by initial evidence and is therefore sometimes called the *anchoring heuristic* (Tversky & Kahneman, 1974), characterizing the familiar phenomenon that first impressions are lasting. The order of information has an effect because people use the information to construct plausible stories or mental models of the world or system. These models differ depending on which information is used first (Bergus et al., 2002). The key point is that, for whatever reason, *information processed early is often most influential,* and this will ultimately affect decision making.

3. *Inattention to later cues.* In contrast to primacy, cues occurring later in time or cues that change over time are often likely to be totally ignored, which may be attributable to attentional factors. In medical diagnosis, this would mean that symptoms, or cues, that are presented first would be more likely to be brought into working memory and remain dominant. It is important to consider that in many dynamic environments with changing information, limitations 2 and 3 can be counterproductive to the extent that older information—recalled when primacy is dominant, may be less accurate as time goes on, more likely to be outdated, and updated by more recent changes.

4. *Cue salience.* Perceptually salient cues are more likely to capture attention and be given more weight (Endsley, 1995; Wickens & Hollands, 2000; see also Chapter 6). As you would expect, *salient cues* in displays are things such as information at the top of a display, the loudest alarm, the largest display, and so forth. Unfortunately, the most salient display cue is not necessarily the most diagnostic.

5. *Overweighting of unreliable cues.* Not all cues are equally reliable. In a trial, some witnesses, for example, will always tell the truth. Others might have faulty memories, and still others might intentionally lie. However, when integrating cues, people often simplify the process by treating all cues as if they are all equally valid and reliable. The result is that people tend to give too much weight to unreliable information (Johnson et al., 1973; Schum, 1975; Wickens & Hollands, 2000).

Heuristics and Biases in Hypothesis Generation, Evaluation and Selection

After a limited set of cues is processed in working memory, the decision maker generates hypotheses by retrieving one or more from long-term memory. There are a number of heuristics and biases that affect this process:

1. *Generation of a limited number of hypotheses.* As we mentioned in our discussion of troubleshooting in Chapter 6, people generate a limited number of hypotheses because of working memory limitations (Lusted, 1976; Mehle, 1982; Rasmussen, 1981). Thus, people will bring in somewhere between one and four hypotheses for evaluation. People consider a small subset of possible hypotheses at one time and often never consider all relevant hypotheses (Elstein et al., 1978; Wickens & Hollands, 2000). Substantial research in real-world decision making under time stress indicates that in these circumstances, decision makers often consider only a single hypothesis (Flin et al., 1996; Klein, 1993). This process degrades the quality of novice decision makers far more than expert decision makers. *The first option considered by experts is likely to be reasonable, but not for novices.*

2. *Availability heuristic.* Memory research suggests that people more easily retrieve hypotheses that have been considered recently or that have been considered frequently (Anderson, 1990). Unusual illnesses are simply not the first things that come to mind to a physician. This is related to another heuristic, the *availability heuristic* (Kahneman et al., 1982; Tversky & Kahneman, 1974). This heuristic assumes that people make certain types of judgment, for example, estimates of frequency, by cognitively assessing how easily the state or event is brought to mind. The implication is that although people try to rationally generate the most likely hypotheses, the reality is that if something comes to mind relatively easily, they assume it is common and therefore a good hypothesis. As an example, if a physician readily thinks of a hypothesis, such as acute appendicitis, he or she will assume it is relatively common, leading to the judgment that it is a likely cause of the current set of symptoms. In actuality, availability to memory may not be a reliable basis for estimating frequency. Availability (to memory) might also be based upon hypotheses that were most *recently* experienced.

3. *Representativeness Heuristic.* Sometimes people diagnose a situation because the pattern of cues "looks like" or is representative of the prototypical example of this situation. This is the *representativeness heuristic* (Kahneman et al., 1982), and usually works well; however the heuristic can be biasing when a perceived situation is slightly different from the prototypical example even though the pattern of cues is similar or representative.

4. *Overconfidence.* Finally, people are often biased in their confidence with respect to the hypotheses they have brought into working memory (Mehle, 1982), believing that they are correct more often than they actually are and reflecting the more general tendency for overconfidence in metacognitive

processes, as described in Chapter 6 (Bjork, 1999). As a consequence, people are less likely to seek out evidence for alternative hypotheses or to prepare for the circumstances that they may be wrong.

Once the hypotheses have been brought into working memory, additional cues are potentially sought to evaluate them. The process of considering additional cue information is affected by cognitive limitations similar to the other subprocesses.

5. *Cognitive tunneling.* As we have noted above in the context of anchoring, once a hypothesis has been generated or chosen, people tend to underutilize subsequent cues. We remain stuck on our initial hypothesis, a process introduced in the previous chapter as cognitive tunneling (Cook & Woods, 1994). Examples of cognitive tunneling abound in the complex systems (e.g., Xiao et al., 1995). Consider the example of the Three Mile Island disaster in which a relief valve failed and caused some of the displays to indicate a rise in the level of coolant (Rubinstein & Mason, 1979). Operators mistakenly thought that that emergency coolant flow should be reduced and persisted to hold this hypothesis for over two hours. Only when a supervisor arrived with a fresh perspective did the course of action get reversed. Notice that cognitive tunneling is a different effect than the cue primacy effect when the decision maker is first generating hypotheses.

Cognitive tunneling can sometimes be avoided by looking at the functionality of objects in terms beyond their normal use. The episode in the moon mission, well captured by the movie *Apollo 13* demonstrated the ability of people to move beyond this type of functional fixedness. Recall that the astronauts were stranded without an adequate air purifier system. To solve this problem, the ground control crew assembled all of the "usable" objects known to be on board the spacecraft (tubes, articles of clothing, etc.). Then they did free brainstorming with the objects in various configurations until they had assembled a system that worked.

6. *Confirmation bias.* Closely related to cognitive fixation are the biases when people consider additional cues to evaluate working hypotheses. First, they tend to seek out only confirming information and not disconfirming information, even when the disconfirming evidence can be more diagnostic (Einhorn & Hogarth, 1978; Schustack & Sternberg, 1981). It is hard to imagine an engineer doing tests for various hardware malfunctions that he thinks are not related to the problem being observed (an exception to this general bias would be when police detectives ask their suspects if they have an alibi). In a similar vein, people tend to underweight, or fail to remember, disconfirming evidence (Arkes & Harkness, 1980; Wickens & Hollands, 2000) and fail to use the absence of important cues as diagnostic information (Balla, 1980). The confirmation bias is exaggerated under conditions of high stress and mental workload (Cook & Woods, 1994; Janis, 1982; Sheridan, 1981; Wright, 1974). Cognitive fixation can occur for any number of reasons, but one reason is the tendency to seek only information that confirms existing belief, which is known as confirmation bias.

The main difference between cognitive fixation and confirmation bias is one of degree. With cognitive fixation, people have adopted and fixated on a single hypothesis, assumed that it is correct, and proceeded with a solution. With confirmation bias, people have a hypothesis that they are trying to evaluate and seek only confirming information in evaluating the hypothesis.

Heuristics and Biases in Action Selection

Choice of action is also subject to a variety of heuristics or biases. Some are based on basic memory processes that we have already discussed.

1. *Retrieve a small number of actions.* Long-term memory may provide many possible action plans, but people are limited in the number they can retrieve and keep in working memory.

2. *Availability heuristic for actions.* In retrieving possible courses of action from long-term memory, people retrieve the most "available" actions. In general, the availability of items from memory are a function of recency, frequency, and how strongly they are associated with the hypothesis or situational assessment that has been selected through the use of "if-then" rules. In high-risk professions like aviation, emergency checklists are often used to insure that actions are available, even if they may not be frequently performed (Degani & Wiener, 1993).

3. *Availability of possible outcomes.* Other types of availability effects will occur, including the generation/retrieval of associated outcomes. As discussed, when more than one possible action is retrieved, the decision maker must select one based on how well the action will yield desirable outcomes. Each action often has more than one associated consequence, which are probabilistic. As an example, a worker might consider adhering to a safety procedure and wear a hardhat versus ignoring the procedure and going without one. Wearing the hardhat has some probability of saving the worker from death due to a falling object. A worker's estimate of this probability will influence the decision to wear the hardhat. The worker's estimate of these likelihoods will not be objective based on statistics, but are more likely to be based on the availability of instances in memory. It is likely that the worker has seen many workers not wearing a hardhat who have not suffered any negative effects, and so he or she is likely to think the probability of being injured by falling objects is less than it actually is. Thus, the availability heuristic will bias retrieval of some outcomes and not others. Chapter 14 describes how warnings can be created to counteract this bias by showing the potential consequences of not complying, thus making the consequences more available.

After someone is injured because he or she did not wear a hardhat, people are quick to criticize because it was such an obvious mistake. The tendency for people to think "they knew it along" is called the *hindsight bias*. This process is evident in the "Monday morning quarterback phenomena" where people believe they would not have made the obvious mistakes of the losing quarterback. More importantly, hindsight bias often plagues accident investigators who, with the benefit of hindsight and the very available (to their memory) example of a bad outcome, inappropriately blame operators for committing errors that are obvious only in hindsight (Fischhoff, 1975).

The decision maker is extremely unlikely to retrieve all of the possible outcomes for an action, particularly under stress. Thus, selection of action suffers from the same cognitive limitations as other decision activities we have discussed (retrieval biases and working-memory limitations). Because of these cognitive limitations, selection of action tends to follow a satisficing model: If an alternative action passes certain criteria, it is selected. If the action does not work, another is considered. Again, this bias is much more likely to affect the performance of novices than experts (Lipshitz et al., 2001).

4. *Framing bias.* The framing bias is the influence of the framing or presentation of a decision on a person's judgment (Kahneman & Tversky, 1984). According to the normative utility theory model, the way the problem is presented should have no effect on the judgment. For example, when people are asked the price they would pay for a pound of ground meat that is 10 percent fat or a pound that is 90 percent lean, they will tend to pay 8.2 cents per pound more for the option presented as 90 percent lean even though they are equivalent (Levin et al., 2002). Likewise, students would likely feel that they are performing better if they are told that they answered 80 percent of the questions on the exam correctly compared to being told that they answered 20 percent of the questions incorrectly. Similarly, people tend to view a certain treatment as more lethal if its risks are expressed as a 20 percent mortality rate than if expressed as 80 percent life saving and are thereby less likely to choose the treatment when expressed in terms of mortality (McNeil et al., 1982). Thus, the *way a decision is framed can bias decisions.*

This has important implications for how individuals and corporations view investments. People judge an investment differently if it is framed as a gain or as a loss. People tend to make conservative decisions when presented with a choice between gains and risky decisions when presented with a choice between losses. For example, when forced to choose between a certain loss of $50 and an equal chance of losing $100 or breaking even, people tend to gamble by preferring the risky option with the hope of breaking even. They tend to make this choice even though the expected utility of each action is equal. In contrast, when presented with a choice between a certain gain of $50 and an equal chance of making nothing or $100, people tend to choose the conservative option of the certain $50. Each example demonstrates the framing bias as a preference for an uncertain loss of greater negative utility compared to a certain loss of a lesser negative utility.

A common manifestation of framing is known as the *sunk cost* bias (Arkes & Hutzel, 2000). This bias affects individual investors who hesitate to sell losing stocks (a certain loss) but tend to sell winning stocks to lock in a gain. Likewise, when you have invested a lot of money in a project that has "gone sour," there is a tendency to keep supporting it in the hopes that it will turn around rather than to give it up. After you have sunk a lot of money into the project, to give up on it is a sure loss. To stay with it is a risky choice that may eventually pay off with some probability but will more likely lead to an even greater cost. Similarly, managers and engineers tend to avoid admitting a certain cost when replacing obsolete equipment. The sunk cost bias describes the tendency to choose the risky loss over the sure one, even when the rational, expected value choice should be to abandon the project. Because people tend to incur greater risk in situations involving losses, *decisions should be framed in terms of gains to counteract this tendency.*

Benefit of Heuristics and the Costs of Biases

This section has focused on the costs of decision making heuristics as defined by the biases that sometimes undermine their effectiveness. In general, decision-making heuristics can be very powerful in simplifying decisions so that a response can be made in a timely manner (Gigerenzer & Todd, 1999). This becomes not only desirable but essential under extreme time pressure, such as the decision a pilot must make before he or she runs out of fuel. However, in some circumstances, the tendency for inexperienced decision makers to generate a limited number of alternatives can result in poor decisions because the best alternatives get overlooked. However, experts with many years of experience might use similar heuristics and avoid the biases because they are able to bring many years of experience to the decision. The one alternative that comes to mind of an expert after assessing the representativeness of the situation is likely to be a good choice. As described in the next section, *experts can also adapt their decision making and avoid heuristics when heuristics might lead to poor decisions.*

DEPENDENCY OF DECISION MAKING ON THE DECISION CONTEXT

The long list of decision-making biases and heuristics above may suggest that people are not very effective decision makers in everyday situations. In fact, however, this is not the case. Most people do make good decisions most of the time, but the list can help account for the infrequent circumstances, like the decision makers in the Three Mile Island nuclear plant, when decisions produce bad outcomes. One reason that most decisions are good, is that heuristics are accurate most of the time. A second reason is that people have a profile of resources: information-processing capabilities, experiences, and decision aids (e.g., a decision matrix) that they can adapt to the situations they face. *To the extent that people have the appropriate resources and can adapt them, they make good decisions.* When people are not able to adapt, such as in some highly constrained laboratory conditions where people have little experience with the situations, poor decisions can result.

One way people adapt to different decision circumstances is by moving from an analytical approach, where they might try to maximize utility, to the use of simplifying heuristics, such as satisficing (Hammond, 1993; Payne, 1982; Payne et al., 1988). Time stress, cognitive resource limitations, and familiarity lead people to use simplifying decision-making heuristics (Janis, 1982). This is commonly found in complex and dynamic operational control environments, such as hospitals, power or manufacturing plant control rooms, air traffic control towers, and aircraft cockpits. Naturalistic decision situations lead people to adopt different strategies than what might be observed in controlled laboratory situations. Understanding how decision making adapts to the characteristics of the person and situation is critical in improving human decision making.

Skill-, Rule-, and Knowledge-Based Behavior

The distinctions of *skill-, rule-,* and *knowledge*-based behavior describe different decisions-making processes that people can adopt depending on their level of expertise and the decision situation (Rasmussen, 1983, 1986, 1993). Rasmussen's SRK (skill, rule, knowledge) model of behavior has received increasing attention in the field of human factors (Vicente, 1999). It is consistent with accepted and empirically supported models of cognitive information processing, such as the three-stage model of expertise proposed by Fitts (1964) and Anderson (1983) and has also been used in popular accounts of human error (Reason, 1990; see also discussions of human error in Chapter 14). These distinctions are particularly important because the ways to improve decision making depend on supporting effective skill-, rule-, and knowledge-based behavior.

Figure 7.3 shows the three levels of cognitive control: skill-based behavior, rule-based behavior, and knowledge-based behavior. Sensory input enters at the

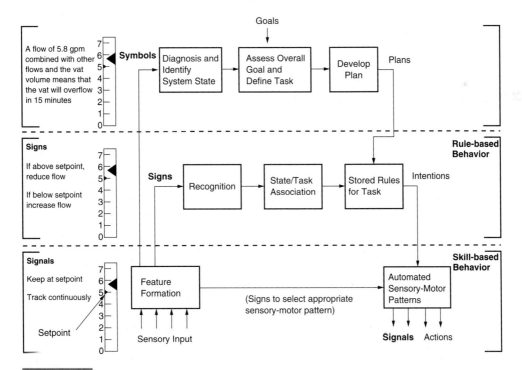

FIGURE 7.3

Ramussen's SRK levels of cognitive control. The same physical cues (e.g., the meter in this figure) can be interpreted as signals, signs, or symbols. (Adapted from Rasmussen (1983). Skills, rules, and knowledge: Signals, signs, and symbols, and other distinctions in human performance models. *SMC-13*(3), 257–266.)

lower left, as a function of attentional processes. This input results in cognitive processing at either the skill-based level, the rule-based level, or the knowledge-based level, depending on the operator's degree of experience with the particular circumstance. (Hammond et al., 1987; Rasmussen, 1993). *People who are extremely experienced with a task tend to process the input at the skill-based level,* reacting to the raw perceptual elements at an automatic, subconscious level. They do not have to interpret and integrate the cues or think of possible actions, but only respond to cues as *signals* that guide responses. Figure 7.3 also shows *signs* at this level of control; however, they are used only indirectly. Signs are used to select the appropriate motor pattern for the situation. For example, my riding style (skill-based behavior) when I come to work on my bike is shifted by signs (ice on the road) to a mode where I am "more careful" (skill-based behavior with a slightly different motor pattern). Because the behavior is automatic, the demand on attentional resources is minimal (Chapter 6). For example, an operator might turn a valve in a continuous manner to counteract changes in flow shown on a meter (see bottom left of Figure 7.3).

When people are familiar with the task but do not have extensive experience, they process input and perform at the rule-based level. The input is recognized in relation to typical system states, termed *signs,* which trigger rules accumulated from past experience. This accumulated knowledge can be in the person's head or written down in formal procedures. Following a recipe to bake bread is an example of rule-based behavior. The rules are "if-then" associations between cue sets and the appropriate actions. For example, Figure 7.3 shows how the operator might interpret the meter reading as a sign and reduce the flow because the procedure is to reduce the flow when the meter is above the setpoint.

When the situation is novel, decision makers do not have any rules stored from previous experience to call on. They therefore have to operate at the knowledge-based level, which is essentially analytical processing using conceptual information. After the person assigns meaning to the cues and integrates them to identify what is happening, he or she processes the cues as *symbols* that relate to the goals and an action plan. Figure 7.3 shows how the operator might reason about the meter reading of 5.8 gallons per minute and think that the flow must be reduced because the flow has reached a point that, when combined with the other flows entering a holding tank, will lead to an overflow in 15 minutes. It is important to note that the same sensory input, the meter in Figure 7.3, for example, can be interpreted as a signal, sign, or symbol.

The SRK levels can describe different levels of expertise. A novice can work only at the analytical knowledge-based level or, if there are written procedures, at the rule-based level. At an intermediate point of learning, people have some rules in their repertoire from training or experience. They work mostly at the rule-based level but must move to knowledge-based processing when encountering new situations. The expert has a greatly expanded rule base and a skill base as well. Thus, the *expert tends to use skill-based behavior,* but moves between the three levels depending on the task. When a novel situation arises, such as a system disturbance not previously experienced, lack of familiarity with the situation moves even the expert back to the analytical knowledge-based level. Effective decision making depends on all three levels of behavior.

Recognition-Primed Decision Making

Recognition primed decision (RPD) making provides a more refined description of how the SRK distinctions interact when experts make complex decisions in difficult situations, such as those associated with naturalistic decision making (Klein, 1989). Experts draw on a huge background of experience to avoid typical decision-making biases. In most instances experts simply recognize a pattern of cues and recall a single course of action, which is then implemented (Klein, 1989; Klein & Calderwood, 1991). The recognition of the situation is similar to the representativeness heuristic described earlier and the selection of an action is similar to rule-based behavior. The biases associated with the representativeness heuristic are avoided if the expert has a sufficiently large set of experiences and is vigilant for small changes in the pattern of cues that might suggest a diagnosis other than the likely one. Simon (1987) describes this type of decision process as "intuition" derived from a capability for rapid recognition linked to a large store of knowledge.

There are three critical assumptions of the RPD model: First, experts use their experience to generate a plausible option the first time around. Second, time pressure should not cripple performance because experts can use rapid pattern matching, which, being almost like perceptual recognition, described in Chapter 6, is resistant to time pressure. Finally, experienced decision makers know how to respond from past experience.

In spite of the prevalence of rapid pattern-recognition decisions, there are cases where decision makers will use analytical methods. In situations where the decision maker is unsure of the appropriate course of action, the action is evaluated by imagining the consequences of what might happen if a course of action is adopted: a *mental simulation,* where the decision maker thinks: "if I do this, what is likely to happen" (Klein & Crandall, 1995). Also, if uncertainty exists and time is adequate, additional analyses are performed to evaluate the current situation assessment, modify the retrieved action plan, or generate alternative actions (Klein et al. 1993). Experts adapt their decision-making strategy to the situation. Table 7.4 summarizes some of the factors that lead to intuitive rule-based decision making and those that lead to analytical knowledge-based decision making.

FACTORS AFFECTING DECISION-MAKING PERFORMANCE: AN INTEGRATED DESCRIPTION OF DECISION MAKING

It is useful to synthesize the different perspectives on decision making into an integrated model that describes the decision-making process. Such a model begins with Rasmussen's three levels of cognitive control, as shown in Figure 7.3. The SRK model is expanded and combined with Figure 7.2 to highlight some of the critical information processing resources, such as selective attention (lower left), long-term memory (bottom of figure), working memory (right of figure), and metacognition (top of figure). As in Figure 7.2, selective attention is needed for cue reception and integration, long-term memory affects the available hypotheses and alternate actions. Importantly, this model shows that metacognition influences the decision-making process by guiding how people adapt to the

TABLE 7.4 **Factors that Lead to Different Decision-Making Processes**

Induces intuitive rule-based decisions	*Induces analytical knowledge-based decisions*
Experience	Unusual situations
Time pressure	Abstract problems
Unstable conditions	Alphanumeric rather than graphic representation
Ill-defined goals	Requirement to justify decision
Large number of cues	Integrated views of multiple stakeholders
Cues displayed simultaneously	Few relationships among cues
Conserve cognitive effort	Requires precise solution

particular decision situation. Metacognition includes the anticipated effort and accuracy of a particular decision making approach.

In this model, people interpret environmental cues at one of three levels: *automatic* skill-based processing, *intuitive* rule-based processing, and *analytical* knowledge-based processing. Automatic processing occurs when environmental cues are sensed (affected by selective attention), but beyond that, there is no demand on cognitive resources.

When the skill- and rule-based processes do not provide a satisfactory solution or decision and time is available, the decision process moves upward in the model; that is, uncertainty coupled with available time leads to a more careful analytical process. Metacognition plays a critical role in recognizing the appropriate decision-making strategy.

The analytical process relies heavily on mental simulation to help assess the hypothesis, action, or plan under consideration (Orasanu, 1993). In this process, the decision maker uses the mental simulations to identify information needed to evaluate his or her understanding and searches the environment for this information. The use of cognitive simulations to generate ideas about additional information to be obtained explains why people tend to look for confirming evidence. The simulation also generates expectations for other cues not previously considered and guides the observation of changes in system variables (Roth, 1997). For example, you might use your mental model of how your car works to diagnose why your car doesn't start by turning on your headlights to confirm your hypothesis that your battery is dead.

Mental models make mental simulation possible and support the evaluation processes. Development of accurate mental models is critical for good decision making. For example, Passaro and colleagues found that inadequate mental models were responsible for decision errors leading to critical mine gas explosions (Passaro et al., 1994), and Lehner & Zirk (1987) found that use of poor mental models can cause a drop in decision performance of anywhere between 30 percent and 60 percent. For example, if you had a poor mental model of your car that did not include the role of the battery, then your ability to diagnose the problem would be greatly limited.

Because recognition of the situation plays such a critical role in expert decision making, adequate awareness of the situation is critical. As discussed earlier, there are 3 levels of situation awareness (Endsley, 1995). Figure 7.4 shows that not everyone needs to, or is able to, achieve all three levels for every decision-making situation. The level of SA required for adequate performance depends on the degree to which the person depends on skill-, rule-, or knowledge-based behavior for a particular decision.

The bottom of the Figure 7.4 shows the importance of monitoring the effects of decisions, a particularly critical part of decision making. In many real-world decisions, a person may iterate many times through the steps we have described. *With clear and diagnostic feedback people can correct poor decisions.* For example, in driving a car, a poor decision to steer to the right is made obvious as the car starts to drift off the road. This process of anticipating the effect of actions also plays a critical role in decision making. People do not passively respond to cues from the system; instead, they actively monitor the effects of their

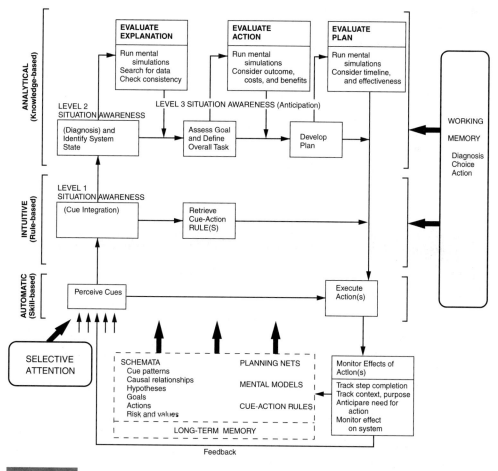

FIGURE 7.4
Integrated model: Adaptive decision making.

actions and look for expected changes in the system (Mumaw et al., 2000). In the case of driving, drivers apply the brakes and expect the car to slow; any failure to slow is quickly recognized.

Over the long term, poor feedback can lead to poor learning and inaccurate mental models (Brehmer, 1980). Although drivers receive good feedback concerning the immediate control of their car, they receive poor feedback about decisions they make regarding speed choice and risk taking. For example, drivers surviving a fatal car crash change their driving habits in only those circumstances which led to the accident, and return to their "normal" driving within a few months (Rajalin & Summala, 1997). In driving, like many other situations, learning is difficult because *people often receive poor feedback about risky situations* due to the great number of probabilistic relationships in these systems (Brehmer, 1980). Chapter 17 discusses the challenges faced by drivers in more detail.

If we consider the activities depicted in Figure 7.4, it is apparent that there are a variety of factors and cognitive limitations that strongly influence decision making. These include the following factors, some of which were identified by Cook and Woods (1994) and Reason (1990) as well as from conclusions drawn earlier in the chapter:

- Inadequate cue integration. This can be due to environmental constraints (such as poor, or unreliable data) or to cognitive factors, that disrupt selective attention, as discussed in Chapter 6, and biases that lead people to weigh cues inappropriately.
- Inadequate or poor-quality knowledge the person holds in long-term memory that is relevant to a particular activity (possible hypotheses, courses of action, or likely outcomes). This limited knowledge results in systematic biases when people use poorly refined rules, such as those associated with representativeness and availability heuristics.
- Tendency to adopt a single course of action and fail to consider the problem space broadly, even when time is available. Working-memory limits make it difficult to consider many alternatives simultaneously, and the tendency towards cognitive fixation leads people to neglect cues after identifying an initial hypothesis.
- Incorrect or incomplete mental model that leads to inaccurate assessments of system state or the effects of an action.
- Working-memory capacity and attentional limits that result in a very limited ability to consider all possible hypotheses simultaneously, associated cues, costs and benefits of outcomes, and so forth.
- Poor awareness of a changing situation and the need to adjust the application of a rule—for example, failing to adjust your car's speed when the road becomes icy.
- Inadequate metacognition leading to an inappropriate decision strategy for the situation. For example, persisting with a rule-based approach when a more precise analytic approach is needed.
- Poor feedback regarding past decisions makes error recovery or learning difficult.

These factors represent important challenges to effective decision making. The following section outlines some strategies to address these challenges and improve decision making.

IMPROVING HUMAN DECISION MAKING

Figure 7.4 shows that decision making is often an iterative cycle in which decision makers are often adaptive, adjusting their response according to their experience, the task situation, cognitive-processing ability, and the available decision-making aids. It is important to understand this adaptive decision process because system design, training, and decision aids need to support it. Attempts to improve decision making without understanding this process tend to fail. In this section, we briefly discuss some possibilities for improving human decision making: *task redesign, decision-support systems,* and *training.*

Task Redesign

We often jump to the conclusion that poor performance in decision making means that we must do something "to the person" to make him or her a better decision maker. However, sometimes a change in the system can support better decision making, eliminating the need for the person to change. As described in Chapter 1, decision making may be improved by task design. Changing the system should be considered before changing the person through training or even providing a computer-based decision aid. For example, consider the situation in which the removal of a few control rods led to a runaway nuclear reaction and the deaths of three people and exposure of 23 others to high levels of radioactivity. Learning from this experience, reactor designers now create reactors that remain stable even when several control rods are removed (Casey, 1998). Creating systems with greater stability leaves a greater margin for error in decisions and can also make it easier to develop accurate mental models.

Decision-Support Systems

Help for decision makers can take many forms, ranging from simple tables to elaborate expert systems. Some decision aids use computers to support working memory and perform calculations. Many decision aids fall in the category of *decision-support systems.* According to Zachary (1988), a decision-support system is "any interactive system that is specifically designed to improve decision making of its user by extending the user's cognitive decision-making abilities." Because this often requires information display, it can be difficult to distinguish between a decision-support system and an advanced information display. Often, the most effective way to support decisions is to provide a good display, as described in Chapter 8. Decision-support systems also share many similarities with automation, discussed in detail in Chapter 16.

Two design philosophies describe decision-support systems. One philosophy tries to reduce poor decisions by eliminating the defective or inconsistent decision making of the person. Decision aids developed using this approach are termed *cognitive prostheses* (Roth et al., 1987). This approach places the person in a subservient role to the computer, in which the person is responsible for data entry and interpretation of the computer's decision. An alternative philosophy tries to support adaptive human decision making by providing useful instruments to support rather than replace the decision maker. Decision aids developed using this approach are termed *cognitive tools*. The cognitive prosthesis philosophy can work quite well when the decision-making situation is well defined and does not include unanticipated conditions; however, the prosthesis approach does not have the flexibility to accommodate unexpected conditions.

Traditional expert systems have not been particularly successful in complex decision environments because they have often been developed using the prosthesis philosophy (Leveson, 1995; Smith et al., 1997). One reason for this lack of success and user enthusiasm is that having a computer system doing the whole task and the human playing a subordinate role by gathering information is not appealing to people (Gordon, 1988). The person has no basis for knowing whether his or her decision is any better or worse than that of the expert system. To make matters worse, there is usually no way to communicate or collaborate the way one might with a human expert. Interestingly, Alty & Coombs (1980) showed that similar types of consultations with highly controlling human advisers were also judged unsatisfactory by "users." Finally, the cognitive prostheses approach can fail when novel problems arise or even when simple data entry mistakes are made (Roth et al., 1987). In other words, the prosthesis approach results in a *brittle* human–computer decision-making system that is inflexible in the face of unforeseen circumstances. For these reasons, the cognitive prosthesis approach is most appropriate for routine situations where decision consistency is more important than the most appropriate response to unusual situations. Decision-support systems that must accommodate unusual circumstances should adapt a cognitive tool perspective that complements rather than replaces human decision making.

Decision Matrices and Trees. One widely used approach has been designed to support the traditional "decision-analysis" cognitive process of weighing alternative actions (see top of Fig. 7.3). This method is popular with engineers and business managers and uses a decision table or decision matrix. It supports the normative multiattribute utility theory described at the start of this chapter and in Chapter 3. Decision tables are used to list the possible outcomes, probabilities, and values of the action alternatives. The decision maker enters estimated probabilities and values into the table. Computers are programmed to calculate and display the utilities for each possible choice (Edwards, 1987; White, 1990). Use of a decision table is helpful because it reduces the working-memory load. By deflecting this load to a computer, it encourages people to consider the decision space more broadly.

Decision trees are useful for representing decisions that involve a sequence of decisions and possible consequences (Edwards, 1987). With this method, a branching point is used to represent the decision alternatives; this is followed by branching points for possible consequences and their associated probabilities. This sequence is repeated as far as necessary for decision making, so the user can see the overall probability for each entire action-consequence sequence. An important challenge in implementing these techniques is user acceptance (Cabrera & Raju, 2001). The multiattribute approach is not how people typically make decisions, and so the approach can seem foreign. However, for those tasks where choices involve high risk and widely varying probabilities, such as types of treatment for cancer, it can be worth training users to be more comfortable with this type of aid.

Spreadsheets. Perhaps one of the most important issues in the design of decision-support systems is the development and use of spreadsheet-based systems. Spreadsheets have emerged as one of the most common decision-support tools, used in a wide range of organizations and created by an equally wide range of developers, many of whom are also the users. Spreadsheets reduce the cognitive load of decisions by performing many tedious calculations. For example, a complex budget for a company can be entered on a spreadsheet, and then managers can perform what-if calculations to evaluate potential operating scenarios or investments. These calculations make examining many outcomes as easy as using a simpler, but less accurate, heuristic that considers only a few outcomes. Because the spreadsheet greatly reduces cognitive load of what-if analysis, people are likely to naturally adopt the easier, more accurate strategy and improve decision quality (Todd & Benbasat, 2000). Unfortunately, spreadsheets are often poorly designed, misused, and contain errors, all of which can undermine decision-making performance.

The surprisingly large number of errors contained in spreadsheets is an important concern. Audits of spreadsheets developed in both laboratory and operational situations show that between 24 percent and 91 percent of spreadsheets contain errors (Panko, 1998). Large spreadsheets tend to contain more errors. One audit of spreadsheets used by businesses found 90 percent of spreadsheets with 150 or more lines containing at least one error (Freeman, 1996). These errors are not due to inherent flaws in the spreadsheet software or the computer processor, even though the Pentium processing error was highly publicized. Instead, the errors include incorrectly entered data and incomplete or inaccurate formulas caused by human error. Spreadsheet errors can induce poor decisions. As an example, one error led to an erroneous transfer of $7 million between divisions of a company (Panko, 1998).

Users' poor understanding of the prevalence of spreadsheet errors compounds this problem. In one study, users rated large spreadsheets as more accurate than small spreadsheets, even though large spreadsheets are much more likely to contain errors. They also rated well-formatted spreadsheets as more accurate than plainly formatted spreadsheets (Reithel et al., 1996). A related concern is that *what-if analyses* performed with a spreadsheet greatly increases users' confidence in their decisions but do not always increase accuracy of their decisions (Davis & Kottemann, 1994). Thus, spreadsheets may actually make some decision biases, such as the over-confidence bias, worse. Because of this, even if spreadsheets are error-free, they may still fail to improve decision-making performance.

Although a somewhat mundane form of decision support, the popularity of spreadsheets makes them an important design challenge. One solution to this challenge is to have several people inspect the formulas (Panko, 1999). Color coding of spreadsheet cells can show data sources and highlight inconsistencies in equations between adjacent cells (Chadwick et al., 2001). Locking cells to prevent inadvertent changes can prevent errors from being introduced when the spreadsheet is being used.

Simulation.   Although spreadsheets can include simple simulations for what-if analysis, more sophisticated, dedicated simulation tools can be useful. Figure 7.4 shows that mental simulation is an important part of the decision process. Since mental simulations can fail because of inaccurate mental models and demands on working memory, it is useful for computers to do the simulation of people. Dynamic simulations can help people evaluate their current working hypotheses, goals, and plans (Roth, 1994; Yoon & Hammer, 1988). These systems can show information related to alternative actions such as resource requirements, assumptions, and required configurations (Rouse & Valusek, 1993). For example, Schraagen (1997) describes a support system for decisions related to naval firefighting. Novices had difficulty predicting (or even considering) the compartments to which fires were most likely to spread. A support system that included a simulation identified compartments most likely to be affected and made recommendations regarding the actions needed to mitigate the effects.

Just as with spreadsheets, simulations do not always enhance decision quality. What-if analyses do not always improve decisions but often increase confidence in the decision. In addition, just like mental models, *computer simulations are incomplete and can be inaccurate.* Any model is a simplification of reality, and people using simulations sometimes overlook this fact. One example is the Hartford Coliseum in which engineers inappropriately relied on a computer model to test the strength of the structure. Shortly after completion, the roof collapsed because the computer model included several poor assumptions (Ferguson, 1992). In addition, these simulations must consider how it supports the adaptive decision process in Figure 7.4.

Expert Systems.   Other decision aids directly specify potential actions. One example of such a computer-based decision aid is the *expert system,* a computer program designed to capture one or more experts' knowledge and provide answers in a consulting type of role (Grabinger et al., 1992; White, 1990).

In most cases, expert systems take situational cues as input and provide either a diagnosis or suggested action as an output. As an example, a medical expert system takes symptoms as input and gives a diagnosis as the output (e.g., Shortliffe, 1976). Expert systems also help filter decisions, such as a financial expert system that identifies and authorizes loans for routine cases, enabling loan officers to focus on more complex cases (Talebzadeh et al., 1995). In another example, a manufacturing expert system speeds the make or buy evaluation and enhances its consistency (Humphreys et al., 2002). As discussed before, this type of decision aid is a cognitive prosthesis and is most effective when applied to routine and well-defined situations, such as the loan approval and manufacturing examples.

Expert systems act as a cognitive tool that provides feedback to the decision maker to improve decision making. Because people sometimes inappropriately rely on rapid, intuitive decisions rather than perform the more difficult deliberate analyses, decision aids might support human decision making by counteracting this "shortcut" or satisficing tendency—at least when it is important and there is ample time for analytical processing (e.g., life-threatening decisions). *Critiquing,* in which the computer presents alternate interpretations, hypotheses, or choices is an extremely effective way to improve decision making (Guerlain et al., 1999; Sniezek et al., 2002). A specific example is a decision-support system for blood typing (Guerlain et al., 1999). Rather than using the expert system as a cognitive prosthesis and identifying blood types, the critiquing approach suggests alternate hypotheses regarding possible interpretations of the cues. The critiquing approach is an example of how expert systems can be used as cognitive tools and help people deal with unanticipated situations. Expert systems are closely related to the issue of automation, a topic covered in detail in Chapter 16.

Displays. Whereas expert systems typically do a lot of "cognitive work" in processing the environmental cues, to provide the decision maker with advice, there are many other forms of decision aids that simply address the **display representation** of those cues. As a consequence, they reduce the cognitive load of information seeking and integration. As discussed in Chapter 4, and again in Chapter 8, *alerts* serve to aid the decision as to whether a variable deserves greater attention (Woods, 1995). As we will discuss in Chapter 8, and again in Chapter 16, *configural displays* can arrange the raw data or cues for a decision in a way that these can be more effectively integrated for a diagnosis, an approach that appears to be particularly valuable when the operator is problem solving at the knowledge based level of Figure 7.4 (Vicente, 2002).

Summary of decision support systems. We have reviewed a variety of decision support tools. Some are more "aggressive" in terms of computer automation and replacement of cognitive activity than others. Which tools will be used? It is apparent that, to some extent, the decisions of users to rely or not upon a particular tool depends upon the metacognitive choice, weighing the anticipated benefit versus the cost (effort and time) of tool use. This cost is directly related to the complexity of the tool, and inversely related to the quality of the interface and of the instructions. Thus it is well established that potentially effective aids will not be used if their perceived complexity is too high (Cook & Woods, 1996; Kirlik, 1993).

Training

Training can address decision making at each of the three levels of control shown in Figure 7.3. First, one method for improving analytical decision making has been to train people to overcome the heuristics/biases described earlier. Some of these efforts focused on teaching the analytical, normative utility methods for decision making (Zakay & Wooler, 1984). Although people can learn the methods, the training efforts were largely unsuccessful simply because people found the methods cumbersome and not worth the cognitive effort. Other training efforts have focused on counteracting specific types of bias, such as the confirmation bias (Tolcott et al., 1989) and overconfidence (Su & Lin, 1998). This type of training has sometimes reduced decision biases, but many studies show little to no effect (Means et al., 1993). A more effective approach might be to allow the natural use of varying strategies, but to teach people when to use them and the shortcomings of each.

As another approach, Cohen, Freeman, and Thompson (1997) suggest training people to do a better job at metacognition, teaching people how to (1) consider appropriate and adequate cues to develop situation awareness, (2) check situation assessments or explanations for completeness and consistency with cues, (3) analyze data that conflict with the situation assessment, and (4) recognize when too much conflict exists between the explanation or assessment and the cues. Training in metacognition also needs to consider when it is appropriate to rely on the automation and when it is not. Automation bias is the tendency to rely on the decision aid too much and can undermine decision quality. Training can reduce automation bias and improve decision making (Skitka et al., 2000).

Analytical decision making can also benefit from training skills such as development of mental models and management of uncertainty and time pressure (Satish & Streufert, 2002). In general, these skills should be taught in the decision-making context. People are better at learning to problem solve or make decisions in a particular area rather than simply learning to do it in general (Lipshitz et al., 2001). Unless the training is carefully structured to present concepts in relation to the particular situation, people fail to connect theoretical knowledge with practical knowledge of the situation (Wagemann, 1998). It is said that their knowledge is "inert." For example, one could teach people a large store of knowledge to use for decision making, but much of it might still remain inert and un-retrieved in the actual decision context (Woods & Roth, 1988).

At the intuitive rule-based level, operators can be provided with training to enhance their perceptual and pattern-recognition skills. Flin and colleagues (1996) and Bass (1998) suggest focusing on situation assessment, where trainees learn to recognize critical situational cues and to improve their ability to maintain their awareness of the situation. This can be achieved by having people either explicitly memorize the cue-action rules or practice a broad variety of trials to implicitly acquire the rules (Lipshitz et al., 2001). For example, Kirlik and colleagues (1996) enhanced perceptual learning and pattern recognition by either (a) having trainees memorize rules or (b) alternating trainee-practice scenarios with modeling scenarios in which the critical situational cues and correct actions were highlighted. Both of these training methods were effective. The broad selection of examples help avoid biases associated with the representativeness heuristic.

To support better processing at the automatic level, training should focus on the relevant cues in *raw data form*. Training skill-based processing takes hundreds of repetition for the associations to become strong enough for automatic processing or automaticity (e.g., Schneider, 1985; see Chapters 6 and 18). In addition, this approach works only for situations where a cue set *consistently* maps onto a particular action. For both rule-based and skill-based training, simulation is often a better medium for extensive practice because it can allow more varied scenarios, and often in less time, than the real-life context (Salas et al., 1998; Salas & Burke, 2002). Finally, for any of the training approaches described, the decision maker should receive feedback, preferably for each cognitive step in addition to feedback of the outcome of the decision as a whole (Gordon, 1994). Additional suggestions for training decision making in complex environments can be found in Means et al. (1993) and Chapter 18. Also, we should realize that training can only do so much and that the task redesign and decision-support systems should also be considered.

CONCLUSION

We discussed decision making and the factors that make it more and less effective. Normative mathematical models of utility theory describe how people should compare alternatives and make the "best" decision. However, limited cognitive resources, time pressure, and unpredictable changes often make this approach unworkable, and people use simplifying heuristics, which make decisions easier but also lead to systematic biases. In real-world situations people often have years of experience that enables them to refine their decision rules and avoid many biases. Real-world decision makers also adapt their decision making by moving from skill- and rule-based decisions to knowledge-based decisions according to the degree of risk, time pressure, and experience. This adaptive process must be considered when improving decision making through task redesign, decision-support systems, or training. The concepts in this chapter have important implications for safety and human error, discussed in Chapter 14. In many ways the decision-support systems described in this chapter can be considered as displays or automation—Chapter 16 addresses automation, and we turn to displays in the next chapter.

Chapter 8

Displays

The operator of an energy-generating plant is peacefully monitoring its operation when suddenly an alarm sounds to indicate that a failure has occurred. Looking up at the top panel of display warning indicators, he sees several warning tiles flashing, some in red, some in amber. Making little sense out of this "Christmas tree" pattern, he looks at the jumbled array of steam gauges and strip charts that present the continuously changing plant variables. Some of the indicators appear to be out of range, but present no coherent pattern, and it is not easy to see which ones are associated with the warning tiles, arrayed in the separate display region above. He turns to the operating manual, which contains a well-laid-out flow diagram of the plant on the early pages. However, he must search for a page at the back to find information on the emergency warning indicators and locate still a different page describing the procedures to follow. Scanning rapidly between these five disconnected sources of information in an effort to understand what is happening within the plant, he finally despairs and shuts down the plant entirely, causing a large loss in profit for the company.

Our unfortunate operator could easily sense the changes in display indicators and read the individual text and diagrams in the manual. He could perceive individual elements. But his ability to perceive the overall meaning of the information was hindered by the poor integration of the displays. In Chapters 4 and 5 we described how the various sensory systems (primarily the eyes and ears) process the raw sensory information and use this information as the bottom-up basis of *perception*, that is, an interpretation of the *meaning* of that information, with the assistance of expectancies and knowledge driving top-down processing. In Chapters 6 and 7 we described the manner in which perceived information is processed further and stored temporarily in working memory, or more permanently in long-term memory, and used for diagnosis and decision making. This chapter focuses on *displays,* which are typically human-made artifacts designed to support the perception of relevant system variables and facilitate the further processing of that information (Fig. 8.1). A speedometer in a car; a warning tone in an aircraft, a message on the phone-based menu system, an instruction panel on an automatic teller, a steam gauge in an industrial plant, and fine print on an application form are all examples of displays, in various modalities, conveying various forms of information used in various tasks.

The concept of the display is often closely linked with that of the *graphical user interface* (GUI), although the former often includes text, while the GUI typically describes graphics and often includes the controls and responses used to manipulate the display, as will be discussed in Chapter 9.

The nature of displays is represented schematically in Figure 8.1: The display acts as a medium between some aspects of the actual information in a system (or action requested of the operator) and the operator's perception and awareness of what the system is doing, what needs to be done, and how the system functions (the mental model). We first describe 13 key human factors principles in the design of displays. Then we describe different categories of tasks for which displays are intended, illustrating various applications of the 13 principles.

WAYS OF CLASSIFYING DISPLAYS

It is possible to classify displays along at least three different dimensions: their physical properties, the tasks they are designed to support, and the properties of the human user that dictate the best mapping between display and task. First,

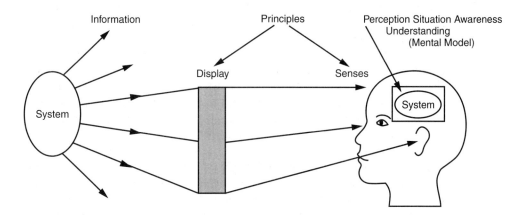

FIGURE 8.1

Key components in display design. A system generates information, some of which must be processed by the operator to perform a task. That necessary information (but *only* that information) is presented on a display and formatted according to principles in such a way that it will support perception, situation awareness, and understanding. Often, this understanding is facilitated by an accurate mental model of the displayed process.

there are differences in the **physical implementation** of the display device. One may think of these as the **physical tools** that the designer has to work with in creating a display. For example, a display may use color or monochrome, visual or auditory modality; a 3-D display may use stereo; the relative location of display elements may be changed and so on.

Such tools are mentioned at various points in the chapter. However, before fabricating a display, the designer must ascertain the nature of the **task** the display is intended to support: Is it navigating, controlling, decision making, learning, and so forth? The chapter is organized around displays to support these various tasks, as we see how different display tools may be optimally suited for different tasks. However, defining the task is only a first step. Once the task and its goals are identified (e.g., designing a map to help a driver navigate from point A to point B) we must do a detailed *information analysis* to identify what the operator needs to know to carry out the task.

Finally, and most important, no single display tool is best suited for all tasks because of characteristics of the human user who must perform those tasks. For example, a digital display that is best for reading of the exact value of an indicator, is not good for assessing at a quick glance the approximate rate of change and value of the indicator. As Figure 8.1 shows, the key mediating factor that determines the best mapping between the physical form of the display and the task requirements is a series of *principles* of human perception and information processing. These principles are grounded in the strengths and weaknesses of human perception, cognition, and performance (Wickens & Hollands, 2000; Boff et al., 1986, see also Chapters 4, 5, and 6), and it is through the careful application of these principles to the output of the information analysis that the best displays emerge.

THIRTEEN PRINCIPLES OF DISPLAY DESIGN

One of the basic tenets of human factors is that lists of longer than five or six items are not easily retained unless they are given with some organizational structure. To help retention of the otherwise daunting list of 13 principles of display design, we associate them into four distinct categories: (1) those that directly reflect *perceptual* operations, (2) those that can be traced to the concept of the *mental model*, (3) those that relate to *human attention,* and (4) those that relate to *human memory.* Some of these principles have been introduced in previous chapters (4, 5, and 6) and others will be discussed more fully later in this chapter.

Perceptual Principles

1. *Make displays legible (or audible).* This guideline is not new. It integrates nearly all of the information discussed in Chapters 4 and 5, relating to issues such as contrast, visual angle, illumination, noise, masking, and so forth. Legibility is so critical to the design of good displays that it is essential to restate it here. Legible displays are necessary, although not sufficient, for creating usable displays. The same is true for audible displays. Once displays are legible, additional perceptual principles should be applied. The following four perceptual principles are illustrated in Figure 8.2.

    2. *Avoid absolute judgment limits.* As we noted in Chapters 4 and 5 when discussing alarm sounds, we do not require the operator to judge the level of a represented variable on the basis of a single sensory variable, like color, size, or loudness, which contains more than five to seven possible levels. To require greater precision as in a color-coded map with nine hues is to invite errors of judgment.

    3. *Top-down processing.* People perceive and interpret signals in accordance with what they *expect* to perceive on the basis of their past experience. If a signal

(a) **Absolute Judgment:**

    "If the light is amber, proceed with caution."

Amber light is one of six possible hues

(b) **Top-Down Processing:**
    A Checklist

> **A** should be on
>
> **B** should be on
>
> **C** should be on
>
> **D** should be off

(c) **Redundancy Gain:** The Traffic Light

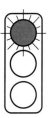

Position and hue are redundant

(d) **Similarity:** Confusion

Figure X············
·························

·························
Altitude ··············
·························

Figure Y ·············
·························

·························
Attitude ··············
·························

FIGURE 8.2

Four *perceptual* principles of display design: (a) absolute judgment; (b) top-down processing (a tendency to perceive as "D should be on"); (c) redundancy gain; and (d) similarity → confusion.

is presented that is contrary to expectations, like the warning or alarm for an unlikely event, then *more physical evidence* of that signal must be presented to guarantee that it is interpreted correctly. Sometimes expectancies are based on long-term memory. However, in the example shown in Figure 8.2b, these expectations are based on the **immediate context** of encountering a series of "on" messages, inviting the final line to also be perceived as on.

4. *Redundancy gain.* When the viewing or listening conditions are degraded, a message is more likely to be interpreted correctly when the same message is expressed more than once. This is particularly true if the same message is presented in *alternative* physical forms (e.g., tone and voice, voice and print, print and pictures, color and shape); that is, redundancy is not simply the same as repetition. When alternative physical forms are used, there is a greater chance that the factors that might degrade one form (e.g., noise degrading an auditory message) will not degrade the other (e.g., printed text). The traffic light (Figure 8.2c) is a good example of redundancy gain.

5. *Discriminability. Similarity causes confusion: Use discriminable elements.* Similar appearing signals are likely to be confused either at the time they are perceived or after some delay if the signals must be retained in working memory before action is taken. What causes two signals to be similar is the **ratio** of similar features to different features (Tversky, 1977). Thus, AJB648 is more similar to AJB658 than is 48 similar to 58, even though in both cases only a single digit is different. Where confusion could be serious, the designer should delete unnecessary similar features and highlight **dissimilar** (different) ones in order to create distinctiveness. Note, for example, the high degree of confusability of the two captions in Figure 8.2d. You may need to look very closely to see its discriminating feature ("l" versus "t" in the fourth word from the end). In Figure 4.11 we illustrated another example of the danger of similarity and confusion in visual information, leading to a major airline crash. Poor legibility (P1) also amplifies the negative effects of poor discriminability.

Mental Model Principles

When operators perceive a display, they often interpret what the display looks like and how it moves in terms of their expectations or *mental model* of the system being displayed (Figure 8.1), a concept discussed in Chapter 6 (Norman, 1988; Gentner & Stevens, 1983). The information presented to our energy system monitor in the opening story was not consistent with the mental model of the operator. Hence, it is good for the format of the display to capture aspects of a user's **correct** mental model, based on the user's experience of the system whose information is being displayed. Principles 6 and 7 illustrate how this can be achieved.

6. *Principle of pictorial realism* (Roscoe, 1968). A display should *look like* (i.e., be a picture of) the variable that it represents. Thus, if we think of temperature as having a high and low value, a thermometer should be oriented vertically. If the display contains multiple elements, these elements can sometimes be *configured* in a manner that looks like how they are configured in the environment that is represented (or how the operator conceptualizes that environment).

7. *Principle of the moving part* (Roscoe, 1968). The moving element(s) of any display of dynamic information should move in a spatial pattern and direction that is compatible with the user's mental model of how the represented element actually moves in the physical system. Thus, if a pilot thinks that the aircraft moves upward when altitude is gained, the moving element on an altimeter should also move upward with increasing altitude.

Principles Based on Attention

Complex multielement displays require three components of attention to process (Parasuraman et al., 1984). As discussed in Chapter 6, *selective attention* may be necessary to choose the displayed information sources necessary for a given task. *Focused attention* allows those sources to be perceived without *distraction* from neighboring sources, and *divided attention* may allow parallel processing of two (or more) sources of information concurrently if a task requires it. All four of the attentional principles described next characterize ways of capitalizing on attentional strengths or minimizing their weaknesses in designing displays.

8. *Minimizing information access cost.* There is typically a cost in time or effort to "move" selective attention from one display location to another to access information. The operator in the opening story wasted valuable time going from one page to the next in the book and visually scanning from there to the instrument panel. The information access cost may also include the time required to proceed through a computer menu to find the correct "page." Thus, good designs are those that minimize the net cost by keeping frequently accessed sources in a location in which the cost of traveling between them is small. This principle was not supported in the maintenance manual in the episode at the beginning of the chapter. We discuss it again in the context of workplace layout in Chapter 10. One direct implication of minimizing access cost is to keep displays small so that little scanning is required to access all information. Such a guideline should be employed carefully however, because very small size can degrade legibility (Kroft & Wickens, 2003) (P1).

9. *Proximity compatibility principle* (Wickens & Carswell, 1995). Sometimes, two or more sources of information are related to the same task and must be **mentally integrated** to complete the task (e.g., a graph line must be related to its legend, or the plant layout must be related to the warning indicator meanings in our opening story); that is, **divided attention** between the two information sources for the one task is necessary. These information sources are thereby defined to have close mental proximity. As described in principle 8, good display design should provide the two sources with close display proximity so that their information access cost will be low (Wickens & Carswell, 1995). However, there are other ways of obtaining close display proximity between information sources besides nearness in space. For example, close proximity can also be obtained by displaying them in a common color, by linking them together with lines or by configuring them in a pattern, as discussed in principle 6. Four of these techniques are shown in Figure 8.3a.

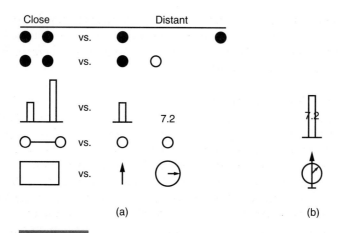

FIGURE 8.3

The proximity compatibility principle. If mental integration is required, close spatial proximity is good. If focused attention is required, close spatial proximity may be harmful. (a) Five examples of close display proximity on the left that will be helpful for tasks requiring integration of information in the two sources shown. These are contrasted with examples of separated, or distant, display pairs on the right. In the five examples, separation is defined by (1) space, (2) color (or intensity), (3) format, (4) links, and (5) object configuration. (b) Two examples of close spatial proximity (overlay) that will hurt the ability to focus on one indicator and ignore the other.

However, as Figure 8.3b shows, too much close display proximity is not always good, particularly if one of the elements must be the subject of focused attention. The clutter of overlapping images makes their individual perception hard. In this case of focused attention, close proximity may be harmful, and it is better for the sources to be more separated. The "lower mental proximity" of the focused attention task is then best served by the "low display proximity" of separation. Thus, the two types of proximity, display and mental, are compatibly related: If mental proximity is high (divided attention for integration), then display proximity should also be high (close). If mental proximity is low (focused attention), the display proximity can, and sometimes should, be lower.

10. *Principle of multiple resources.* As we discussed in Chapter 6, sometimes processing a lot of information can be facilitated by dividing that information across resources—presenting visual and auditory information concurrently, for example, rather than presenting all information visually or auditorily.

Memory Principles

Human memory is vulnerable, particularly working memory because of its limited capacity: We can keep only a small number of "mental balls" in the air at one time, and so, for example, we may easily forget a phone number before we have had a chance to dial it or write it down. Our operator in the opening story had a hard time remembering information on one page of the manual while he was reading the other. Our long-term memory is vulnerable because we forget certain things or sometimes because we remember other things *too well* and persist in doing them when we should not. The final three principles address different aspects of these memory processes.

   11. *Replace memory with visual information: knowledge in the world.* The importance of presenting knowledge in the world of what to do (Norman, 1988) is the most general memory principle, echoing guidelines presented in Chapter 6. People ought not be required to retain important information solely in working memory or retrieve it from long-term memory. There are several ways that this is manifest: the visual echo of a phone number (rather than reliance on the fallible phonetic loop), the checklist (rather than reliance on prospective memory), and the simultaneous rather than sequential display of information to be compared. Of course, sometimes too much knowledge in the world can lead to clutter problems, and systems designed to rely on knowledge in the head are not necessarily bad. For example, in using computer systems, experts might like to be able to retrieve information by direct commands (knowledge in the head) rather than stepping through a menu (knowledge in the world). Good design must balance the two kinds of knowledge. One specific example of replacing memory with perception becomes a principle in its own right, which defines the importance of predictive aiding.

   12. *Principle of predictive aiding.* Humans are not very good at predicting the future. In large part this limitation results because prediction is a difficult cognitive task, depending heavily on working memory. We need to think about current conditions, possible future conditions, and then "run" the mental model by which the former may generate the latter. When our mental resources are consumed with other tasks, prediction falls apart and we become *reactive,* responding to what has already happened, rather than *proactive,* responding in anticipation of the future. Since proactive behavior is usually more effective than reactive, it stands to reason that displays that can explicitly predict what will (or is likely to) happen are generally quite effective in supporting human performance. A predictive display removes a resource-demanding cognitive task and replaces it with a simpler perceptual one. Figure 8.4 shows some examples of effective predictor displays.

   13. *Principle of consistency.* When our long term-memory works too well, it may continue to trigger actions that are no longer appropriate, and this a pretty instinctive and automatic human tendency. Old habits die hard. Because there is no way to avoid this, good designs should try to accept it and design displays in a manner that is consistent with other displays that the user may be perceiving concurrently (e.g., a user alternating between two computer systems) or may have perceived in the recent past. Hence, the old habits from those other displays will transfer positively to support processing of the new displays. Thus, for example, color coding should be consistent across a set of displays so that red always means the same thing. As another example, a set of different display panels should be consistently organized, thus reducing information access cost (P8) each time a new set is encountered.

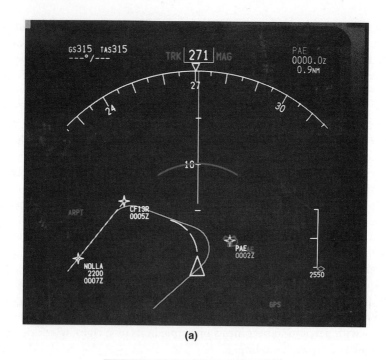

(a)

LEFT TURN
1 MILE
AHEAD

(b)

FIGURE 8.4

Two predictive displays. (a) an aircraft flight predictor, shown by the curved, dashed line extending from the triangular aircraft symbol at the bottom. This predicts the turn and future heading of the aircraft. (*Source:* Courtesy of the Boeing Corporation.) (b) a highway sign.

Conclusion

In concluding our discussion of principles, it should be immediately apparent that principles sometimes conflict or "collide." Making all displays consistent, for example, may sometimes cause certain displays to be less compatible than others, just as making all displays optimally compatible may make them inconsistent. Putting too much knowledge in the world or incorporating too much redundancy can create very cluttered displays, thereby making focused attention more difficult. Minimizing information access effort by creating very small displays may reduce legibility. Alas, there is no easy solution to say what principles are more important than others when two or more principles collide. But clever and creative design can sometimes enable certain principles to be more effectively served without violating others. We now turn to a discussion of various categories of displays, illustrating the manner in which certain principles have been applied to achieve better human factors. As we encounter each principle in application, we place a reminder of the principle number in parentheses, for example, (P10) refers to the principle of multiple resources.

ALERTING DISPLAYS

We discussed alerting displays to some extent in Chapter 5 in the context of auditory warnings, and shall do so again when we discuss both safety and automation. If it is critical to *alert* the operator to a particular condition, then the *omnidirectional* auditory channel is best. However, there may well be several different levels of seriousness of the condition to be alerted, and not all of these need or should be announced auditorily. For example, if my car passes a mileage level in which a particular service is needed, I do not need the time-critical and intrusive auditory alarm to tell me that.

Conventionally, system designers have classified three levels of alerts—warnings, cautions, and advisories—which can be defined in terms of the severity of consequences of failing to heed their indication. Warnings, the most critical category, should be signaled by salient auditory alerts; cautions may be signaled by auditory alerts that are less salient (e.g., softer voice signals); advisories need not be auditory at all, but can be purely visual. Both warnings and cautions can clearly be augmented by redundant visual signals as well (P4). When using redundant vision for alerts, flashing lights are effective because the onsets that capture attention occur repeatedly. Each onset is itself a redundant signal. In order to avoid possible confusion of alerting severity, the aviation community has also established explicit guidelines for *color coding,* such that warning information is always red; caution information is yellow or amber; advisory information can be other colors (e.g., white), clearly discriminable (P5) from red and amber.

Note that the concept of defining three levels of condition severity is consistent with the guidelines for "likelihood alarms" discussed in Chapter 5 (Sorkin et al., 1988), in which different degrees of danger or risk are explicitly signaled to the user.

LABELS

Labels may also be thought of as displays, although they are generally static and unchanging features for the user. Their purpose is to unambiguously signal the identity or function of an entity, such as a control, display, piece of equipment, entry on a form, or other system component; that is, they present knowledge in the world (P11) of what something is. Labels are usually presented as print but may sometimes take the form of icons (Fig. 8.5). The four-key design criteria for

FIGURE 8.5
Some typical icons.

labels, whether presented in words or pictures, are visibility, discriminability, meaningfulness, and location.

1.   *Visibility/legibility.* This criterion (P1) relates directly back to issues of contrast sensitivity, discussed in Chapter 4. Stroke width of lines (in text or icons) and contrast from background must be sufficient so that the shapes can be discerned under the poorest expected viewing conditions. This entails some concern for the shape of icons, an aspect delivered at low spatial frequencies.

2.   *Discriminability (P5).* This criterion dictates that any feature that is necessary to discriminate a given label from an alternative *that may be inferred by the user to exist in that context* is clearly and prominently highlighted. We noted that confusability increases with the ratio of shared to distinct features between potential labels. So, two figure legends that show a large amount of identical (and perhaps redundant) text are more confusable than two in which this redundancy is deleted (Fig. 8.2d).

As described in Chapter 6, a special "asymmetrical" case of confusion is the tendency to confuse negative labels ("no exit") with positive ones (exit). Unless the negative "no," "do not," "don't," and so on is clearly and saliently displayed, it is very easy for people to miss it and assume the positive version, particularly when viewing the label (or hearing the instructions) under degraded sensory conditions.

3.   *Meaningfulness.* Even if a word or icon is legible and not confusable, this is no guarantee that it triggers the appropriate meaning in the mind of the viewer when it is perceived. What, for example, do all the icons in Figure 8.5 mean? Or, for the English viewer of the sign along the German Autobahn, what does the word *anfang* mean? Unfortunately, too often icons, words, or acronyms that are highly meaningful in the mind of the designer, who has certain expectations of the mindset that the user *should* have when the label is encountered, are next to meaningless in the mind of some proportion of the actual users. Because this unfortunate situation is far more likely to occur with the use of abbreviations and icons than with words, we argue that labels based *only* on icons or abbreviations should be avoided where possible (Norman, 1981). Icons may well be advantageous where the word labels may be read by those who are not fluent in the language (e.g., international highway symbols) and sometimes under degraded viewing conditions; thus, the *redundancy gain* (P4) that such icons provide is usually of value. But the use of icons *alone* appears to carry an unnecessary risk when comprehension of the label is important. The same can be said

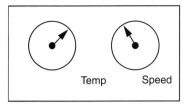

FIGURE 8.6
The importance of unambiguous association between displays and labels.

for abbreviations. When space is small—as in the label of a key that is to be pressed, effort should be made to perceptually "link" the key to a verbal label that may be presented next to the key.

4. *Location.* One final obvious but sometimes overlooked feature of labels: They should be physically close to and unambiguously associated with the entity that they label, thereby adhering to the proximity compatibility principle (P9). Note how the placement of labels in Figure 8.6 violates this. While the display indicating temperature is closest to the temperature label, the converse cannot be said. That is, the temperature label is just as close to the speed display as it is to the temperature display. If our discussion concerned the location of buttons, not displays and labels, then the issue would be one of *stimulus-response compatibility,* discussed in Chapter 9.

As described in Chapter 5, computer designers are applying the concept of icons to sound in the generation of *earcons,* synthetic sounds that have a direct, meaningful association with the thing they represent. In choosing between icons and earcons, it is important for the designer to remember that earcons (sound) are most compatible for representing **events** that play out over time (e.g., informing that a computer command has been accomplished), whereas icons are better for representing the identity of **locations** that exist in space.

MONITORING

Displays for monitoring are those that support the viewing of potentially changing quantities, usually represented on some analog or ordered value scale, such as a channel frequency, speed, temperature, noise level, or changing job status. A variety of tasks may need to be performed on the basis of such displays. A monitored display may need to be *set,* as when an appropriate frequency is dialed in to a radio channel. It may simply need to be *watched* until it reaches a value at which some discrete action is taken, or it may need to be *tracked,* in which case another variable must be manipulated to follow the changing value of the monitored variable. (Tracking is discussed in considerably more detail in Chapter 9.) Whatever the action to be taken on the basis of the monitored variable, discrete or continuous, immediate or delayed, four important guidelines can be used to optimize the monitoring display.

1.   *Legibility.* Display legibility (P1) is of course the familiar criterion we re-visited in the previous section, and it relates to the issues of contrast sensitivity discussed in Chapter 4. If monitoring displays are digital, the issues of print and character resolution must be addressed. If the displays are analog dials or point-ers, then the visual angle and contrast of the pointer and the legibility of the scale against which the pointer moves become critical. A series of guidelines may be found in Sanders and McCormick (1993) and Helander (1987) to assure such legibility. But designers must be aware of the possible degraded viewing condi-tions (e.g., low illumination) under which such scales may need to be read, and they must design to accommodate such conditions.

2.   *Analog versus digital.* Most variables to be monitored are continuously changing quantities. Furthermore, users often form a mental model of the chang-ing quantity. Hence, adhering to the principle of pictorial realism (P6, Roscoe, 1968) would suggest the advantage of an analog (rather than digital) representa-tion of the continuously changing quantity. The data appear to support this guideline (Boff & Lincoln, 1988). In comparison to digital displays (Fig. 8.7a), analog displays like the moving pointer in Figure 8.7b can be more easily read at a short glance; the value of an analog display can be more easily estimated when the display is changing, and it is also easier to estimate the rate and direction of that change. At the same time, digital displays *do* have an advantage if very precise "check reading" or setting of the exact value is required. But unless these are the *only* tasks required of a monitoring display, and the value changes slowly, then if a digital display is used, it should be redundantly provided with its analog counter-part (P4), like the altitude display shown in Figure 8.7c.

3.   *Analog form and direction.* If an analog format is chosen for display, then the principle of pictorial realism (P6; Roscoe, 1968) would state that the orienta-tion of the display scale should be in a form and direction congruent with the op-erator's mental model of the displayed quantity. Cyclical or circular variables (like compass direction or a 24-hour clock) share an appropriate circular form for a round dial or "steam gauge" display, whereas linear quantities with clearly defined

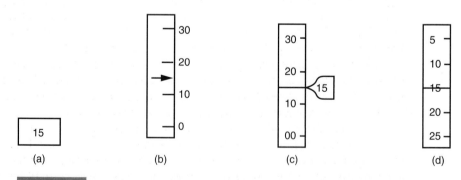

FIGURE 8.7
(a) digital display; (b) moving pointer analog display; (c) moving scale analog display with redundant digital presentation. Both (a) and (b) adhere to the principle of pictorial realism. (d) Inverted moving scale display adheres to principle of the moving part.

high and low points should ideally be reflected by linear scales. These scales should be vertically arrayed so that high is up and low is down. This orientation feature is easy to realize when employing the fixed-scale moving pointer displays (Figure 8.7b) or the moving scale fixed-pointer display shown in Figure 8.7c.

However, many displays are fairly dynamic, showing visible movement while the operator is watching or setting them. The principle of the moving part (P7) suggests that displays should move in a direction consistent with the user's mental model: An increase in speed or any other quantity should be signaled by a movement *upward* on the moving element of the display (rightward and clockwise are also acceptable, but less powerful movement stereotypes for increase). While the moving pointer display in Figure 8.7b clearly adheres to this stereotype, the moving scale display in Figure 8.7c does not. Upward display movement will signal a decrease in the quantity. The moving scale version in Figure 8.7d, with the scale inverted, can restore the principle of the moving part, but only at the expense of a violation of the principle of pictorial realism (P6) because the scale is now inverted. *Both* moving scale displays suffer from a difficulty of reading the scale value while the quantity is changing rapidly.

Despite its advantages of adhering to the principles of both pictorial realism and the moving part, there is one cost with a linear moving pointer display (Figure 8.7b). It cannot present a wide range of scale values within a small range of physical space. If the range of scale over which the variable travels is large and the required reading precision is also high (a pilot's altimeter, for example), this can present a problem. One answer is to revert to the moving scale display, which can present high numbers at the top. If the variable does not change rapidly (i.e., there is little motion), then the principle of the moving part has less relevance, and so its violation imposes less of a penalty. A second option is to use circular moving pointer displays that are more economical of space. While these options may destroy some adherence to the principle of pictorial realism (if displaying linear quantities), they still possess a reasonable stereotype of increase clockwise. A third possibility is to employ a frequency-separated concept of a hybrid scale in which high-frequency changes of the displayed variable drive a moving pointer against a stable scale, while sustained low-frequency changes can gradually shift the scale quantities to the new (and appropriate) range of values as needed (maintaining high numbers at the top) (Roscoe, 1968; Wickens & Hollands, 2000).

Clearly, as in any design solution, there is no "magic layout" that will be cost-free for all circumstances. As always, task analysis is important. The analysis should consider the rate of change of the variable, its needed level of precision, and its range of possible values before a display format is chosen.

One final factor influencing the choice of display concerns the nature of *control* that may be required to set or to track the displayed variable. Fortunately for designers, many of the same laws of display expectations and mental models apply to control; that is, just as the user expects (P3) that an upward (or clockwise) movement of the display signals an increasing quantity, so the user also expects that an upward (or clockwise) movement of the control will be required to *increase* the displayed quantity. We revisit this issue in more detail in Chapter 9 when we address issues of display-control compatibility.

4. *Prediction and sluggishness.* Many monitored variables in high-inertia systems, like ships or chemical processes, are sluggish in that they change relatively slowly. But as a consequence of the dynamic properties of the system that they represent, the slow change means that their future state can be known with some degree of certainty. Such is the case of the supertanker, for example: Where the tanker is now in the channel and how it is moving will quite accurately predict where it will be several minutes into the future. Another characteristic of such systems is that efforts to control them which are executed now will also not have an influence on their state until much later. Thus, the shift in the supertanker's rudder will not substantially change the ship's course until minutes later, and the adjustment of the heat delivered to a chemical process will not change the process temperature until much later (Chapter 16). Hence, control should be based on the operator's prediction of *future* state, not present conditions. But as we discussed in Chapter 6, prediction is not something we do very well, particularly under stress; hence, good predictor displays (P11) can be a great aid to human monitoring and control performance (Fig. 8.4).

Predictive displays of physical systems are typically driven by a computer model of the dynamics of the system under control and by knowledge of the current and future inputs (forces) acting on the system. Because, like the crystal ball of the fortune-teller, these displays really are driven by automation making inferences about the future, they may not always be correct and are less likely to be correct the further into the future the prediction. Hence, the designer should be wary of predicting forward further than is reasonable and might consider depicting limits on the degree of certainty of the predicted variable. For example, a display could predict the most likely state and the 90 percent confidence interval around possible states that could occur a certain time into the future. This confidence interval will grow as that time—the *span of prediction*—is made longer.

MULTIPLE DISPLAYS

Many real-world systems are complex. The typical nuclear reactor may have at least 35 variables that are considered critical for its operation, while the aircraft is assumed to have at least seven that are important for monitoring in even the most routine operations. Hence, an important issue in designing multiple displays is to decide where they go, that is, what should be the *layout* of the multiple displays (Wickens et al., 1997). In the following section we discuss several guidelines for display layout, and while these are introduced in the context of monitoring displays, you should realize that the guidelines apply to nearly any type of display, such as the layout of windows on a Web page. We use the term "guidelines" to distinguish them from the 13 principles, although many of the guidelines we describe are derived from the principles. We then address similar issues related to head-up displays and configural displays.

Display Layout

In many work environments, the designer may be able to define a *primary visual area* (PVA) (see Chapter 10). For the seated user, this maybe the region of forward view as the head and eyes look straight forward. For the vehicle operator, it may be the direction of view of the highway (or runway in an aircraft approach). Defining this region (or point in space) of the PVA is critical because the first of six guidelines of display layout, *frequency of use,* dictates that frequently used displays should be adjacent to the PVA. This makes sense because their frequent access dictates a need to "minimize the travel time" between them and the PVA (P8). Note that sometimes a very frequently used display can itself define the PVA. With the conventional aircraft display suite shown in Figure 8.8, this

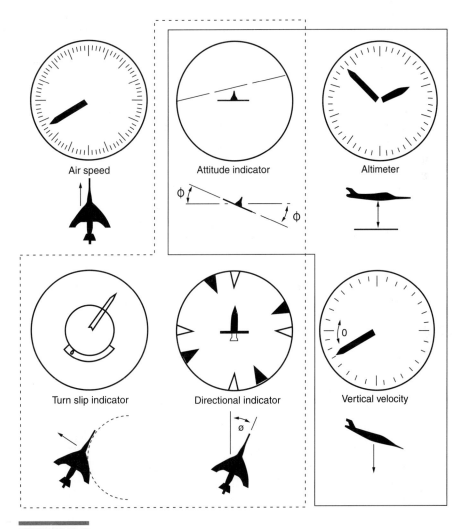

FIGURE 8.8

Conventional aircraft instrument panel. The attitude directional indicator is in the top center. The outlines surround displays that are related in the control of the vertical (solid outline) and lateral (dashed box) position of the aircraft. Note that each outline surrounds physically proximate displays. The three instruments across the top row and that in the lower center form a T shape, which the FAA mandates as a consistent layout for the presentation of this information across all cockpit designs.

principle is satisfied by positioning the most frequently used instrument, the attitude indicator, at the top and center, closest to the view out the windshield on which the pilot must fixate to land the aircraft and check for other traffic.

Closely related to frequency of use is *importance of use,* which dictates that important information, even if it may not be frequently used, be displayed so that attention will be captured when it is presented. While displaying such information within the PVA often accomplishes this, other techniques, such as auditory alerts coupled with guidance of where to look to access the information, can accomplish the same goal.

Display relatedness or *sequence of use* dictates that related displays and those pairs that are often used in sequence should be close together. (Indeed, these two features are often correlated. Displays are often consulted sequentially *because* they are related, like the commanded setting and actual setting of an indicator.) This principle captures the key feature of the proximity compatibility principle (P9) (Wickens & Carswell, 1995). We saw the manner in which it was violated for the operator in our opening story. As a positive example, in Figure 8.8, the vertical velocity indicator and the altimeter, in close spatial proximity on the right side, are also related to each other, since both present information about the vertical behavior of the aircraft. The figure caption also describes other examples of related information in the instrument panel.

Consistency is related to both memory and attention. If displays are always consistently laid out with the same item positioned in the same spatial location, then our memory of where things are serves us well, and memory can easily and automatically guide selective attention to find the items we need (P8, P13). Stated in other terms, top-down processing can guide the search for information in the display. Thus, for example, the Federal Aviation Administration provides strong guidelines that even as new technology can revolutionize the design of flight instruments, the basic form of the four most important instruments in the panel in Figure 8.8—those forming a T—should always be preserved (FAA, 1987).

Unfortunately, there are many instances in which the guideline of consistency conflicts with those of frequency of use and relatedness. These instances define *phase-related* operations, when the variables that are frequently used (or related and used in sequence) during one phase of operation may be very different from those during another phase. In nuclear power-plant monitoring, the information that is important in startup and shutdown is different from what is important during routine operations. In flying, the information needed during cruise is quite different from that needed during landing, and in many systems, information needed during emergency is very different from that needed during routine operations. Under such circumstances, a totally consistent layout for all phases may be unsatisfactory, and current, "soft" computer-driven displays allow flexible formats to be created in a phase-dependent layout. However, if such flexibility is imposed, then three key design guidelines must be kept in mind: (1) It should be made very clear to the user by *salient visible* signals which configuration is in effect; (2) where possible, some consistency (P13) across all formats should be sought; (3) the designer should resist the temptation to create excessive flexibility (Andre & Wickens, 1992). Remember that as long as a display design *is* consistent, the user's memory will help guide attention to find the needed information rapidly, even if that information may not be in the very best location for a particular phase.

Organizational grouping is a guideline that can be used to contrast the display array in Figure 8.9a with that in Figure 8.9b. An organized, "clustered" display, such as that seen in Figure 8.9a, provides an aid that can easily guide visual attention to particular groups as needed (P8), as long as all displays within a group are functionally related and their relatedness is clearly known and identified to the user. If these guidelines are *not* followed, however, and unrelated items belong to a common spatial cluster, then such organization may actually be counterproductive (P9).

Two final guidelines of display layout are related to *stimulus-response compatibility,* which dictates that displays should be close to their associated controls, and *clutter avoidance,* which dictates that there should ideally be a minimum visual angle between all pairs of displays. We discuss stimulus-response compatibility in Chapter 9 and clutter avoidance in the following sections.

Head-Up Displays and Display Overlay

We have already seen that one important display layout guideline involves moving important information sources close to the PVA. The ultimate example of this approach is to actually *superimpose* the displayed information on top of the PVA creating what is known as the *head-up display,* or HUD (Weintraub & Ensing, 1992; Newman, 1995; Wickens et al., 2003). These are often proposed (and used) for vehicle control but may have other uses as well when the PVA can be clearly specified. For example, a HUD might be used to superimpose a computer graphics designer's palette information over the design workspace (Harrison & Vicente, 1996). Two examples of HUDs, one for aircraft and one for automobiles, are shown in Figure 8.10.

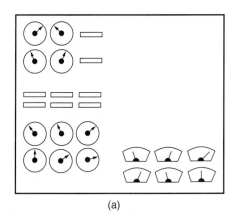
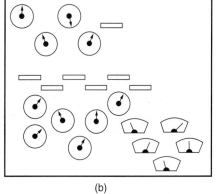

(a)   (b)

FIGURE 8.9

Differences in display organization: (a) high; (b) low. All displays within each physical grouping and thus have higher display proximity must be somehow related to each other in order for the display layout on the left to be effective (P9).

(a)

(b)

FIGURE 8.10

Head-up displays: (a) for automobile (*Source:* Kaptein, N. A. Benefits of In-car Head-up Displays. Report TNO-TM 1994 B-20. Soesterberg, TNO Human Factors Research Institute.); (b) for aircraft. (*Source:* Courtesy of *Flight Dynamics.*)

The proposed advantages of HUDs are threefold. First, assuming that the driver or pilot should spend most of the time with the eyes directed outward, then overlapping the HUD imagery should allow both the far-domain environment and the near-domain instrumentation to be monitored in parallel with little information access cost (P8). Second, particularly with aircraft HUDs, it is possible to present imagery that has a direct spatial counterpart in the far domain. Such imagery, like a schematic runway or horizon line that overlays its counterpart, seen in Figure 8.10b, is said to be *conformal.* By positioning this imagery in the HUD overlaying the far domain, divided attention between the two domains is supported (P9). Third, many HUDs are projected via *collimated imagery,* which essentially reorients the light rays from the imagery in a parallel fashion, thereby making the imagery to appear to the eyes to be at an accommodative distance of *optical infinity.* The advantage of this is that the lens of the eyeball accommodates to more distant viewing than the nearby windshield and so does not have to reaccommodate to shift between focus on instruments and on far domain viewing (see Chapter 4).

Against these advantages must be considered one very apparent cost. Moving imagery too close together (i.e., superimposed) violates the seventh guideline of display layout: creation of excessive *clutter* (P9. See Figure 8.3b). Hence, it is possible that the imagery may be difficult to read against the background of varied texture and that the imagery itself may obscure the view of critical visual events in the far domain. The issue of overlay-induced clutter is closely related to that of map overlay, discussed later in this chapter.

Evaluation of HUDs indeed suggests that the three overall benefits tend to outweigh the clutter costs. In aircraft, flight control performance is generally better when critical flight instruments are presented head-up (and particularly so if they are conformal; Wickens & Long, 1995; Fadden et al., 2001). In driving, the digital speedometer instrument is sampled for a shorter time in the head-up location (Kiefer, 1991), although in both driving and flying, speed control is not substantially better with a HUD than with a head-down display (Kiefer, 1991; Sojourner & Antins, 1990; Wickens & Long, 1995). There is also evidence that relatively expected discrete events (like the change in a digital display to be monitored) are better detected when the display is in the head-up location (Sojourner & Antin, 1990; Fadden et al., 2001; Horrey & Wickens 2003).

Nevertheless, the designer should be aware that there are potential costs from the HUD of overlapping imagery. In particular, these clutter costs have been observed in the detection of unexpected events in the far domain, such as the detection of an aircraft taxiing out onto the runway toward which the pilot is making an approach (Wickens & Long, 1995; Fischer et al., 1980; Fadden et al., 2001).

Head-Mounted Displays

A close cousin to the HUD is the head-mounted or helmet-mounted display in which a display is rigidly mounted to the head so that it can be viewed no matter which way the head and body are oriented (Melzer & Moffett, 1997). Such a display has the advantage of allowing the user to view superimposed imagery across a much wider range of the far domain than is possible with the HUD. In an aircraft or helicopter, the head-mounted displays (HMDs) can allow the pilot to retain a view of HMD flight instruments while scanning the full range of the outside world for threatening traffic or other hazards (National Research Council, 1995). For other mobile operators, the HMD can be used to minimize information access costs while keeping the hands free for other activities. For example, consider a maintenance worker, operating in an awkward environment in which the head and upper torso must be thrust into a tight space to perform a test on some equipment. Such a worker would greatly benefit by being able to consult information on how to carry out the test, displayed on an HMD, rather than needing to pull his head out of the space every time he must consult the test manual. The close proximity between the test space and the instructions thus created assists the integration of these two sources of information (P9). The use of a head-orientation sensor with conformal imagery can also present information on the HMD specifying the direction of particular locations in space relative to the momentary orientation of the head; for example, the location of targets, the direction to a particular landmark, or due north (Yeh et al., 1999).

HMDs can be either monocular (presented to a single eye), biocular (presented as a single image to both eyes), or binocular (presented as a separate image to each eye); furthermore, monocular HMDs can be either opaque (allowing only the other eye to view the far domain) or transparent (superimposing the monocular image on the far domain). Opaque binocular HMDs are part of virtual reality systems. Each version has its benefits and costs (National Research Council, 1995). The clutter costs associated with HUDs may be mitigated somewhat by using a monocular HMD, which gives one eye unrestricted view of the far domain. However, presenting different images to the two eyes can sometimes create problems of *binocular rivalry* or *binocular suppression* in which the two eyes compete to send their own image to the brain rather than fusing to send a single, integrated image (Arditi, 1986).

To a greater extent than is the case with HUDs, efforts to place conformal imagery on HMDs can be problematic because of potential delays in image updating. When conformal displays, characterizing *augmented reality,* are used to depict spatial positions in the outside world, they must be updated each time the display moves (ie., head rotates) relative to that world. Hence, conformal image updating on the HMD must be fast enough to keep up with potentially rapid head rotation. If it is not, then the image can become disorienting and lead to motion sickness (Durlach & Mavor, 1995); alternatively, it can lead users to adopt an unnatural strategy of reducing the speed and extent of their head movements (Seagull & Gopher, 1995; Yeh et al., 1999).

At present, the evidence is mixed regarding the relative advantage of presenting information head-up on an HMD versus head-down on a handheld display (Yeh et al., 1999; Yeh et al., 2003). Often, legibility issues (P1) may penalize the small-sized image of the handheld display, and if head tracking is available, then the conformal imagery that can be presented on the HMD can be very valuable for integrating near- and far-domain information (P9). Yet if such conformal imagery or augmented reality cannot be created, the HMD value diminishes, and diminishes still further if small targets or high detail visual information must be seen through a cluttered HMD in the world beyond (Yeh et al., 2003).

Configural Displays

Sometimes, multiple displays of single variables can be arrayed in both space and format so that certain properties relevant to the monitoring task will emerge from the combination of values on the individual variables. Figure 8.11a shows an example, a patient-respiration monitoring display developed by Cole (1986). In each rectangle the height indicates the volume or depth of patient breathing, and the width indicates the rate. Therefore, the total area of the rectangle indicates the total amount of oxygen respired by the patient (right rectangle) and imposed by the respirator (left rectangle). This relationship holds because the amount = depth × rate and the rectangle area = height × width. Thus, the display has been configured to produce an *emergent feature* (Pomerantz & Pristach, 1989; Sanderson et al., 1989); that is, a property of the configuration of individual variables (in this case depth and rate) emerges on the display to signal a significant, task-relevant, integrated variable (the rectangle area or amount of oxygen (P9). Note also in the figure that a second emergent feature may be perceived as the *shape* of the rectangle—the ratio of height to width that signals either shallow rapid breathing or slow deep breathing (i.e., different "styles" of breathing, which may indicate different states of patient health).

The rectangle display can be fairly widely used because of the number of other systems in which the product of two variables represent a third, important variable. Examples are distance = speed × time, amount = rate × time, value (of information) = reliability × diagnosticity, and expected value (in decision making) = probability × value.

Another example of a configural display, shown in Figure 8.11b, is the safety-parameter monitoring display developed by Woods, Wise, and Hanes (1981) for a nuclear power control room. The eight critical safety parameters are configured in an octagon such that when all are within their safe range, the easily perceivable emergent feature of *symmetry* is observed. Furthermore, if a parameter departs from its normal value as the result of a failure, the distorted *shape* of the polygon can uniquely signal the nature of the underlying fault, a feature that was sadly lacking for our operator in the story at the beginning of the chapter. Such a feature would also be lacking in more conventional arrays of displays like those shown in Figure 8.9.

In the case of the two displays in Figure 8.11, configuring the to-be-integrated variables as dimensions of a single object creates a sort of attentional "glue" that fuses them together, thus adhering to the proximity compatibility principle (P9). But configural displays and their emergent features do not have to come from a single object. Consider Figure 8.12, the proposed design for a boiler power plant supervisory display (Rantanen & Gonzalez de Sather, 2003). The 13 bar graphs, representing critical plant parameters, configure to define an imagined straight line across the middle of the display to signal the key state that all are

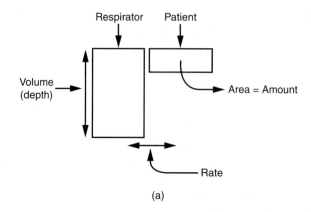

(a)

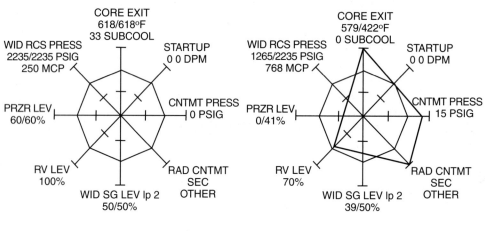

(b)

FIGURE 8.11

(a) Configural respiration monitoring display (*Source:* Developed by Cole, W. 1986 "Medical Cognitive Graphics." Proceedings of CHI. Human Factors in Computing Systems. New York: Association for Computing Machinery). (b) Integrated spoke or polar display for monitoring critical safety parameters in nuclear power. Left: normal operation; right: wide-range iconic display during loss-of-coolant accident. (*Source:* Woods, D. D., Wise, J., and Hanes, L. "An Evaluation of Nuclear Power Plant Safety Parameter Display Systems." Proceedings of the 25th Annual Meeting of the Human Factors Society; 1981, p. 111. Santa Monica, CA: Human Factors Society. Copyright 1981 by the Human Factors Society, Inc. Reproduced by permission.)

operating within normal range. In Figure 8.12, the "break" of the abnormal parameter (FW Press) is visually obvious.

Configural displays generally consider space and spatial relations in arranging dynamic displayed elements. Spatial proximity may help monitoring performance, and object integration may also help, but neither is sufficient or necessary to support information integration from emergent features. The key to such support lies in emergent features that *map to task-related variables*

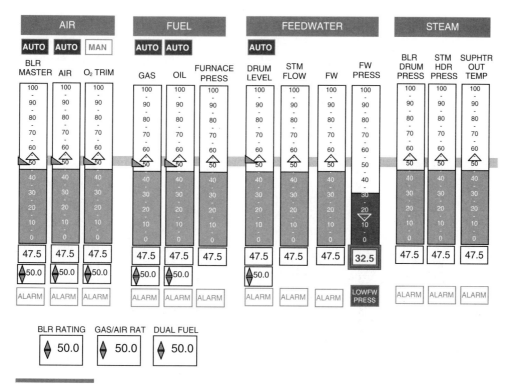

FIGURE 8.12

Monitoring display for a boiler power plant (Rantanen & Gonzalez de Sather, 2003). The emergent feature of the straight line, running across the display at the top of the gray bars, is salient.

(Bennett & Flach, 1992). The direct perception of these emergent features can replace the more cognitively demanding computation of derived quantities (like amount in Figure 8.11a). Will such integration hinder focused attention on the individual variables? The data in general suggest that it does not (Bennett & Flach, 1992). For example, in Figure 8.12, it remains relatively easy to perceive the particular value of a variable (focused attention) even as it is arrayed within the configuration of the 13 parallel bars.

Putting It All Together: Supervisory Displays

In many large systems, such as those found in the industrial process–control industry (see Chapter 16), dynamic supervisory displays are essential to guarantee appropriate situation awareness and to support effective control. As such, several of the display principles and guidelines discussed in this chapter should be applied and harmonized. Figure 8.12 provides such an example. In the figure, we noted the alignment of the parallel monitoring displays to a common baseline to make their access easy (P8) and their comparison or integration (to assure normality) also easy by providing the emergent feature (P9). The display provides redundancy (P4) with the digital indicator at the bottom and a color change in the bar when it moves out of acceptable range. A predictor (P12), the white triangle, shows the trend. The fixed-scale moving pointer display conforms to mental model principles P6 and P7. Finally, the display replaced a separate, computer-accessible window display of alarm information with a design that positioned each alarm directly under its relevant parameter (P9).

One of the greatest challenges in designing such a display is to create one that can simultaneously support monitoring in routine or modestly nonroutine circumstances as well as in abnormal circumstances requiring diagnosis, problem solving, and troubleshooting, such as those confronting the operator at the beginning of the chapter. The idea of presenting totally different display suites to support the two forms of behavior is not always desirable, because in complex systems, operators may need to transition back and forth between them; and because complex systems may fail in many ways, the design of a display to support management of one form of failure may harm the management of a different form.

In response to this challenge, human factors researchers have developed what are called *ecological interfaces* (Vicente, 2002; Vicente & Rasmussen, 1992). The design of ecological interfaces is complex and well beyond the scope of this textbook. However, their design capitalizes in part upon graphical representation of the process, which can produce emergent features that will perceptually signal the departure from normality, and in some cases help diagnose the nature of a failure (Figures 8.11b and 8.12 provide examples). Ecological interface design also capitalizes upon spatial representations of the system, or useful "maps," as we discuss in the following section. However, a particular feature of ecological interfaces is their incorporation of flexible displays that allow the operator/supervisor to reason at various **levels of abstraction** about the problem (Rasmussen, 1983). Where is a fault located? Is it creating a loss of energy or buildup of excessive pressure in the plant? What are its implications for production and safety? These three questions represent different levels of abstraction, ranging from the physical (very concrete, like question 1) to the much more conceptual or abstract (question 3), and an effective manager of a fault in a high-risk system must be able to rapidly switch attention or "move" cognition between various levels. A recent review of the research on ecological interfaces (Vicente, 2002) suggests that they are more effective in supporting fault management than other displays, while not harming routine supervision.

If there are different forms of displays that may support different aspects of the task, or different levels of abstraction which must be compared, it is important to strive, if possible, to keep these visually available at the same time, thereby keeping knowledge in the world (P11), rather than forcing a great deal of sequential paging or keyboard interaction to obtain screens (Burns, 2000).

NAVIGATION DISPLAYS AND MAPS

A navigational display (the most familiar of which is the map) should serve four fundamentally different classes of tasks: (1) provide guidance about how to get to a destination, (2) facilitate planning, (3) help recovery if the traveler becomes lost, and (4) maintain situation awareness regarding the location of a broad range of objects (Garland & Endsley, 1995). For example, a pilot map might depict other air traffic or weather in the surrounding region, or the process controller might view a "mimic diagram" or map of the layout of systems in a plant (Chapter 16). The display itself may be paper or electronic. Environments in which these tasks should be supported range from cities and countrysides to buildings and malls. Recently, these environments have also included spatially defined "electronic environments" such as databases, hypertext, and large menu systems (see Chapter 15). Navigational support also may be needed in multitask conditions while the traveler is engaged in other tasks, like driving the vehicle.

Route Lists and Command Displays

The simplest form of navigational display is the route list or *command display.* This display typically provides the traveler with a series of commands (turn left, go straight, etc.) to reach a desired location. In its electronic version, it may provide markers or pointers of where to turn at particular intersections. The command display is easy to use. Furthermore, most navigational commands can be expressed in words, and if commands are issued verbally through synthesized voice they can be easily processed while the navigator's visual/spatial attention is focused on the road (Streeter et al., 1985), following the attention principle of multiple resources (P10) described in Chapter 6.

Still, to be effective, command displays must possess an accurate knowledge of where the traveler is as each command is issued so that it will be given at the right place and time. Thus, for example, a printed route list is vulnerable if the traveler strays off the intended route, and any sort of electronically mediated command display will suffer if navigational choice points (i.e., intersections) appear in the environment that were not in the database (our unfortunate traveler turns left into the unmarked alley). Thus, command displays are not effective for depicting where one is (allowing recovery if lost), and they are not very useful for planning and maintaining situation awareness. In contrast, spatially configured maps do a better job of supporting these services (planning and situation awareness). There are many different possible design features within such maps, and we consider them in turn.

Maps

Legibility.   To revisit a recurring theme (P1), maps must be legible to be useful. For paper maps, care must be taken to provide necessary contrast between labels and background and adequate visual angle of text size. If color-coded maps are used, then low-saturation coding of background areas enables text to be more visible (Reynolds, 1994). However, colored text may also lead to poor contrast (Chapter 4). In designing such features, attention should also be given to the conditions in which the maps may need to be read (e.g., poor illumination, as discussed in Chapter 4). Unfortunately, legibility may sometimes suffer because of the need for detail (a lot of information) or because limited display size forces the use of a very small map. With electronic maps, detail can be achieved without sacrificing legibility if *zooming* capabilities are incorporated.

Clutter and Overlay.   Another feature of detailed maps is their tendency to become cluttered. Clutter has two negative consequences: It slows down the time to access information (P8) (i.e., to search for and find an item) and it slows the time to read the items as a consequence of masking by nearby items (the focused attention disruption resulting from close proximity, P9). Besides the obvious solution of creating maps with minimal information, three possible solutions avail themselves. First, effective color coding can present different classes of information in different colors. Hence, the human selective attention mechanism is more readily able to focus on features of one color (e.g., roads), while filtering out the temporarily unneeded items of different colors (e.g., text symbols, rivers, terrain; Yeh & Wickens, 2001). Care should be taken to avoid an extensive number of colors (if absolute judgment is required, P2) and to avoid highly saturated colors (Reynolds, 1994). Second, with electronic maps, it is possible for the user to highlight (intensify) needed classes of information selectively while leaving others in the background (Yeh & Wickens, 2001). The enhanced intensity of target information can be a more effective filter for selective and focused attention than will be the different color. Third, carrying the concept of highlighting to its extreme, *decluttering* allows the user to simply turn off unwanted categories of information (Stokes et al., 1990; Mykityshyn et al., 1994). One problem with both highlighting and decluttering is that the more flexible the options are, the greater is the degree of choice imposed on the user, and this may impose unnecessary decision load (Yeh & Wickens, 2001). Furthermore, in some environments, such as a vibrating vehicle, the control interface necessary to accomplish the choice is vulnerable.

Position Representation.   Users benefit in navigational tasks if they are presented with a direct depiction of where they are on the map. This feature can be helpful in normal travel, as it relieves the traveler of the mental demands of inferring the direction and rate of travel. In particular, however, this feature is extremely critical in aiding recovery from getting lost. This, of course, is the general goal of providing "you are here" maps in malls, buildings, and other medium-scale environments (Levine, 1982).

Map Orientation.   A key feature of good maps is their ability to support the navigator's rapid and easy cross-checking between features of the environment (the forward view) and the map (Wickens, 1999). This can be done most easily if the map is oriented in the direction of travel so that "up" on the map is forward and, in particular, left on the map corresponds to left in the forward view. Otherwise, time-consuming and error-prone mental rotation is required (Aretz, 1991). To address this problem, electronic maps can be designed to rotate so that up on the map is in the direction of travel (Wickens et al., 1996; Wickens, 2000B), and "you are here" maps can be mounted so that the top of the map corresponds to the direction of orientation as the viewer observes the map (Levine, 1982), as shown in Figure 8.13. When this correspondence is achieved, the principle of pictorial realism (P6) is satisfied.

Despite the advantages of map rotation for navigation, however, there are some costs associated. For paper maps, the text will be upside down if the traveler

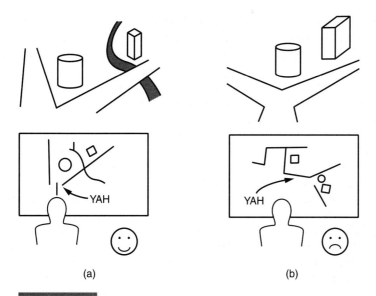

(a)   (b)

FIGURE 8.13

Good (a) and poor (b) mounting of "you are here" map. Note in (b) that the observer must mentally rotate the view of the map by 90° so that left and right in the world correspond to left and right in the map.

is headed south. For electronic maps containing a lot of detail, vector graphics will be needed to preserve upright text (Wickens, 2000B). Furthermore, for some aspects of planning and communications with others, the stability and universal orientation of a fixed north-up map can be quite useful (Baty et al., 1974; Aretz, 1991). Thus, electronic maps should be designed with a fixed-map option available.

Scale. In general, we can assume that the level of detail, scale, or availability with which information needs to be presented becomes less of a concern in direct proportion to the distance away from the traveler and falls off more rapidly in directions behind the traveler than in front (because the front is more likely to be in the future course of travel). Therefore, electronic maps often position the navigator near the bottom of the screen (see Figure 8.4a). The map scale should be user-adjustable if possible, not only because of clutter but because the nature of the traveler's needs can vary from planning, in which the location of a route to very distant destinations may need to be visualized (small scale), to guidance, in which only detailed information regarding the next choice point is required (large scale).

One possible solution to addressing the issue of scale is in the creation of dual maps in which local information regarding one's momentary position and orientation is presented alongside more global large-scale information regarding the full environment. The former can be ego-referenced and correspond to the direction of travel, and the latter can be world-referenced. Figure 8.14 shows some examples. Such a dual map creation is particularly valuable if the user's

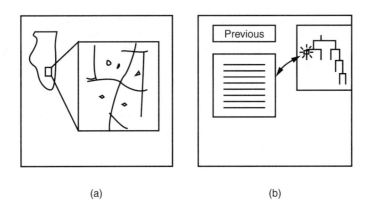

(a)   (b)

FIGURE 8.14

Examples of global and local map presentation: (a) from typical state quadrangle map;
(b) map of a hierarchical database, on the right, flashing the page that is viewed on the left.
Note that the region depicted by the local map is also depicted in the global map. These
examples illustrate visual momentum to assist the viewer in seeing how one piece of
information fits into the context of the other.

momentary position and/or orientation is highlighted on the wide-scale, world-
referenced map (Aretz, 1991; Wickens et al., 2000), thereby capturing the principle of *visual momentum,* which serves to visually and cognitively link two related
views (P9)-(Woods, 1984). Both maps in Figure 8.14 indicate the position of the
local view within the global one.

Three-Dimensional Maps.   Increasing graphics capabilities have enabled the creation of effective and accurate 3-D or perspective maps that depict terrain and
landmarks (Wickens, 2000a & b). If it is a rotating map, then such a map will
nicely adhere to the principle of pictorial realism (P6; Roscoe, 1968). But are
3-D maps helpful? The answer depends on the extent to which the vertical information, or the visual identity of 3-D landmark objects, is necessary for navigation. For the pilot flying high over flat terrain or for the driver navigating a
gridlike road structure, vertical information is likely to play little role in navigation. But for the hiker or helicopter pilot in mountainous terrain, for the pilot
flying low to the ground, or the vehicle driver trying to navigate by recognizing
landmark objects in the forward field of view, the advantages of vertical (i.e.,
3-D) depiction become far more apparent (Wickens, 1999). This is particularly
true given the difficulties that unskilled users have reading 2-D contour maps.
Stated simply, the 3-D display usually looks more like a picture of the area that is
represented (P6), and this is useful for maintaining navigational awareness.
More guidance on the use of 3-D displays is offered in the following section.

Planning Maps and Data Visualization.   Our discussion of maps has assumed the
importance of a *traveler* at a particular location and orientation in the map-
depicted database. But there are several circumstances in which this is not the
case; the user does not "reside" within the database. Here we consider examples

such as air traffic control displays, vehicle dispatch displays, process-control mimic diagrams, construction plans, wiring diagrams, and the display of 3-D scientific data spaces. The user is more typically a "planner" who is using the display to understand the spatial relations between its elements.

Many of the features we have described apply to these "maps for the non-traveler" as well (e.g., legibility and clutter issues, flexibility of scale). But since there typically is no direction of travel, map rotation is less of an issue. For geographic maps, north-up is typically the fixed orientation of choice. For other maps, the option of flexible, user-controlled orientation is often desirable.

The costs and benefits of 3-D displays for such maps tend to be more task-specific. For maps to support a good deal of 3-D visualization (like an architect's plan), 3-D map capabilities can be quite useful (Wickens et al., 1994). In tasks such as air traffic control, where very precise separation along lateral and vertical dimensions must be judged, however, 3-D displays may impose costs because of the ambiguity with which they present this information (see Chapter 4). Perhaps the most appropriate guidance that should be given is to stress the need for careful task and information analysis before choosing to implement 3-D maps: (1) How important *is* vertical information in making decisions? (2) Does that information need to be processed at a very precise level (in which case 3-D representations of the vertical dimensions are not good (Wickens, 2000a & b; St. John et al., 2001), or can it be processed just to provide some global information regarding "above" or "below," in which case the 3-D displays can be more effective?

If a 3-D (perspective) map is chosen, then two important design guidelines can be offered (Wickens et al., 1989). First, as noted in Chapter 4, the greater number of natural depth cues that can be rendered in a synthetic display, the more compelling will be the sense of depth or three dimensionality. Stereo, interposition and motion parralex (which can be created by allowing the viewer to rotate the display) are particularly valuable cues. (Wickens et al., 1989; Sollenberger & Milgram, 1993). Second, if display viewpoint rotation is an option, it is worthwhile to have a 2-D viewpoint (i.e., overhead lookdown) available as a default option.

QUANTITATIVE INFORMATION DISPLAYS: TABLES AND GRAPHS

Some displays are designed to present a range of numbers and values. These may be as varied as tables depicting the nutrition and cost of different products for the consumer, the range of desired values for different maintenance testing outcomes, a spreadsheet, or a set of economic or scientific data. The format of depiction of such data has a strong influence on its interpretability (Gillan et al., 1998). An initial choice can be made between representation of such values via tables or graphs. As with our discussion of dynamic displays, when the comparison was between digital and analog representation, one key consideration is the *precision* with which a value must be read. If high precision is required, the table may be a wise choice. Furthermore, unlike dynamic digital displays, tables do not suffer the problems of reading digital information while it is changing. However, as shown in Figure 8.15a, tables do not support a very good perception

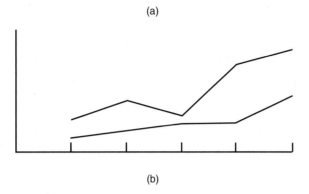

| 22 | 25 | 26 | 24 | 28 |
| 26 | 32 | 29 | 38 | 42 |

(a)

(b)

FIGURE 8.15

(a) Tabular representation of trend variables; (b) graphical representation of the same trend variables as (a). Note how much easier it is to see the trend in (b).

of change over space; that is, the increasing or decreasing trend of values across the table is not very discernible compared to the same data presented in line-graph form in Figure 8.15b. Tables are even less supportive of perception of the *rate* of trend change (acceleration or deceleration) across space and less so still for trends that exist over two dimensions of space (e.g., an interaction between variables), which can be easily seen by the divergence of the two lines on the right side of the graph of Figure 8.15b.

Thus, if absolute precision is *not* required and the detection or perception of trend information is important, the graph represents the display of choice. If so, then the questions remain: What kind of graph? Bar or line? Pie? 2-D or 3-D? and so on. While you may refer to Kosslyn (1994), or Gillan et al. (1998) for good treatments of human factors of graphic presentation, a number of fairly straightforward guidelines can be offered as follows.

Legibility (P1)

The issues of contrast sensitivity are again relevant. However, in addition to making lines and labels of large enough visual angle to be readable, a second critical point relates to *discriminability* (P5). Too often, lines that have very different meanings are distinguished only by points that are highly confusable (Fig. 8.16a). Here is where attention to incorporating salient and *redundant* coding (P4) of differences (Figure 8.16b) can be quite helpful. In modern graphics packages, color is often used to discriminate lines. In this case, it is essential to use color coding redundantly with another salient cue. Why? As we noted in Chapter 4, not all viewers have good color vision, and a non-redundant colored graph printed from a noncolor printer or photocopied may be useless.

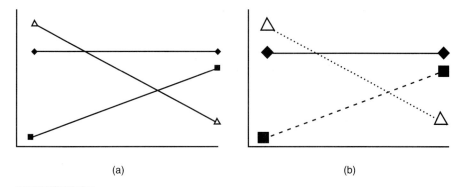

(a)   (b)

FIGURE 8.16

(a) Confusable lines on a graph; (b) discriminable lines created in part by use of redundancy. (*Source:* Wickens, CD., 1992b. The human factors of graphs at HFS annual meetings. *Human Factors Bulletin, 35* [7], 1–3.)

Clutter

Graphs can easily become cluttered by presenting a lot more lines and marks than the actual information they convey. As we know, excessive clutter can be counterproductive (Lhose, 1993), and this has led some to argue that the *data-ink ratio* should always be maximized (Tufte, 1983, 1990); that is, the greatest amount of data should be presented with the smallest amount of ink. While adhering to this guideline is a valuable safeguard against the excessive ink of "boutique" graphs, such as those that unnecessarily put a 2-D graph into 3-D perspective (Fig. 8.17a; Carswell, 1992). The guideline of minimizing ink can however be counterproductive if carried too far. Thus, for example, the "minimalist" graph in Figure 8.17b, which maximizes data-ink ratio, gains little by its decluttering and loses a lot in its representation of the trend, compared to the

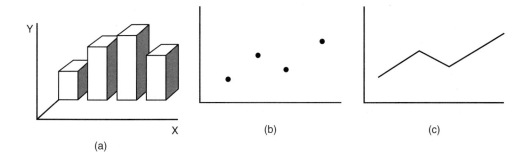

(a)   (b)   (c)

FIGURE 8.17

(a) Example of a boutique graph with a very low data-ink ratio. The 3-D graph contains the unnecessary and totally noninformative representation of the depth dimension; (b) minimalist graph with very high data-ink ratio; (c) line graph with intermediate data-ink ratio. Note the redundant trend information added by the line.

line graph of Figure 8.17c. The lines of Figure 8.17c contain an emergent feature—their slope—which is not visible in the dot graph of 8.17b. The latter is
also much more vulnerable to the conditions of poor viewing (or the misinter-
pretation caused by the dead bug on the page!).

Proximity

Visual attention must sometimes do a lot of work, traveling from place to place
on the graph (P8), and if this visual search effort is excessive, it can hinder graph
interpretation, competing for perceptual-cognitive resources with the cognitive
processes required to understand what the graph means. Hence, it is important
to construct graphs so things that need to be compared (or integrated) are either
close together in space or can be easily linked perceptually by a common visual
code. This, of course, is a feature for the proximity compatibility principle (P9)
and can apply to keeping legends close to the lines that they identify (Fig. 8.18a;

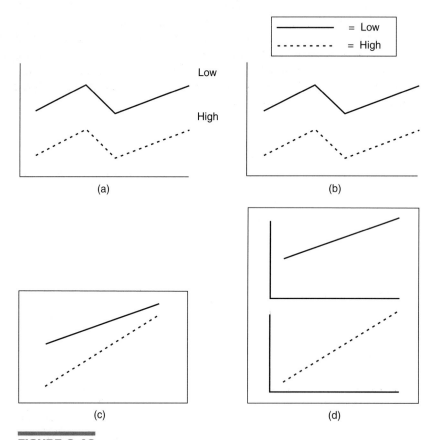

FIGURE 8.18

Graphs and proximity. (a) Close proximity of label to line; a good design feature. (b) Low
proximity of label to line; a poor design feature. (c) Close proximity of lines to be compared
(good). (d) Low proximity of lines to be compared (poor).

rather than in remote captions or boxes (Fig. 8.18b) and keeping two lines that need to be compared on the same panel of a graph (Fig. 8.18c) rather than on separate panels (8.18d). The problems of low proximity will be magnified as the graphs contain more information—more lines).

Format

Finally, we note that as the number of data points in graphs grows quite large, the display is no longer described as a graph but rather as one of *data visualization,* some of whose features were described in the previous section on maps. Others are discussed in Chapter 15.

CONCLUSION

We presented a wide range of display principles designed to facilitate the transmission of information from the senses, discussed in Chapters 4 and 5, to cognition, understanding, and decision making, discussed in Chapters 6 and 7. There is no single "best" way to do this, but consideration of the 13 principles presented above can certainly help to rule out bad displays. Much of the displayed information eventually leads to *action*—to an effort to *control* some aspect of a system or the environment or otherwise to respond to a displayed event. In the next chapter, we discuss some of the ways in which the human factors engineer can assist with that control process.

Chapter

Control

The rental car was new, and as he pulled onto the freeway entrance ramp at dusk, he started to reach for what he thought was the headlight control. Suddenly, however, his vision was obscured by a gush of washer fluid across the windshield. As he reached to try to correct his mistake, his other hand twisted the very sensitive steering wheel and the car started to veer off the ramp. Quickly, he brought the wheel back but overcorrected, and then for a few terrifying moments the car seesawed back and forth along the ramp until he brought it to a stop, his heart pounding. He cursed himself for failing to learn the location of controls before starting his trip. Reaching once more for the headlight switch, he now activated the flashing hazard light—fortunately, this time, a very appropriate error.

Our hapless driver experienced several difficulties in control that can be placed in the context of the human information-processing model discussed in Chapter 6. This model can be paraphrased by "knowing the state of affairs, knowing what to do, and then doing it." Control is the "doing it" part of this description. It is both a noun (a control) and an action verb (to control). Referring to the model of information processing presented in Chapter 6, we see that control primarily involves the selection and execution of responses—that is, the last two stages of the model—along with the feedback loop that allows the human to determine that the control response has been executed in the manner that was intended. In this chapter, we first describe some important principles concerning the selection of responses. Then we discuss various aspects of response execution that are influenced by the nature of the control device, which is closely intertwined with the task to be performed. We address discrete activation of controls or switches, controls used as setting or pointing devices, controls used for verbal or symbolic input (e.g., typing), and continuous control used in tracking and traveling.

PRINCIPLES OF RESPONSE SELECTION

The difficulty and speed of selecting a response or an action is influenced by several variables (Fitts & Posner, 1967; Wickens & Hollands, 2000), of which five are particularly critical for system design: decision complexity, expectancy, compatibility, the speed-accuracy tradeoff, and feedback.

Decision Complexity

The speed with which an action can be selected is strongly influenced by the number of possible alternative actions that *could* be selected in that context. This is called the *complexity* of the decision of what action to select. Thus, each action of the Morse code operator, in which only one of two alternatives is chosen (*dit* or *dah*) follows a much simpler choice than each action of the typist, who must choose between one of 26 letters. Hence, the Morse code operator can generate a greater number of keystrokes per minute. Correspondingly, users can select an action more rapidly from a computer menu with two options than from the more complex menu with eight options.

Engineering psychologists have characterized this dependency of response selection time on decision complexity by the *Hick-Hyman law* of reaction time (RT), shown in Figure 9.1 (Hick 1952; Hyman, 1953). When reaction time or response time is plotted as a function of $Log_2(N)$ rather than N (see Figure 9.1b), the function is generally linear. Because $Log_2(N)$ represents the amount of information, in bits, conveyed by a choice, in the formal information theory the linear relation of RT with bits, conveyed by the Hick-Hyman law, suggests that humans process information at a constant rate.

The Hick-Hyman law does not imply that systems designed for users to make simpler decisions are superior. In fact, if a given amount of information

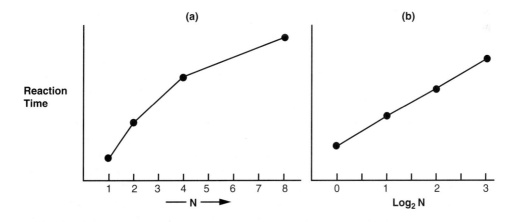

FIGURE 9.1

The Hick-Hyman law of reaction time. (a) The figure shows the logarithmic increase in RT as the number of possible stimulus-response alternatives (N) increases. This can sometimes be expressed by the formula: $RT = a + bLog_2N$. This linear relation is shown in (b).

needs to be transmitted by the user, it is generally more efficient to do so by a smaller number of complex decisions than a larger number of simple decisions. This is referred to as the *decision complexity advantage* (Wickens and Hollands 2000). For example, a typist can convey the same message more rapidly than can the Morse code operator. Although keystrokes are made more slowly, there are far fewer of them. Correspondingly, as we learn in Chapter 15, "shallow" menus with many items (i.e., eight in the example above) are better than "deep" menus with few items.

Response Expectancy

In chapters 4, 6, and 8 we learned that we perceive rapidly (and accurately) information that we *expect*. In a corresponding manner, we select more rapidly and accurately those actions we expect to carry out than those that are surprising to us. We do not, for example, expect the car in front of us to come to an abrupt halt on a freeway. Not only are we slow in perceiving its expansion in the visual field, but we are much slower in applying the brake (selecting the response) than we would be when the light expectedly turns yellow at an intersection that we are approaching.

Compatibility

In Chapter 8 we discussed the concept of display compatibility between the orientation and movement of a display and the operator's expectancy of movement, or mental model, of the displayed system in the context of *the principle of the moving part. Stimulus-response compatibility* (or display control compatibility) describes the expected relationship between the location of a control or movement of a control response and the location or movement of the stimulus or display to which the control is related (Fitts & Seeger, 1953).

Two subprinciples characterize a compatible (and hence good) mapping between display and control (or stimulus and response). (1) *Location compatibility:* The control location should be close to (and in fact closest to) the entity being controlled or the display of that entity. Figure 8.6 showed how location compatibility is applied to bad label placement. If the labels in that figure were instead controls, it would represent poor (ambiguous) location compatibility. (2) *Movement compatibility:* The direction of movement of a control should be congruent with the direction both of movement of the feedback indicator and of the system movement itself. A violation of movement compatibility would occur if the operator needed to move a lever to the left, in order to move a display indicator to the right.

The Speed-Accuracy Tradeoff

For the preceding three principles, the designer can assume that factors that make the selection of a response longer (complex decisions, unexpected actions, or incompatible responses) will also make errors more likely. Hence, there is a positive correlation between response time and error rate or, in other terms, a positive correlation between speed and accuracy. These variables do not trade off. However, there are some circumstances in which the two measures *do* trade off: For example, if we try to execute actions very rapidly (carrying out procedures under a severe time deadline), we are more likely to make errors. In contrast, if we must be very cautious because the consequences of errors are critical, we will be slow. Hence, in these two examples there is a **negative** correlation, or a *speed-accuracy tradeoff.* In these examples, the tradeoff was caused by user strategies. As we will see below, sometimes control devices differ in the speed-accuracy tradeoff because one induces faster but less precise behavior and the other more careful but slower behavior.

Feedback

Most controls and actions that we take are associated with some form of visual feedback that indicates the *system* response to the control input. For example, in a car the speedometer offers visual feedback from the control of the accelerator. However, good control design must also be concerned with more direct feedback of the control state itself. As we learned in Chapter 5, this feedback may be kinesthetic/tactile (e.g., the feel of a button as it is depressed to make contact or the resistance on a stick as it is moved). It may be auditory (the click of the switch or the beep of the phone tone), or it may be visual (a light next to a switch to show it is on or even the clear and distinct visual view that a push button has been depressed).

Through whatever channel, we can state with some certainty that more feedback of both the current control state (through vision) and the change in control state is good as long as the feedback is nearly instantaneous. However, feedback that is delayed by as little as 100 msec can be harmful if rapid sequences of control actions are required. Such delays are particularly harmful if the operator is less skilled (and therefore depends more on the feedback) or if the feedback cannot be filtered out by selective attention mechanisms (Wickens & Hollands, 2000). A good example of such harmful delayed feedback is a voice feedback delay while talking on a radio or telephone.

DISCRETE CONTROL ACTIVATION

Our driver in the opening story was troubled, in part, because he simply did not know, or could not find, the right controls to *activate* the wipers. Many such controls in systems are designed primarily for the purpose of activating or changing the discrete state of some system. In addition to making the controls easily visible (Norman, 1988), there are several design features that make the activation of such controls less susceptible to errors and delays.

Physical Feel

Feedback is a critical, positive feature of discrete controls. Some controls offer more feedback channels than others. The toggle switch is very good in this regard. It changes its state in an obvious *visual* fashion and provides an auditory click and a tactile snap (a sudden *loss* of resistance) as it moves into its new position. The auditory and tactile feedback provide the operator with instant knowledge of the toggle's change in state, while the visual feedback provides continuous information regarding its new state. A push button that remains depressed when on has similar features, but the visual feedback may be less obvious, particularly if the spatial difference between the button at the two positions is small.

Care should be taken in the design of other types of discrete controls that the feedback (indicating that the system has received the state change) is obvious. Touch screens do not do this so well; neither do push-button phones that lack an auditory beep following each keypress. Computer-based control devices often replace the auditory and tactile state-change feedback with artificial visual feedback (e.g., a light that turns on when the switch is depressed). If such visual feedback is meant to be the only cue to indicate state change (rather than a redundant one), then there will be problems associated both with an increase in the *distance* between the light and the relevant control (this distance should be kept as short as possible) and with the possible electronic failure of the light or with difficulties seeing the light in glare. Hence, feedback lights ideally should be redundant with some other indication of state change; of course, any visual feedback should be immediate.

Size.   Smaller keys are usually problematic from a human factors standpoint. If they are made smaller out of necessity to pack them close together in a miniaturized keyboard, they invite "blunder" errors when the wrong key (or two keys) are inadvertently pressed, an error that is particularly likely for those with large fingers or wearing gloves. If the spacing between keys is not reduced as they are made smaller, however, the time for the fingers to travel between keys increases.

Confusion and Labeling.   Keypress or control activation errors also occur if the identity of a key is not well specified to the novice or casual user (i.e., one who does not "know" the location by touch). This happened to our driver at the beginning of the chapter. These confusions are more likely to occur (a) when large sets of identically appearing controls are unlabeled or poorly labeled and (b) when labels are physically displaced from their associated controls, hence violating the proximity compatibility principle (see Figure 8.6 in Chapter 8).

POSITIONING CONTROL DEVICES

A common task in much of human–machine interaction is the need to **position** some entity in space. This may involve moving a cursor to a point on a screen, reaching with a robot arm to contact an object, or moving the setting on a radio dial to a new frequency. Generically, we refer to these spatial tasks as those involving positioning or *pointing* (Baber, 1997). A wide range of control *devices*, such as the mouse, joystick, and thumbpad are available to accomplish such tasks. Before we compare the properties of such devices, however, we consider the important nature of the human performance skill underlying the pointing task: movement of a controlled entity, which we call a *cursor,* to a destination, which we call a *target.* We describe a model that accounts for the time to make such movements.

Movement Time

Controls typically require movement of two different sorts: (1) movement is often required for the hands or fingers to reach the control (not unlike the movement of attention to access information, discussed in Chapter 8), and (2) the control may then be moved in some direction, often to position a cursor. Even in the best of circumstances in which control location and destination are well learned, these movements take time. Fortunately for designers, such times can be relatively well predicted by a model known as *Fitts's law* (Fitts, 1954; Jagacinski & Flach, 2003):

$$MT = a + b \log_2(2A/W)$$

where A = amplitude of the movement and W = width of the target or the desired precision with which the cursor must land. This means that movement time is linearly related to the logarithm of the term $(2A/W)$, which is the *index of difficulty* of the movement. We show three examples of Fitts's law in Figure 9.2, with the index of difficulty calculated to the right. As shown in rows a and b, each time the distance to the key doubles, the index of difficulty and therefore movement time increases by a constant amount. Correspondingly, each time the required precision of the movement is doubled (the target width or allowable

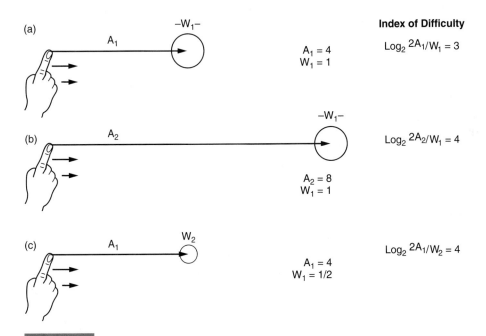

FIGURE 9.2
Fitts's law of movement time. Comparing (a) and (b) shows the doubling of movement amplitude from $A_1 \rightarrow A_2$; comparing (a) to (c) shows halving of target width $W_1 \rightarrow W_2$ (or doubling of target precision); (b) and (c) will have the same movement time. Next to each movement is shown the calculation of the index of difficulty of the movement to which movement time will be directly proportional.

precision is halved; compare rows a and c), the movement time also increases by a constant amount unless the distance is correspondingly halved (compare rows b and c, showing the same index of difficulty and therefore the same movement time). As we saw in the previous section, making keys smaller (reducing W) increases movement time unless they are proportionately moved closer together. Another implication of Fitts's law is that if we require a movement of a given amplitude, A, to be made within a shorter time constraint, MT, then the precision of that movement will decrease as shown by an increase in the variability of movement endpoints, represented by W. This characterizes a speed-accuracy tradeoff in pointing movements. The value of W in this case characterizes the distribution of endpoints of the movement. Higher W means higher error.

The mechanisms underlying Fitts's law are based heavily on the visual feedback aspects of controlled aiming, and hence the law is equally applicable to the actual physical movement of the hand to a target (i.e., reaching for a key) as to the movement of a displayed cursor to a screen target achieved by manipulation of some control device (e.g., using a mouse to bring a cursor to a particular item in a computer menu; Card et al., 1978). It is also applicable to movements as coarse as a foot reaching for a pedal (Drury, 1975) and as fine as assembly and manipulation under a microscope (Langolf et al., 1976). This generality gives the law great value in allowing designers to predict the costs of different keyboard layouts and target sizes in a wide variety of circumstances (Card et al., 1983). In particular, in comparing rows (b) and (c) of Figure 9.2, the law informs that miniaturized keyboards—reduced distance between keys—will not increase the speed of keyboard use.

Device Characteristics

The various categories of control devices that can be used to accomplish these pointing or position tasks may be grouped into four distinct categories. In the first category are *direct position controls* (light pen and touch screen) in which the position of the human hand (or finger) directly corresponds with the desired location of the cursor. The second category contains *indirect position controls*—the mouse or touch pad—in which changes in the position of the limb directly correspond to changes in the position of the cursor, but the limb is moved on a surface different from the display cursor surface.

The third category contains *indirect velocity controls*, such as the joystick and the cursor keys. Here, typically an activation of control in a given direction yields a **velocity** of cursor movement in that direction. For cursor keys, this may involve either repeated presses or holding it down for a long period. For joystick movements, the magnitude of deflection typically creates a proportional velocity. Joysticks may be of three sorts: *isotonic*, which can be moved freely and will rest wherever they are positioned; *isometric* (see Chapter 5), which are rigid but produce movement proportional to the force applied; or *spring-loaded*, which offer resistance proportional to both the force applied and the amount of displacement, springing back to the neutral position when pressure is released. The spring-loaded stick, offering both proprioceptive and kinesthetic feedback of movement extent, is typically the most preferred. (While joysticks can be configured as position controls, these are not generally used, for reasons discussed later.) The fourth category is that of voice control.

Across all display types, there are two important variables that affect usability of controls for pointing (and they are equally relevant for controls for tracking). First, feedback of the current state of the cursor should be salient, visible, and as applied to indirect controls, immediate. Thus, system lags greatly disrupt pointing activity, particularly if this activity is at all repetitive. Second, performance is affected in a more complex way by the system *gain*. Gain may be described by the ratio:

$$G = (\text{change of cursor})/(\text{change of control position}).$$

Thus, a high-gain device is one in which a small displacement of the control produces a large movement of the cursor or produces a fast movement in the case of a velocity control device. (This variable is sometimes expressed as the reciprocal of gain, or the control/display ratio.) The gain of direct position controls, such as the touch screen and light pen, will obviously be 1.0.

There is some evidence that the ideal gain for indirect control devices should be in the range of 1.0 to 3.0 (Baber, 1997). However, two characteristics partially qualify this recommendation. First, humans appear to adapt successfully to a wider range of gains in their control behavior (Wickens, 1986). Second, the ideal gain tends to be somewhat task-dependent because of the differing properties of low-gain and high-gain systems. Low-gain systems tend to be effortful, since a lot of control response is required to produce a small cursor movement; however, high-gain systems tend to be imprecise, since it is very easy to overcorrect when trying to position a cursor on a small target. Hence, for example, to the extent that a task requires a lot of repetitive and lengthy movements to large targets, a higher gain is better. This might characterize the actions required in the initial stages of a system layout using a computer-aided design tool where different elements are moved rapidly around the screen. In contrast, to the extent that small, high-precision movements are required, a low-gain system is more suitable. These properties characterize tasks such as uniquely specifying data points in a very dense cluster or performing microsurgery in the operating room, where an overshoot could lead to serious tissue damage. Many factors can influence the effectiveness of control devices (see Baber, 1997; Bullinger et al., 1997) and we describe these below.

Task Performance Dependence

For the most critical tasks involved in pointing (designating targets and "dragging" them to other locations), there is good evidence that the best overall devices are the two direct position controls (touch screen and light pen) and the mouse (as reflected in the speed, accuracy and preference data shown in Fig. 9.3; Baber, 1997; Epps, 1987; Card et al., 1978). Analysis by Card and Colleagues (1978), using Fitts's law to characterize the range of movement distances and precisions, suggests that the mouse is superior to the direct pointing devices. However, Figure 9.3 also reveals the existence of a speed-accuracy tradeoff between the direct position controls, which tend to be very rapid but less accurate,

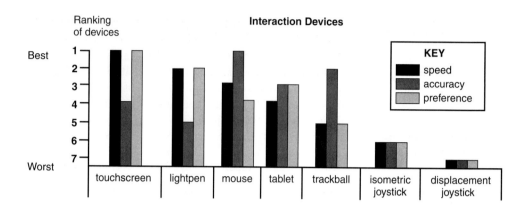

FIGURE 9.3

A comparison of performance of different control devices, based on speed, accuracy, and user preference. (*Source:* Baber, C., 1997. *Beyond the Desktop.* San Diego, CA: Academic Press.)

and the mouse, which tends to be slower but generally more precise. Problems in accuracy with the direct positioning devices arise from several factors: parallax errors in which the position where the hand or light pen is seen to be does not correspond to where it is if the surface is viewed at an angle, instability of the hand or fingers (particularly on a vertical screen), and in the case of touch screens, the imprecision of the finger area in specifying small targets. In addition to greater accuracy, indirect position devices like the mouse have another clear advantage over the direct positioning devices. Their gain may be adjustable, depending on the required position accuracy (or effort) of the task.

When pointing and positioning is required for more complex spatial activities, like drawing or handwriting, the advantages for the indirect positioning devices disappear in favor of the most natural feedback offered by the direct positioning devices.

Cursor keys, not represented in Figure 9.3, are adequate for some tasks, but they do not produce long movements well and generally are constrained by "city block" movement, such as that involved in text editing. Voice control may be feasible in designating targets by nonspatial means (e.g., calling out the target identity rather than its location), but this is feasible only if targets have direct, visible, and unambiguous symbolic labels.

Closely related to performance effects are the effects of device on workload (see Chapter 13). These are shown in Table 9.1.

The Work Space Environment

An important property of the broader workspace within which the device is used is the display, which presents target and cursor information. As we have noted, display size (or the physical separation between display elements) influences the extent of device-movement effort necessary to access targets. Greater display size places a greater value on efficient high-gain devices. In contrast,

TABLE 9.1   Interaction Devices Classified in Terms of Workload

Interaction Device	Cognitive Load	Perceptual Load	Motor Load	Fatigue
Light pen	Low	Low	Medium	Medium
Touch panel	Low	Low	Low	Low
Table (stylus)	High	Medium	Medium	High
Alphanumeric keyboard	High	High	High	High
Function keyboard	Low	Medium	Low	Low
Mouse	Low	Medium	Medium	Medium
Trackball	Low	Medium	Medium	Medium

Source: Baber, C., 1997. *Beyond the Desktop.* San Diego, CA: Academic Press.

smaller, more precise targets (or smaller displays) place a greater need for precise manipulation and therefore lower gain.

The physical characteristics of the display also influence usability. Vertically mounted displays or those that are distant from the body impose greater costs on direct positioning devices where the hand must move across the display surface. Frequent interaction with keyboard editing creates a greater benefit of devices that are physically integrated with the keyboard (i.e., cursor keys or a thumb touch pad rather than the mouse) or can be used in parallel with it (i.e., voice control). Finally, the available workspace *size* may constrain the ability to use certain devices. In particular, devices like joysticks or cursor keys that may be less effective in desktop workstations become relatively more advantageous for control in mobile environments, like the vehicle cab or small airplane cockpit, in which there is little room for a mouse pad. Here the thumb pad, in which repeated movement of the thumb across a small surface moves the cursor proportionately, is an advantage (Bresley, 1995).

Finally, the environment itself can have a major impact on usability For example, direct position control devices suffer greatly in a vibrating environment, such as a vehicle cab. Voice control is more difficult in a noisy environment.

The preceding discussion should make clear that it is difficult to specify in advance what the best device will be for a particular combination of task, workspace, and environment. It should, however, be possible to eliminate certain devices from contention in some circumstances and at the same time to use the factors discussed above to understand why users may encounter difficulties during early prototype testing. The designer is referred to Baber (1997) and Bullinger et al. (1997) regarding more detailed treatment of the human factors of control device differences.

VERBAL AND SYMBOLIC INPUT DEVICES

Spatial positioning devices do not generally offer a compatible means of inputting or specifying much of the symbolic, numerical, or verbal information that is involved in system interaction (Wickens et al., 1983). For this sort of information, keyboards or voice control have generally been the interfaces of choice.

Numerical Data Entry

For numerical data entry, numerical keypads or voice remain the most viable alternatives. While voice control is most compatible and natural, it is hampered by certain technological problems that slow the rate of possible input. Numeric keypads, are typically represented in one of three forms. The linear array, such as found at the top of the computer keyboard is generally not preferred because of the extensive movement time required to move from key to key. The 3×3 square arrays minimize movement distance (and therefore time). General design guidelines suggest that the layout with 123 on the top row (telephone) is preferable (Baber, 1997), to that with 789 on top (calculator) although the advantage is probably not great enough to warrant redesign of the many existing "7-8-9" keyboards.

Linguistic Data Entry

For data entry of linguistic material, the computer keyboard has traditionally been the device of choice. Although some alternatives to the traditional QWERTY layout have been proposed, it is not likely that this design will be changed.

An alternative to dedicated keys that require digit movement is the *chording keyboard* in which individual items of information are entered by the simultaneous depression of combinations of keys, on which the fingers may remain (Seibel, 1964; Gopher & Raij, 1988). Chording works effectively in part by allowing a single complex action to convey a large amount of information and hence benefit from the decision complexity advantage, discussed earlier in this chapter. A single press with a 10-key keyboard can, for example, designate any of $2^{10} - 1$ (or 1,023) possible actions/meanings.

Such a system has three distinct advantages. First, since the hands never need to leave the chord board, there is no requirement for visual feedback to monitor the correct placement of a thumb or finger digit. Consider, for example, how useful this feature would be for entering data in the high-visual-workload environment characteristic of helicopter flight or in a continuous visual inspection task. Second, because of the absence of a lot of required finger movement, the chording board is less susceptible to repetitive stress injury or carpal tunnel syndrome (Chapter 11). Finally, after extensive practice, chording keyboards have been found to support more rapid word transcription processing than the standard typewriter keyboard, an advantage due to the absence of movement-time requirements (Seibel, 1964; Barton, 1986; Wickens & Hollands, 2000).

The primary cost of the chording keyboard is in the extensive learning required to associate the finger combinations with their meaning (Richardson et al., 1987). In contrast, typewriter keyboards provide knowledge in the world regarding the appropriate key, since each key is labeled on the top and each letter is associated with a unique location in space (Norman, 1988). For the chord board there is only knowledge in the head, which is more difficult to acquire and may be easier to lose through forgetting. Still, however, various chording systems have found their way into productive use; examples are both in postal mail sorting (Barton, 1986) and in court transcribing (Seibel, 1964), where specialized users have invested the necessary training time to speed the flow of data input.

VOICE INPUT

Within the last several years, increasingly sophisticated voice recognition technology has made this a viable means of control, although such technology has both costs and benefits.

Benefits of Voice Control

While chording is efficient because a single action can select one of several hundred items (the decision complexity advantage), an even more efficient linguistic control capability can be obtained by voice, where a single utterance can represent any of several thousand possible meanings. Furthermore, as we know, voice is usually a very "natural" communications channel for symbolic linguistic information and one with which we have had nearly a lifetime's worth of experience. This naturalness may be (and has been) exploited in many control interfaces when the benefits of voice control outweigh their technological costs.

Particular benefits of voice control may be observed in dual-task situations. When the hands and eyes are busy with other tasks, like driving (which prevents dedicated manual control on a keyboard and the visual feedback necessary to see if the fingers are properly positioned), designs in which the operator can time-share by talking to the interface using separate resources are of considerable value. Some of the greatest successes have been realized, for example, in using voice to enter radio-frequency data in the heavy visual-manual load environment of the helicopter. "Dialing" of cellular phones by voice command while driving is considered a useful application of voice recognition technology. So also is the use of this technology in assisting baggage handlers to code the destination of a bag when the hands are engaged in "handling" activity. There are also many circumstances in which the combination of voice and manual input for the same task can be beneficial (Baber, 1997). Such a combination, for example, would allow manual interaction to select objects (a spatial task) and voice to convey symbolic information to the system about the selected object (Martin, 1989).

Costs of Voice Control

Against these benefits may be arrayed four distinct costs that limit the applicability of voice control and/or highlight precautions that should be taken in its implementation. These costs are related closely to the sophistication of the voice recognition technology necessary for computers to translate the complex four-dimensional analog signal that is voice (see Chapter 5) into a categorical vocabulary, which is programmed within the computer-based voice recognition system (McMillan et al., 1997).

Confusion and Limited Vocabulary Size. Because of the demands on computers to resolve differences in sounds that are often subtle even to the human ear, and because of the high degree of variability (from speaker to speaker and occasion to occasion) in the physical way a given phrase is uttered, voice recognition systems are prone to make confusions in classifying similar-sounding utterances (e.g., "cleared to" versus "cleared through"). How such confusions may be dealt with can vary (McMillan et al., 1997). The recognizing computer may simply take its "best guess" and pass it on as a system input. This is what a computer keyboard would do if you hit the wrong letter. Alternatively, the system may provide feedback if it is uncertain about a particular classification or if an utterance is not even close to anything in the computer's vocabulary. The problem is that if the recognition capabilities of the computer are still far from perfect, the repeated occurrences of this feedback will greatly disrupt the smooth flow of voice communications if this feedback is offered in the auditory channel. If the feedback is offered visually, then it may well neutralize the dual-task benefit (i.e., keeping the eyes free). These costs of confusion and misrecognition can be addressed only by reducing the vocabulary size and constructing the vocabulary in such a way that acoustically similar items are avoided.

Constraints on Speed.   Most voice recognition systems do not easily handle the continuous speech of natural conversation. This is because the natural flow of our speech does not necessarily place physical pauses between different words (see Chapter 5; Fig. 5.8d). Hence, the computer does not easily know when to stop "counting syllables" and demarcate the end of a word to look for an association of the sound with a given item in its vocabulary. To guard against these limitations, the speaker may need to speak unnaturally slowly, pausing between each word.

A related point concerns the time required to "train" many voice systems to understand the individual speaker's voice prior to the system's use. This training is required because there are so many physical differences between the way people of different gender, age, and dialect may speak the same word. Hence, the computer can be far more efficient if it can "learn" the pattern of a particular individual (called a *speaker-dependent* system) than it can if it must master the dialect and voice quality of all potential users (*speaker-independent* system). For this reason, speaker-dependent systems usually can handle a larger vocabulary.

Acoustic Quality and Noise and Stress.   Two characteristics can greatly degrade the acoustic quality of the voice and hence challenge the computer's ability to recognize it. First, a noisy environment is disruptive, particularly if there is a high degree of spectral overlap between the signal and noise (e.g., recognizing the speaker's message against the chatter of other background conversation). Second, under conditions of stress, one's voice can change substantially in its physical characteristics, sometimes as much as doubling the fundamental frequency (the high-pitched "Help, emergency!"; Sulc, 1996). As we will see in Chapter 13, stress appears to occur often under emergency conditions, and hence great caution should be given before designing systems in which voice control must be used as part of emergency procedures.

Compatibility.   Finally, we have noted that voice control is less suitable for controlling continuous movement than are most of the available manual devices (Wickens et al., 1985; Wickens et al., 1984). Consider, for example, the greater difficulties of trying to steer a car along a curvy road by saying "a little left, now a little more left" than by the more natural manual control of the steering wheel.

Conclusion. Clearly all of these factors—costs, benefits, and design cautions (like restricting vocabulary)—play off against each other in a way that makes it hard to say precisely when voice control will be better or worse than manual control. The picture is further complicated because of the continued improvement of computer algorithms that are beginning to address the two major limitations of many current systems (continuous speech recognition and speaker-dependence). However, even if such systems do successfully address these problems, they are likely to be expensive, and for many applications, the cheaper, simpler systems can be useful within the constraints described above. For example, one study has revealed that even with excellent voice recognition technology, the advantages for voice control over mouse and keyboard data entry are mixed (Mitchard & Winkes, 2002). For isolated words, voice control is faster than typing only when typing speed is less than 45 words/minute, and for numerical data entry, the mouse or keypad are superior.

CONTINUOUS CONTROL AND TRACKING

Our discussion of the positioning task focused on guiding a cursor to a fixed target either through fairly direct hand movement (the touch screen or light pen) or as mediated by a control device (the trackball, joystick, or mouse). However, much of the world of both work and daily life is characterized by making a cursor or some corresponding system (e.g., vehicle) output follow or "track" a *continuously moving dynamic* target. This may involve tasks as mundane as bringing the fly swatter down on the moving pest or riding the bicycle around the curve, or as complex as guiding an aircraft through a curved flight path in the sky, guiding your viewpoint through a virtual environment, or bringing the temperature of a nuclear reactor up to a target value through a carefully controlled trajectory. These cases and many more are described by the generic task of *tracking* (Jagacinski & Flach, 2003; Wickens, 1986); that is, the task of making a system output (the cursor) correspond in time and space to a time-varying command target input.

The Tracking Loop: Basic Elements

Figure 9.4 presents the basic elements of a tracking task. Each element receives a time-varying input and produces a corresponding time-varying output. Hence, every signal in the tracking loop is represented as a function of time, $f(t)$. These elements are described here within the context of automobile driving (Chapter 17), although it is important to think about how they may generalize to any number of different tracking tasks.

When driving an automobile, the *human operator* perceives a discrepancy or *error* between the desired state of the vehicle and its actual state. As an example, the car may have deviated from the center of the lane or may be pointing in a direction away from the road. The driver wishes to reduce this error function of time, $e(t)$. To do so, a force (actually a torque), $f(t)$, is applied to the steering wheel or *control* device. This force in turn produces a rotation, $u(t)$, of the steering wheel itself, called *control output*. (Note that our frame of reference is the

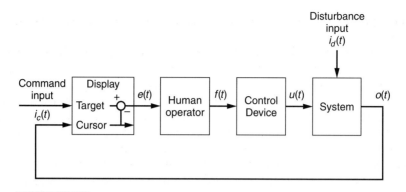

FIGURE 9.4

The tracking loop.

human. Hence, we use the term *output* from the human rather than the term *input* to the system.) The relationship between the force applied and the steering wheel control output is defined as the *control dynamics,* which are responsible for the proprioceptive feedback that the operator receives (see Chapter 5).

Movement of the steering wheel or control device according to a given time function, $u(t)$, then causes the vehicle's actual position to move laterally on the highway, or more generally, the controlled system to change its state. This movement is called the *system output,* $o(t)$. As noted earlier, when presented on a display, the representation of this output position is often called the *cursor.* The relationship between control output, $u(t)$, and system response, $o(t)$, is defined as the *system dynamics.* In discussing positioning control devices, we described the difference between position and velocity system dynamics. If the driver is successful in the correction applied to the steering wheel, then the discrepancy between vehicle position on the highway, $o(t)$ and the desired or "commanded" position at the center of the lane, $i_c(t)$ is reduced. That is, the error, $e(t)$, is reduced to zero. On a display, the symbol representing the input is called the *target.* The difference between the output and input signals (between target and cursor) is the error, $e(t)$, which was the starting point of our discussion. A good driver responds in such a way as to keep $o(t) = i(t)$ or, equivalently, $e(t) = 0$. The system represented in Figure 9.4 is called a closed-loop control system (Powers, 1973). It is sometimes called a negative feedback system because the operator corrects in the opposite direction from (i.e., "negates") the error.

Because errors in tracking stimulate the need for corrective responses, the operator need never respond at all as long as there is no error. This might happen while driving on a straight smooth highway on a windless day. However, errors typically arise from one of two sources. *Command inputs,* $i_c(t)$, are changes in the target that must be tracked. For example, if the road curves, it generates an error for a vehicle traveling in a straight line and so requires a corrective response. *Disturbance inputs,* $i_d(t)$, are those applied directly to the system for which the operator must compensate. For example, a wind gust that blows the car off the center of the lane is a disturbance input. So is an accidental movement of the steering wheel by the driver, as happened in the story at the beginning of the chapter.

The source of all information necessary to implement the corrective response is the *display* (see Chapter 8). For an automobile driver, the display is the field of view seen through the windshield, but for an aircraft pilot making an instrument landing, the display is represented by the instruments depicting pitch, roll, altitude, and course information (see Chapter 17). An important distinction may be drawn between *pursuit* and *compensatory* tracking displays, as shown in Figure 9.5. A pursuit display presents an independent representation of movement of both the target and the cursor against the frame of the display. Thus, the driver of a vehicle sees a pursuit display, since movement of the automobile can be distinguished and viewed independently from the curvature of the road (the command input; Fig. 9.5a). A compensatory display presents only movement of the error relative to a fixed reference on the display. The display provides no indication of whether this error arose from a change in system output or command input (Roscoe et al., 1981). Flight navigation instruments are typically compensatory displays (Fig. 9.5b).

As we noted in Chapter 8, displays may contain *predictive* information regarding the future state of the system, a valuable feature if the system dynamics are sluggish. The automobile display is a kind of predictor because the current direction of heading relative to the vanishing point of the road provides a prediction of the future lateral deviation. The preview is provided by the future curvature of the road in Figure 9.5a.

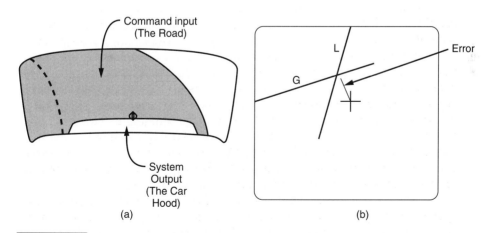

(a)   (b)

FIGURE 9.5

(a) A pursuit display (the automobile); the movement of the car (system output), represented as the position of the hood ornament, can be viewed independently of the movement of the road (command input); (b) a compensatory display (the aircraft instrument landing system). G and L respectively represent the glideslope (commanded vertical input) and localizer (commanded horizontal input). The + is the position of the aircraft. The display will look the same whether the plane moves or the command inputs move.

Finally, tracking performance is typically measured in terms of *error, e(t)*. It may be calculated at each point in time as the absolute deviation and then cumulated and averaged (divided by the number of sample points) over the duration of the tracking trial. This is the mean absolute error (MAE). Sometimes, each error sample may be squared, the squared samples summed, the total divided by the number of samples, and the square root taken. This is the root mean squared error (RMSE). Kelley (1968) discusses different methods of calculating tracking performance.

Now that we have seen the elements of the tracking task, which characterize the human's efforts to make the system output match the command target input, we can ask what characteristics of the human–system interaction make tracking difficult (increased error or increased workload). With this knowledge in mind, the designer can intervene to improve tracking systems. As we will see, some of the problems lie in the tracking system itself, some lie within the human operator's processing limits, and some involve the interaction between the two.

The Input

Drawing a straight line on a piece of paper or driving a car down a straight stretch of road on a windless day are both examples of tracking tasks. There is a command target input and a system output (the pencil point or the vehicle position). But the input does not vary; hence, the task is easy. After you get the original course set, there is nothing to do but move forward, and you can drive fast (or draw fast) about as easily as you can drive (or draw) slowly. However, if the target line follows a wavy course, or if the road is curvy, you have to make corrections, and there is uncertainty to process; as a result, both error and workload can increase if you try to move faster. This happens because the frequency of corrections you must make increases with faster movement and your ability to generate a series of rapid responses to uncertain or unpredictable stimuli (wiggles in the line or highway) is limited. Hence, driving too fast on the curvy road, you will begin to deviate more from the center of the lane, and your workload will be higher if you attempt to stay in the center. We refer to the properties of the tracking input, which determine the frequency with which corrections must be issued, as the *bandwidth* of the input. While the frequency of "wiggles" in a command input is one source of bandwidth, so too is the frequency of disturbances from a disturbance input like wind gusts (or drawing a straight line on the paper in a bouncing car).

In tracking tasks, we typically express the bandwidth in terms of the cycles per second (Hz) of the highest input frequency present in the command or disturbance input. It is very hard for people to perform tracking tasks with random-appearing input having a bandwidth above about 1 Hz. In most naturally occurring systems that people are required to track (cars, planes), the bandwidth is much lower, less than about 0.5 Hz. High bandwidth inputs keep an operator very busy with visual sampling and motor control, but they do not involve very much cognitive complexity. This complexity, however, is contributed by the *order* of a control system, to which we now turn.

Control Order

Position Control. We introduced the concept of control order in our discussion of positioning controls, when position and velocity control systems were contrasted (e.g., the mouse and the joystick). Thus, the *order* of a control system refers to whether a change in the position of the control device (by the human operator) leads to a change in the **position** (zero-order), **velocity** (first-order), or **acceleration** (second-order) of the system output. Consider moving a pen across the paper or a pointer across the blackboard, or moving the computer mouse to position a cursor on the screen. In each case, a new position of the control device leads to a new position of the system output. If you hold the control still, the system output will also be still. This is zero-order control (see Figure 9.6a).

$$O(t) = \int i(t)dt$$

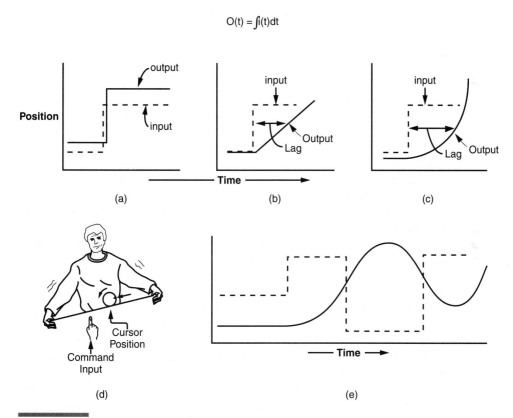

FIGURE 9.6

Control order. The solid line represents the change in position of a system output in response to a sudden change in position of the input (dashed line), both plotted as a function of time. (a) Response of a zero-order system; (b) response of a first-order system. Note the lag. (c) Response of a second-order system. Note the greater lag in (c) than in (b). (d) A second-order system: Tilt the board so the pop can (the cursor) lines up with the command-input finger. (e) Overcorrection and oscillations typical of control of second-order systems.

Velocity Control.　　Now consider the scanner on a typical digital car radio. Depressing the button (a new position) creates a constant rate of change or *velocity* of the frequency setting. In some controls, depressing the button harder or longer leads to a proportionately greater velocity. This is a first-order control. As noted earlier, most pointing-device joysticks use velocity control. The greater the joystick is deflected, the faster will be the cursor motion. An analogous first-order control relation is between the *position* of your steering wheel (input) and the *rate of change* (velocity) of heading of your car (output). As shown in Figure 9.6b, a new steering wheel angle (position) brings about a constant rate-of-change of heading. A greater steering wheel angle leads to a tighter turn (greater rate-of-change of heading). In terms of integral calculus, the order of control corresponds to the number of time integrals between the input and output; that is, for first-order control or velocity control,

$$O(t) = \int i(t)dt$$

This relation holds because the integration of position over time produces a velocity.

For zero-order control,

$$O(t) = i(t)$$ There are no (zero) time integrals.

Both zero-order (position) and first-order (velocity) controls are important in designing manual control devices. Each has its costs and benefits. To some extent, the "which is best?" question has an "it depends" answer. In part, this depends on the goals. If, on the one hand, accurate positioning is very important (like positioning a cursor at a point on a screen), then position control (with a low gain) has its advantages, as we saw in Figure 9.3. On the other hand, if following a moving target or traveling (moving forward) on a path is the goal (matching velocity), then one can see the advantages of first-order velocity control. An important difference is that zero-order control often requires a lot of physical effort to achieve *repeated* actions. Velocity control can be more economical of effort because you just have to set the system to the appropriate velocity (e.g., rounding a curve) and let it go on until system output reaches the desired target.

Any control device that uses first-order dynamics should have a clearly defined and easily reachable **neutral point** at which no velocity is commanded to the cursor. This is because stopping is a frequent default state. This is the advantage of spring-loaded joysticks for velocity control because the natural resting point is set to give zero velocity. It represents a problem when the mouse is configured as a first-order control system, since there is no natural zero point on the mouse tablet. While first-order systems are effort conserving, as shown in Figure 9.6b, first-order systems tend to have a little more lag between when the human commands an output to the device (applies a force) and when the system reaches its desired target position. The amount of lag depends on the gain, which determines how rapid a velocity is produced by a given deflection.

Acceleration Control.　　Consider the astronaut who must maneuver a spacecraft into a precise position by firing thrust rockets. Because of the inertia of the craft, each rocket thrust produces an *acceleration* of the craft for as long as the engine is firing. The time course looks similar to that shown in Figure 9.6c. This, in general, is a second-order acceleration control system, described in the equation

$$o(t) = \int \quad i(t)\, dt \qquad\qquad \int$$

To give yourself an intuitive feel for second-order control, try rolling a pop can to a new position or command input, *i,* on a board, as shown in Figure 9.6d. Second-order systems are generally very difficult to control because they are both *sluggish* and *unstable.* The sluggishness can be seen in the greater lag in Figure 9.6c compared to that in first- and zero-order control (9.6b and a respectively). Both of these properties require the operator to *anticipate* and *predict* (control based on the future, not the present), and, as we learned in Chapters 6 and 8, this is a cognitively demanding source of workload for the human operator.

Because second-order control systems are hard to control, they are rarely if ever intentionally designed into systems. However, a lot of systems that humans are asked to control have a sluggish acceleration-like response to a position input because of the high mass and inertia of controlled elements in the physical world. As we saw, applying a new position to the thrust control on a spacecraft causes it to accelerate endlessly. Applying a new position to the steering wheel via a fixed lateral rotation causes the car's position, with regard to the center of a straight lane, to accelerate, at least initially. In some chemical or energy conversion processes, application of the input (e.g., added heat) yields a second-order response to the controlled variable. Hence, second-order systems are important for human factors practitioners to understand because of the things that designers or trainers can do to address their harmful effects (increased tracking error and workload) when humans must control them.

Because of their long lags, second order systems can only be successfully controlled if the tracker anticipates, inputting a control now, for an error that will be predicted to occur in the future. Without such anticipation, unstable behavior will result. As we learned in Chapters 6 and 8, such anticipation is demanding of mental resources and not always done well.

Sometimes anticipation or prediction can be gained by paying attention to the trend in error. One of the best cues about where things will be in the future is for the tracker to perceive *trend* information of where they are going right now— that is, attend to the current rate of change. For example, in driving, one of the best clues to where the vehicle will be with regard to the center of the lane is where and how fast it is heading *now.* This trend information can be gained better by looking down the roadway to see if the direction of heading corresponds with the direction of the road than it can be by looking at the deviation immediately in front of the car. Predictive information can also be obtained from explicit *predictor displays* as described in Chapter 8 (see Figure 8.4a). Finally, as we discuss in Chapter 16, designers often *automate* the control of higher order systems with lags.

Time Delays and Transport Lags

We saw that higher-order systems (and particularly second-order ones) have a lag (see Figs. 9.6b and c). Lags may sometimes occur in systems of lower order as well. When navigating through virtual environments that must be rendered with time-consuming computer graphics routines, there is often a delay between moving the control device and updating the position or viewpoint of the displays (see Chapter 15; Sherman & Craig, 2003). These time delays, or *transport lags,* produce the same problems of anticipation that we saw with higher-order systems: Lags require anticipation, which is a source of human workload and system error.

Gain

As we noted in discussing input devices, system gain describes how much output the system provides from a given amount of input. Hence, gain may be formally defined as the ratio $\Delta O / \Delta I$, where Δ is a given change or difference in the relevant quantity. In a high-gain system, a lot of output is produced by a small change of input. A sports car is typically high gain because a small movement of the steering wheel produces a large change in output (change in heading). Note that gain can be applied to any order system, describing the amount of *change* in position (zero), speed (first), or acceleration (second) produced by a given deflection of the control.

Just as we noted in our discussion of the pointing task, whether high, low, or medium gain is best is somewhat task-dependent. When system output must travel a long distance (or change by a large amount), high-gain systems are best because the large change can be achieved with little control effort (for a position control system) or in a rapid time (for a velocity control system). However, when precise positioning is required, high-gain systems present problems of overshooting and undershooting, or *instability.* Hence, low gain is preferable. As might be expected, gains in the midrange of values are generally best, since they address both issues— reduce effort and maintain stability—to some degree (Wickens, 1986).

Stability

Now that we have introduced concepts of lag (due to higher system order or transport delay), gain, and bandwidth, we can discuss briefly one concept that is extremely important in the human factors of control systems: *stability.* Novice pilots sometimes show unstable altitude control as they oscillate around a desired altitude. Our unfortunate driver in the chapter's beginning story also suffered instability of control. This is an example of unstable behavior known as *closed-loop instability.* It is sometimes called *negative feedback instability* because of the operator's well-intentioned but ineffective efforts to correct in a direction that will reduce the error (i.e., to *negate* the error). Closed-loop instability results from a particular combination of three factors:

1. There is a lag somewhere in the total control loop in Figure 9.4, either from the system lag or from the human operator's response time.
2. The gain is too high. This high gain can represent either the system's gain—too much heading change for a given steering wheel deflection— or the human's gain—a tendency to overcorrect if there is an error (our unfortunate driver).

3. The human is trying to correct an error too rapidly and is not wa⟩ until the lagged system output stabilizes before applying another c⟨ rective input. Technically, this third factor results when the input band width is high relative to the system lag, and the operator chooses to respond with corrections to all of the input "wiggles" (i.e., does not fil ter out the high-frequency inputs).

Exactly how much of each of these quantities (lag, gain, bandwidth) are re sponsible for producing the unstable behavior is beyond the scope of this chap ter, but there are good models of both the machine and the human that have been used to predict the conditions under which this unstable behavior will occur (McRuer, 1980; Wickens, 1986; Wickens & Hollands, 2000; Jagacinski & Flach, 2003). This is, of course, a critical situation for a human performance model to be able to predict.

Human factors engineers can offer five solutions that can be implemented to reduce closed-loop instability: (1) Lower the gain (either by system design or by instructing the operator to do so). (2) Reduce the lags (if possible). This might be done, for example, by reducing the required complexity of graphics in a virtual reality system (Pausch, 1991; Sherman & Craig, 2003). (3) Caution the operator to change strategy in such a way that he or she does not try to correct every input but filters out the high-frequency ones, thereby reducing the bandwidth. (4) Change strategy to seek input that can anticipate and pre dict (like looking farther down the road when driving and attending to head ing, or paying more attention to rate-of-change indicators). (5) Change strategy to go "open loop." This is the final tracking concept we shall now discuss.

Open-Loop Versus Closed-Loop Systems

In all of the examples we have described, we have implicitly assumed that the operator is perceiving an error and trying to correct it; that is, the loop depicted in Figure 9.6 is closed. Suppose, however, that the operator did not try to correct the error but just "knew" where the system output needed to be and responded with the precise correction to the control device necessary to produce that goal. Since the operator does not then need to perceive the error and therefore will not be looking at the system output, this is a situation akin to the loop in Figure 9.6 being broken (i.e., opening the loop). In open-loop behavior the operator is not trying to correct for outputs that may be visible only after system lags. As a result, the operator will not fall prey to the evils of closed-loop instability. Of course, open-loop behavior depends on the operator's knowledge of (1) where the target will be and (2) how the system output will respond to his or her con trol input; that is, a well-developed mental model of the system dynamics (Ja gacinski & Miller, 1978; Chapter 6). Hence, open-loop behavior is typical only of trackers who are highly skilled in their domain.

Open-loop tracking behavior might typify the process control operator (Chapter 16) who knows exactly how much the heat needs to be raised in a process to reach a new temperature, tweaks the control by precisely that amount, and walks away. Such behavior must characterize a skilled baseball hitter who takes one quick look at the fast ball's initial trajectory and knows exactly how to swing the bat to connect. In this case there is no time for closed-loop feedback to guide the response. It also characterizes the skilled computer user who does not need to wait for screen readout prior to depressing each key in a complex sequence of commands. Of course, such users still receive feedback *after* the skill is performed, feedback that will be valuable in learning or "fine tuning" the mental model (Chapter 18).

REMOTE MANIPULATION OR TELEROBOTICS

There are many circumstances in which continuous and direct human control is desirable but not feasible. Two examples are *remote manipulation,* such as when operators control an undersea explorer or an unmanned air vehicle (UAV), and *hazardous manipulation,* such as is involved in the manipulation of highly radioactive material. This task, sometimes known as *telerobotics* (Sheridan, 1997, 2002), possesses several distinct challenges because of the absence of direct viewing. The goal of the designer of such systems is often to create a sense of "telepresence," that is, a sense that the operator is actually immersed within the environment and is directly controlling the manipulation as an extension of his or her arms and hands. Similar goals of creating a sense of presence have been sought by the designers of virtual reality systems (Durlach & Mavor, 1995; Sherman & Craig, 2003; Barfield & Furness, 1995). Yet there are several control features of the situation that prevent this goal from being easily achieved in either telerobotics or virtual reality (Stassen & Smets, 1995).

Time Delay

Systems often encounter time delays between the manipulation of the control and the availability of visual feedback for the controller. In some cases these may be transmission delays. For example, the round-trip delay between earth and the moon is 5 seconds for an operator on earth carrying out remote manipulation on the moon. High-bandwidth display signals that must be transmitted over a low-bandwidth channel also suffer such a delay. Sometimes the delays might simply result from the inherent sluggishness of high-inertial systems that are being controlled. In still other cases, the delays might result from the time it takes for a computer system to construct and update elaborate graphics imagery as the viewpoint is translated through or rotated within the environment. In all cases, such delays present challenges to effective control.

Depth Perception and Image Quality

Teleoperation normally involves tracking or manipulating in three dimensions. Yet, as we saw in Chapter 4, human depth perception in 3-D displays is often less than adequate for precise judgment along the viewing axis of the display. One solution that has proven quite useful is the implementation of stereo. The problem with stereo teleoperation, however, lies in the fact that two cameras must be mounted and two separate dynamic images must be transmitted over what may be a very limited bandwidth channel, for example, a tethered cable connecting a robot on the ocean floor to an operator workstation in the vessel above. Similar constraints on the bandwidth may affect the quality or fuzziness of even a monoscopic image, which could severely hamper the operator's ability to do fine, coordinated movement. It is apparent that the tradeoff between image quality and the speed of image updating grows more severe as the behavior of the controlled robot becomes more dynamic (i.e., its bandwidth increases).

Proprioceptive Feedback

While visual feedback is absolutely critical to remote manipulation tasks, there are many circumstances in which proprioceptive or tactile feedback is also of great importance (Durlach & Mavor, 1995; Sherman & Craig, 2003). This is true because the remote manipulators are often designed so that they can produce extremely great forces, necessary, for example, to move heavy objects or rotate rusted parts. As a consequence, they are capable of doing great damage unless they are very carefully aligned when they come in contact with or apply force to the object of manipulation. Consider, for example, the severe consequences that might result if a remote manipulator accidentally punctured a container of radioactive material by squeezing too hard, or stripped the threads while trying to unscrew a bolt. To prevent such accidents, designers would like to present the same tactile and proprioceptive sensations of touch, feel, pressure, and resistance that we experience as our hands grasp and manipulate objects directly (see Chapter 5). Yet it is extremely challenging to present such feedback effectively and intuitively, particularly when there are substantial loop delays. In some cases, visual feedback of the forces applied must be used to replace or augment the more natural tactile feedback.

The Solutions

Perhaps the most severe problem in many teleoperator systems is the time delay. As we have seen, the most effective solution is to reduce the delay. When the delay is imposed by graphics complexity, it may be feasible to sacrifice some complexity. While this may lower the reality and sense of presence, it is a move that can improve usability (Pausch, 1991).

A second effective solution is to develop predictive displays that are able to anticipate the future motion and position of the manipulator on the basis of present state and the operator's current control actions and future intentions (see Chapter 8). While such prediction tools have proven to be quite useful (Bos et al., 1995), they are only as effective as the quality of the control laws of system dynamics that they embody. Furthermore, the system cannot achieve effective prediction (i.e., preview) of a randomly moving target, and without reliable preview, many of the advantages of prediction are gone.

A third solution is to avoid the delayed feedback problem altogether by implementing a computer model of the system dynamics (without the delay), allowing the operator to implement the required manipulation in "fast time" off line, relying on the now instant feedback from the computer model (Sheridan,

1997, 2002; see Chapter 6). When the operator is satisfied that he or she has created the maneuver effectively, this stored trajectory can be passed on to the real system. This solution has the problem that it places fairly intensive demands on computer power and of course will not be effective if the target environment itself happened to change before the planned manipulation was implemented.

Clearly, as we consider designs in which the human plans an action but the computer is assigned responsibility for carrying out those actions, we are crossing the boundary from manual control to automated control, an issue we discuss in depth in Chapter 16. We also note other important aspects of control that are covered in other chapters: process control because of its high levels of automation and its many facets that have little to do with actual control (e.g., monitoring and diagnosis) are also covered in Chapter 16. Many aspects of ground and air vehicle control are addressed in (Chapter 17). Finally, many characteristics of telerobotics are similar to those being addressed in the implementation of *virtual reality* systems, which is discussed again in Chapter 15.

Chapter 10

Engineering Anthropometry and Workspace Design

John works in a power plant. As part of his daily job duties, he monitors several dozen plant status displays. Some of the displays are located so high that he has to stand on a stool in order to read the displayed values correctly. Being 6 feet 6 inches tall himself, he wonders how short people might do the same job. "Lucky me, at least I don't have to climb a ladder," he calms himself down every time he steps on the stool.

Susan is a "floater" at a manufacturing company. That means she goes from one workstation to another to fill in for workers during their breaks. She is proud that she is skilled at doing different jobs and able to work at different types of workstations. But she is frustrated that most of the workstations are too high for her. "One size fits all!? How come it doesn't fit me, a short person!" She not only feels uncomfortable working at these stations, but worries every day that she may hurt herself someday if she overextends her shoulder or bends forward too much when reaching for a tool.

We do not have to go to a power plant or a manufacturing company to find these types of scenarios. In daily life, we do not like to wear clothes that do not fit our body. We cannot walk steadily if our shoes are of the wrong size. We look awkward and feel terrible when we sit on a chair that is either too wide or too narrow. We cannot reach and grasp an object if it is too high on a wall or too far across a table.

These descriptions seem to offer no new insight to us because they all are common sense. We all seem to know that the physical dimensions of a product or workplace should fit the body dimensions of the user. However, some of us may be surprised to learn that inadequate dimensions are one of the most common causes of error, fatigue, and discomfort because designers often ignore or forget this requirement or do not know how to put it into design.

In many power plants and chemical-processing plants, displays are located so high that operators must stand on stools or ladders in order to read the displayed values. In the cockpits of some U.S. Navy aircrafts, 10 percent of the controls could not be reached even by the tallest aviators, and almost 70 percent of the emergency controls were beyond the reach of the shortest aviators. To find everyday examples, simply pay attention to the desks, chairs, and other furnishings in a classroom or a home. Are they well designed from the human factors point of view? Try to answer this question now, and then answer it again after studying this chapter.

In this chapter we introduce the basic concepts of a scientific discipline called anthropometry, which provides the fundamental basis and quantitative data for matching the physical dimensions of workplaces and products with the body dimensions of intended users. We also describe some general principles and useful rules of thumb for applying anthropometric information in design.

Anthropometry is the study and measurement of human body dimensions. Anthropometric data are used to develop design guidelines for heights, clearances, grips, and reaches of workplaces and equipments for the purpose of accommodating the body dimensions of the potential workforce. Examples include the dimensions of workstations for standing or seated work, production machinery, supermarket checkout counters, and aisles and corridors. The workforce includes men and women who are tall or short, large or small, strong or weak, as well as those who are physically handicapped or have health conditions that limit their physical capacity.

Anthropometric data are also applied in the design of consumer products such as clothes, automobiles, bicycles, furniture, hand tools, and so on. Because products are designed for various types of consumers, an important design requirement is to select and use the most appropriate anthropometric database in design. Grieve and Pheasant (1982) note that "as a rule of thumb, if we take the smallest female and the tallest male in a population, the male will be 30–40 percent taller, 100 percent heavier, and 500 percent stronger." Clearly, products designed on the basis of male anthropometric data would not be appropriate for many female consumers. When designing for an international market, applying the data collected from one country to other regions with significant size differences is inappropriate.

In ergonomics, another use of anthropometric information is found in occupational biomechanics, discussed in Chapter 11. Anthropometric data are used in biomechanical models in conjunction with information about external loads to assess the stress imposed on worker's joints and muscles during the performance of work.

Because of the importance of considering human variability in design, this chapter starts with a discussion of the major sources of human variability and how statistics can help designers analyze human variability and use this information in design. We then describe briefly some of the devices and methods used for anthropometric measurements and the major types of anthropometric data. Some general procedures of applying anthropometric data in design are then introduced, followed by a discussion of the general principles for workspace design. Design of standing and seated work areas is discussed in the last section.

HUMAN VARIABILITY AND STATISTICS
Human Variability

Age Variability.   Everyone knows that the stature of a person changes quickly from childhood to adolescence. In fact, a number of studies have compared the stature of people at each year of age. The data indicate stature increases to about age 20 to 25 (Roche & Davila, 1972; VanCott & Kinkade, 1972) and starts to decrease after about age 35 to 40, and women show more shrinkage than men (Trotter & Gleser, 1951; VanCott & Kinkade, 1972). Unlike stature, some other body dimensions such as weight and chest circumference may increase through age 60 before declining.

Sex Variability.   Adult men are, on average, taller and larger than adult women. However, 12-year-old girls are, on average, taller and heavier than their male counterparts because girls see their maximum growth rate from ages 10 to 12 (about 2.5 in./year), whereas boys see theirs around ages 13 to 15 (about 2.7 in./year). Girls continue to show noticeable growth each year until about age 17, whereas the growth rate for boys tapers off gradually until about age 20 (Stout et al., 1960). On average, adult female dimensions are about 92 percent of the corresponding adult male values (Annis, 1978). However, significant differences exist in the magnitude of the differences between males and females on the various dimensions. Although adult men are generally larger than adult women on most dimensions, some dimensions, such as hip and thigh measurements, do not show major differences between men and women, and women exceed men on a number of dimensions, such as skinfold thickness.

Racial and Ethnic Group Variability.   Body size and proportions vary greatly between different racial and ethnic groups. Anthropometric surveys of black and white males in the U.S. Air Force show that their average height was identical, but blacks tended to have longer arms and legs and shorter torsos than whites (Long & Churchill, 1965; NASA, 1978). Comparisons of the U.S. Air Force data with the Japanese Air Force data (Yokohori, 1972) found that the Japanese were shorter in stature, but their average sitting height did not differ much from the American data. Similar differences were also found between the American, the French, and the Italian anthropometric data. On the basis of these differences, Ashby (1979) states that if a piece of equipment was designed to fit 90 percent of the male U.S. population, it would fit roughly 90 percent of Germans, 80 percent of Frenchmen, 65 percent of Italians, 45 percent of Japanese, 25 percent of Thai, and 10 percent of Vietnamese.

Occupational Variability.   Differences in body size and dimensions can be easily observed between people working in different occupational groups. Professional basketball players are much taller than most American males. Ballet dancers tend to be thinner than average. Existing data show that truck drivers tend to be taller and heavier than average (Sanders, 1977), and coalminers appear to have larger torso and arm circumferences (Ayoub et al., 1982). Occupational variability can result from a number of factors, including the type and amount of physical activity involved in the job, the special physical requirements of certain occupations, and the self-evaluation and self-selection of individuals in making career choices.

Generational or Secular Variability.   Annis (1978) graphed the trend of change in stature of the American population since 1840 and noted that there has been a growth in stature of about 1 cm per decade since the early 1920s. Improved nutrition and living conditions are offered as some of the possible reasons for this growth. However, it appears that this trend toward increasing stature and size is leveling off (Hamil et al., 1976). Griener and Gordon (1990) examined the secular trends in 22 body dimensions of male U.S. Army soldiers and found that some dimensions still show a clear trend of growth (e.g., body weight and shoulder breath), while others are not changing considerably (e.g., leg length).

Transient Diurnal Variability.   Kroemer (1987) notes that a person's body weight varies by up to 1 kg per day because of changes in body water content. The stature of a person may be reduced by up to 5 cm at the end of the day, mostly because of the effects of gravitational force on a person's posture and the thickness of spinal disks. Measuring posture in different positions also may yield different results. For example, leaning erect against a wall may increase stature by up to 2 cm as opposed to free standing. Chest circumference changes with the cycle of breathing. Clothes can also change body dimensions.

Statistical Analysis

In order to deal with these variabilities in engineering design, an anthropometric dimension is analyzed as a statistical distribution rather than a single value. Normal distribution (also called Gaussian distribution in some science and engineering disciplines) is the most commonly used statistical distribution because it approximates most anthropometric data quite closely.

Normal Distribution.   The normal distribution can be visualized as the normal curve, shown in Figure 10.1 as a symmetric, bell-shaped curve. The mean and the standard deviation are two key parameters of the normal distribution. The mean is a measure of central tendency that tells us about the concentration of a group of scores on a scale of measurement. The mean (most often referred to as the average in our everyday conversations) is calculated as the sum of all the individual measurements divided by the sample size (the number of people measured). To put it in a formula form, we have,

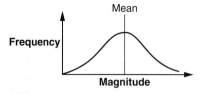

FIGURE 10.1
A graphical representation of the normal distribution.

$$M = \Sigma \, (X_i)/N,$$

where M is the mean of the sample, X_i represents the ith measurement, and N is the sample size.

The standard deviation is a measure of the degree of dispersion or scatter in a group of measured scores. The standard deviation, s, is calculated with the following formula:

$$s = \sqrt{\frac{\Sigma \, (X_i - M)^2}{N - 1}}$$

In Figure 10.1 the value of the mean determines the position of the normal curve along the horizontal axis, and the value of the standard deviation determines whether the normal curve has a more peaked or flat shape. A normal curve with a smaller mean is always located to the left of a normal curve with a larger mean. A small value of the standard deviation produces a peaked normal curve, indicating that most of the measurements are close to the mean value. Conversely, a large value of the standard deviation suggests that the measured data are more scattered from the mean.

Percentiles.   In engineering design, anthropometric data are most often used in percentiles. A percentile value of an anthropometric dimension represents the percentage of the population with a body dimension of a certain size or smaller. This information is particularly important in design because it helps us estimate the percentage of a user population that will be accommodated by a specific design. For example, if the width of a seat surface is designed using the 50th-percentile value of the hip breadth of U.S. males, then we can estimate that about 50 percent of U.S. males (those with narrower hips) can expect to have their hips fully supported by this type of seat surface, whereas the other 50 percent (those with wider hips) cannot.

For normal distributions, the 50th-percentile value is equivalent to the mean of the distribution. If a distribution is not normally distributed, the 50th-percentile value may not be identical to the mean. However, for practical design purposes, we often assume that the two values are identical or approximately the same, just as we assume that most anthropometric dimensions are normally distributed, though they may not be so in reality.

For normal distributions, percentiles can be easily calculated by using Table 10.1 and the following formula together:

$$X = M + F \times s,$$

where X is the percentile value being calculated, M is the mean (50th-percentile value) of the distribution, s is the standard deviation, F is the multiplication factor corresponding to the required percentile, which is the number of standard deviations to be subtracted from or added to the mean. F can be found in Table 10.1.

TABLE 10.1   Multiplication Factors for Percentile Calculation

Percentile	F
1st	−2.326
5th	−1.645
10th	−1.282
25th	−0.674
50th	−0
75th	+0.674
90th	+1.282
95th	+1.645
99th	+2.326

ANTHROPOMETRIC DATA

Measurement Devices and Methods

Many body dimensions can be measured with simple devices. Tapes can be used to measure circumferences, contours, and curvature as well as straight lines. An anthropometer, which is a straight, graduated rod with one sliding and one fixed arm, can be used to measure the distance between two clearly identifiable body landmarks. The spreading caliper has two curved branches joined in a hinge. The distance between the tips of the two branches is read on a scale attached on the caliper. A small sliding compass can be used for measuring short distances, such as hand length and hand breadth. Boards with holes of varying diameters drilled on it can be used to measure finger and limb diameters. Figure 10.2 contains a set of basic anthropometric instruments.

Anthropometric data collected by different measures usually requires clearly identifiable body landmarks and fixed points in space to define the various measurements. For example, *stature* is defined as the distance between the standing surface (often the floor) and the top of the head, whereas *hand length* is the distance from the tip of the middle finger of the right hand to the base of the thumb. The person being measured is required to adopt a standard posture specified by a measurer, who applies simple devices on the body of the subject to obtain the measurements. For most measurements, the subject is asked to adopt an upright straight posture, with body segments either in parallel with each other or at 90° to each other. For example, the subject may be asked to "stand erect, heels together; butt, shoulder blades, and back of head touching a wall . . ." (Kroemer, 1987). The subject usually does not wear clothes and shoes. For seated measurements, the subject is asked to sit with thighs horizontal, lower legs vertical, and feet flat on their horizontal support.

The Morant technique is a commonly used conventional measurement technique that uses a set of grids that are usually attached on two vertical surfaces meeting at right angles. The subject is placed in front of the surfaces, and the body landmarks are projected onto the grids for anthropometric measurements. Photographic methods, filming and videotaping techniques, use of multiple cam-

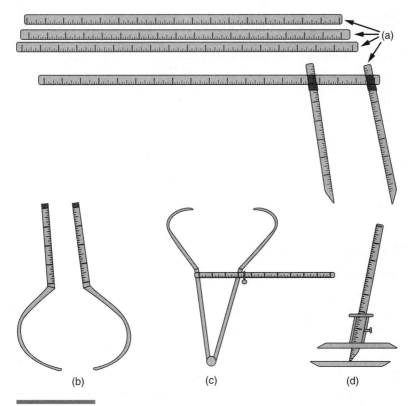

FIGURE 10.2

Basic anthropometric measuring instruments. (a) Anthropometer with straight branches, (b) curved branches for anthropometer, (c) spreading calipers, and (d) sliding compass.

eras and mirrors, holography, and laser techniques are some of the major measurement techniques that have appeared in the past few decades. They continue to be used and improved for various design and research purposes.

To avoid potential ambiguity in interpretation, the following terms are defined and used in anthropometry (Kroemer, 1987):

Height: A straight-line, point-to-point vertical measurement.

Breadth: A straight-line, point-to-point horizontal measurement running across the body or segment.

Depth: A straight-line, point-to-point horizontal measurement running fore-aft the body.

Distance: A straight-line, point-to-point measurement between body landmarks.

Circumference: A closed measurement following a body contour, usually not circular.

Curvature: A point-to-point measurement following a body contour, usually neither circular nor closed.

Civilian and Military Data

Large-scale anthropometric surveys are time consuming, labor-intensive, and expensive. Not surprisingly, significant gaps exist in the world anthropometric database. Most anthropometric surveys were done with special populations, such as pilots or military personnel. Civilian data either do not exist for some populations or are very limited in scope. Much of the civilian data from the United States and other countries were collected many years ago and thus may not be representative of the current user population.

Several large-scale surveys of civilian populations were carried out a few decades ago. O'Brien and Sheldon (1941) conducted a survey of about 10,000 civilian women for garment sizing purposes. The National Center for Health Statistics conducted two large-scale surveys of civilian men and women; the first was conducted from 1960 to 1962 and measured 3,091 men and 3,581 women, and the second was from 1971 to 1974 and measured 13,645 civilians. Two relatively small-scale surveys were carried out recently: the Eastman Kodak Company's (1983) survey of about 100 men and 100 women, and the Marras and Kim's (1993) survey of 384 male and 125 female industrial workers.

The most recent reported civilian anthropometric effort is the Civilian American and European Surface Anthropometry Resource (CAESAR) project, which measured 2,500 European and 2,500 U.S. civilian men and women of various weights, between the ages of 18 and 65. This project used the U.S. Air Force's whole body scanner to digitally scan the human body to provide more comprehensive data than was previously available through traditional measurement methods and to produce 3-D data on the size and shape of human body (Society of Automotive Engineers, 2002).

Surveys of civilian populations were usually limited in scope. Although measurements of body dimensions of military personnel are most extensive and up to date, there may exist significant differences between the military and civilian populations. For example, Marras and Kim (1993) found that significant differences exist in weight and abdominal dimensions between the industrial and military data. An industrial worker of 95th-percentile weight is much heavier than the 95th-percentile U.S. Army soldier. However, 5th-percentile female industrial workers are slightly lighter than U.S. Army women at the same percentile value.

Due to the lack of reliable anthropometric information on civilian populations in the United States and worldwide, the current practice in ergonomic design is to use military data as estimates of the body dimensions of the civilian population. However, the documented differences between civilian and military anthropometric data suggest that designers need to be cautious of any potential undesirable consequences of using these estimates and be ready to make necessary adjustments accordingly in design. Table 10.2 contains a sample of the anthropometric data obtained largely on U.S. Air Force and Army men and women (Clauser et al., 1972; NASA, 1978; White & Churchill, 1971). The dimensions in Table 10.2 are depicted in Figure 10.3 and Figure 10.4.

TABLE 10.2 **Anthropometric Data (unit: inches)**

Measurement	Males 50th percentile	±1S.D	Females 50th percentile	±1S.D.	Population Percentiles, 50/50 Males/Females 5th	50th	95th
Standing							
1. Forward Functional Reach							
a. includes body depth	32.5	1.9	29.2	1.5	27.2	30.7	35.0
at shoulder	(31.2)	(2.2)	(28.1)	(1.7)	(25.7)	(29.5)	(34.1)
b. acromial process to	26.9	1.7	24.6	1.3	22.6	25.6	29.3
function pinch							
c. abdominal extension	(24.4)	(3.5)	(23.8)	(2.6)	(19.1)	(24.1)	(29.3)
to functional pinch							
2. Abdominal Extension Depth	9.2	0.8	8.2	0.8	7.1	8.7	10.2
3. Waist Height	41.9	2.1	40.0	2.9	37.4	40.9	44.7
	(41.3)	(2.1)	(38.8)	(2.2)	(35.8)	(39.9)	(44.5)
4. Tibial Height	17.9	1.1	16.5	0.9	15.3	17.2	19.4
5. Knuckle Height	29.7	1.6	2.80	1.6	25.9	28.8	31.9
6. Elbow Height	43.5	1.8	40.4	1.4	38.0	42.0	45.8
	(45.1)	(2.5)	(42.2)	(2.7)	(38.5)	(43.6)	(48.6)
7. Shoulder Height	56.6	2.4	51.9	2.7	48.4	54.4	59.7
	(57.6)	(3.1)	(56.3)	(2.6)	(49.8)	(55.3)	(61.6)
8. Eye Height	64.7	2.4	59.6	2.2	56.8	62.1	67.8
9. Stature	68.7	2.6	63.8	2.4	60.8	66.2	72.0
	(69.9)	(2.6)	(64.8)	(2.8)	(61.1)	(67.1)	(74.3)
10. Functional Overhead Reach	82.5	3.3	78.4	3.4	74.0	80.5	86.9
Seated							
11. Thigh Clearance Height	5.8	0.6	4.9	0.5	4.3	5.3	6.5
12. Elbow Rest Height	9.5	1.3	9.1	1.2	7.3	9.3	11.4
13. Midshoulder Height	24.5	1.2	22.8	1.0	21.4	23.6	26.1
14. Eye Height	31.0	1.4	29.0	1.2	27.4	29.9	32.8
15. Sitting Height, Normal	34.1	1.5	32.2	1.6	32.0	34.6	37.4
16. Functional Overhead Reach	50.6	3.3	47.2	2.6	43.6	48.7	54.8
17. Knee Height	21.3	1.1	20.1	1.9	18.7	20.7	22.7
18. Popliteal Height	17.2	1.0	16.2	0.7	15.1	16.6	18.4
19. Leg Length	41.4	1.9	39.6	1.7	37.3	40.5	43.9
20. Upper-Leg Length	23.4	1.1	22.6	1.0	21.1	23.0	24.9
21. Buttocks-to-Popliteal Length	19.2	1.0	18.9	1.2	17.2	19.1	20.9
22. Elbow-to-Fit Length	14.2	0.9	12.7	1.1	12.6	14.5	16.2
	(14.6)	(1.2)	(13.0)	(1.2)	(11.4)	(13.8)	(16.2)
23. Upper-Arm Length	14.5	0.7	13.4	0.4	12.9	13.8	15.5
	(14.6)	(1.0)	(13.3)	(0.8)	(12.1)	(13.8)	(16.0)
24. Shoulder Breadth	17.9	0.8	15.4	0.8	14.3	16.7	18.8

(continued)

TABLE 10.2 (continued)

Measurement	Males		Females		Population Percentiles, 50/50 Males/Females		
	50th percentile	±1S.D	50th percentile	±1S.D.	5th	50th	95th
Foot							
25. Hp Breadth	14.0	0.9	15.0	1.0	12.8	14.5	16.3
26. Foot Length	10.5	0.5	9.5	0.4	8.9	10.0	11.2
27. Foot Breadth	3.9	0.2	3.5	0.2	32	3.7	4.2
Hand							
28. Hand Thickness Metacarpal III	1.3	0.1	1.1	0.1	1.0	1.2	1.4
29. Hand Length	7.5	0.4	7.2	0.4	6.7	7.4	8.0
30. Digit Two Length	3.0	0.3	2.7	0.3	2.3	2.8	3.3
31. Hand Breadth	3.4	0.2	3.0	0.2	2.8	3.2	3.6
32. Digit One Length	5.0	0.4	4.4	0.4	3.8	4.7	5.6
33. Breadth of Digit One Interphalangeal Joint	0.9	0.05	0.8	0.05	0.7	0.8	1.0
34. Breadth of Digit Three Interphalangeal Joint	0.7	0.05	0.6	0.04	0.6	0.7	0.8
35. Grip Breadth, Inside Diameter	1.9	0.2	1.7	0.1	1.5	1.8	2.2
36. Hand Spread, Digit One to to Two, 1st Phalangeal Joint	4.9	0.9	3.9	0.7	3.0	4.3	6.1
37. Hand Spread, Digit One to Two, 2nd Phalangeal Joint	4.1	0.7	3.2	0.7	2.3	3.6	5.0
Head							
38. Head Breadth	6.0	0.2	5.7	0.2	5.4	5.9	6.3
39. Interpupillary Breadth	2.4	0.2	2.3	0.2	2.1	2.4	2.6
40. Biocular Breadth	3.6	0.2	3.6	0.2	3.3	3.6	3.9
Other Measurements							
41. Flexion-Extension, Range of Motion of Wrist, Degrees	134	19	141	15	108	138	166
42. Ulnar-Radial Range of Motion of Wrist, Degrees	60	13	67	14	41	63	87
43. Weight, in Pounds	183.4	33.2	146.3	30.7	105.3	164.1	226.8

Source: Eastman Kodak Company, 1983.

Structural and Functional Data

Depending on how they are collected, anthropometric data can be classified into two types: structural (or static) data and functional (or dynamic) data. The two types of data serve different purposes in engineering design.

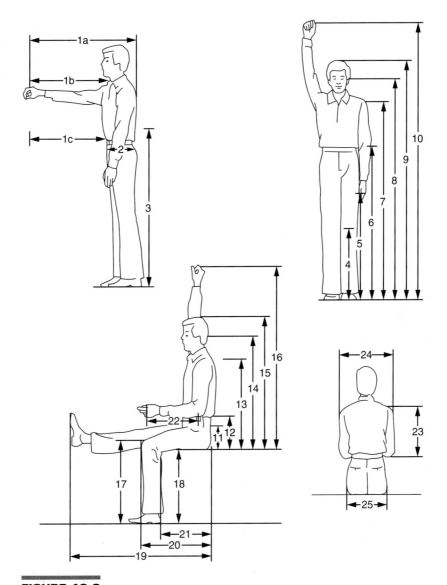

FIGURE 10.3

Anthropometric measures: standing and sitting. (*Source:* Eastman Kodak Company, 1986. *Ergonomic Design for People at Work,* Vol. 1. New York: Van Nostrand Reinhold.)

Structural anthropometric data are measurements of the body dimensions taken with the body in standard and still (static) positions. Examples include stature, shoulder breadth, waist circumference, length of the forearm, and width of the hand.

Functional anthropometric data are obtained when the body adopts various working postures (i.e., when the body segments move with respect to standard

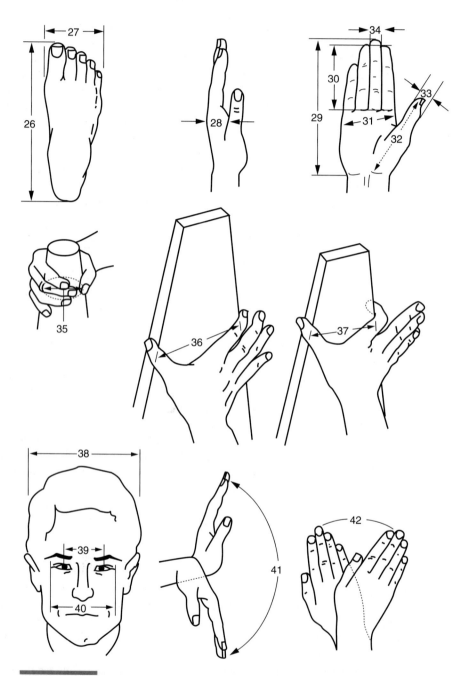

FIGURE 10.4

Anthropometric measures: hand, face, and foot. (*Source:* Eastman Kodak Company, 1986. *Ergonomic Design for People at Work,* Vol. 1. New York: Van Nostrand Reinhold.)

reference points in space). The flexion-extension range of wrist motion and the ulnar-radial range of wrist motion (measures 41 and 42 in Figure 10.4) are examples of functional data. Another example is the *reach envelope,* described later in this chapter. For example, the area that can be reached by the right hand of a standing person defines a standing reach envelope of the right hand, which provides critical information for workspace design for right-handed standing workers. Detailed anthropometric tables, including both static and dynamic data, can be found in Birt, Snyder, and Duncanson (1996) and Roebuck (1995).

Most anthropometric data are static, although work activities can be more accurately represented by dynamic data. Because standard methods do not exist that allow one to convert static into dynamic data, the following procedure suggested by Kroemer (1983) may be useful for designers to make estimates:

1. Heights (stature, eye, shoulder, hip) should be reduced by 3 percent.
2. Elbow height requires no change or an increase of up to 5 percent if elbow needs to be elevated for the work.
3. Forward and lateral reach distances should be decreased by 30 percent if easy reach is desirable, and they can be increased by 20 percent if shoulder and trunk motions are allowed.

Some anthropometric dimensions are highly correlated with each other. For example, a tall person is likely to have long legs and be heavier than a short person. But some dimensions are not highly correlated. It appears, for example, that a person's stature says little about the breadth of that person's head. Detailed information about the correlation among various body dimensions can be found in Roebuck, Kroemer, and Thomson (1975).

Note that it is very unlikely that one can find an "average person" in a given population who is average (50th-percentile value) on all body dimensions. A person with average stature may have a long or short hand, large or small shoulder breath, or wide or narrow feet.

Note also that when designing for people with special needs, e.g., wheelchair users, anthropometric data collected from the corresponding populations should be used (Curtis et al., 1995; Das & Kozey, 1999).

Use of Anthropometric Data in Design

Data contained in anthropometric tables provide critical information with which designers can design workplaces and products. Use of the data, however, requires a thorough analysis of the design problem. The following procedure provides a systematic approach for the use of anthropometric data in design:

1. Determine the user population (the intended users). The key question is, Who will use the product or workplace? People of different age groups have different physical characteristics and requirements. Other factors that must also be considered include gender, race, and ethnic groups; military or civilian populations.
2. Determine the relevant body dimensions. The key question is, Which body dimensions are most important for the design problem? For example, the

design of a doorway must consider the stature and shoulder width of the intended users. The width of a seat surface must accommodate the hip breadth of the users.

3. Determine the percentage of the population to be accommodated. Although a simple answer to this problem is that we should accommodate 100 percent of the population, this answer is not practical or desirable in many design situations because of various financial, economical, and design constraints. For example, there may be limits on how far a seat can be adjusted in a vehicle to accommodate the smallest and largest 1 percent of drivers because to do so would force changes in the overall structure of the design—at a tremendous expense. For most design problems, designers try to accommodate as large a proportion of the intended user population as possible within these constraints. There are three main approaches to this problem.

The first approach is called *design for extremes,* which means that for the design of certain physical dimensions of the workplace or living environment, designers should use the anthropometric data from extreme individuals, sometimes at one end and sometimes at both ends of the anthropometric scale in question. One example is the strength of supporting devices. Designers need to use the body weight of the heaviest users in designing the devices to ensure that the devices are strong enough to support all potential users of the devices.

The second approach, called *design for adjustable range,* suggests that designers should design certain dimensions of equipment or facilities in a way that they can be adjusted to the individual users. Common examples include seats and steering wheels of automobiles and office chairs and desks.

According to the third approach, *design for the average,* designers may use average anthropometric values in the design of certain dimensions if it is impractical or not feasible to design for extremes or for adjustability because of various design constraints. Many checkout counters in department stores and supermarkets, for example, are designed for customers of average height. Although they are not ideal for every customer, they are more convenient to use for most customers than those checkout counters that are either too low or too high. Clearly, it is impractical to adjust the height of a counter for each customer. However, design for the average should be used only as a last resort after having seriously considered the other two design approaches.

4. Determine the percentile value of the selected anthropometric dimension. The key design questions are, Which percentile value of the relevant dimension should be used: 5th, 95th, or some other value? Should the percentile value be selected from the male data or the female data? The percentage of the population to be accommodated determines the percentile value of the relevant anthropometric dimension to be used in design. However, a design decision to accommodate 95 percent of the population does not always mean that the 95th-percentile value should be selected. Designers need to be clear whether they are designing a lower or an upper limit for the physical dimensions of the system or device.

Lower-limit refers to the physical size of the system, not the human user; that is, lower-limit means that the system cannot be smaller, or else it will be unusable by the largest users. Therefore, designers must use a high percentile for the design of lower-limit physical dimensions. For example, if a stool should be strong enough to support a very heavy person, then the 95th or 99th percentile of male body weight should be used as its minimum strength requirement. The logic is simple: If the heaviest (or tallest, largest, widest, etc.) people have no problem with this dimension, then almost everyone can use it. Another example of lower-limit dimensions is the height of a doorway in public places.

In contrast to the lower-limit dimensions, an upper-limit dimension requires the designers to set a maximum value (the upper limit) for the dimension so that a certain percentage of a population can be accommodated. Here, upper limit means that the physical size of the system cannot be bigger than this limit, or else it will not be usable by smallest users. Thus, designers should use a low percentile for the design of upper-limit dimensions. In other words, in order to accommodate 95 percent of the population, the 5th percentile (most often from the female data) should be used in design. The logic is simple: If the shortest (or smallest, lightest, etc.) people have no problem with this dimension, then most people can use it. For example, the size and weight of a tray to be carried by workers should be small enough so that the smallest workers can carry it without any problem. Other examples of upper-limit dimensions include the height of steps in a stairway or the reach distance of control devices.

5. Make necessary design modifications to the data from the anthropometric tables. Most anthropometric measures are taken with nude or nearly nude persons, a method that helps standardize measurements but does not reflect real-life situations. Clothing can change body size considerably. A light shirt for the summer is very different from a heavy coat for winter outdoor activities. Therefore, necessary adjustments must be made in workplace design to accommodate these changes. Allowance for shoes, gloves, and headwear must also be provided if the workers are expected to wear them at work.

Another important reason for data adjustment is that most anthropometric data are obtained with persons standing erect or sitting erect. Most of us do not assume these types of body postures for long. In order to reflect the characteristics of a person's "natural" posture, necessary adjustments must be made. For example, the "natural standing" (slump-posture) eye height is about 2 cm lower than the erect standing eye height, and the "natural sitting" eye height is about 4.5 cm lower than the erect sitting eye height (Hertzberg, 1972). These considerations are critical for designing workplaces that have high viewing requirements.

The use of anthropometric tables to develop and evaluate various possible layouts is often a slow and cumbersome process when several physical dimensions are involved (e.g., a vehicle cab, which involves visibility setting adjustments and several different kinds of reach). Advanced computer graphics now enable the use of more interactive anthropometric models, like Jack or COMBIMAN, in which dynamic renderings of a human body can be created with varying percentile dimensions and then moved through the various dimensions of a computer-simulated workspace in order to assess the adequacy of design (Badler et al., 1990; Chaffin et al., 2001; Karwowski et al., 1990).

6. Use mock-ups or simulators to test the design. Designers often need to evaluate whether the design meets the requirements by building mock-ups or simulators with representative users carrying out simulated tasks. This step is important because various body dimensions are measured separately in a standardized anthropometric survey, but there may exist complicated interactions between the various body dimensions in performing a job. Mock-ups can help reveal potential interactions and help designers make necessary corrections to their preliminary design. A limitation of mock-ups is often encountered because the available human users for evaluation may not span the anthropometric range of potential users. This limitation points again to the potential advantages of anthropometric models, where such users can be simulated.

GENERAL PRINCIPLES FOR WORKSPACE DESIGN

The goal of human factors is to design systems that reduce human error, increase productivity, and enhance safety and comfort. Workplace design is one of the major areas in which human factors professionals can help improve the fit between humans and machines and environments. This section summarizes some general principles of workspace design. Although we describe workspace design only from the human factors perspective, these human factors concerns should be considered in the context of other critical design factors, such as cost, aesthetics, durability, and architectural characteristics. Design is an art as well as a science. There are no formulas to ensure success. But the general guidelines described here may help remind workplace designers of some basic requirements of a workplace and prevent them from designing workplaces that are clearly nonoptimal.

Clearance Requirement of the Largest Users

Clearance problems are among the most often encountered and most important issues in workspace design. The space between and around equipments, the height and width of passageways, and the dimensions provided for the knees, legs, elbows, feet, and head are some examples of clearance design problems. Some workers may not be able to access certain work areas if there is not enough clearance provided. Inadequate clearance may also force some workers to adopt an awkward posture, thus causing discomfort and reducing productivity.

As mentioned earlier, clearance dimensions are lower-limit dimensions and should be adequate for the largest users (typically 95%) who are planning to use the workplace, and then often adjusted upward to reflect the increased space needs of a person with heavy clothes. While design for lower-limit dimensions such as clearance spaces always means that high percentiles are used in design, it does not always mean that male data should be used all the time. Clearly, for female-only workplaces, data from the female population should be used. What is not so obvious is that female data should also be used sometimes for mixed-sex workplaces. For example, the body width of a pregnant woman may need to be used to set the lower limit for some design dimensions.

Reach Requirements of the Smallest Users

Workers often need to extend their arms to reach and operate a hand-operated device or to use their feet to activate a foot pedal. In contrast to the clearance problem, which sets the design limits at the largest users, reach dimensions should be determined on the basis of the reach capabilities of the *smallest* users, typically 5th-percentile. Because heavy clothing reduces a person's reach capability, raw data from an anthropometric table need to be adjusted downward to reflect the reduced reach capacity of a person with heavy clothes.

An important concept here is reach envelope (also called reach area), which is the 3-D space in front of a person that can be reached without leaning forward or stretching. The seated reach envelope for a fifth-percentile female is shown in Figure 10.5, as an example of reach envelopes. The figure show only the right arm's reach area. For practical purposes, the left arm's reach can be approximated as the mirror image of the right arm's. Establishing the shape and size of the reach envelopes for various work situations is an ongoing research area (Sengupta & Das, 2000).

Clearly, objects that must be reached frequently should be located within the reach area and as close to the body as possible. If these objects have different sizes and weights, large and heavy ones should be placed closer to the front of the worker. A worker may be allowed to lean forward occasionally to reach something outside the work area, but such activities should not become a frequent and regular part of jobs with short work cycles.

In considering the issues of object location, manipulation, and reach, issues of strength and fatigue must also be addressed. The same physical layout for two workers of the same physical proportions will have very different long-term health and safety implications if the workers differ substantially in their strength or if, for example, the parts to be lifted and moved from one point in the work space to another differ substantially in their weight. The role of these critical issues is addressed in the next chapter.

Special Requirements of Maintenance People

A well-designed workplace should consider not only the regular functions of the workplace and the workers who work there everyday, but also the maintenance needs and special requirements of maintenance personnel. Because maintenance people often must access areas that do not have to be accessed by regular workers, designers must analyze the special requirements of the maintenance people and design of the workplace accordingly. Because regular workers and maintenance people often have different needs, an adjustable workplace becomes particularly desirable.

Adjustability Requirements

People vary in many anthropometric dimensions, and their own measurements may change as a function of factors such as the clothes they wear on a particular day. Because of the conflicting needs of different people, it is often impossible to have "one size fits all." In considering adjustments as discussed above, designers

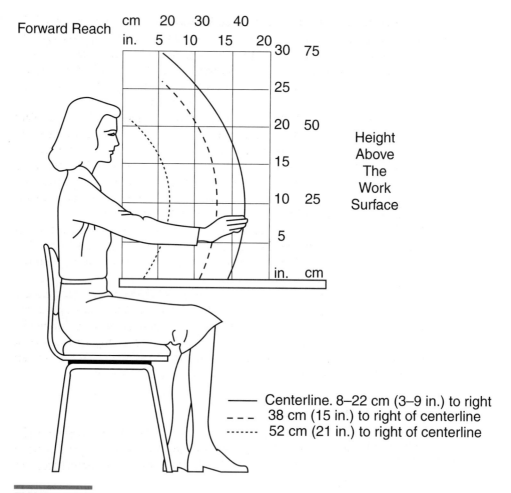

FIGURE 10.5

The seated forward reach of a small female's right hand. (*Source:* Eastman Kodak Company, 1986. *Ergonomic Design for People at Work,* Vol. 1. New York: Van Nostrand Reinhold; developed from data in Faulkner & Day, 1970.)

should also make sure that the adjustment mechanisms are easy to use; otherwise, users are often intimidated by the complexity of the adjustment methods and refuse to use them. For example, the ease of adjusting automobile seating parameters can be greatly influenced both by placing those controls in a location where they can be easily reached and by paying attention to issues of movement compatibility (discussed in Chapter 9) so that the direction in which a control should be moved to adjust the seat in a particular direction is obvious.

There are many ways in which a workplace can be adjusted. The following summarizes four general approaches to workplace adjustment that should be considered in workplace design (Eastman Kodak Company, 1986).

1. *Adjusting the workplace.* The shape, location, and orientation of the workplace may be adjusted to achieve a good fit between the worker and the task. For example, front surface cutouts can be used to allow the worker to move closer to the reach point so that reach requirement can be minimized. Reach distance may also be reduced by height and orientation adjustments relative to the worker and other equipments involved in the same task.

2. *Adjusting the worker position relative to the workplace.* When workplace adjustments are not feasible because they conflict with the requirements of other vital equipment or services or because they exceed budget constraints, designers may consider various ways of adjusting the working position relative to the workplace. Change in seat height and use of platforms or step-up stools are some of the means of achieving vertical adjustability. A swing chair may be used to change the orientation of the worker relative to the equipment.

3. *Adjusting the workpiece.* Lift tables or forklift trucks can be used to adjust the height of a workpiece. Jigs, clamps, and other fixtures can be used to hold a workpiece in a position and orientation for easy viewing and operation. Parts bins can help organize items for easier access.

4. *Adjusting the tool.* An adjustable-length hand tool can allow people with different arm lengths to reach objects at different distances. In an assembly plant, such tools can allow a worker to access an otherwise inaccessible workpiece. Similarly, in a lecture hall, a changeable-length pointing stick allows a speaker to point to items displayed on varying locations of a projection screen without much change in his or her standing position and posture.

Visibility and Normal Line of Sight

Designers should ensure that the visual displays in a workplace can be easily seen and read by the workers. This requires that the eyes are at proper positions with respect to viewing requirements. In this regard, the important concept of "normal" line of sight is of particular relevance.

The normal line of sight is the preferred direction of gaze when the eyes are at condition. It is considered by most researchers to be about 10° to 15° below the horizontal plane (see Figure 10.6). Grandjean, Hunting, and Pidermann (1983) reported the results of a study that showed that the normal line of sight is also the preferred line of sight of computer users watching a screen. Bhatnager, Drury, and Schiro (1985) studied how the height of a screen affected the performance, discomfort, and posture of the users. They found that the best performance and physical conform were observed for the screen height closest to the normal line of sight. Therefore, visual displays should be placed within ±15° in radius around the normal line of sight. When multiple visual displays are used in a workplace, primary displays should be given high priority in space assignment and should be placed in the optimal location.

Of course, presenting visual material within 15° around the normal line of sight is not sufficient to ensure that it will be processed. The visual angle and the contrast of the material must also be adequate for resolving whatever information is presented there, a prediction that also must take into account the viewing

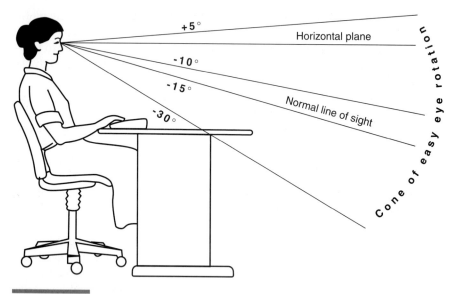

FIGURE 10.6

The normal line of sight and the range of easy eye rotation. (*Source:* Grandjean, E., 1988. *Fitting the Task to the Man* (4th ed.). London: Taylor and Francis. Reprinted by permission of Taylor and Francis.)

distance of the information as well as the visual characteristics of the user. Visibility analysis may also need to address issues of whether critical signals will be seen if they are away from the normal line of sight. Can flashing lights in the periphery be seen? Might other critical warning signals be blocked by obstructions that can obscure critical hazards or information signs in the outside world?

Component Arrangement

Part of a workplace designer's task is to arrange the displays and controls, equipment and tools, and other parts and devices within some physical space. Depending on the characteristics of the user and the tasks in question, optimum arrangements can help a user access and use these components easily and smoothly, whereas a careless arrangement can confuse the user and make the jobs harder. The general issue is to increase overall movement efficiency and reduce total movement distance, whether this is movement of the hands, of the feet, or of the total body through locomotion.

Principles of display layout, discussed in Chapter 8, can be extended to the more general design problem of component arrangements. These principles may be even more critical when applied to components than to displays, since movement of the hands and body to reach those components requires greater effort than movement of the eyes (or attention) to see the displays. In our discussion, the components include displays, controls, equipment and tools, parts and supplies, and any device that a worker needs to accomplish his or her tasks.

1. *Frequency of use principle.* The most frequently used components should be placed in most convenient locations. Frequently used displays should be positioned in the primary viewing area, shown in Figure 10.6; frequently used hand tools should be close to the dominant hand, and frequently used foot pedals should be close to the right foot.

2. *Importance principle.* Those components that are more crucial to the achievement of system goals should be located in the convenient locations. Depending on their levels of importance for a specific application, displays and controls can be prioritized as primary and secondary. Primary displays should be located close to the primary viewing area, which is the space in front of an operator and 10° to 15° within the normal line of sight. Secondary displays can be located at the more peripheral locations. One suggested method of arranging controls according to their priority is shown in Figure 10.7 (Aeronautical Systems Division, 1980).

3. *Sequence of use principle.* Components used in sequence should be located next to each other, and their layout should reflect the sequence of operation. If an electronic assembly worker is expected to install an electronic part on a device immediately after picking the part up from a parts bin, then the parts bin should be close to the device if possible.

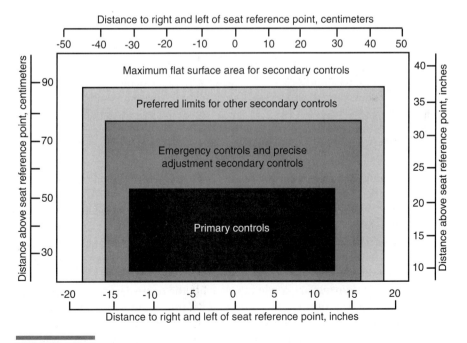

FIGURE 10.7
Preferred vertical surface areas for different classes of control devices. (*Source:* Sanders, M. S., and McCormick, E. J., 1993. *Human Factors in Engineering and Design* (7th ed.). New York: McGraw-Hill. Adapted from Aeronautical Systems Division, 1980.)

4. *Consistency principle.* Components should be laid out with the same component located in the same spatial locations to minimize memory and search requirements. Consistency should be maintained both within the same workplace and across workplaces designed for similar functions. For example, a person would find it much easier to find a copy machine in a university library if copy machines are located at similar locations (e.g., by the elevator) in all the libraries on a campus.

Standardization plays an important role in ensuring that consistency can be maintained across the borders of institutions, companies, and countries. Because arrangements of automobile components are rather standardized within the United States, we can drive cars made by different companies without much problem.

5. *Control-display compatibility principle of colocation.* This is a specific form of stimulus-response compatibility discussed in earlier chapters. In the context of arrangement, this principle states that control devices should be close to their associated displays, and in the case of multiple controls and displays, the layout of controls should reflect the layout of displays to make visible the control-display relationship.

6. *Clutter-avoidance principle.* We discussed the importance of avoiding display clutter; clutter avoidance is equally important in the arrangement of controls. Adequate space must be provided between adjacent controls such as buttons, knobs, and pedals to minimize the risk of accidental activation.

7. *Functional grouping principle.* Components with closely related functions should be placed close to each other. Displays and controls associated with power supply, for example, should be grouped together, whereas those responsible for communications should be close to each other. Various groups of related components should be easily and clearly identifiable. Colors, shapes, sizes, and separation borders are some of the means to distinguish the groups.

Ideally, we would like to see all seven principles satisfied in a design solution. Unfortunately, it is often the case that some of the principles are in conflict with each other and thus cannot be satisfied at the same time. For example, a warning display may be most important for the safe operation of a system, but it may not be the component that is most frequently used. Similarly, a frequently used device is not necessarily the most crucial component. Such situations call for careful tradeoff analysis to decide the relative importance of each principle in the particular situation. Some data suggests that functional grouping and sequence of use principles are more critical than the importance principle in positioning controls and displays (Fowler et al., 1968; Wickens et al., 1997).

Applications of these principles require subjective judgments. For example, expert judgments are needed to evaluate the relative importance of each component and to group various components into functionally related groups. However, quantitative methods such as link analysis and optimization techniques are available that can be used in conjunction with these subjective approaches.

Link analysis is a quantitative and objective method for examining the relationships between components, which can be used as the database for optimizing component arrangements. A link between a pair of components represents a relationship between the two components. The strength of the relationship is reflected by link values. For example, a link value of three for the A-B link (connecting A to B) means that component B has been used three times immediately following (or preceding) the use of A. This is called a sequential link. It may be applied to movement of the eyes across displays in visual scanning, of the hands in a manual task, or of the whole body within a workspace.

Clearly, data about sequential links are useful for the application of sequence of use principle in workplace design. Link analysis also yields a measure of the number of times that each component is used per unit of time. This measure is called functional links. If these component-use data are known for a particular application, then these values can be used to apply the frequency of use principle.

One goal of link analysis is to support a design that minimizes the total travel time across all components; that is, to make the most traveled links the shortest. Figure 10.8 illustrates this process with a simple four-component system. The width of a link represents its strength. The system on the left shows the analysis before redesign, and that on the right shows the analysis after.

With simple systems that have a small number of components, such as that shown in Figure 10.8, designers may adopt a simple trial-and-error procedure in using link data to arrange components. Designers can develop a number of design alternatives and see how the link values change when the arrangements change and finally adopt the design option that best meet the needs of the design. With complex systems that have many components, however, designers may use mathematical methods to help them attack the problem. For example, designers may treat component layout as an optimization problem and use well-developed operations research methods such as linear programming to arrange the components in a way that optimizes some design criterion. The design criterion could be defined as some operational cost, which is expressed as a mathematical function of variables that define the spatial layout of the components.

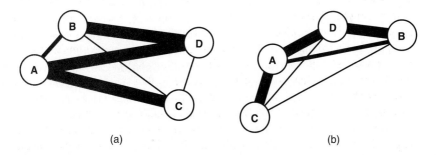

(a)   (b)

FIGURE 10.8

Applying link analysis in system design. The width of a link represents the frequency of Travel (or the strength of connection) between two components. The purpose of the design is to minimize the total travel time across all components. (a) Before reposition of components. Note that thick lines are long. (b) After reposition. Note that the thick lines are shorter.

DESIGN OF STANDING AND SEATED WORK AREAS
Choice Between Standing and Seated Work Areas

In most job environments, workers either stand or sit during work. Standing workplaces are usually used where the workers need to make frequent movements in a large work area, handle heavy or large objects, or exert large forces with their hands. Long-duration standing duty is also observed in the service industry, such as the jobs of the airline or hotel reservation clerks and bank tellers. Because prolonged standing is a strainful posture that puts excessive load on the body and may lead to body fluid accumulation in the legs, a worker should not be required to stand for long time without taking a break. Use of floor mats and shoes with cushioned soles may also help increase a standing worker's comfort.

Whenever possible, a seated workplace should be used for long-term duration jobs, because a seated posture is much easier to maintain and much less of a strain to the body. It also allows for better controlled arm movements, provides a stronger sense of balance and safety, and improves blood circulation. Workplace designers must make sure, however, that leg room (leg and knee clearance) is provided for the seated worker. Furthermore, prolonged sitting can be harmful to the lower back. Seated workplaces should also be provided with adjustable chairs and footrests, and workers should be allowed to stand up and walk around after a period of seated work.

A sit-stand workplace is sometimes used as a compromise or tradeoff between the standing and sitting requirements of a job. This type of workplace may be used when some of the job components are best done standing and others are best done sitting. Designers must analyze the job components involved and decide which type of workplace is best for each.

Work Surface Height

The nature of the tasks being performed should determine the correct work surface height for standing or seated work. A simple but useful rule of thumb to determine the work surface height is to design standing working heights at 5 to 10 cm (2–4 in.) below elbow level and to design seated working heights at elbow level unless the job requires precise manipulation or great force application (Ayoub, 1973; Grandjean, 1988; Eastman Kodak Company, 1986).

Whether seated or standing, precise manipulation calls for working heights above the elbow level; the work surface must be raised to a level at which the worker can see clearly without bending his or her back forward. Great force application or coarse work involving much movement requires working heights lower than that specified by the rule of thumb but should not be so low that there is not enough knee or leg room left under the work surface. Figure 10.9 provides a schematic illustration of this rule of thumb for determining the surface height for standing work.

If feasible, working surface height should be adjustable to suit the workers of varying sizes. If it is impossible to do so for financial or various other practical reasons, then working heights should be set according to the anthropometric values of the tallest workers. Shorter workers should be provided with something to stand on.

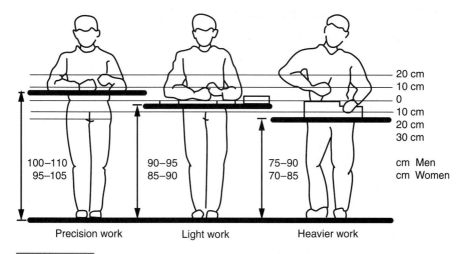

20 cm
10 cm
0
10 cm
20 cm
30 cm

| 100–110 | 90–95 | 75–90 | cm Men |
| 95–105 | 85–90 | 70–85 | cm Women |

Precision work   Light work   Heavier work

FIGURE 10.9

Recommended work surface height for standing work. The reference line (0 cm) is the height of the elbows above the floor. (*Source:* Grandjean, 1988. *Fitting the Task to the Man* [4th ed.]. London: Taylor and Francis.)

Work Surface Depth

An important concept in determining work surface depth is normal and maximum work areas. These areas were first proposed by Farley (1955) and Barnes (1963). The areas defined by Barnes are shown in Figure 10.10, in which the normal work area in horizontal plane is the area covered by a sweep of the forearm without extending the upper arm, and the maximum work area is the area defined by a sweep of the arm by extending the arm from the shoulder. In defining the normal work area, Barnes assumes that the elbow stays at a fixed point. The normal work area defined by Squires (1956) is also shown in Figure 10.10; it does not make this fixed-elbow assumption.

Clearly, normal and maximum work areas must be considered in determining work surface depth. Items that must be reached immediately or frequently should be located within the normal work area and as close to the body as possible, while other items can be located within the maximum work area. It may be permissible to have a worker occasionally lean forward to reach an item outside the maximum work area, but such reaches should not occur regularly and frequently.

Work Surface Inclination

Most work surfaces are designed as horizontal surfaces. However, a number of studies have shown that slightly slanted surfaces (about 15°) should be used for reading. Eastman and Kamon (1976) and Bridger (1988) found that slant surfaces improve body posture, involve less trunk movement, require less bending of the neck, and produce less worker fatigue and discomfort. However, for other types of visual tasks, such as extensive writing, a slanted surface may not be the

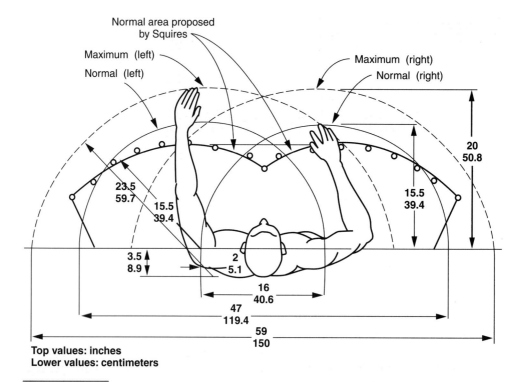

Top values: inches
Lower values: centimeters

FIGURE 10.10

Normal and maximum working areas (in inches and centimeters) proposed by Barnes and normal work area proposed by Squires. (*Source:* Sanders, M. S., and McCormick, E. J., 1993. *Human Factors in Engineering and Design* [7th ed.]. New York: McGraw-Hill. Copyright 1993. Reprinted by permission of the McGraw-Hill Companies.)

best choice. Bendix and Hagberg (1984) found that users preferred horizontal desks for writing, although the same users preferred the slanted desks for reading.

CONCLUSION

Matching the physical layout of the workspace to the physical dimensions and constraints of the user is a necessary but not sufficient task to create a well-human-factored workspace. As we noted, just because a worker can reach a component does not mean that he or she can easily manipulate it or lift it without doing damage to the lower back. To address this second dynamic aspect of workspace design, we must consider the biomechanics of the human body, the issue to which we now turn.

Chapter

11

Biomechanics of Work

Mary is the CEO of a package-shipping company. She and her management team recently decided to increase the package weight limit from 80 pounds per package to 145 pounds, hoping to increase productivity and competitiveness of the company. This decision immediately stirred an uproar among the workers, and the union is planning to organize a strike. The union believes that the new package weight limit puts workers at a great risk of physical injury. "Actually, the current weight limit of 80 pounds is already too high!" some workers complain.

Mary does not wish to put the workers in a dangerous work environment. She does not want to see a strike in her company. She is also afraid of any lawsuits against the company if a worker gets injured in the workplace. But at the same time, Mary wants to see the company survive and succeed in a competitive market, and to do so, she has to constantly improve the productivity. She wonders, "Is the limit of 145 pounds too high? Is it true that 80 pounds is already too heavy? Is there any scientific answer to these questions?"

In the previous chapter, we discussed the importance of ensuring the fit between the physical dimensions of products and workplaces and the body dimensions of the users. Products and workplaces that are not designed according to the anthropometric characteristics of the users will either prevent the worker from using them or force them to adopt awkward postures that are hard to maintain and stressful to the body.

Awkward postures are not the only factor that can cause physical stress to the body. In this chapter, we bring another important factor into our discussion about ergonomic design of workplaces and devices. This factor is concerned with the mechanical forces exerted by a worker in performing a task such as lifting a load or using a hand tool. In fact, awkward postures and heavy exertion forces are two major causes of musculoskeletal problems, whose prevalence and severity can be illustrated with the following statistics.

According to a report of the National Institute for Occupational Safety and Health (NIOSH, 1981), about half a million workers in the United States suffer some kind of overexertion injury each year. The two most prevalent musculoskeletal problems are low back pain and upper-extremity (fingers, hands, wrists, arms, and shoulders) cumulative trauma disorders. About 60 percent of the overexertion injuries reported each year involve lifting and back pain. The National Council on Compensation Insurance estimates that low-back-pain-related worker compensation payments and indirect costs total about $27 billion to $56 billion in the United States (Pope et al., 1991). Armstrong and Silverstein (1987) found that in industries where the work requires repetitive hand and arm exertions, more than one in 10 workers annually reported upper-extremity cumulative trauma disorders (UECTDs).

In this chapter we introduce the scientific discipline of occupational biomechanics, which plays a major role in studying and analyzing human performance and musculoskeletal problems in manual material handling and provides the fundamental scientific basis for ergonomic analysis of physical work. As defined by Chaffin, Andersson, and Martin (1999, p. xv), occupational biomechanics is "a science concerned with the mechanical behavior of the musculoskeletal system and component tissues when physical work is performed. As such, it seeks to provide an understanding of the physics of manual activities in industry."

Occupational biomechanics is an interdisciplinary science that integrates knowledge and techniques from diverse physical, biological, and engineering disciplines. In essence, biomechanics analyzes the human musculoskeletal system as a mechanical system that obeys laws of physics. Thus, the most basic concepts of occupational biomechanics are those concerning the structure and properties of the musculoskeletal system and the laws and concepts of physics. These two aspects of biomechanics are covered first in this chapter. We then discuss low back pain and UECTDs in detail because they are the musculoskeletal problems that occur most often in work environments and incur greatest danger and cost.

THE MUSCULOSKELETAL SYSTEM

The musculoskeletal system is composed of the bones, muscles, and connective tissues, which include ligaments, tendons, fascia, and cartilage. Bone can also be considered a connective tissue. The main functions of the musculoskeletal system are to support and protect the body and body parts, to maintain posture and produce body movement, and to generate heat and maintain body temperature.

Bones and Connective Tissues

There are 206 bones in a human body, and they form the rigid skeletal structure, which plays the major supportive and protective roles in the body. The skeleton establishes the body framework that holds all other body parts together. Some bones protect internal organs, such as the skull, which covers and protects the brain, and the rib cage, which shields the lungs and heart from the outside. Some bones, such as the long bones of the upper and lower extremities, work with the attached muscles to support body movement and activities.

Each of the other four types of connective tissues has its own special functions. Tendons are dense, fibrous connective tissues that attach muscles to bones and transmit the forces exerted by the muscles to the attached bones. Ligaments are also dense, fibrous tissues, but their function is to connect the articular extremities of bones and help stabilize the articulations of bones at joints. Cartilage is a translucent elastic tissue that can be found on some articular bony surfaces and in some organs, such as the nose and the ear. Fascia covers body structures and separates them from each other.

Two or more bones are linked with each other at joints, which can be classified into three types. Most joints are synovial joints, where no tissue exists between the highly lubricated joint surfaces. The other two types of joints are fibrous joints, such as those connecting the bones of the skull through fibrous tissues, and cartilaginous joints, such as those bridging vertebral bones and intervertebral discs. Depending on the type of movement allowed, joints can also be classified as no-mobility joints, hinge joints, pivot joints, and ball-and-socket joints. No-mobility joints, such as the seams in the skull of an adult, do not support movement. A hinge joint, such as the elbow, permits motion in only one plane. A pivot joint, such as the wrist joint, allows two degrees of freedom in movement. A ball-and-socket joint, such as the hip and shoulder, has three degrees of freedom.

Bones change their structure, size, and shape over time as a result of the mechanical loads placed on them. Wolff (1892) suggests that bones are deposited where needed and resorbed where not needed. However, the precise relationships between bone changes and mechanical loads remain unknown. More important, it should be realized that bones can fracture when they are exposed to excess or repetitive loading in the form of bending forces, torsional forces, or combined forces. The amount of load, the number of repetitions, and the frequency of loading are the three most important factors that can cause bone fracture. Further, bone is capable of repairing small fractures if adequate recovery time is given. Thus, the repetition rate of manual exertions or the recovery period after exertions can become significant factors (Chaffin et al., 1999). Connective tissues may also be damaged after excessive or repeated use. For example, heavy loads may increase tension in tendons and cause tendon pain. Excessive use of tendons may also cause inflammation of tendons.

Muscles

The musculoskeletal system has about 400 muscles, which make up about 40 to 50 percent of the body weight. Muscles consume almost half of the body's metabolism, which not only supplies the energy for maintaining body posture and producing body motion but is also used to generate heat and maintain body temperature. The energy metabolism of muscles is discussed in Chapter 12. Here we describe the basic structures and mechanical properties of muscles.

Muscles are composed of bundles of muscle fibers, connective tissue, and nerves. Muscle fibers are long, cylindrical cells consisting largely of contractile elements called myofibrils. Muscles with larger cross-sections are able to exert larger forces. The connective tissue of muscle provides a channel through which nerves and blood vessels enter and leave the muscle. Muscles contain sensory and motor nerve fibers. Information about the length and tension of the muscle is transmitted through sensory nerve fibers to the central nervous system. Muscle activities are regulated by motor nerve fibers, which transmit impulses from the central nervous system to the muscles. Each motor nerve fiber regulates a group of related muscle fibers through its branches. The group of muscle fibers regulated by the branches of the same motor nerve is called *a motor unit,* which is the basic functional unit of the muscle.

Muscles can contract concentrically, eccentrically, and isometrically in response to motor nerve impulses. A concentric contraction is also called an isotonic contraction, in which the muscle shortens while contracting and producing a constant internal muscle force. An eccentric contraction is one in which the muscle lengthens while contracting, which occurs when the external force is greater than the internal muscle force. In an isometric contraction, the muscle length remains unchanged during the contraction process. Concentric contractions can be observed in the arm flexor muscles when an object is lifted upward. Eccentric contractions can be seen when a person picks up a heavy object and is unable to hold it in the desired position, and the muscles are forcibly lengthened (Eastman Kodak Company, 1986). Isometric contractions occur when a person pauses during lifting and holds the object in a static position. Muscle contraction produces muscle force or tension, which is transmitted to bones through tendons and is used to maintain body posture and perform physical work.

Currently, no measuring device exists that can measure the tensions within the muscle directly. Hence, muscle "strength" is inferred from the amount of force or torque it exerts. Torque, also called moment, is the product of force and the perpendicular distance from its line of action to the axis of rotation. The movement of an arm is an example of torque; the axis of rotation is at the center of the joint at the elbow or the shoulder. The torque generated by arm movement transforms arm muscle contraction into physical work, such as pulling or pushing an object. Similarly, torques generated by movements of other body parts allow one to accomplish a variety of physical tasks.

Muscle strength is the amount and direction of force or torque measured by a measuring device under standardized measuring procedures (Chaffin et al., 1999; Kroemer et al., 1994). Muscle strength can be classified as static strength and dynamic strength. Static strength is also called isometric strength, which is the maximal voluntary isometric muscle exertion level. More specifically, static strength is measured when a group of static exertions is performed. Each lasts about 4 to 6 sec, with 30 to 120 sec rests provided between exertions. The mean exertion levels of the first 3 sec of the steady exertions are used as the measured strength level.

Dynamic muscle strength is more difficult to measure than static strength, because body accelerations have significant effects on the muscle force measured. Therefore, dynamic strength data can vary considerably depending on the dynamics of the task and the way in which the subjects perform it. Several methods have been developed to help standardize the measurement of dynamic strength. One method uses specially designed isokinetic equipments to ensure fixed-speed body motion by providing a variable resistance to the motion.

Another method, called psychophysical method, requires the subjects to adjust the load upward or downward after each trial in a simulated task situation until they believe the load has reached their maximum capacity. Clearly, a number of factors such as a person's motivation and cooperation may affect the measurement of a person's dynamic strength using the psychophysical method. However, until more comprehensive methods are developed, psychophysical method based on simulations of task situations may be the most accurate method of estimating a person's acceptable strength limit (Chaffin et al., 1999).

Muscle strength data have been collected for some muscle groups. For example, Kamon and Goldfuss (1978) found that the average male worker has a forearm flexion and extension strength of about 276 Newtons when one arm is used, and the average female worker has a forearm strength of about 160 Newtons. Asmussen and Heebol-Nielsen (1961) found that the torque-generating capability of an average male is about 14.1 Newton-meters when turning a handle and about 4.1 Newton-meters when turning a key. The corresponding strength data for an average female are 8.6 Newton-meters and 3.2 Newton-meters respectively (Eastman Kodak Company, 1986).

In performing physical work, excessive loading can cause musculoskeletal problems such as bone fracture and muscle fatigue. To determine whether a load is excessive for a body segment, we need to quantify the magnitude of physical stress imposed on the body segment in performing the task. How do we obtain these quantitative estimates? Biomechanical modeling provides an important method for answering this question.

BIOMECHANICAL MODELS

Biomechanical models are mathematical models of the mechanical properties of the human body. In biomechanical modeling, the musculoskeletal system is analyzed as a system of mechanical links, and the bones and muscles act as a series of levers. Biomechanical models allow one to predict the stress levels on specific musculoskeletal components quantitatively with established methods of physics and mechanical engineering and thus can serve as an analytical tool to help job designers identify and avoid hazardous job situations.

The fundamental basis of biomechanical modeling is the set of three Newton's laws:

1. A mass remains in uniform motion or at rest until acted on by an unbalanced external force.
2. Force is proportional to the acceleration of a mass.
3. Any action is opposed by reaction of equal magnitude.

When a body or a body segment is not in motion, it is described as in static equilibrium. For an object to be in static equilibrium, two conditions must be met: The sum of all external forces acting on an object in static equilibrium must be equal to zero, and the sum of all external moments acting on the object must be equal to zero. These two conditions play an essential role in biomechanical modeling.

The following is a description of a planar, static model of isolated body segments based on Chaffin, Andersson, and Martin (1999). Planar models (also called 2-D models) are often used to analyze symmetric body postures with forces acting in a single plane. Static models assume that a person is in a static position with no movement of the body or body segments. Although the model is elementary, it illustrates the methods of biomechanical modeling. Complex 3-D, whole-body models can be developed as expansions of elementary models.

Single-Segment Planar Static Model

A single-segment model analyzes an isolated body segment with the laws of mechanics to identify the physical stress on the joints and muscles involved. As an illustration, suppose a person is holding a load of 20-kg mass with both hands in front of his body and his forearms are horizontal. The load is equally balanced between the two hands.

The distance between the load and elbow is 36 cm, as shown in the schematic diagram in Figure 11.1. Only the right hand, right forearm, and right elbow are shown in Figure 11.1 and analyzed in the following calculations. The left hand, left forearm, and left elbow follow the same calculation method and yield the same results, because the load is equally balanced between the two hands.

The forces and rotational moments acting on the person's elbow can be determined using the laws of mechanics. First, load weight can be calculated with the equation

$$W = mg$$

where

W is the weight of object measured in Newtons (N),
m is the mass of object measured in kilograms (kg),
g is the gravitational acceleration (a constant of 9.8 m/s^2).

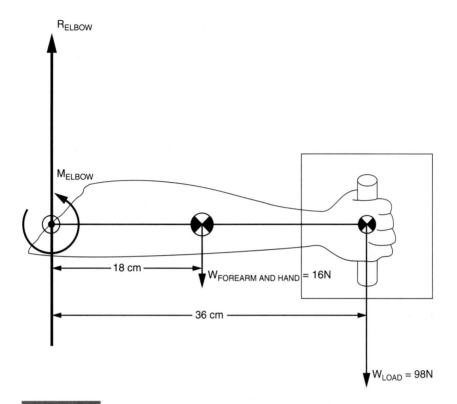

R_{ELBOW}

M_{ELBOW}

18 cm

$W_{FOREARM\ AND\ HAND} = 16N$

36 cm

$W_{LOAD} = 98N$

FIGURE 11.1

A single segment biomechanical model of a forearm and a hand holding a load in the horizontal position. (*Source:* Adapted from Chaffin, D. B., Andersson, G. B. J., and Martin, B. J., 1999. *Occupational Biomechanics* [3rd ed.]. New York: Wiley, Copyright 1999. Reprinted by permission of John Wiley & Sons, Inc.)

For the current problem, we have

$$W = 20 \text{ kg} \times 9.8 \text{ m/s}^2 = 196 \text{ N.}$$

When the center of mass of the load is located exactly between the two hands and the weight is equally balanced between both hands, each hand supports half of the total weight. We have

$$W_{on\text{-}each\text{-}hand} = 98 \text{ N}$$

Furthermore, for a typical adult worker, we assume that the weight of the forearm-hand segment is 16 N, and the distance between the center of mass of the forearm-hand segment and the elbow is 18 cm, as shown in Figure 11.1.

The elbow reactive force R_{elbow} can be calculated using the first condition of equilibrium described above. For the current problem, it means that R_{elbow} must be in the upward direction and large enough to resist the downward weight forces of the load and the forearm-hand segment. That is,

$$\Sigma \text{ (forces at the elbow)} = 0$$
$$- 16 \text{ N} - 98 \text{ N} + R_{elbow} = 0$$
$$R_{elbow} = 114 \text{ N}$$

The elbow moment M_{elbow} can be calculated using the second condition of equilibrium. More specifically, the clockwise moments created by the weight forces of the load and the forearm-hand segment must be counteracted by an equal-magnitude, counterclockwise M_{elbow}. That is,

$$\Sigma \text{ (moments at the elbow)} = 0$$
$$(- 16 \text{ N})(0.18 \ m) + (- 98 \text{ N})(0.36 \text{ m}) + M_{elbow} = 0$$
$$M_{elbow} = 38.16 \text{ N-m}$$

The force on the elbow, described above, will be different on the shoulder. To compute this, one must extend to a two-segment model whose details may be found in Chaffin et al (1999).

LOW-BACK PROBLEMS

As mentioned earlier, low-back pain is perhaps the most costly and prevalent work-related musculoskeletal disorder in industry. According to the estimates of the National Council on Compensation Insurance, low-back pain cases account for approximately one-third of all workers' compensation payments. When indirect costs are included, the total cost estimates range from about $27 to $56 billion in the United States (Pope et al., 1991). About 60 percent of overexertion injuries reported each year in the United States are related to lifting (NIOSH, 1981). Further, it is estimated that low-back pain may affect as much as 50 to 70 percent of the general population due to occupational and other unknown factors (Andersson, 1981; Waters et al., 1993).

Manual material handling involving lifting, bending, and twisting motions of the torso are a major cause of work-related low-back pain and disorders, both in the occurrence rate and the degree of severity. However, low-back problems are not restricted to these situations. Low-back pain is also common in sedentary work environments requiring a prolonged, static sitting posture. Thus, manual handling and seated work become two of the primary job situations in which the biomechanics of the back should be analyzed.

Low-Back Biomechanics of Lifting

The lower back is perhaps the most vulnerable link of the musculoskeletal system in material handling because it is most distant from the load handled by the hands, as shown in Figure 11.2. Both the load and the weight of the upper torso create significant stress on the body structures at the low back, especially at the disc between the fifth lumbar and the first sacral vertebrae (called the L5/S1 lumbosacral disc).

A more accurate determination of the reactive forces and moments at the L5/S1 disc requires the use of a multisegment model, as illustrated when we estimated forces and moments at the shoulder. It also requires the consideration of abdominal pressure, created by the diaphragm and abdominal wall muscles (Morris et al., 1961). However, a simplified single-segment model can be used to obtain a quick estimate of the stress at the low back (Chaffin et al., 1999).

When a person with an upper-body weight of W_{torso} lifts a load with a weight of W_{load}, the load and the upper torso create a combined clockwise rotational moment that can be calculated as

$$M_{load\text{-}to\text{-}torso} = W_{load} \times h + W_{torso} \times b$$

where

h is the horizontal distance from the load to the L5/S1 disc, and
b is the horizontal distance from the center of mass of the torso to the L5/S1 disc.

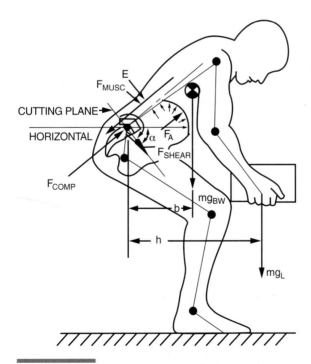

FIGURE 11.2

A low-back biomechanical model of static coplanar lifting. (*Source:* Chaffin, D. B., Andersson, G. B. J., and Martin, B. J., 1999. *Occupational Biomechanics* [3rd ed.]. New York: Wiley. Copyright 1999. Reprinted by permission of John Wiley & Sons, Inc.)

This clockwise rotational moment must be counteracted by a counterclockwise rotational moment, which is produced by the back muscles with a moment arm of about 5 cm. That is,

$$M_{back\text{-}muscle} = F_{back\text{-}muscle} \times 5 \text{ (N-cm)}.$$

According to the second condition of static equilibrium, we have,

$$\Sigma \text{ (moments at the L5/S1 disc)} = 0.$$

That is,

$$F_{muscle} \times 5 = W_{load} \times h + W_{torso} \times b$$
$$F_{muscle} = W_{load} \times h/5 + W_{torso} \times b/5.$$

Because h and b are always much larger than 5 cm, F_{muscle} is always much greater than the sum of the weights of the load and torso. For example, if we assume that $h = 40$ cm and $b = 20$ cm for a typical lifting situation, we have

$$F_{muscle} = W_{load} \times 40/5 + W_{torso} \times 20/5$$
$$= 8 \times W_{load} + 4 \times W_{torso}$$

This equation indicates that for a lifting situation discussed here, which is typical of many lifting tasks, the back muscle force is eight times the load weight and four times the torso weight combined. Suppose a person has a torso weight of 350 N and is lifting a load of 300 N (about 30 kg). The above equation tells us that the back muscle force would be 3,800 N, which may exceed the capacity of some people. If the same person lifts a load of 450 N, the equation indicates that the muscle force would reach 5,000 N, which is at the upper limit of most people's muscle capability. Farfan (1973) estimates that the normal range of strength capability of the erector spinal muscle at the low back is 2,200 to 5,500 N.

In addition to the muscle strength considerations, we must also consider the compression force on the L5/S1 disc, which can be estimated with the following equation on the basis of the first condition of equilibrium:

$$\Sigma \text{ (forces at the L5/S1 disc)} = 0.$$

As a simple approximation, we can ignore the abdominal force, f_a, shown in Figure 11.2, and we have

$$F_{compression} = W_{load} \times \cos \alpha + W_{torso} \times \cos \alpha + F_{muscle}$$

where α is shown in Figure 11.2 as the angle between the horizontal plane and the sacral cutting plane, which is perpendicular to the disc compression force.

This equation suggests that disc compression force can be even greater than the muscle force. For example, suppose $\alpha = 55°$. When a person with a torso weight of 350 N lifts a load of 450 N, we have

$$F_{compression} = 450 \times \cos 55° + 350 \times \cos 55° + 5000$$
$$= 258 + 200 + 5000 = 5458N$$

Disc compression at this level can be hazardous to many workers.

In carrying out a lifting task, several factors influence the load stress placed on the spine. Our analysis considers explicitly two of the factors—the weight and the position of the load relative to the center of the spine. A number of other factors are also important in determining the load on the spine, including the degree of twisting of the torso, the size and shape of the object, and the distance the load is moved. Developing a comprehensive and accurate biomechanical model of the low back that includes all these factors is beyond the scope of this book. For practical ergonomics analysis purposes, the lifting guide developed by the National Institute for Occupational Safety and Health is of great value (described in detail in the next section).

NIOSH LIFTING GUIDE

NIOSH developed an equation in 1981 to help ergonomists and occupational safety and health practitioners analyze lifting demands on low back (NIOSH, 1981). The purpose is to help prevent or reduce the occurrence of lifting-related low-back pain and injuries. The equation, known as the NIOSH lifting equation, provides a method for determining two weight limits associated with two levels of back injury risk. More specifically, the first limit is called an action limit (AL), which represents a weight limit above which a small portion of the population may experience increased risk of injury if they are not trained to perform the lifting task. The second limit, called the maximum permissible limit (MPL), is calculated as three times the action limit. This weight limit represents a lifting condition at which most people would experience a high risk of back injury. Lifting jobs must be redesigned if they are above the MPL. The NIOSH lifting equation can be used to identify high-risk lifting jobs and evaluate alternative job designs; it has received wide acceptance among ergonomics and safety practitioners.

The 1981 equation could only be applied to symmetrical lifting tasks that do not involve torso twisting. It was revised and expanded in 1991 to apply to a greater variety of lifting tasks. The equation allows one to compute an index called the recommended weight limit (RWL), which represents a load value for a specific lifting task that nearly all healthy workers could perform for a substantial period of time without an increased risk of developing lifting-related low-back pain (Waters et al., 1993).

The lifting equation is based on three criteria established on the basis of research results and expert judgments from the perspectives of biomechanics, psychophysics, and work physiology. The biomechanical criterion selects 3.4 kN as the compressive force at the L5/S1 disc that defines an increased risk of low-back injury. In setting the biomechanical criterion, it is realized that lifting tends to incur the greatest stress at the L5/S1 disc and compressive force is likely to be the critical stress vector responsible for disc injuries such as disc herniation, vertebral end-plate fracture, and nerve root irritation. Although shear force and torsional force are also transmitted to the L5/S1 disc during lifting, their effects on back tissues remain unclear and thus are not considered in designing the NIOSH lifting equation.

The 3.4 kN limit was established on the basis of epidemiological data and cadaver data. Epidemiological data from industrial studies provide quantitative evidence linking lifting-related low-back pain and injury incidence with estimated disc compressive force on the L5/S1 disc. For example, Herrin, Taraiedi, and Anderson (1986) traced the medical reports of 6,912 incumbent workers employed in 55 industrial jobs involving 2,934 potentially stressful manual material handling tasks. They found that the rate of reported back problems for jobs with predicted compressive force between 4.5 kN and 6.8 kN was more than 1.5 times greater than that for jobs with compressive force below 4.5 kN. Cadaver data have also been used to evaluate the compressive strength of the spine. For example, Jager and Luttman (1989) found a mean value of 4.4 kN with a standard deviation of 1.88 kN. In general, the studies show that spine specimens are more likely to show damage as the compressive force increases.

Physiological and psychophysical criteria were also used in developing the lifting equation. The physiological criterion was selected to limit loads for repetitive lifting. Activities such as walking, load carrying, and repeated load lifting use more muscle groups than infrequent lifting tasks. These kinds of activities require large energy expenditures, which should not exceed the energy producing capacity of a worker. The physiological criterion sets the limit of maximum energy expenditure for a lifting task at 2.2 to 4.7 kcal/min. The meaning and the importance of these terms will be discussed in the next chapter on work physiology.

The psychophysical criterion is developed on the basis of measurements of the maximum-acceptable-weight-of-lift, which is the amount of weight a person chooses to lift for a given task situation. The maximum-acceptable-weight-of-lift is obtained in experiments in which workers are asked to "work as hard as you can without straining yourself, or without becoming unusually tired, weakened, overheated, or out of breath" (Snook & Ciriello, 1991; Waters et al., 1993). Studies have shown that low-back pain and injuries are less likely to occur for lifting tasks that are judged acceptable by workers than those that are not. The psychophysical criterion of the NIOSH lifting equation was selected to ensure that the lifting demands would not exceed the acceptable lifting capacity of about 99 percent of male workers and 75 percent of female workers, which include about 90 percent of a 50-50 mixed-sex working population.

Based on these three criteria, the following lifting equation was developed for calculating the recommended weight limit (Waters et al., 1993):

$$RWL = LC \times HM \times VM \times DM \times AM \times FM \times CM$$

RWL is the recommended weight limit.

LC is the load constant. It defines the maximum recommended weight for lifting under optimal conditions, which refers to lifting tasks satisfying the following conditions: symmetric lifting position with no torso twisting, occasional lifting, good coupling, ≤ 25cm vertical distance of lifting.

HM is the horizontal multiplier, which reflects the fact that disc compression force increases as the horizontal distance between the load and the spine increases, and thus the maximum acceptable weight limit should be decreased from *LC* as the horizontal distance increases.

VM is the vertical multiplier. The NIOSH lifting equation assumes that the best originating height of the load is 30 inches (or 75 cm) above the floor. Lifting from near the floor (too low) or high above the floor (too high) is more stressful than lifting from 30 inches above the floor. Thus, the allowable weights for lifts should be a function of the absolute distance of the originating height of the load from 30 inches. *VM* accommodates this consideration by using a $|V - 30|$ term in its calculation.

DM is the distance multiplier, established on the basis of results of empirical studies that suggest physical stress increases as the vertical distance of lifting increases.

AM is the asymmetric multiplier. Asymmetric lifting involving torso twisting is more harmful to the spine than symmetric lifting. Therefore, the allowable weight of lift should be reduced when lifting tasks involve asymmetric body twists. *AM* incorporates this consideration into the lifting equation.

CM is the coupling multiplier, which takes on different values depending on whether it is easy to grab and lift the loads. If the loads are equipped with appropriate handles or couplings to help grab and lift the loads, it is regarded as good coupling. If the loads are not equipped with easy-to-grab handles or couplings but are not hard to grab and lift, (e.g., they do not have a large or awkward shape and are not slippery), it is regarded as fair coupling. If the loads are hard to grab and lift, it is regarded as poor coupling.

FM is the frequency multiplier, which is used to reflect the effects of lifting frequency on acceptable lift weights.

The values of the first five components can be determined with the formulas in the Table 11.1. The values of *FM* and *CM* can be found in Tables 11.2 and 11.3 respectively.

H is the horizontal distance between the hands lifting the load and the midpoint between the ankles. Note that although the biomechanical model shown in Figure 11.2 uses the horizontal distance between the hands lifting the load and the L5/S1 in its analysis, the NIOSH lifting equation was established

TABLE 11.1   Definition of Components of NIOSH Lifting Equation (1991)

Component	Metric System	U.S. System
LC (load constant)	23 kg	51 lb
HM (horizontal multiplier)	(25/H)	(10/H)
VM (vertical multiplier)	$(1-0.003\,|V-75|)$	$(1-0.0075\,|V-30|)$
DM (distance multiplier)	(0.82 + 4.5/D)	(0.82 + 1.8/D)
AM (asymmetric multiplier)	(1–0.0032A)	(1–0.0032A)
FM (frequency multiplier)	from Table 10.2	from Table 10.2
CM (coupling multiplier)	from Table 10.3	from Table 10.3

on the basis of using the horizontal distance between the hands lifting the load and the midpoint between the ankles in its calculations, because this distance is much easier to measure in real-world applications than the one shown in Figure 11.2.

V is the vertical distance of the hands from the floor.

TABLE 11.2   Frequency Multiplier (FM) (Note: 75cm = 30 inches)

Frequency lifts/min	Work Duration					
	≤1h		≤2h		≤8h	
	V < 75cm	V ≥ 75cm	V < 75cm	V ≥ 75cm	V < 75cm	V ≥ 75cm
0.2	1.00	1.00	0.95	0.95	0.85	0.85
0.5	0.97	0.97	0.92	0.92	0.81	0.81
1	0.94	0.94	0.88	0.88	0.75	0.75
2	0.91	0.91	0.84	0.84	0.65	0.65
3	0.88	0.88	0.79	0.79	0.55	0.55
4	0.84	0.84	0.72	0.72	0.45	0.45
5	0.80	0.80	0.60	0.60	0.35	0.35
6	0.75	0.75	0.50	0.50	0.27	0.27
7	0.70	0.70	0.42	0.42	0.22	0.22
8	0.60	0.60	0.35	0.35	0.18	0.18
9	0.52	0.52	0.30	0.30	0.00	0.15
10	0.45	0.45	0.26	0.26	0.00	0.13
11	0.41	0.41	0.00	0.23	0.00	0.00
12	0.37	0.37	0.00	0.21	0.00	0.00
13	0.00	0.34	0.00	0.00	0.00	0.00
14	0.00	0.31	0.00	0.00	0.00	0.00
15	0.00	0.28	0.00	0.00	0.00	0.00
>15	0.00	0.00	0.00	0.00	0.00	0.00

Source: Waters, T.R., Putz-Anderson, V., Garg, A., and Fine, L. (1993). Revised NIOSH equation for the design and evaluation of manual lifting tasks, *Ergonomics,* 36, 7, 749-76. Copyright © 1993. Reprinted by permission of Taylor & Francis.

TABLE 11.3 Coupling Multiplier

Couplings	V < 75 cm (30 in.)	V ≥ 75 cm (30 in.)
	Coupling multipliers	
Good	1.00	1.00
Fair	0.95	1.00
Poor	0.90	0.90

D is the vertical travel distance between the origin and the destination of the lift.

A is the angle of asymmetry (measured in degrees), which is the angle of torso twisting involved in lifting a load that is not directly in front of the person.

F is the average frequency of lifting measured in lifts/min (see Table 11.2).

The NIOSH lifting equation allows us to calculate the *RWL* for specific task situations as an index of the baseline capacity of workers. Clearly, the risk of back injury increases as the load lifted exceeds this baseline. To quantify the degree to which a lifting task approaches or exceeds the *RWL*, a lifting index *(LI)* was proposed for the 1991 NIOSH lifting equation, which is defined as the ratio of the load lifted to the *RWL*. The *LI* can be used to estimate the risk of specific lifting tasks in developing low-back disorders and to compare the lifting demands associated with different lifting tasks for the purpose of evaluating and redesigning them (Waters et al., 1993). The current belief is that lifting tasks with an *LI* > 1 are likely to pose an increased risk for some workers. When *LI* > 3, however, many or most workers are at a high risk of developing low-back pain and injury. A recent study of the relationship between the *LI* and one-year prevalence of low-back pain showed a higher low-back pain prevalence in jobs with the *LI* between 2 and 3 than those with no lifting requirements (Waters et al., 1999).

An example of a lifting job that can be analyzed with the NIOSH lifting equation is illustrated in Figure 11.3. The job requires the worker to move tote boxes from an incoming flat conveyor to an outgoing J-hook conveyor at a rate of about three boxes per minute. Each tote box weighs 15 lbs, and the worker performs this job for 8 hours each day. The worker can grasp the tote box quite comfortably. The physical dimensions of the workplace that are relevant for using the NIOSH lifting equation are shown in Figure 11.3. More specifically, the horizontal distance between the hands and the midpoint between the ankles is 16 inches, which is assumed to stay relatively constant during lifting. The vertical distance of the hands from the floor at the starting position of lifting is 44 inches. The vertical distance of the hands from the floor at the destination is 62 inches, and thus the distance lifted is 18 inches (62 − 44 = 18). Although it is not shown in the figure, it is estimated that the worker needs to twist his or her torso about 80° while transferring a tote box

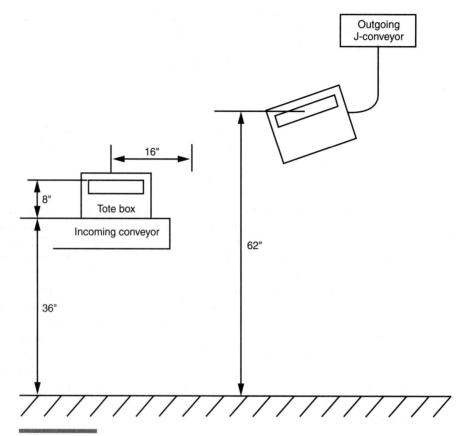

FIGURE 11.3

A schematic representation of the workplace for tote box transfer.

from the incoming to the outgoing conveyor. These parameters can be summarized as follows:

$H = 16''$
$V = 44''$
$D = 18''$
$A = 80°$
$F = 3$ lifts/minute
C: Good coupling
Job duration: 8 hours per day
Weight lifted: 15 lbs

The six multipliers can be calculated as follows:

$HM = 10/H = 10/16 = 0.625$
$VM = 1 - 0.0075 \times |V - 30| = 1 - 0.0075 \times |44 - 30| = 0.895$

$$DM = 0.82 + 1.8/D + 0.82 + 1.8/18 + 0.92$$
$$AM = 1 - 0.0032 \times A = 1 - 0.0032 \times 80 = 0.744$$
$$FM = 0.55 \text{ (from Table 11.2, 3 lifts/min, 8 hours, V > 30'')}$$
$$CM = 1.0 \text{ (from Table 11.3, good coupling)}$$

So we have

$$RWL = 51 \times HM \times VM \times DM \times AM \times FM \times CM$$
$$= 51 \times 0.625 \times 0.895 \times 0.92 \times 0.744 \times 0.55 \times 1.0$$
$$= 10.74 \text{ (lbs)}$$
$$LI = \text{Weight of tote}/RWL = 15/10.74 = 1.40$$

The result of this analysis suggests that some workers would experience an increased risk of back injury while performing this lifting task because the lifting index (*LI*) of 1.4 associated with this job is slightly higher than 1.0. Necessary precautions must be taken to minimize the risk of injury, and the job may need to be redesigned to lower the *LI*.

Although the 1991 NIOSH lifting equation represents a major advancement over the 1981 NIOSH lifting equation, it still has many limitations in its usability. For example, this equation is restricted to analyzing static lifting jobs and is not intended for analyzing jobs with pushing, pulling, or carrying tasks (Dempsey, 1999; Dempsey et al., 2000). Current and future research in ergonomics and occupational biomechanics will undoubtedly provide job analysis methods that are more comprehensive and more widely applicable.

Manual Materials Handling

The 1991 NIOSH lifting equation not only provides a job analysis tool for evaluating lifting demands, it also suggests a list of seven major design parameters that job designers should try to optimize in designing workplaces and devices for material handling.

The horizontal and vertical multipliers in the NIOSH equation remind job designers that loads or material handling devices (MHDs) should be kept close to the body and located at about thigh or waist height if possible. Large packages located on or near the floor are particularly hazardous because they cannot be easily kept close to the body, and a person must lean the torso forward, resulting in a significant increase in low-back disc compression force, as illustrated in the low-back biomechanical model. Thus, large packages should not be presented to a worker at a height lower than about midthigh, or about 30 in. above the floor (Chaffin, 1997). For example, adjustable lift tables can be used to assist workers when handling large or heavy objects, as illustrated in Figure 11.4. Lift tables can also help reduce the vertical travel distance that an object needs to be lifted, which is suggested by the distance multiplier.

The asymmetric multiplier reminds the designers that torso twisting should be minimized in materials handling. Figure 11.5 shows that a simple and careful redesign of workplace layout can help eliminate unnecessary torso twisting

(a)   (b)

FIGURE 11.4

Use of adjustable lift tables to avoid stooped lifting of heavy materials: (a) A lift and tilt table, (b) a pallet lift table. (*Source:* Adapted from the United Auto Workers—Ford Job Improvement Guide, 1988.)

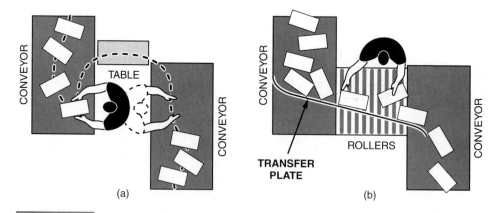

(a)   (b)

FIGURE 11.5

Workplace redesign: (a) Old workplace design requiring lifting and torso twisting; (b) redesigned workplace minimizing these requirements. (*Source:* Adapted from the United Auto Workers—Ford Job Improvement Guide, 1988.)

movements and significantly reduce the risk of worker discomfort and injury. To minimize torso twisting, a lifting task should be designed in a way that requires the use of both hands in front of the body and balances the load between the hands. Extra caution should be exercised in lifting bags of powdered materials because the contents of the bag may shift during lifting. This type of lifting should be avoided if possible.

The NIOSH lifting equation also reminds the job designers that the frequency of lifting should be minimized by adopting adequate lifting and work-rest schedules. Much of the frequent and heavy lifting in a workplace should be done with the assistance of MHDs, and the loads or MHDs should be easy to grasp and handle. Every effort should be made to minimize the weight of the load by selecting lightweight materials if possible.

Clearly, these design parameters do not constitute a complete list of the causes of musculoskeletal problems in manual materials handling. Other factors, such as whole body vibration, psychosocial factors, age, health, physical fitness, and nutrition conditions of a person, are also important in determining the incidence rate and severity of low-back pain in material handling. Furthermore, lifting-related low-back pain comprise only a portion of all cases of low-back pain in the workplace (Frymoyer et al., 1980; National Safety Council, 1990). The following discussion of seated work illustrates another common cause of low-back problems.

Seated Work and Chair Design

In Chapter 10 we mentioned that, whenever possible, a seated workplace should be used for long-duration jobs because a seated posture is much easier to maintain and less strainful to the body. It also allows for better-controlled arm movements, provides a stronger sense of balance and safety, and improves blood circulation. However, the sitting posture has its own cost: It is particularly vulnerable to low-back problems. In fact, low-back pain is common in seated work environments where no lifting or manual handling activities occur.

Low-back disorders in seated work are largely due to a loss of lordotic curvature in the spine and a corresponding increase in disc pressure for the sitting posture. The lumbar (low-back) spine of an adult human when standing erect is curved forward—a spinal posture called *lordosis,* while the thoracic spine is curve backward, known as *kyphosis.* When a person sits down, the pelvis rotates backward and the lumbar lordosis is changed into a kyphosis, particularly when a person sits with a slumped posture. Without proper body support, most people adopt a slumped sitting posture soon after sitting down, in which the front part of the intervertebral discs is compressed and the back part stretched. These forces cause the discs to protrude backward, pressurizing the spinal soft tissues and possibly the nerve roots, which may result in back pain (Bridger, 1995; Keegan, 1953).

Loss of lumbar lordosis in a sitting posture increases the load within the discs because the trunk load moment increases when the pelvis rotates backward and the lumbar spine and torso rotate forward. A number of studies have shown that the disc pressures for upright standing postures were at least 35 to 40 percent lower than those for sitting (Nachemson & Morris, 1964; Chaffin et al., 1999). In different unsupported sitting postures, the lowest pressure was found when sitting with the back straight. As shown in Figure 11.6, disc pressure is much lower in an erect sitting posture than in slumped sitting. Further, disc pressure varies considerably depending on the sitting posture.

To reduce the incidence rate and severity of low-back pain in seated work, workplace designers must pay special attention to the design of seats. A properly designed seat can support a person to adopt a less strainful posture and reduce the loads placed on the spine. Several seat-design parameters are effective in achieving this purpose, including the backrest inclination angle, lumbar support, and arm rest.

Backrest is effective in reducing low-back stress. The most important parameter of back rest design is its inclination angle, which is the angle between the backrest and the seat surface. A 90° back-inclination angle (a seat with a straight back) is inappropriate because it forces a person to adopt a slumped posture. An increase in backrest inclination results in an increase in the transfer of body weight to the backrest and a reduced disc pressure. The optimal inclination angle should be between 110° and 120° (Hosea et al., 1986; Andersson et al., 1974).

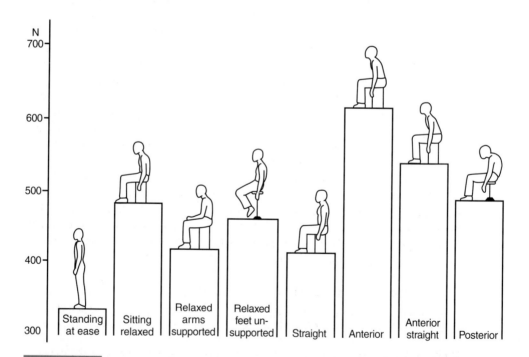

FIGURE 11.6

Disc pressure measurements in standing and unsupported sitting. (*Source:* Andersson, G. B. J., 1974. Biomechanical aspects of sitting: An application to VDT terminals. *Behavior and Information Technology* 6(3), 257–269. Copyright 1974. Reprinted by permission of Taylor & Francis.)

The backrest should also have a pad in the lumbar region (called a lumbar support), which can greatly reduce the low-back stress because it helps a seated person maintain lordosis. Lumbar support is particularly important when the back inclination angle is small. There is also evidence that a lumbar support is as effective as a full back support (Chaffin et al., 1999). The thickness of lumbar support should be about 5 cm. It is desirable, however, that the lumbar support is adjustable in height and size to maximize the comfort for people of different sizes.

Arm rests can help support part of the body weight of a seated person and thus reduce the load on the spine. A tiltable seat surface is also desirable in that it allows variations in posture, although there is no clear evidence that tiltable seats can change the spinal load significantly (Bendix et al., 1985). Properly adjusted seat height, use of cushioned seat surfaces, and adequate leg space can all help reduce back stress. Further, it should be emphasized that no matter how well seats are designed, a person should not adopt a static sitting posture for long. Sedentary workers should have regular breaks in which they should stand up and walk around.

UPPER-EXTREMITY CUMULATIVE TRAUMA DISORDERS

In some industries where repetitive hand and arm exertions are prevalent, cumulative trauma disorders (CTDs) of the upper extremities are common and can be even more costly than low-back problems. Since the early 1980s, there has been a sharp rise in reported CTD cases. Armstrong and Silverstein (1987) found that in workplaces involving frequent hand and arm exertions, more than 1 in 10 workers annually reported CTDs. According to CTD News (1995), the U.S. Bureau of Labor Statistics' most recent report shows that 302,000 CTD-related injuries and illnesses were reported in 1993, which was up more than 7 percent from 1992 and up 63 percent from 1990. CTD News estimates that American employers spend more than $7.4 billion a year in workers' compensation costs and untold billions on medical treatment and other costs such as litigation.

Several other terms have been used to describe upper-extremity cumulative trauma disorders, including *cumulative effect trauma, repetitive motion disorders,* and *repetitive strain injury* (RSI). RSI is commonly used in Europe, and CTD is used in the United States. These terms all emphasize that the disorders are largely due to the cumulative effects of repetitive, prolonged exposures to physical strain and stress.

Common Forms of CTD

CTDs are disorders of the soft tissues in the upper extremities, including the fingers, the hand and wrist, the upper and lower arms, the elbow, and the shoulder.

Tendon-Related CTD. Tendons attach muscles to bones and transfer muscle forces to bones. When an increased blood supply is needed in repetitive work, the muscles may "steal" blood from tendons, particularly in static work in which there is an increased tension in tendons. These conditions may cause *tendon pain.* Excessive and repetitive use of tendons can cause inflammation of tendons, which is a common CTD known as *tendonitis*. The sheaths surrounding tendons provide the necessary nutrition and lubrication to the tendons. When the sheaths also show inflammation and secret excess synovial fluid, the condition is called *tenosynovitis*.

Neuritis. Sensory and motor nerves enter and leave the muscles and connect the muscles to the central nervous system. Repeated use of the upper extremities in awkward posture can stretch the nerves or rub the nerves against bones and cause nerve damage, leading to neuritis. This ailment is accompanied by tingling and numbness in the affected areas of the body.

Ischemia. The sensations of tingling and numbness can also occur when there is a localized tissue anemia due to an obstruction of blood flow. Repeated exposures of the palm to pressure forces from the handle of a hand tool, for example, can cause obstructions of blood flow to fingers, leading to ischemia at the fingers.

Bursitis. Bursitis is the inflammation of a bursa, which is a sac containing synovia or viscous fluid. Bursae can be found near the joints, and they protect tendons from rubbing against bones and help reduce friction between tissues. Bursitis is usually accompanied by a dull pain in the affected part of the body.

CTDs can also be classified according to specific body parts affected, that is, the fingers, hand and wrist, elbow, and shoulder.

CTDs of the Fingers. Repeated and prolonged use of vibrating hand tools may cause numbness, tingling, or pain when the hands are exposed to cold, which is an ailment known as *vibration-induced white fingers* or *Raynaud's phenomenon*. Excessive use of digit fingers against resistance or sharp edges and repeated use of index finger with pistol type hand tools may cause a condition called *trigger finger* in which the affected finger cannot straighten itself once flexed. Forceful extensions of the thumb may cause impaired thumb movement, a condition called *gamekeeper's thumb*.

CTDs of the Hand and Wrist. Carpal tunnel syndrome (CTS) is a common CTD affecting the wrist and hand. Several types of soft tissues pass through a narrow channel in the wrist known as the carpal tunnel. Finger movements are controlled by the muscles in the forearm, which are connected to the fingers by the long tendons passing through the carpal tunnel. Nerves and blood vessels also pass through this channel between the hand and the forearm.

CTS can have many occupational causes, including rapid and repetitive finger movements, repeated exertions with a bent wrist, static exertion for a long time, pressure at the base of the palm, and repeated exposure to hand vibration. CTS has been reported by typists and users of conventional computer keyboards, whose jobs require rapid finger movements and bent wrists (Hedge et al., 1996). Use of conventional keyboards bend the wrists outward; it may also bend the wrist upward if a wrist-rest is not provided, because the surfaces of the keys and the desk are at different heights. As shown in Figure 11.7, bending the wrist causes the finger tendons to rub against adjacent structures of the carpal tunnel and produces large intrawrist forces. Large forces and pressure in the

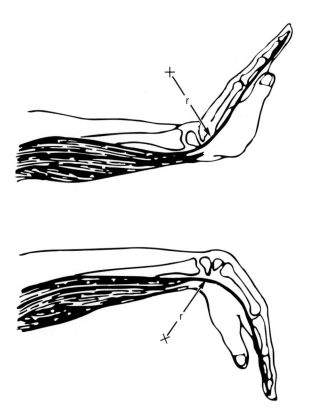

FIGURE 11.7

Bending the wrist causes the finger flexor tendons to rub on adjacent nerves and other tissues of the carpal tunnel. (*Source:* Armstrong, T. J., *1983. An ergonomics guide to carpal tunnel syndrome.* Akron, OH: American Industrial Hygiene Association. Copyright 1983. Reprinted by permission of Industrial Hygiene Association, Fairfax, VA.)

carpal tunnel can cause tendon inflammation and swelling. Carpal tunnel syndrome develops if the median nerve in the carpal tunnel is affected, resulting in tingling and numbness in the palm and fingers.

CTDs at the Elbow.   Many of the muscles of the forearm start from the elbow. Thus, wrist activities may affect the elbow. Repeated forceful wrist activities such as frequent use of a hammer can cause overexertion of the extensor muscles on the outside of the elbow, which leads to tendon irritation, an ailment known as *tennis elbow* or *lateral epicondylitis.* When the flexor muscles and their tendons on the inside of the elbow are affected, the ailment is called *golfer's elbow* or *medial epicondylitis.* Another well-known CTD at the elbow is called *telephone operator's elbow,* which is often found in workplaces where workers rest their elbows on a sharp edge of a desk or a container. The constant pressure from the sharp edge may irritate the nerve and cause tingling and numbness in the vicinity of the little finger.

CTDs at the Shoulder. Working with fast or repetitive arm movements or with static elevated arms may cause shoulder pain and injuries, particularly when the hands are raised above the shoulder height. Such activities may cause CTDs at the shoulder, such as tenosynovitis and bursitis, often known as *impingement syndrome, rotator cuff irritation, swimmer's shoulder, or pitcher's arm.*

Causes and Prevention of CTDs

It is clear that CTDs can have many work-related causes, including repetitive motion, excessive force application, unnatural posture, prolonged static exertion, fast movement, vibration, cold environment, and pressure of tools or sharp edges on soft tissues.

Rapid, repetitive movements of hand or fingers can irritate the tendons and cause the sheaths surrounding tendons to produce excess synovial fluid, leading to tenosynovitis and tendonitis. These problems are more likely to occur when forceful exertions are involved because of the increased tensions in muscles and tendons. Unnatural joint postures such as bent wrists, elevated elbows, or raised shoulders preload and stretch the soft tissues and may press the tendons against the bones and increase their frictions with each other. Using a short tool handle against the base of the palm, grasping sharp objects in the hand, or resting the arm on a sharp edge can cause obstructions of blood flow and possibly irritate the nerves, which may also occur in vibrational or cold environments. These factors often combine in a job situation and increase the risk of CTDs.

A number of nonoccupational factors, including health condition, wrist size, pregnancy, use of oral contraceptives, sex, age, and psychosocial factors, have also been identified as potential causes for CTDs. (Armstrong, 1983; Armstrong et al., 1993; Barton et al., 1992; Posch & Marcotte, 1976). People with preexisting health conditions such as arthritis, diabetes, and peripheral circulatory impairments are particularly vulnerable to the development of CTDs, which also appear to be more common among individuals with a small hand or wrist. Pregnancy, menopause, and use of oral contraceptives are also linked to the development of CTDs, which partially explains why women may be more prone to them. Elderly people have a greater risk of developing CTDs, particularly those with poor general health conditions. Further, psychosocial factors such as job satisfaction, self-esteem, and tolerance of discomfort are important factors in determining a person's vulnerability to developing CTDs.

The existence of the various occupational and nonoccupational causes calls for a comprehensive approach to the prevention of CTDs in workplaces through administrative and engineering methods. Administrative methods include worker education and training and the provision of appropriate work-rest schedules. Engineering methods refer to the use of engineering techniques to redesign the workplace and tools.

Human factors professionals and ergonomists need to work with management and related worker organizations to establish continuing education programs to increase the workers' knowledge of the risks, causes, and preventive methods of CTDs. Attention to worker health conditions, establishment of regular exercise programs and facilities, and creation of a desirable social environment are some of the approaches that management can adopt to minimize the risk of work-related CTDs.

Job schedules should be carefully evaluated and designed to reduce time and pace pressure and provide flexibility. Warm-up exercises before the start of the work and the adoption of adequate work-rest cycles are effective ways of conditioning and relaxing the body in a work environment. Task rotation can increase task variety and help minimize the repetitive components of a job. As discussed in the previous chapter, workers are forced to adopt an awkward posture when the workplace is not designed according to the anthropometric characteristics of workers. Elevated elbows and raised arms are required when using a high work surface. Static postures are unavoidable when the work space is too small to allow any movement. Neck and shoulder pain are likely to develop when the visual displays are located either too high or too low. Therefore, anthropometric design of workplaces is an important method for preventing work-related CTDs.

Use of automated equipments, provision of supporting devices, and careful design of work tools can also help reduce CTD risks. For example, highly repetitive tasks or tasks requiring forceful exertions should be done by automated equipment if possible. Arm rests to support the weight of the arms can help reduce the load on the elbow and shoulder. Design of a work tool should be based on a careful analysis of the joint postures required in using the tool, and every effort should be made to avoid unnatural postures such as bent, twisted, or overextended joint positions. For computer keyboard users, wrist rests with a proper surface contour and soft cloth material can help the wrists maintain a more natural posture and minimize the wrist contact with a potentially cold and sharp table edge.

Hand-Tool Design

Hand tools can be seen everywhere. Screwdrivers, handsaws, hammers, pliers, scissors, forks, knives, and chopsticks constitute only a small sample of the hand tools used by millions of people every day. Hand tools extend the capabilities of the human hands to accomplish tasks that are otherwise impossible or dangerous. However, poorly designed hand tools not only jeopardize task performance and productivity but are a major cause of CTDs. Four guidelines have been developed for the design of hand tools to reduce the risk of developing CTDs (Armstrong, 1983; Chaffin et al., 1999; Greenberg & Chaffin, 1976; Pheasant, 1986; Tichauer, 1978).

    1. *Do not bend the wrist.* Unnatural postures are harmful to the musculoskeletal structures involved. When using a hand tool, the wrist should remain straight rather than bent or twisted. In other words, the hand, wrist, and forearm should remain in alignment when using a hand tool. Straight-handled hand tools often require a bent-wrist posture for certain task situations, while a bent handle may help the worker maintain a straight wrist. As shown in Figure 11.8, the proper shape of the handle should be determined by a careful analysis of the task situation. Figure 11.8 shows that pistol-grip handles are desirable for

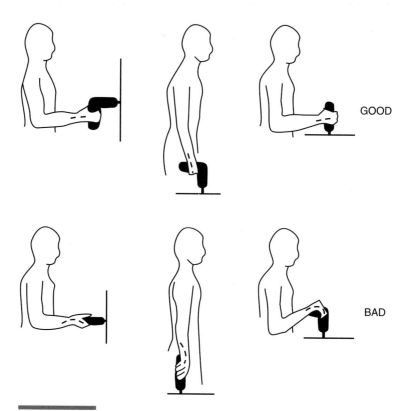

FIGURE 11.8

Wrist posture is determined by the height and orientation of the work surface and the shape of the hand tool. The three "good designs" illustrated in the figure allow the worker to maintain a good posture, that is, a straight wrist. The "bent wrist" shown in the three "bad designs" indicate bad postures, which should be avoided in hand tool and workplace design. (*Source:* Adapted from Armstrong, T. J. 1983. *An ergonomics guide to carpal tunnel syndrome.* Akron, OH: AIHA Ergonomics Guide Series, American Industrial Hygiene Association. Copyright 1983. Reprinted by permission of American Industrial Hygiene Association, Fairfax, VA.)

powered drivers when working with a vertical surface at elbow height or a horizontal surface below waist height, whereas straight handles are better when working with a horizontal surface at elbow height.

2. *Shape tool handles to assist grip.* The center of the palm is vulnerable to force applications because the median nerve, the arteries, and the synovium for the finger flexor tendons are located in the area. Tool handles should be padded, be sufficiently long, and have a small curvature to help distribute the forces on either side of the palm and the fingers.

3. *Provide adequate grip span.* As shown in Figure 11.9, grip strength is a function of grip span, which is the distance between the two points where the hand contacts the two open handles of a hand tool. The grip strength of men is

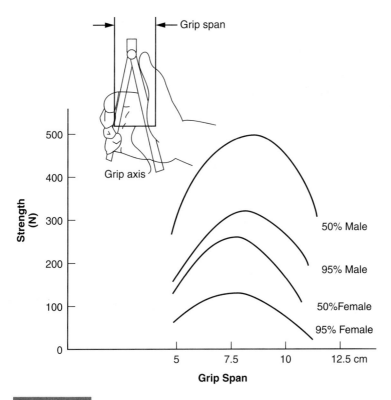

FIGURE 11.9

Maximum grip strength as a function of the width of a handle opening (grip span). (*Source:* Chaffin, D. B., Andersson, G. B. J., and Martin, B. J., 1999. *Occupational Biomechanics.* New York: Wiley. Copyright 1999. Reprinted by permission of John Wiley & Sons, Inc.)

about twice that of women, and both men and women achieve the maximum grip strength when the grip span is about 7 to 8 cm (Greenberg & Chaffin, 1976).

For round tool handles such as those for screwdrivers, the grip span is defined as the diameter of the handles. Ayoub and Lo Presti (1971) found that the maximum grip strength was observed when the grip span was about 4 cm. In general, the handle diameter should not be greater than 4 to 5 cm and should allow slight overlap of the thumb and fingers of the user (Pheasant & O'Neill, 1975; Bridger, 1995).

4. *Provide finger and gloves clearances.* Adequate finger clearance must be provided to ensure a full grip of an object and to minimize the risk of squeezing and crushing the fingers. Similarly, sufficient clearance for gloves should be provided if the workers are expected to wear them, such as in cold workplaces or when handling hazardous materials. Because gloves reduce both the sensory and the motor capabilities of the hands, extra caution must be exercised in tool and job design to avoid tool slippage or accidental activation of neighboring devices.

CONCLUSION

We have seen in this chapter how the human musculoskeletal system can be analyzed with biomechanical methods and how these analyses can give us deeper and quantitative insights into real-world physical stress problems such as low-back pain and CDT problems. These analyses can also help us identify methods of improving workplaces and reducing physical injury risks.

Biomechanical methods discussed in this chapter focus on the mechanical aspects of physical work. Workers can perform a job only if they have enough energy to support their job activities. A person's energy is generated through a complex physiological system, the topic of the next chapter.

Chapter 12

Work Physiology

Judy works as a greeter in a large supermarket. During her 8-hour shift, she stands roughly at the same spot at the entrance of the supermarket, maintaining an upright posture and a constant smile, while greeting shoppers. Although she gets regular breaks, she feels she needs more frequent breaks. But she hesitates to bring it up to the manager, because her manager and coworkers think she already has the easiest job. Being a very sweet lady, Judy does not like to carry any negative thought about anything, and she feels, "Maybe it is because I am old that I get this easy job just standing here." But only she herself knows how terribly tired she feels at the end of each day.

Joe is a construction worker, healthy, strong, and proud of his skills. When his wife received a nice job offer in southern Florida, they left Minnesota, where they had grown up, and moved to the Sunshine State. Joe quickly found a construction job, but for the first time in his life, he found himself easily tiring and not as swift and strong as his coworkers. Under the scorching sun and suffocating humidity, he had to take frequent breaks that slowed down the whole crew's progress. Joe felt badly, but his boss and coworkers were very understanding: "Don't worry. You will get used to it very soon. And you don't have to shovel snow any more. Think about that!"

The human body can maintain the body posture, walk and run, and lift and carry other objects because it has a musculoskeletal system of bones, muscles, and connective tissues. In Chapter 11 we focused on the mechanical aspects of physical work and described how awkward postures and heavy exertion forces can lead to severe musculoskeletal problems such as low-back pain and upper-extremity disorders. We also described how biomechanical methods can be applied to analyze the mechanical behavior of the musculoskeletal system.

In this chapter we shift the focus of discussion from the mechanical to the physiological aspects of muscular work. Physical work is possible only when there is enough energy to support muscular contractions. A central topic of this chapter is how various physiological systems work together to meet the energy-expenditure requirements of work and how these requirements can be measured quantitatively and considered in the analysis of physical work.

This chapter starts with a description of the physiological structure of muscles and how energy is generated and made available for use by the muscles. We then describe how the raw materials for energy production are supplied and its waste products removed by the circulatory and respiratory systems. Energy expenditure requirements of various types of activities are then described, together with a discussion about how the levels of energy expenditure can be measured quantitatively. Clearly, there are upper limits of energy production and muscular work for each individual. The implications of these work capacity limits for ergonomic job design are discussed in the last section of the chapter.

MUSCLE STRUCTURE AND METABOLISM
Muscle Structure

The primary function of muscle is to generate force and produce movement. The body has three types of muscle cells (also known as muscle fibers): smooth muscle, cardiac muscle, and skeletal muscle. *Smooth muscle* is found in the stomach and intestines, blood vessels, urinary bladder, and uterus. Smooth muscle is involved in the digestion of food and the regulation of the internal environment of the body. The contraction of smooth muscle is not normally under conscious control. *Cardiac muscle*, as the name implies, is the muscle of the heart and, like smooth muscle, is not normally under direct conscious control. This chapter is primarily concerned with the third type of muscle, skeletal muscle, which is directly responsible for physical work.

Skeletal muscle is the largest tissue in the body, accounting for about 40 percent of the body weight. It is attached to the bones of the skeleton, and its contraction enables bones to act like levers. The contraction of most skeletal muscles is under direct conscious control, and the movements produced by skeletal muscle make physical work possible.

Each skeletal muscle is made up of thousands of cylindrical, elongated muscle fibers (muscle cells). The individual fibers are surrounded by a network of connective tissues through which blood vessels and nerve fibers pass to the muscle fibers. Each fiber consists of many cylindrical elements arranged in parallel, called *myofibrils*, each of which is further divided longitudinally into a number of *sarcomeres* that are arranged in series and form a repeating pattern along the length of the myofibril. The sarcomeres are the contractile unit of skeletal muscle.

The sarcomere is comprised of two types of protein filaments—a thick filament called *myosin* and a thin one called *actin*. The two types of filaments are layered over each other in alternate dark and light bands, as shown in Figure 12.1. The layers of thick filaments are found in the central region of the sarcomere,

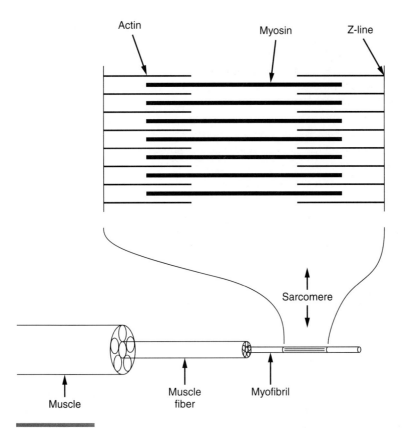

FIGURE 12.1
The structure of muscle.

forming the dark bands, known as the A bands. The layers of thin filaments are connected to either end of the sarcomere to a structure called the Z line. Two successive Z lines define the two ends of one sarcomere.

Aerobic and Anaerobic Metabolism

Physical work is possible only when there is energy to support muscular contraction. Figure 12.2 illustrates the various physiological systems that work together to meet the energy expenditure demands of work. These systems are described in this section on metabolism and the next section on circulatory and respiratory systems.

The energy required for muscular contraction (and for many other physiological functions of the body) comes in the form of high-energy phosphate compounds known as ATP (adenosine triphosphate) and CP (creatine phosphate). These compounds are derived from metabolism of nutrients either in the presence of oxygen *(aerobic metabolism)* or without oxygen *(anaerobic metabolism)*, and the process of creating high-energy phosphate compounds is called *phosphorylation*.

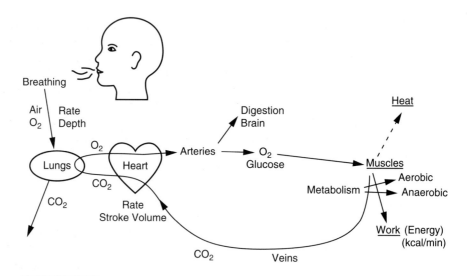

FIGURE 12.2

The various systems that work together to meet the energy expenditure requirements of work.

The ATP and CP compounds are energy carriers and are found in all body cells, where they are formed and used to fuel activities of the body and to sustain life. When energy is required for a reaction such as muscle contraction and relaxation, ATP is converted to ADP (adenosine diphosphate) by splitting off one of the phosphate bonds, and energy is made available for use in this process. In this respect, ATP behaves like a rechargeable battery, which provides a short-term storage of directly available energy (Astrand & Rodahl, 1986).

The body has a very limited capacity for ATP storage. For example, a 75-kg (165-lb) person has about 1 kilocalorie of ATP-stored energy available at any one time. Thus, if a muscle had to rely on its ATP storage for contraction, it would run out of this energy supply in a few seconds. To maintain the contractile activity of a muscle, ATP compounds must be continuously synthesized and replenished at the same rate as they are broken down. There are three sources for supplying ATP: creatine phosphate, oxidative phosphorylation (aerobic metabolism), and anaerobic glycolysis (anaerobic metabolism).

The molecules of CP contain energy that can be transferred to the molecules of ADP to recharge the ADP back to ATP. In this regard, the CP system acts like a backup storage for ATP and provides the most rapid means of replenishing ATP in the muscle cell. However, although the CP system has an energy storage capacity that is about four times that of the ATP system, it is still of very limited capacity. The total energy supply from the ATP and CP systems can only support either heavy work for about 10 seconds or moderately heavy work for about 1 minute.

If muscle activities are to be sustained for a longer period of time, the muscle cells must be able to form ATP from sources other than CP. When enough oxygen is available and muscle activity is at moderate levels (moderate rates of

ATP breakdown), most of the required ATP can be supplied by the process of oxidative phosphorylation. In this process, nutrients (carbohydrates and fatty acids derived from fat) are burned in the presence of oxygen and energy is released to form ATP for muscle work. The nutrients are obtained from the food we eat, and oxygen is obtained from the air we breathe. The nutrients and oxygen are transported to the muscle cells by the blood through the circulatory system. The nutrients can also be obtained from storage in the cells. The liver and muscle cells store the carbohydrates in the form of glycogen, which is derived from glucose in the blood stream. The muscle protein myoglobin allows the muscle to store a very small amount of oxygen, which can be used in short, intense muscle contractions. This oxidative phosphorylation process releases energy for use by the muscles but also produces carbon dioxide as a waste byproduct, which must be removed from the tissues by the circulatory system.

Because it usually requires about 1 to 3 minutes for the circulatory system to respond to increased metabolic demands in performing physical tasks, skeletal muscles often do not have enough oxygen to carry out aerobic metabolism (oxidative phosphorylation) at the beginning of physical work. During this period of time, part of the energy is supplied through *anaerobic glycolysis*, which refers to the generation of energy through the breakdown of glucose to lactic acid in the absence of oxygen.

Although anaerobic glycolysis can produce ATP very rapidly without the presence of oxygen, it has the disadvantage of producing lactic acid as the waste product of this process. Lactic acid causes the acidity of the muscle tissue to increase and is believed to be a major cause of muscle pain and fatigue. The removal of lactic acid requires oxygen, and when oxygen is not available, lactic acid diffuses out the muscle cells and accumulates in the blood, causing an "oxygen debt," which must be paid back when the muscle activity ceases. In other words, to remove these waste products, the muscle must continue to consume oxygen at a high rate after it has stopped contraction so that its original state can be restored. Another disadvantage of anaerobic glycolysis is that it is not efficient in its use of glucose to produce energy. It requires much larger quantities of glucose to produce the same amount of ATP as compared to aerobic metabolism.

When enough oxygen is available, *aerobic metabolism* can supply all the energy required for light or moderate muscular work. Under these circumstances, the body is considered to be in the "steady state." For very heavy work, however, even when adequate oxygen is available, aerobic metabolism may not be able to produce ATP quickly enough to keep pace with the rapid rate of ATP breakdown. Thus, for very heavy work, anaerobic glycolysis serves as an additional source for producing ATP, and fatigue can develop rapidly as lactic acid accumulates in the muscle cells and in the blood.

The overall efficiency with which muscle converts chemical energy to muscular work is only about 20 percent. Metabolic heat accounts for the remaining 80 percent of the energy released in metabolism (Edholm, 1967). The heavier the work, the greater the amount of heat produced. This increased heat production may severely affect the body's ability to maintain a constant body temperature, especially in hot environments.

CIRCULATORY AND RESPIRATORY SYSTEMS

Muscular work can be sustained only when adequate amounts of nutrients and oxygen are continuously supplied to the muscle cells and when the waste products of metabolism such as carbon dioxide can be quickly removed from the body. It is the duty of the circulatory and respiratory systems to perform these functions and to meet these requirements. The circulatory system serves as the transportation system of the body; it delivers oxygen and nutrients to the tissues and removes carbon dioxide and waste products from the tissues. The respiratory system exchanges oxygen and carbon dioxide with the external environment.

The Circulatory System

The circulatory system is composed of the blood and the cardiovascular system, which is the apparatus that transports the blood to the various parts of the body.

The Blood.   Blood consists of three types of blood cells and plasma. Red blood cells transport oxygen to the tissues and help remove carbon dioxide from them. White blood cells fight invading germs and defend the body against infections. Platelets help stop bleeding. Plasma, in which the blood cells are suspended, contains 90 percent water and 10 percent nutrient and salt solutes.

Of the three types of specialized blood cells, red blood cells are of most interest to work physiology because of their oxygen-carrying property. Red blood cells are formed in bone marrow and carry a special type of molecule known as the hemoglobin molecule (Hb). A hemoglobin molecule can combine with four molecules of oxygen to form oxyhemoglobin, allowing it to carry oxygen in the blood efficiently.

The total blood weight of an average adult is about 8 percent of his or her body weight. Because one kilogram of blood has a volume of about 1 liter (L), the total blood volume of an average adult, as measured in liters, is about 8 percent of his or her body weight, as measured in kilograms. Therefore, a 65-kg adult would have a total blood volume of about 5.2 liters ($0.08 \times 65 = 5.2$), of which about 2.85 liters consist of plasma and 2.35 liters of blood cells.

The ability of the blood to deliver oxygen and nutrients to the tissues and remove carbon dioxide from them is reduced if an individual has a low blood volume or a low red-cell count, or if an individual works in a polluted or poorly ventilated environment or at high altitudes where the air has a low oxygen content. Working in these environments increases the stress on the circulatory system because it has to work harder to compensate for the reduced ability of the blood to perform its functions.

The Structure of the Cardiovascular System.   The cardiovascular system is composed of blood vessels through which blood flows, and the heart, which is the pump that generates this flow.

The heart is a four-chambered muscular pump located in the chest cavity. It is divided into right and left halves, each consisting of two chambers, an *atrium* and a *ventricle* (Figure 12.3). Between the two chambers on each side of the

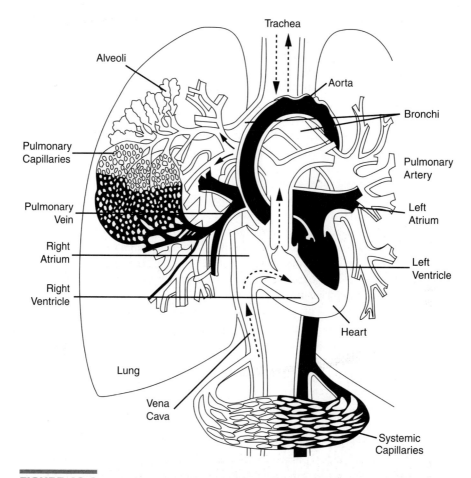

FIGURE 12.3

The anatomy of the circulatory and respiratory systems. The figure shows the major elements of the two systems and the two circuits of blood circulation: systemic (or general body) circulation and the pulmonary (or lung) circulation. (*Source:* Comroe, J. H., Jr., 1966. The lung. *Scientific American*, 220, 56–68. Copyright February 1966 by Scientific American. All rights reserved.)

heart are the atrioventricular valves (AV valves), which force one-directional blood flow from atrium to ventricle but not from ventricle to atrium. Furthermore, the right chambers do not send blood to the left chambers, and vice versa.

The cardiovascular system actually consists of two circuits of blood circulation, both originating and ending in the heart. In both circuits, the vessels carrying blood away from the heart are called *arteries*, and the vessels bringing blood back to the heart are called *veins*.

In the first circulation, known as the *systemic circulation*, fresh blood rich in nutrients and oxygen is pumped out of the left ventricle via a large artery called the aorta. From the aorta a series of ever-branching arteries conduct blood to the tissues and organs of the body. These arteries split into progressively smaller branches, and within each organ or tissue, the arteries branch into the next series of vessels called the *arterioles*. The arterioles further split into a network of tiny, thin blood vessels called *capillaries* that permeates the tissues and organs. It is through this network of capillaries that the fresh blood delivers oxygen and nutrients to the tissues, collects carbon dioxide and waste products from the tissues and carries them away on its way back to the heart.

On its way back to the heart, the blood in the capillaries first merges into larger vessels called *venules*, and the venules are further combined into still larger vessels, veins. Ultimately, the veins from the upper half of the body are joined into a large vein called the superior vena cava, and the veins from the lower half of the body are combined into another large vein called the inferior vena cava. Via these two veins blood is returned to the right atrium of the heart, completing a cycle of the systemic circulation.

In the second circulation, known as the *pulmonary circulation*, blood rich in carbon dioxide is pumped out of the right ventricle via the pulmonary artery, which splits into two arteries, one for each lung. Similar to the systemic circulation, the arteries branch into arterioles, which then split into capillaries. Through the bed of capillaries in the lungs, blood expels carbon dioxide and absorbs oxygen (a process called *oxygenation*). On its way back to the heart, the oxygenated blood in the capillaries first merges into venules and then into progressively larger veins. Finally, via the largest of these veins, the pulmonary veins, the oxygenated blood leaves the lungs and returns to the left atrium of the heart, completing a cycle of the pulmonary circulation.

Blood Flow and Distribution.   The heart generates the pressure to move blood along the arteries, arterioles, capillaries, venules, and veins. The heart pumps blood through its rhythmic actions of contraction and relaxation and at a rate that is adjusted to physical workload as well as other factors such as heat and humidity. Although the heart plays the critical role in producing the sustained blood flow, the role of the blood vessels is much more sophisticated than that of simple inert plumbing. The blood flow encounters resistance in the blood vessels between the heart and the tissues, and the blood vessels can change their resistance to blood flow significantly to match the oxygen demands of various organs and tissues.

The resistance to flow is a function of the blood vessel's radius which can be changed significantly to alter the flow of blood to the muscles according to their need.

Each type of blood vessel makes its own unique contribution to achieving adequate blood distribution. Because the arteries have large radii, they offer little resistance to blood flow. Their role is to serve as a pressure tank to help move the blood through the tissues. The arteries show the maximum arterial pressure during peak ventricular contraction and the minimum pressure at the end of ventricular relaxation. The maximum arterial pressure is called the *systolic pressure*, and the minimum pressure is called the *diastolic pressure*. They are recorded as systolic/diastolic, for example, 135/70 mm Hg. The difference between systolic and diastolic pressure is called the *pulse pressure*.

In contrast to the negligible resistance offered by arteries, the radii of arterioles are small enough to provide significant resistance to blood flow. Furthermore, the radii of arterioles can be changed precisely under physiological control mechanisms. Therefore, arterioles are the major source of resistance to blood flow and are the primary site of control of blood-flow distribution.

Although capillaries have even smaller radii than arterioles, the huge number of capillaries provide such a large area for flow that the total resistance of all the capillaries is much less than that of the arterioles. Capillaries are thus not considered the main source of flow resistance. However, there does exist in the capillary network another mechanism for controlling blood flow distribution—*thoroughfare channels*, small blood vessels that provide direct links or shortcuts between arterioles and venules. These shortcuts allow the blood in the arterioles to reach the venules directly without going through the capillaries and are used to move blood away from resting muscles quickly when other tissues are in more urgent need of blood supply.

The veins also contribute to the overall function of blood flow. They contain oneway valves, which allow the blood in the veins to flow only toward the heart. Furthermore, the rhythmic pumping actions of dynamic muscle activities can massage the veins and serve as a "muscle pump" (also called "secondary pump") to facilitate the blood flow along the veins back to the heart.

The amount of blood pumped out of the left ventricle per minute is called the *cardiac output* (Q). It is influenced by physiological, environmental, psychological, and individual factors. The physiological demands of muscular work changes cardiac output greatly. At rest the cardiac output is about 5 liters per minute (L/min). In moderate work the cardiac output is about 15 L/min. During heavy work it may increase as much as fivefold to 25 L/min. Work in hot and humid environments also increases cardiac output when the body must supply more blood to the skin to help dissipate excess body heat. Cardiac output may also increase when an individual is excited or under emotional stress. Age, gender, health, and fitness conditions may also influence the cardiac output of an individual under various job situations.

The heart has two ways to increase its cardiac output: Increase the number of beats per minute (called heart rate, or HR) or increase the amount of blood per beat (called stroke volume, or SV). In fact, cardiac output is the product of heart rate and stroke volume, as shown in the following formula:

$$Q \text{ (L/min)} = HR \text{ (beats/min)} \times SV \text{ (L/beat)}$$

In a resting adult stroke volume is about 0.05 to 0.06 L/beat. For moderate work stroke volume can increase to about 0.10 L/min. For heavy work, increased cardiac output is accomplished largely through heart rate increases. Heart rate is one of the primary measurements of physical workload at all workload levels.

Each tissue or organ receives a portion of the cardiac output. The blood-flow distribution for a resting adult is given in the left column of Table 12.1. At rest, the digestive system, brain, kidneys, and muscles each receive about 15 to 20 percent of the total cardiac output. In moderate work in a hot environment of

TABLE 12.1 **Blood Flow Distribution in Different Resting and Working Conditions**

		Blood Flow Distribution (%)	
Organs	Resting	Moderate Work (environment: 38° C)	Heavy Work (environment: 21° C)
Muscles	15–20	45	70–75
Skin	5	40	10
Digestive system	20–25	6–7	3–5
Kidney	20	6–7	2–4
Brain	15	4–5	3–4
Heart	4–5	4–5	4–5

This table shows the blood flow distribution at several organs or tissues in three situations. For example, at rest condition, muscles receive about 15–20% of the total cardiac output, but during moderate work in a hot environment (38° C) they receive about 45% of the total cardiac output. During heavy work in a moderate environment (21° C) this percentage increases to about 70–75%.

Source: Adapted from Astrand & Rodahl, 1986; Brouha, 1967; Eastman Kodak, 1986.

38° C, as shown in the middle column of Table 12.1, about 45 percent of cardiac output goes to the working muscles to meet their metabolic requirements. During very heavy work, this percentage increases to about 70 to 75 percent, even in a moderate environment of 21° C, as shown in the right column of Table 12.1. In hot environments more blood is distributed to the skin to dissipate the excess body heat. The fraction of blood that goes to the digestive system and the kidneys falls sharply with increased workload. An interesting aspect of blood-flow distribution is the remarkable stability of brain blood flow. The brain receives the same amount of blood under all situations, although it represents a smaller fraction of the total cardiac output in heavy work than at rest. As mentioned, blood-flow distribution is made possible primarily by dilating and constricting arterioles in different organs and tissues on a selective basis.

The Respiratory System

The respiratory system is the gas-exchanger of the body. It obtains oxygen from and dispels carbon dioxide to the environment.

The Structure of the Respiratory System. The respiratory system is composed of the nose, pharynx, larynx, trachea, bronchi, lungs, the muscles of the chest wall, and the diaphragm, which separates the chest cavity from the abdomen. The nose and the airway from the nose to the lungs conduct air to the lungs and filter it to prevent dust and harmful substances from reaching the lungs. They also moisturize the inspired air and adjust its temperature before it reaches the lungs.

The lungs consist of a huge number of *alveoli* (between 200 million and 600 million of them), which provide a large surface for the gas exchange to take place in the lungs. Blood flowing through the pulmonary capillaries absorbs oxygen from the alveoli and dispels carbon dioxide. The amount of gas exchanged per minute in the alveoli is called the *alveolar ventilation*. The respiratory system adjusts the alveolar ventilation according to the level of physical workload and demands of metabolism.

Air is breathed into the lungs when the muscles of the chest wall work with the abdominal muscles to expand the chest and lower the diaphragm. These muscle actions increase the chest volume and makes the lung pressure smaller than the atmospheric pressure, so air is brought into the lungs. Similarly, when the chest muscles relax and the diaphragm moves up, air is breathed out of the lungs.

Lung Capacity.   Not all the air in the lungs is exhaled even after a person tries his or her best to breathe out all the air in his or her lungs (called a maximum expiration). The amount of air that remains in the lungs after a maximum expiration is called the *residual volume*. The amount of air that can be breathed in after a maximum inspiration is called the *vital capacity*. The total lung capacity is the sum of the two volumes, as illustrated in Figure 12.4.

Maximum inspiration or maximum expiration rarely occurs in life. The amount of air breathed in per breath (called *tidal volume*) is less than the vital capacity, leaving an inspiratory reserve volume (IRV) and an expiratory reserve volume (ERV). A resting adult has a tidal volume of about 0.5 L, which can increase to about 2 L for heavy muscular work. The increase in tidal volume is realized by using portions of the inspiratory and expiratory reserve volumes.

The respiratory system adjusts the amount of air breathed per minute (called the *minute ventilation or minute volume*) by adjusting the tidal volume

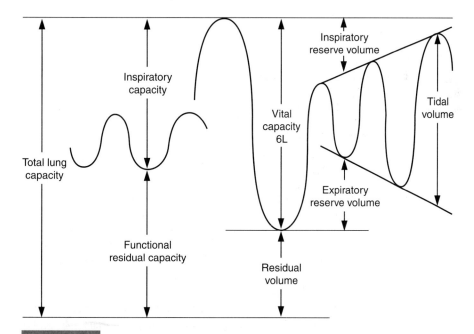

FIGURE 12.4

Respiratory capacities and volumes. (*Source:* Kroemer, K. et al., 1990. *Engineering Physiology: Bases of Human Factors/Ergonomics,* 2nd ed. New York: Van Nostrand Reinhold. Copyright 1990. Reprinted by permission of Van Nostrand Reinhold.)

and the frequency of breathing. In fact, minute ventilation is calculated as the product of tidal volume and breathing frequency. The body carefully controls the two parameters to maximize the efficiency of breathing in meeting the needs of alveolar ventilation. A resting adult breathes about 10 to 15 times per minute. The tidal volume increases for light work, but the breathing frequency does not. This is because there is a constant anatomical space in the air pathways between the nose and the lungs that is ventilated on each breath and the air in that space does not reach the alveoli. The deeper the breath (the larger the tidal volume), the larger is the percentage of air that reaches the alveoli. Therefore, increasing the tidal volume is more efficient than increasing the breathing frequency. As workload further increases, however, increasing tidal volume alone is not sufficient to meet the ventilation needs, and thus the frequency of breathing also increases rapidly with increasing workload. For heavy work, the respiratory frequency can increase threefold over its resting level to about 45 breaths per minute.

The environmental air we breathe is normally composed of 21 percent oxygen, 0.03 percent carbon dioxide, the remaining being mostly nitrogen. Clearly, if the working environment has poor ventilation or is polluted with smoke or other chemical substances, then the respiratory and the circulatory systems must work harder to compensate for the reduced oxygen supply. The respiratory and the circulatory systems are also under increased stress when working at high altitudes above sea level because of the lower oxygen content in the air and the reduced difference between the atmospheric pressure and the lung pressure.

ENERGY COST OF WORK AND WORKLOAD ASSESSMENT
Energy Cost of Work

The human body must consume energy to maintain the basic life functions even if no activities are performed at all. The lowest level of energy expenditure that is needed to maintain life is called the *basal metabolism*. The basal metabolic rate is measured in a quiet and temperature-controlled environment for a resting person after he or she has been under dietary restrictions for several days and had no food intake for twelve hours. There are individual differences in their basal metabolic rate. Gender, age, and body weight are some of the main factors that influence a person's basal metabolic rate. Human energy expenditure is measured in kilocalories. The average basal metabolic rate for adults is commonly considered to be about 1,600 to 1,800 kcal per 24 hours (Schottelius & Schottelius, 1978), or about 1 kcal per kilogram of body weight per hour (Kroemer et al., 1994).

Even for low-intensity sedentary or leisure activities, the human body needs more energy than that supplied at the basal metabolic level. Various estimates have been made about the energy costs of maintaining a sedentary, nonworking life. For example, it is estimated that the *resting metabolism* measured before the start of a working day for a resting person is about 10 to 15 percent higher than basal metabolism (Kroemer et al., 1994). Luehmann (1958) and Schottelius and

Schottelius (1978) estimate that the energy requirement is about 2,400 kcal per day for basal metabolism and leisure and low-intensity everyday nonworking activities.

With the onset of physical work, energy demand of the body rises above that of the resting level. The body increases its level of metabolism to meet this increased energy demand. The term *working metabolism*, or *metabolic cost of work*, refers to this increase in metabolism from the resting to the working level. The metabolic or energy expenditure rate during physical work is the sum of the basal metabolic rate and the working metabolic rate. Estimates of energy expenditure rates for some daily activities and certain types of work have been made, ranging from 1.6 to 16 kcal/min. For example, Durnin and Passmore (1967) report that the work of a male carpenter has an energy requirement of about 2.9 to 5.0 kcal/min, and a female worker doing laundry work has an energy cost of about 3.0 to 4.0 kcal/min. Table 12.2 provides a sample list of energy expenditure rates for various activities.

As shown in Figure 12.5, it usually takes some time for the body to increase its rate of metabolism and meet the energy requirements of work imposed by the muscles at the end of the loop in Figure 12.2. In fact, it usually takes about 1 to 3 minutes for the circulatory and respiratory systems to adjust to the increased metabolic demands and reach the level at which the energy requirements of work are met. During this initial warm-up period at the start of physical work, the amount of oxygen supplied to the tissues is less than the amount of oxygen needed, creating an oxygen deficit. Due to this oxygen deficit or the inadequate oxygen supply, anaerobic metabolism is a main source of energy. If the physical work is not too heavy, a steady state can be reached in which

TABLE 12.2   Estimates of Energy Expenditure Rates for Various Activities

Activity	Estimates of Energy Expenditure Rates (kcal/min)
Sleeping	1.3
Sitting	1.6
Standing	2.3
Walking (3 km/hr)	2.8
Walking (6 km/hr)	5.2
Carpenter-assembling	3.9
Woodwork-packaging	4.1
Stockroom work	4.2
Welding	3.4
Sawing wood	6.8
Chopping wood	8.0
Athletic activities	10.0

Source: Based on Durnin & Passmore, 1967; Edholm, 1967; Passmore & Durnin, 1955; Vos, 1973; Woodson, 1981.

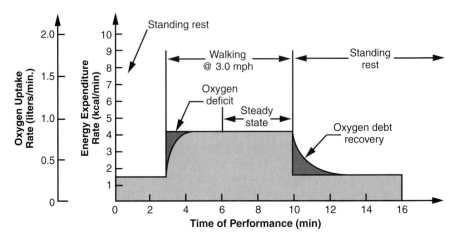

*Assumes 5 kcal of energy expended
per liter of oxygen used

FIGURE 12.5

The change in total energy expenditure rate as activity level changes. (*Source:* Garg, A.,
Herrin, G., and Chaffin, D., 1978. Prediction of metabolic rates from manual materials
handling jobs. *American Industrial Hygiene Association Journal, 39*[8], 661–674.)

oxidative metabolism produces sufficient energy to meet all energy requirements. The oxygen deficit incurred at the start of work must be repaid at some
time, either during work if the work is light or during the recovery period immediately after work ceases if the work is moderate or heavy. This is why the respiratory and circulatory systems often do not return to their normal activity
levels immediately on completion of a moderate or heavy work.

The physical demands of work can be classified as light, moderate, heavy,
very heavy, and extremely heavy according to their energy expenditure requirements (Astrand & Rodahl, 1986; Kroemer et al., 1994). In *light work*, the energy
expenditure rate is fairly small (under 2.5 kcal/min) and the energy demands
can be met easily by oxidative metabolism of the body. *Moderate work* has energy requirements of about 2.5 to 5.0 kcal/min, which are still largely met
through oxidative metabolic mechanisms. *Heavy work* requires energy at expenditure rates between 5.0 and 7.5 kcal/min. Only physically fit workers are able to
carry out this type of work for a relatively long period of time with energy supplied through oxidative metabolism. The oxygen deficit incurred at the start of
work cannot be repaid until the end of work. In *very heavy work* (with energy
expenditure rates between 7.5 and 10.0 kcal/min) and *extremely heavy work*
(greater than 10.0 kcal/min), even physically fit workers cannot reach a steady-
state condition during the period of work. The oxygen deficit and the lactic acid
accumulation continue to increase as the work continues and make it necessary
for the worker to take frequent breaks or even to quit the work completely.

Measurement of Workload

The results of extensive research on work physiology have shown that energy expenditure rate of a work is linearly related to the amount of oxygen consumed by the body and to heart rate. Therefore, oxygen consumption rate and heart rate are often used to quantify the workload of physical work. In this section we describe the two measurements, along with blood pressure and minute ventilation, which are two less commonly used but sometimes useful physiological measures of physical workload. We also describe subjective measures of workload which, when used in conjunction with physiological measures, often provide job analysts with a more comprehensive understanding of the working condition than do physiological measures alone.

Oxygen Consumption. As described earlier, aerobic (oxidative) metabolism is the source of energy for sustained muscular work when the body is in a steady state. Extensive research has shown that there is a linear relationship between oxygen consumption and energy expenditure: For every liter of oxygen consumed, an average of about 4.8 kcal of energy is released. Thus, the amount of aerobic metabolism or energy expenditure of work can be determined by multiplying the oxygen-consumption rate (liters/min) by 4.8 (kcal/liter).

The amount of oxygen consumed can be determined by measuring the amount of air expired per unit of time and the difference between the fraction of oxygen in the expired air and that in the inspired air. For most workplaces, except those at high altitudes or in polluted work environments, the fraction of oxygen in the inspired air can be assumed to be about 21 percent.

To collect the expired air in a workplace, the worker is asked to wear a face mask or a mouthpiece through which the air is inhaled and exhaled. The expired air either is collected in a large bag (called the Douglas bag) and analyzed later for its oxygen content or passes directly through an instrument that analyzes its oxygen content (Astrand & Rodahl, 1986; Harrison et al., 1982). A flow meter installed in the face mask or mouthpiece can be used to determine the volume of inspired or expired air. For the Douglas bag method, the volume of expired air can be determined by measuring the volume of air in the filled bag. Portable devices are available commercially for measuring expired air flow rates and oxygen consumption. An important requirement for these devices is that their usage should cause minimal interference with the worker's job performance. The equipment should not be too bulky for use in the field, and its airway (mask, tube, valves, etc.) should not cause great resistance to breathing during heavy physical work. Continuous efforts are made to improve the instruments and meet these requirements as closely as possible.

Note that measuring the amount of oxygen consumed during work can only help determine the amount of aerobic metabolism involved. To estimate the amount of anaerobic (nonoxidative) metabolism used in a work, we must measure the additional amount of oxygen consumed during the recovery period over that of the resting state. As described earlier, oxygen consumption rate does not return to its resting value immediately upon cessation of work. It remains elevated for a period of time and gradually falls back to the resting level. The excess oxygen used during this recovery period recharges the depleted stores of ATP and CP and repays the oxygen debt incurred at the start and during the period of work. The greater the amount of anaerobic metabolism involved in a work, the greater the amount of excess oxygen needed to pay back the oxygen debt during the recovery period. Therefore, measurement of oxygen consumption during the recovery period provides an estimate of the amount of anaerobic metabolism of a job.

Another important issue that must be noted is that oxygen consumption can only be used to estimate the energy demands of "dynamic" work, such as walking, running, and dynamic lifting, in which muscle contractions alternate with relaxation periods. It is not a good measure of the workload of "static" work, such as holding a heavy object at a fixed position for long. This is because static work usually recruits a small number of localized muscle groups and keeps them in a contracted state continuously. Sustained muscle contraction disrupts blood flow to these muscles because of their continued compression of the blood vessels. Energy supply to the contracted muscles is restricted due to inadequate blood flow. Therefore, although static work is very demanding and leads to fatigue quickly, static work effort is not well reflected in measures of oxygen consumption. Methods of evaluating static work are described in the last section of this chapter.

Heart Rate. Heart rate, the number of heart beats per minute, is another commonly used physiological measure of physical workload. Heart rate usually increases as workload and energy demands increase. It reflects the increased demand for the cardiovascular system to transport more oxygen to the working muscles and remove more waste products from them. Extensive research has shown that for moderate work, heart rate is linearly related to oxygen consumption (Astrand & Rodahl, 1986). Because heart rate is easier to measure than oxygen consumption, it is often used in industrial applications as an indirect measure of energy expenditure.

Heart rate is not as reliable as oxygen consumption as a measure of energy expenditure. It is influenced by many factors, and the linear relationship between heart rate and oxygen consumption can be violated by these factors, which include emotional stress, drinking coffee or tea, working with a static and awkward posture, or working in hot environments. Any of these circumstances can lead to disproportionately high heart rates without an equally significant increase in oxygen consumption. Furthermore, the relationship between heart rate and oxygen consumption varies among individuals. Different individuals can show different heart rates when they have the same level of oxygen consumption.

Despite these complicating factors, because of the convenience of measuring heart rate and its relative accuracy in reflecting workload, heart rate is considered to be a very useful index in physical work evaluation.

Portable telemetry devices, available commercially, allow monitoring and recording the heart rate of a worker unobtrusively and from a distance. To measure the heart rate, the worker wears a set of electrodes on his or her chest that detects the signals from the heart. The signals are transmitted to a receiver for recording and analysis. A simple but somewhat intrusive method to measure heart rate is to use the fingers to count the pulse of the radial artery located at the thumb side of the wrist. Heart rate can also be collected by counting the pulse of the carotid artery on the neck near the angle of the jaw.

Because the relationship between heart rate and oxygen consumption varies for different individuals, this relationship must be established for each worker before heart rate is used alone as an estimate of workload. This process requires the measurement of heart rate and oxygen consumption in controlled laboratory conditions in which several levels of workloads are varied systematically. After the relationship between the two variables are established for a worker, the same worker's energy expenditure rate in the workplace can be estimated by collecting his or her heart rate and converting it to oxygen-consumption and energy-expenditure data. Studies have shown that heart-rate data offer valid estimates of energy-expenditure rate when the heart rate–oxygen consumption relationship is calibrated for each worker (Bridger, 1995).

In general, the change of heart rate before, during, and after physical work follows the same pattern as that of oxygen consumption or energy expenditure, shown in Figure 12.5. A resting adult has a typical heart rate of about 60 to 80 beats/min, although large differences exist among different individuals. During physical work, the heart rate first rises and then levels off at the steady state, and it does not return to its resting value immediately on cessation of work. The amount of increase in heart rate from the resting to the steady state is a measure of physical workload, and so also is the heart rate recovery time. The heavier the physical work, the greater is the increase in heart rate, and the longer is the heart rate recovery time.

There is a maximum heart rate for each individual, which is affected by many factors such as age, gender, and health and fitness level. The primary factor determining the maximum heart rate is age, and the decline of the maximum heart rate as a function of age can be estimated by the following linear equation (Astrand & Rodahl, 1986).

$$\text{maximum heart rate} = 206 - (0.62 \times \text{age}).$$

Another commonly used formula to estimate the maximum heart rate is (Cooper et al., 1975)

$$\text{maximum heart rate} = 220 - \text{age}.$$

Maximum heart rate directly determines the maximum work capacity or the maximum energy expenditure rate of an individual.

Blood Pressure and Minute Ventilation. The term *blood pressure* refers to the pressure in the large arteries. Arteries offer little resistance to blood flow and serve as a pressure tank to help move the blood through the tissues. Arteries show the maximum arterial pressure during peak ventricular contraction and the minimum pressure at the end of ventricular relaxation. The maximum arterial pressure is called *systolic pressure,* and the minimum pressure is called *diastolic pressure.* The two blood pressures can be measured with a blood pressure gauge (sphygmomanometer), cuff, and stethoscope and are recorded as systolic/diastolic, for example, 135/70 mm Hg.

Because blood pressure measurements require workers to stop their work and thus interfere with or alter the regular job process, they are not used as often as oxygen-consumption and heart-rate measurements. However, studies have shown that for work involving awkward static postures, blood pressure may be a more accurate index of workload than the other two measurements (Lind & McNichol, 1967).

Another physiological measurement that is sometimes used in job evaluation is minute ventilation or minute volume, which refers to the amount of air breathed out per minute. It is often measured in conjunction with oxygen consumption and used as an index of emotional stress. When workers are under emotional stress, as in emergency situations or under time pressure, they may show a change in their respiration pattern and an increase in their minute ventilation. However, there is usually not a corresponding increase in the measurement of oxygen consumption, because little additional oxygen is consumed by the body under these situations.

Subjective Measurement of Physical Workload.   Subjective rating scales of physical workload have been developed as simple and easy-to-use measures of workload. A widely used subjective rating scale is the Borg RPE (ratings of perceived exertion) scale (Borg, 1985), which requires workers to rate their perceived level of physical effort on a scale of 6 to 20. The two ends of the scale represent the minimum and maximum heart rate of 60 and 200 beats/min respectively. Subjective scales are cheaper and easier to implement than physiological measures, and they often provide valid and reliable quantification of physical efforts involved in a job. However, subjective measures may be influenced by other factors, such as worker's satisfaction with a workplace, motivation, and other emotional factors. Therefore, caution should be exercised in the use and analysis of subjective measures, and it is often desirable to use subjective ratings in conjunction with physiological measures to achieve a more comprehensive understanding of the work demands.

PHYSICAL WORK CAPACITY AND WHOLE-BODY FATIGUE
Short-Term and Long-Term Work Capacity

Physical work capacity is a person's maximum rate of energy production during physical work, and it varies as a function of the duration of the work. The maximum energy-expenditure rate that can be achieved by an individual for a few minutes is called the *short-term maximum physical work capacity (MPWC)* or *aerobic capacity.* Figure 12.6 shows the linear relationship between energy-expenditure rate and heart rate for a healthy individual with a maximum heart rate of 190 beats/min and a MPWC of about 16 kcal/min for dynamic work. It

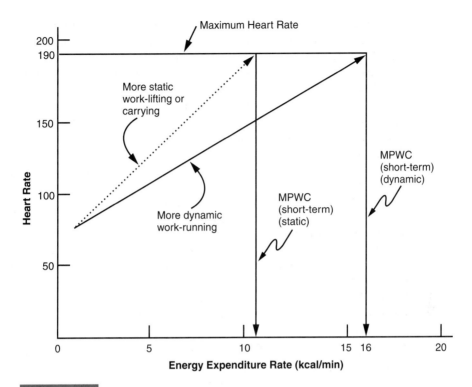

FIGURE 12.6

The relationship between heart rate and energy-expenditure rate for static and dynamic work. At the same maximum heart rate, the maximum physical work capacity is larger for dynamic than for static work. (*Source:* Garg, A., Herrin, G., and Chaffin, D., 1978. Prediction of metabolic rates from manual materials handling jobs. *American Industrial Hygiene Association Journal*, 39[8], 661–674.)

also shows that the MPWC is significantly reduced for static muscular work in which anaerobic metabolism takes place due to restricted blood flow to the muscles (Garg et al., 1978).

The short-term MPWC is also referred to as VO_{2max} in the literature to describe a person's capacity to utilize oxygen. It is believed that the MPWC is determined by the maximum capacity of the heart and lungs to deliver oxygen to the working muscles. During physical work, heart rate and oxygen consumption increase as workload increases. However, they cannot increase indefinitely. As workload further increases, a limit is reached at which the heart cannot beat faster and the cardiovascular system cannot supply oxygen at a faster rate to meet the increasing energy demands of the work. At this point, the person has reached his or her aerobic capacity or VO_{2max}.

There are great individual differences in aerobic capacity. Age, gender, health and fitness level, training, and genetic factors all influence an individual's aerobic capacity. According to the data published by NIOSH (1981), the aerobic capacity for average healthy males and females are approximately 15 kcal/min and 10.5 kcal/min respectively.

Physical work capacity drops sharply as the duration of work increases. The decline of long-term MPWC from the level of short-term MPWC is shown in Figure 12.7 (Bink, 1964). For job design purposes, NIOSH (1981) states that workers should not work continuously over an 8-hour shift at a rate over 33 percent of their short-term MPWC. This means that for continuous dynamic work, healthy male workers should not work at a rate over 5 kcal/min, and healthy female workers should not work at a rate over 3.5 kcal/min. For dynamic jobs performed occasionally (1 hour or less during an 8-hour shift), NIOSH states that the recommended energy-expenditure limit should be 9 kcal/min and 6.5 kcal/min for healthy males and females respectively. Clearly, older and less-fit workers have lower MPWC than young, fit workers and require reduced 8-hour work capacity limits.

In ergonomic job evaluation, the energy cost of different jobs can be measured and compared with the NIOSH recommendations to determine whether a job can be performed by the workforce and whether it must be redesigned to lower the required energy-expenditure rate to make it acceptable to the intended workforce. For example, if a job is identified to require an energy-expenditure rate of about 5 kcal/min, then we know that only healthy male workers can perform this job continuously over an 8-hour shift. To make this job acceptable to a wider range of workers, we need to either redesign the job (e.g., use of auto-

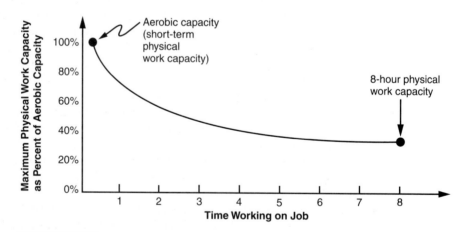

FIGURE 12.7

MPWC as a function of work duration. (*Source:* Bink, B. 1962. The physical working capacity in relation to working time and age. *Ergonomics,* 5[1], 25–28; Bink, B., 1964. Additional studies of physical working capacity in relation to working time and age. *Proceedings of the Second International Congress on Ergonomics,* Dortmund, Germany: International Ergonomics Association.)

mated material-handling devices) or adopt an appropriate work-rest schedule, as discussed in the following section.

Causes and Control of Whole-Body Fatigue

A worker is likely to experience whole-body fatigue during or at the end of an 8-hour shift if the energy demands of work exceed 30 to 40 percent of his or her maximum aerobic capacity and will certainly feel fatigued if the energy cost exceeds 50 percent of the aerobic capacity. Both subjective and physiological symptoms may appear as indicators of fatigue. The fatigued worker may experience a feeling of slight tiredness, weariness, or complete exhaustion, and show impaired muscular performance or difficulties in keeping awake. There may also be an increase in blood lactic acid accumulation and a drop in blood glucose. Prolonged whole-body fatigue may lead to low job satisfaction and even increased risk of health problems such as heart attacks.

One explanation of the cause of whole-body fatigue is that when the energy expenditure rate exceeds 40 to 50 percent of the aerobic capacity, the body cannot reach the steady state in which aerobic metabolism supplies enough oxygen to meet all the energy needs. Consequently, anaerobic metabolism contributes an increasing proportion of the energy supplied and produces an increasing amount of waste products such as lactic acid during the process.

It should be noted, however, that the exact nature and causes of fatigue is still largely unknown (Astrand & Rodahl, 1986; Simonson, 1971; Kroemer et al., 1994). For example, although increased accumulation of lactic acid in the blood is often observed in prolonged heavy work, it is not usually associated with prolonged moderate work, which may also cause fatigue (Astrand & Rodahl, 1986). Depletion of ATP and CP has traditionally been regarded as a main cause for fatigue; however, this view has been challenged as well (Kahn & Monod, 1989; Kroemer et al., 1994). Fatigue may also be a symptom of disease or poor health. Furthermore, the development of fatigue is influenced by a worker's motivation, interest in the job, and other psychological factors. The same worker may develop fatigue more quickly in one job than in another, although the two jobs may have comparable energy requirements. Similarly, two workers of the same health and fitness condition may develop fatigue at different rates for the same job. However, regardless of the causes, complaints of job-related fatigue in a workplace should be treated as important warning signals and dealt with seriously so that related job hazards can be identified and removed.

Engineering and administrative methods can be used to reduce the risk of wholebody fatigue in industrial workplaces. Engineering methods refer to the use of engineering techniques to redesign the job and provide job aids. For example, use of conveyer belts or automated material-handling devices can help reduce the need for load carrying. A better layout of the workplace designed according to the frequency and sequence of use of various workplace components can help reduce the distance of lifting, pushing, or pulling heavy objects and thus greatly reduce the energy-expenditure requirements of work.

When an existing heavy job cannot be redesigned with engineering techniques due to various constraints, work-rest scheduling is the most commonly adopted administrative method to keep the work at acceptable energy-expenditure levels.

When environmental heat load is not present, a work-rest schedule can be determined with the following formula:

Rest period as a fraction of total work time $= (PWC - E_{job})/(E_{rest} - E_{job})$

PWC is the physical work capacity for workers of concern,
E_{job} is the energy-expenditure rate required to perform the job, and
E_{rest} is the energy-expenditure rate at rest. A value of 1.5 kcal/min (90 kcal/hr) is often used to represent the energy expenditure rate for seated rest.

As an example, suppose the energy-expenditure rate of a physical work is 6.5 kcal/min and the work is performed by healthy male and female workers on an 8-hour shift basis. Recall that the NIOSH-recommended 8-hour work capacity limits are 5 kcal/min and 3.5 kcal/min for healthy males and females respectively. It is clear that this job cannot be performed continuously for 8 hours by either group of workers. If this job cannot be redesigned with engineering techniques, then a proper work-rest schedule must be implemented to reduce the risk of whole-body fatigue. Furthermore, the rest schedule should be determined separately for the two groups of workers because of the difference in their physical work capacities.

Using the formula presented above, we have, for male workers,

Rest period as a fraction of total work time
$= (5 - 6.5)/(1.5 - 6.5) = 1.5/5 = 0.30$

For female workers, we have

Rest period as a fraction of total work time
$= (3.5 - 6.5)/(1.5 - 6.5) = 3/5 = 0.60$

Therefore, during an 8-hour shift, male workers should have a total rest period of 2.4 hours ($0.30 \times 8 = 2.4$), and female workers should have a total rest period of 4.8 hours ($0.60 \times 8 = 4.8$) because of the heavy physical demands of the job. The total rest time should be divided into many short breaks and distributed throughout the 8-hour work shift rather than taken as few long breaks.

When environmental heat stress is present in a workplace, such as working in a hot climate or near heat sources, workers may need to take frequent rests even when the energy-expenditure rate required for performing the physical task is not high. About 80 percent of metabolic energy is released in the form of metabolic heat (Edholm, 1967), which must be dissipated from the body so that the body can maintain a constant normal temperature of 98.6° F. Dissipation of metabolic heat can be difficult in a working environment in which large radiant heat or high humidity exist or there is a lack of adequate air flow. For these work situations workers need to take breaks in a cool area to avoid heat-related health risks.

Figure 12.8 contains a set of recommended work-rest schedules for various workloads at different levels of environmental heat conditions. A comprehensive index of the environmental heat load, called wet bulb globe temperature (WBGT), must first be determined with the following equations (NIOSH, 1972) before using these guidelines:

When the level of radiant heat is low in a working environment, the WBGT is

$$WBGT = 0.7 \text{ (natural wet bulb temperature)} + 0.3 \text{ (globe temperature)}$$

When the level of radiant heat is high (e.g., working in sunlight or near a radiant heat source), WBGT is

$$WBGT = 0.7 \text{ (natural wet bulb temperature)}$$
$$+ 0.2 \text{ (globe temperature)} + 0.1 \text{ (dry bulb temperature)}$$

where,

NWBT is the natural wet bulb temperature (WBT), which is the temperature of a wet wick measured with actual air flow present. NWBT is the same as WBT when the air velocity is greater than 2.5 m/sec (8 ft/sec). NWBT = 0.9 WBT + 0.1 (dry bulb temperature) for slower air velocities.

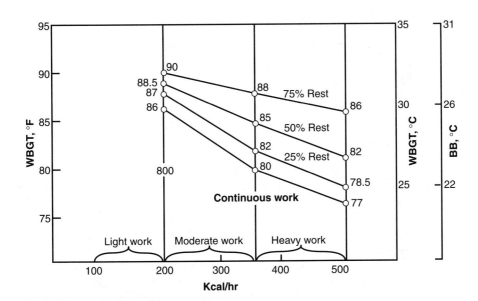

FIGURE 12.8

Recommended WBGT limits for various workload levels and work-rest schedules. (*Source:* American Society of Heating, Refrigerating, and Air-Conditioning Engineers. *ASHRAE Handbook*, 1985 *Fundamentals.* New York: ASHRAE.)

Devices are available to measure and calculate these temperature indexes.

It is clear from Figure 12.8 that when working in a hot or humid workplace, frequent rests in a cool place are often necessary even when the energy cost of performing the physical task is not high. For example, although a light work of 3.4 kcal/min (204 kcal/h) can be performed continuously by most workers when heat stress is not present, the same physical task would require the workers to spend 50 percent of the time resting in a cool environment when the working environment has a WBGT of 88.5 degrees F.

Three cautionary notes must be made with regard to the use of Figure 12.8. First, although significant differences exist between males and females in their physical work capacities, Figure 12.8 does not take into account this difference. Second, the term *continuous work* used in Figure 12.8 does not necessarily mean that a work can be performed continuously for 8 hours. For example, although a light work (< 200 kcal/h or 3.4 kcal/min) can be performed continuously for 8 hours in a workplace with a 75° F WBGT by both male and female workers, a heavy work of 390 kcal/h (6.5 kcal/min) cannot be sustained by many healthy male workers, as we calculated earlier. Most workers cannot perform a very heavy work of 480 kcal/h (8 kcal/min) for long, even when there is no environmental heat stress. Third, Figure 12.8 applies only to heat-acclimatized workers (workers who are not new to a hot working environment). Workers who are new to a hot environment (heat-unacclimatized workers) should be given work at lower energy-expenditure levels. Recommended heat exposure and energy-expenditure limits for heat-unacclimatized workers can be found in NIOSH (1986).

Static Work and Local Muscle Fatigue

While whole-body fatigue is often associated with prolonged dynamic whole-body activities that exceed an individual's MPWC, local muscle fatigue is often observed in jobs requiring static muscle contractions. Dynamic muscle activities provide a "muscle pump" that massages the blood vessels and assists blood flow through the muscle's rhythmic actions. Static muscle contractions, in contrast, impede or even occlude blood flow to the working muscles because the sustained physical pressure on the blood vessels prevents them from dilating as long as the contraction continues. The lack of adequate oxygen supply forces anaerobic metabolism, which can produce local muscle fatigue quickly due to the rapid accumulation of waste products and depletion of nutrients near the working muscles.

The maximum length of time a static muscle contraction can be sustained (muscle endurance time) is a function of the exerted force expressed as a percentage of the muscle's *maximum voluntary contraction (MVC)*, which is the maximal force that the muscle can develop. This relationship is shown in Figure 12.9, which is often called the *Rohmert curve* (Rohmert, 1965). It is clear from Figure 12.9 that the maximal force can be sustained for only a few seconds. A 50 percent force can be sustained for about one minute, but the static contraction can be maintained for minutes and even up to hours if the exerted muscle force is below 15 percent of the MVC (Simonson & Lind, 1971).

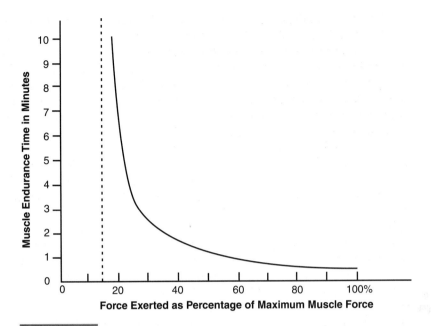

FIGURE 12.9

Relationship between static muscle endurance time and muscle exertion level. (*Source:* Rohmert, W., 1965. Physiologische Grundlagen der Erholungszeitbestimmung, *Zeitblatt der Arbeitswissenschaft*, 19, p. 1. Cited in Simonson, E., ed., 1971. *Physiology of Work Capacity and Fatigue*, Springfield, IL: Charles C. Thomas Publishers, p. 246.). Although this figure suggests that low-level muscle contractions can be sustained indefinitely, recent evidence (Sato, et al., 1984; Sjogaard et al., 1986) indicates muscle fatigue will develop at any contraction level.

Some studies suggest that static contractions can be held almost indefinitely if the exerted force is less than 10 percent of the MVC (Bjorksten & Jonsson, 1977). But other research indicates that muscle fatigue will develop at any contraction level of the MVC (Sato et al., 1984;).

Muscle endurance time drops sharply at levels above 15 percent of the MVC, and muscle fatigue develops quickly (in seconds) if the static work requires more than 40 percent of the MVC. The symptoms of local muscle fatigue include muscle pain or discomfort, reduced coordination of muscle actions, and increased muscle tremor. Reduced motor control may lead to occupational injuries and accidents. Prolonged muscle fatigue may lead to disorders of the adjoining ligaments and tendons.

Two methods are commonly used to measure local muscle fatigue: electromyography (EMG) and subjective rating (psychophysical) scales. *Electromyography* is a technique for measuring the electrical activities of muscles from electrodes taped on the skin over the muscles. Extensive research has found that the EMG signals often shift to lower frequencies and show higher amplitudes as muscle fatigue develops (Hagberg, 1981; Lindstrom et al., 1977). These changes in EMG are often used as objective indicators of the development of local muscle fatigue.

As in the measurement of whole-body fatigue and work capacity, subjective rating scales can be used to measure muscle fatigue. Workers are asked to rate the level of fatigue experienced in a job on a set of rating scales, each of which represents a local muscle group (e.g., left shoulder, right shoulder, left wrist, right wrist). Each scale is marked with numerical markers such as 1 through 7, and the two ends of each scale represent very low and very high levels of muscle fatigue respectively. In ergonomic job analysis of static work and muscle fatigue, it is often desirable to use subjective ratings in conjunction with EMG measurements.

As in the cases of whole-body fatigue, engineering and administrative methods can be used to reduce the risk of local muscle fatigue in industrial workplaces. Engineering methods focus on redesigning the job to eliminate static postures and reduce loads on various joints. This is often accomplished by improving workplace layouts and providing arm rests, body supports, and job aids. The biomechanical methods of job analysis, described in Chapter 11, can be applied in this process to help identify stressful loads and evaluate alternative workplace layouts and work methods.

The most commonly adopted administrative method of reducing the risk of local muscle fatigue is to adopt job procedures that provide adequate muscle rests between exertions and during prolonged static work. The job procedure should allow workers to change their postures periodically and use different muscle groups from time to time during the work. For example, periodic leg activities during prolonged seated work can greatly reduce swelling and discomfort at the lower legs and ankles, compared to continuous sitting during an 8-hour shift (Winkel & Jorgensen, 1985).

CONCLUSION

Physical work is possible only when there is enough energy to support muscular contractions. In this chapter, we saw how the cardiovascular and respiratory systems work together to meet the energy requirements of work and how these requirements can be measured quantitatively and considered in the analysis of physical work.

Although anthropometric, biomechanical, and physiological issues are discussed separately in chapters 10, 11, and 12, a job analyst must consider all three aspects together when designing or analyzing a workplace. Workplaces and workstations must be designed according to the anthropometric characteristics of the users. Otherwise, users will have to adopt awkward postures. From the biomechanics point of view, awkward postures are very likely to create stress on a person's joints and muscles. Biomechanical methods can be used to analyze the user's postures, together with any required exertion forces, to identify the risk of physical injuries. The energy-expenditure demands of a work can be evaluated using physiological methods to reduce the risk of whole-body fatigue. Jobs involving static muscle contractions should be identified and redesigned so as to reduce local muscle fatigue.

Poorly designed workstations and manual material handling may cause both physical and psychological stress, but they are not the only causes of stress in life and work. Other factors, such as noise and vibration, as well as time pressure and anxiety, may cause stress as well. These stressors are the topic of the next chapter.

Chapter 13

Stress and Workload

The proposal must be postmarked no later than 5 P.M., *but as the copying is frantically pursued an hour before, the machine ceases to function, displaying a series of confusing error messages on its computer-driven display. With the panic of the approaching deadline gripping an unfortunate victim, he finds himself unable to decipher the complex and confusing instructions. In another building on campus, a job candidate, giving a talk, has fielded a few difficult questions and now turns to the video demo that should help answer the questions. Nervous and already upset, she finds that the video player machine will not function, and while she fiddles with the various buttons, no one lifts a hand to assist her; instead, the audience waits impatiently for the show to go on.*

Meanwhile, on the other side of the state, the climber has been concentrating on a difficult rock pitch when she suddenly realizes that the clouds have closed in around her. A sudden clap of thunder follows the tingle of electricity on her skin, and the patter of sleet on the now slippery rocks makes the once-challenging climb a truly life-threatening experience. To make matters worse, the cold has crept into her fingers, and as she fumbles with the rope through her protection on the rock, it takes all the concentration she can muster to deal with securing the protective rope. Inexplicably, rather than calling a retreat in the dangerous circumstances, she decided to continue to lead her team upward.

These three anecdotes illustrate some of the varying effects of stress on performance—the stress of time pressure, the stress of threat and anxiety, and the stress imposed by factors in the environment, such as the cold on the rock. The concept of stress is most easily understood in the context of Figure 13.1. On the left of the figure is a set of *stressors*, influences on information availability and processing that are not inherent in the content of that information itself. Stressors may include such influences as noise, vibration, heat, and dim lighting

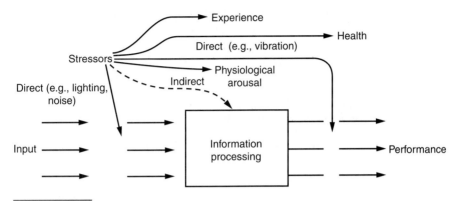

FIGURE 13.1
A representation of stress effects.

as well as such psychological factors as anxiety, fatigue, frustration, and anger. Such forces typically have four effects: (1) They produce a psychological *experience.* For example, we are usually (but not always) able to report a feeling of frustration or arousal as a consequence of a stressor. (2) Closely linked, a change in physiology is often observable. This might be a short-term change— such as the increase in heart rate associated with taking the controls of an aircraft or the stress of air traffic controllers in high-load situations—or it might be a more sustained effect—such as the change in the output of catecholamines, measured in the urine after periods of flying combat maneuvers or actual battlefield events (Bourne, 1971). The psychological experience and physiological characteristics are often, but not invariantly, linked. (3) Stressors affect the efficiency of information processing, generally by degrading performance (Driskell & Salas, 1996). (4) The stressors may have long-term negative consequences for health. To the extent that all four effects are present, the cause can be labeled a *stressor.*

As the figure shows, these effects may be *direct* or *indirect.* Direct effects influence the quality of information received by the receptors or the precision of the response. For example, vibration reduces the quality of visual input and motor output, and noise does the same for auditory input (Poulton, 1976). Time stress may simply curtail the amount of information that can be perceived in a way that quite naturally degrades performance. Hence, many of the negative influences of direct effect stressors on performance can be easily predicted. Most direct effect stressors are categorized as *environmental stressors,* and their physical magnitude can be objectively measured (e.g., the degrees of temperature at a workplace).

Some of these direct-effect physical stressors, like noise or vibration, as well as others for which no direct effect can be observed—like anxiety or fear—appear to show more indirect effects by influencing the efficiency of information processing through mechanisms that have not yet been described. Many of the effects are mediated by *arousal.*

In this chapter, we consider first those *environmental* stressors that typically have clearly defined direct effects (although they may have indirect effects as well). We then consider internal, psychological stressors of threat and anxiety, those stressors associated with job and home, and finally the interrelated effects of stress imposed by work overload, underload, fatigue, and sleep disruption. As we discuss each stressor, we consider both the nature of negative stress effects on performance and the possible system remediations that can reduce those effects.

ENVIRONMENTAL STRESSORS

We have already had an introduction to two of the most important environmental stressors in the form of *lighting* (in Chapter 4) and *noise* (in Chapter 5). Our discussion of both is instructive in setting the stage for the stressors we discuss in this chapter; in both cases, the particular *level* of the variable involved determines whether a degradation of performance occurs, with intermediate levels often producing better performance than levels that are too low or too high. (This is particularly true with lighting, where both low illumination and glare can exert direct detrimental effects on performance.) Furthermore, in both cases, but particularly in the case of noise, the detrimental effects can be partitioned into those that disrupt performance of a task concurrent with the stressor (e.g., the noise masks the conversation) and those that have delayed effects that are more likely to endanger health (e.g., deafness in the case of noise). It is reasonable to argue that any stressor that produces delayed effects should trigger steps to reduce its magnitude, whether or not it also induces effects on concurrent performance. In contrast, those stressors that induce only direct effects may be tolerated as long as the level of performance loss sacrifices neither safety nor performance quality.

Motion

Stress effects of motion can result from either sustained motion or cyclic motion. The effects of sustained motion on motion sickness are discussed along with our treatment of the vestibular system in Chapter 5. In this section we discuss the effects of *cyclic motion,* also called *vibration,* including both high-frequency vibration, which may lead to performance decrements or repetitive motion disorders, and low-frequency vibration, which is another cause of motion sickness.

High-Frequency Vibration. High-frequency vibration may be distinguished in terms of whether it is specific to a particular limb, such as the vibration produced by a handheld power saw, or whether it influences the whole body, such as that from a helicopter or ground vehicle. The aversive long-term health consequences of the former type are well documented in the literature on repetitive stress injuries (see Chapter 11). As a consequence of this danger, standard "dosage" allowances for exposure to different levels of vibration have been established (Wasserman, 1987), not unlike the noise dosages discussed in Chapter 5. It is also obvious that hand vibration from a handheld tool disrupts the *precision* of the hand and arm in operating that tool (i.e., a direct effect), possibly endangering the worker.

In addition to the remediations of limiting dose exposures, efforts can be made to select tools whose vibrations are reduced through design of the engine itself or incorporation of vibration-damping material.

In contrast to the well-documented effects of repetitive motion disorders, the health consequences of full-body vibration are somewhat less well documented, although effects on both body posture and oxygen consumption have been observed (Wasserman, 1987). However, such vibration has clear and noticeable effects on many aspects of human performance (Griffin, 1997a, 1997b). Its presence in a vehicle can, for example, make touch screens extremely unreliable as control input devices and lead instead to the choice of dedicated keypads (see Chapter 9). Vibration may disrupt the performance of any eye-hand coordination task unless the hand itself is stabilized by an external source (Gerard & Martin, 1999). Finally, vibration can disrupt the performance of purely visual tasks through the apparent blurring of the images to be perceived, whether these are words to be read or images to be detected (Griffin, 1997a, 1997b). As might be expected from our discussion in Chapter 4, the effect of any given high-frequency vibration amplitude can be predicted on the basis of the *spatial frequency resolution* necessary for the task at hand; the smaller the line or dot that needs to be resolved (the higher the spatial frequency), the greater will be the disruptive effect of a given vibration amplitude. Similar predictions can be made on the basis of the spatial precision of movement. Hence, one remediation to vibration is to ensure that text fonts are larger than the minimum specified for stable environments and that target sizes for control tasks are larger. Naturally, insulating both user and interface from the source of vibration using cushioning is helpful.

Low-Frequency Vibration and Motion Sickness.   As we discussed in Chapter 5, motion effects at a much lower frequency, such as the regular sea swell on a ship, the slightly faster rocking of a light airplane in flight, or the environment of a closed cab in a tank or ground vehicle, can lead to motion sickness. We discussed the contributing factors of a *decoupling* between the visual and vestibular inputs (in such a way that motion sickness can be induced even where there is no true motion, as in full-screen visual displays). When considered as a stressor, the primary effects of motion sickness seem to be those of a *distractor*. Quite simply, the discomfort of the sickness is sufficiently intrusive that it is hard to concentrate on anything else, including the task at hand.

Thermal Stress

Both excessive heat and excessive cold can produce performance degradation and health problems. A good context for understanding their effects can be appreciated by the representation of a comfort zone, which defines a region in the space of temperature and humidity and is one in which most work appears to be most productive (Fanger, 1977). Regions above the comfort zone produce heat stress; those below produce cold stress. The temperature range is 73° F to 79° F in the summer and 68° F to 75° F in the winter. The zone is skewed such that less humidity is allowed (60 percent) at the upper temperature limit of 79° F than at the lower limit of 68° F (85 percent humidity allowed).

The stress of excessive heat, either from the sun or from nearby equipment such as furnaces or boilers, produces well-documented decrements in performance (Konz, 1997), particularly on perceptual motor tasks like tracking and reaction time (Bowers et al., 1996). The effects of heat are primarily *indirect*, affecting the efficiency of information processing rather than the quality of information available in visual input or the motor stability of hand movement. The long-term consequences of heat exposure to health are not well-documented unless the exposure is one that leads to dehydration, heat stroke, or heat exhaustion.

In predicting the effects of certain levels of ambient heat (and humidity), it is important to realize the influence of three moderating variables of the clothing worn; (Bensel & Santee, 1997). The amount of *air movement,* induced by natural breezes or fans, and the degree of *physical work* carried out by the operator, (see Chapter 12).

Implicit in the discussion of moderating factors are the recommendations for certain kinds of remediations when heat in the workplace is excessive. For example, the choice of clothing can make a difference, the job may be redesigned to reduce the metabolic activity, and fans can be employed appropriately. Furthermore, ample amounts of liquids (and opportunities to consume them) should be provided.

The effects of cold stress are somewhat different from those of heat. Long-term cold exposure can obviously lead to frostbite, hypothermia, and health endangerment. Generally, cold effects on information processing (indirect effects) do not appear to be documented, other than through distraction of discomfort and trying to keep warm. As experienced by the mountain climber at the beginning of the chapter, the most critical performance aspects of cold stress are the direct effects related to the disruption of coordinated motor performance coordinated by the hands and fingers. This disruption results from the joint effects of cold and wind. The remediation for cold stress is, obviously, wearing appropriate clothing to trap body heat. Such clothing varies considerably in its effectiveness in this regard (Bensel & Santee, 1997), and of course there are many circumstances in which the protective value of some clothing, such as gloves and mittens, must be traded off against the loss in manual dexterity that results from their use (see Chapter 9).

Air Quality

Poor air quality is often a consequence of poor ventilation in closed working spaces like mines or ship tanks but also increasingly in environments polluted by smog or carbon monoxide. Included here are the pronounced effects of *anoxia,* the lack of oxygen frequently experience in high altitudes (West, 1985). Any of these reductions in air quality can have relatively pronounced negative influences on perceptual, motor, and cognitive performance (Houston, 1987; Kramer et al., 1993). To make matters worse, some causes of anoxia, like carbon monoxide, can sometimes appear insidiously so the effected operator is unaware of the danger imposed by the degrading air quality. The interacting effects of cold and anoxia at high altitude are evident when the human's physiology, in an effort to preserve the adequate flow of now precious oxygen to the brain and heart, essentially shuts down delivery of blood to the extremities of the fingers and toes. These now become extremely vulnerable to frostbite.

PSYCHOLOGICAL STRESSORS

The environmental stressors that we discussed in the previous section all had in common the characteristic that some *physical* measure in the environment—such as that recorded by a noise meter, vibration or motion indicator, or thermometer—could be used to assess the magnitude of the stress influence. In contrast, consider two of the stressors on the people described at the beginning of the chapter. The candidate giving her job talk was stressed by the threat of embarrassment; the climber was stressed by the potential injury or even loss of life in the hazardous situation. In neither of these cases is it possible to physically measure an environmental quantity that is responsible for the psychological state of stress. Yet in both cases, the negative consequences to performance can be seen, and such consequences are consistent with a great deal of experimental and incident analysis data. Thus, when we talk of *psychological stressors* in this chapter, we are discussing specifically those stressors resulting from the perceived threat of harm or loss of esteem (i.e., potential embarrassment), of something valued, or of bodily function through injury or death.

Cognitive Appraisal

Several factors make the understanding of such psychological stressors more challenging and difficult than is the case with environmental stressors. First, it is difficult to ascertain for each individual what may constitute a threat. The expert climber may perceive circumstances as being an "exciting challenge," whereas the novice may perceive the identical combinations of steep rock and exposure as being a real danger, simply because of the difference in skill level that the two climbers possess to deal with the problem. Second, as noted by Lazarus and Folkman (1984), the amount of stress for a given circumstance is very much related to the person's understanding or *cognitive appraisal* of the situation.

There are several possible reasons for differences in cognitive appraisal. One may fail to perceive the circumstances of risk. For example, the climber may simply be so intent on concentrating on the rock that she fails to notice the deteriorating weather, and she will not feel stressed until she does. One may fail to understand the risk. Here the climber may see the clouds approaching but not appreciate their implications for electrical activity and wet rock. One may be relatively more confident or even *overconfident* (see Chapter 6) in one's ability to deal with the hazard. Finally, if people appraise that they are more in control of the situation, they are less likely to experience stress than if they feel that other agents are in control (Bowers et al., 1996). These facts together thwart the effort to derive hard numbers to predict the amount of stress for such psychological stressors in any particular circumstance (although such numbers may indeed be obtained from correlated physiological measures like heart rate). An added challenge in predicting individual responses to stressors lies in the availability of different strategies (Hockey, 1997).

Ethical Issues

There are also considerable challenges in doing research in the area of psychological stressors. For clear ethical reasons, it is not always appropriate to put participants in psychological research in circumstances in which they may be stressed by the threat of physical or psychological damage (even though the former may be guaranteed never to occur). This has meant, as we discussed in Chapter 2, that research in this area must document *in advance* that the benefits to society of the knowledge gained by the research outweigh the potential psychological risks to the participant of being placed in the stressful circumstance. This documentation is often sufficiently difficult to provide that research knowledge in the area of psychological stressors progresses very slowly. Nevertheless, the collective results of laboratory research and case studies from incident and accident analysis has revealed a general pattern of effects that can be predicted to occur under psychological stress (Broadbent, 1972; Hockey, 1986; Driskell & Salas, 1996, Hancock & Desmond, 2001).

Level of Arousal

Stressful circumstances of anxiety and danger produce an increase in *physiological arousal,* which can be objectively documented by changes in a variety of physiological indicators, such as heart rate, pupil diameter, and hormonal chemistry (Hockey, 1986). Concurrent with this arousal increase, investigators have long noted what is characterized as an *inverted U* function of performance, shown in Figure 13.2; that is, performance first increases up to a point known as the *optimum level of arousal* (OLA) and then subsequently declines as stress-induced arousal increases further. Also note in the figure that the OLA is higher for simpler tasks than for complex ones (or for more highly skilled operators for whom a given task is simpler than for the novice). This function is sometimes referred to as the *Yerkes-Dodson law* (Yerkes & Dodson, 1908). The cause of the performance increase as arousal increases to the optimum (the left side of the curve) can be thought of as the facilitory effect of investing effort—trying harder; for example, the threat of loss caused by a psychological stressor will generally make us more motivated to work harder and perform better. However, the loss in performance above the OLA (the right side of the curve) appears to be due to a more complex set of effects of *overarousal.*

Performance Changes with Overarousal

Several different changes in information-processing characteristics have been noted to occur as different forms of the sense of danger or threat have been imposed on people. *Perceptual* or *attentional narrowing,* sometimes known as *tunneling,* describes the tendency to restrict the range or breadth of attention, to concentrate very hard on only one "thing," and to ignore surrounding information sources (this thing is often the source of stress or information on how to

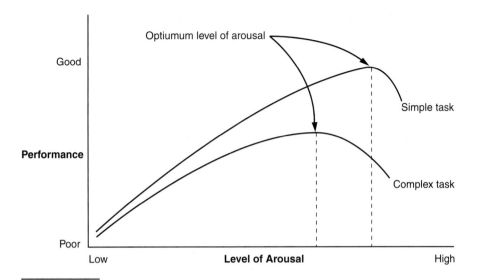

FIGURE 13.2

The Yerkes-Dodson law showing the relation between level of arousal (induced by stress) and performance. The OLA is shown to be higher for less complex tasks.

avoid it). While this strategy of focused attention may be appropriate if the object of tunneling does indeed provide the path to safety, it may be highly inappropriate if safety instead requires considering a broader set of less obvious signals, events, or information channels. Thus, the stressed speaker at the beginning of the chapter may have become so focused on the buttons on the video that she failed to notice that the machine was unplugged. Indeed, there is evidence that the catastrophe at the Three Mile Island nuclear power plant resulted, in part, because the stress caused by the auditory alert in the nuclear power control room and the dangerous condition that it signaled led operators to tunnel on one single indicator (which incorrectly indicated that the water level in the reactor was too high) and fail to perform a wider visual scan that would have allowed attention to be directed to other, correct indicators (suggesting correctly that the water level was too low; Rubinstein & Mason, 1979, Wickens, 1992).

Just as visual attention can be tunneled to a particular part of the visual environment, so *cognitive tunneling* under stress describes the tendency to focus attention exclusively on one hypothesis of what is going on (e.g., only one failure candidate as the cause of an alarm) and ignore a potentially more creative diagnosis by considering a wider range of options. Thus, our climber at the beginning of the chapter may have focused only on the one solution—"climb upward." Such a trend is consistent with findings that increased stress reduces performance on tests of creativity (Shanteau & Dino, 1993).

Working memory loss describes just that. Under stress, people appear to be less capable of using working memory to store or rehearse new material or to perform computations and other attention-demanding mental activities (Wickens et al., 1991; Stokes & Kite, 1994; Hockey, 1986). The stressed pilot, panicked over the danger of a failed engine and lost in bad weather, may be less able to correctly remember the air traffic controller's spoken guidance about where he is and the correct compass heading to turn to.

While working memory may degrade under stress, a person's *long-term memory* for well-known facts and skills will be little hampered and may even be enhanced. Thus, under stress we tend to engage in the most available thoughts and actions. The problem occurs when these actions are different from the appropriate response to the stressful situation, for example, when the appropriate and seldom practiced response in an emergency (a condition that will rarely occur) is *incompatible* with the usual response in (frequently encountered) routine circumstances. An example of this is the appropriate emergency response to a skid while driving on an icy road. Under these stressful circumstances, you should first turn *toward* the direction of skid to bring the car under control, precisely the opposite of your normal response on dry pavement, which is to turn away from the direction you do not want to go. It is because of this tendency to revert to the dominant habit in emergency that it is important to *overlearn* the pattern of behavior appropriate for emergencies, an issue we address in Chapter 18.

Finally, certain *strategic shifts* are sometimes observed in stress-producing emergency circumstances. One is the tendency to "do something, *now*"—that is, to take immediate action (Hockey, 1986). The trouble is, as we learned in Chapter 9, fast action often sacrifices accuracy through the *speed-accuracy trade-off*. Thus, the wrong action might be taken, whereas a more measured and delayed response could be based on more information and more careful reasoning. This is why organizations may wish to caution operators *not* to take any action at all for a few seconds or even minutes following an emergency, until the appropriate action is clearly identified.

Remediation of Psychological Stress

The previous description of performance tendencies following the experience of psychological stress suggests some logical remediations that can be taken (Wickens, 1996). Most appropriately, since these stresses are most likely to occur in emergency conditions, remediations depend on an analysis of the likely circumstances of emergency and actions that should be taken. Remediations should proceed with the design of displays, controls, and procedures in a way that *simplifies* these elements as much as possible. For example, emergency instructions should be easy to locate and salient (so that tunneling will not prevent them from being followed correctly). The actions to be taken should depend as little as possible on holding information in working memory. As we discussed in Chapter 6 and 8, knowledge should be in the world (Norman, 1988). Actions to be taken in emergency should be explicitly instructed when feasible and should be as compatible as possible with conventional, well-learned patterns of action and compatible mapping of displays to controls (Chapter 9). As discussed in Chapter 5, auditory alerts and warnings should be designed to avoid excessively loud and stressful noises.

Finally, training can be employed in two productive directions (Johnston & Cannon-Bowers, 1996). First, extensive (and some might say excessive) training of emergency procedures can make these a dominant habit, readily available to long-term memory when needed. Second, generic training of *emergency stress management* can focus both on guidelines, like inhibiting the tendency to respond immediately (unless this is absolutely necessary), and on techniques, such as breathing control, to reduce the level of arousal to a more optimal value. Such stress training has been validated to have some degree of success and to transfer from one stressor to another (Driskell et al., 2001).

LIFE STRESS

There is another large category of stressors related to stressful circumstances on the job and in the worker's personal life that can lead to disruption in performance (Cooper & Cartwright, 2001; Cooper, 1995). It has been documented, for example, that industries with financial difficulties may have poorer safety records, or alternatively, that workers who are content with labor-management relations (relieving a potential source of job stress) enjoy greater productivity. Correspondingly, stressful life events, like deaths in the family or marital strife (Holmes & Rahe, 1967) have been associated with events such as aircraft mishaps (Alkov et al., 1982), although this relationship is not a terribly strong one; that is, there are lots of people who suffer such life stress events who may be able to cope extremely well on the job.

The cause of both of these types of stress may be related to the different aspects of attention. First, poorer performance by those who are stressed by job-related factors (e.g., poor working conditions, inequitable wages) may be related to the *lack* of attention, resources, or effort put into the job (i.e., low motivation). In contrast, the greater safety hazards of some who suffer life stress may be related to *distraction* or *diversion* of attention; that is, attention diverted from the job-related task to thinking about the source of stress (Wine, 1971).

The full discussion of remediations for such stresses are well beyond the scope of this book, as they pertain to topics such as psychological counseling or industrial relations. In brief, however, the possibility of removing workers from job settings as a consequence of life stress events is questionable, only because so many people are able to cope effectively with those events and would be unfairly displaced. In a comprehensive review of stress in organizations, Cooper and Cartwright (2001) offer three general approaches that organizations can take:

1. Address and remove the source of stress within the organization (i.e., low pay, long working hours, future job uncertainty).
2. Implement stress management programs that can teach workers strategies for dealing with stress.
3. Provide counselors to individuals.

While the first option is preferable, the latter two options have had some success. In one study, absenteeism was found to be reduced by 60 percent following the introduction of stress management training (Cooper & Cartwright, 2001). However, the findings are that the benefits of such programs may be short lived, and they are more likely to address the effects of stress than the attitude toward the job. Cooper and Cartwright conclude that the best solution is to try to eliminate the stress (approach 1) rather than to deal with its consequences.

WORKLOAD OVERLOAD

Stress can be imposed by having too much to do in too little time (Svenson & Maule, 1993). In 1978, an airliner landed far short of the Pensacola Airport runway in Escambia Bay. While flying at night, the flight crew had apparently neglected to monitor their altitude after having to make a faster than usual approach, cramming a lot of the prelanding cockpit tasks into a shorter-than-expected period of time. The high workload apparently caused the pilots to neglect the key task of altitude monitoring. Several years later, an air traffic controller forgot that a commuter aircraft had been positioned on the active runway, a failure of prospective memory, discussed in Chapter 6, and the controller cleared a commercial airliner to land on the same runway. In examining the tragic collision that resulted, the National Transportation Safety Board concluded that, among other causes, the controller had been overloaded by the number of responsibilities and planes that needed to be managed at that time (National Transportation Safety Board, 1991). In the following pages we describe how workload can be predicted and then how it is measured.

The Timeline Model.   The concept of workload can be most easily and intuitively understood in terms of a ratio of the time required (to do tasks) to the time available (to do them in). That is, the ratio TR/TA. We can all relate to the high workload of "so much to do, so little time." The concept of workload is a good deal more sophisticated than this, but the time-ratio concept is a good starting place (Hendy et al., 1997). Thus, when we wish to calculate the workload experienced by a particular operator in a particular environment, we can begin by laying out a *timeline* of when different tasks need to be performed and how long they typically take, as shown in Figure 13.3. Such a time line should be derived on the basis of a careful task analysis. We may then calculate the workload for particular intervals of time as the ratio within that interval of TR/TA (Parks & Boucek, 1989; Kirwan & Ainsworth, 1992). These ratio values are shown at the bottom of the figure for five intervals.

This calculation can be designed to accomplish two objectives. First, it should predict how much workload a human experiences, a subjective state that can be measured. Second, it should predict the extent to which performance will suffer because of overload. However, these two effects are not entirely linked, as shown in Figure 13.4. As the ratio increases, the experience of workload, shown

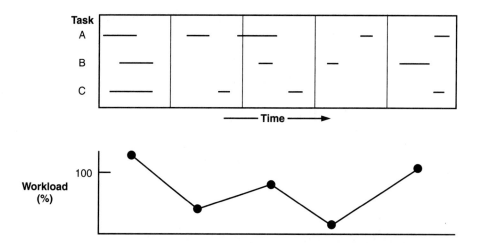

FIGURE 13.3
Timeline analysis. The percentage of workload at each point is computed as the average number of tasks per unit time, within each window. Shown at the bottom of the figure is the computed workload value TR/TA.

by the solid line, also increases relatively continuously. However, human performance decrements due to overload occur only at or around the breakpoint of the dashed line, where TR/TA = 1.0, and above where people are required to time-share two or more tasks, producing dual-task decrements, discussed in Chapter 6. Figure 13.4 therefore defines two qualitatively different regions of

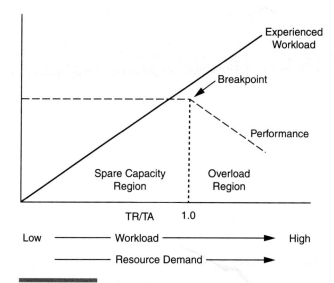

FIGURE 13.4
Hypothetical relation between workload imposed by a task, measured by TR/TA, and workload experienced and performance.

workload, an overload region to the right of the breakpoint, and a **spare capacity** region to the left. Designers have sometimes suggested that it is a good idea to create job environments with a workload of less than 0.8 from the time/line analysis in order to provide a margin of spare capacity should unexpected circumstances arise (Kirwan & Ainsworth, 1992; Parks & Boucek, 1989).

While it might seem quite feasible to construct task timelines of the sort shown in Figure 13.3 and use them to derive workload estimates, in fact, four factors make this endeavor somewhat challenging.

1.   *Identification of task times.* The lengths of lines in Figure 13.3 must be derived. Some investigators provide these in terms of table lookups (Luczak, 1997) or in software packages such as the Army's IMPRINT program. Sometimes the values are provided by the workload analyst or subject matter expert (Sarno & Wickens, 1995), and sometimes they may be observed by watching and recording people performing the tasks in question. In estimating these tasks times, it is critically important to include *covert* tasks, like planning, diagnosis, rehearsing, or monitoring; even though they may not be reflected in any direct behavioral activity, they are still a major source of workload.

2.   *Scheduling and prioritization.* Figure 13.3 indicates that there is overload in the first time period. However, the wise operator might choose to postpone performance of one or two of the overloading tasks to the second time period, when workload is relatively light, in order to better distribute the workload. Such *task management strategies* were discussed in Chapter 6, and a timeline model must account for these possibilities.

3.   *Task resource demands and automaticity.* Figure 13.3 suggests that all tasks are equal in their contribution to task overload. As we learned in Chapter 6, this is not the case. If one of two overlapping tasks are automated (e.g., walking), it will impose very little overload on a concurrent task. Even if two overlapping tasks are not fully automated, if they are relatively easy and demand few resources for their performance, they are not likely to produce an overload performance decrement. The issue of where task resource demand values come from is similar to that associated with task time values. Some authors have offered explicit measures of the demands of specific tasks (McCracken & Aldrich, 1989), values that are embedded in certain software packages like IMPRINT. It is also possible to reasonably estimate resource demands to be at one of two levels, 1.0 or 0.5, while considering specific task factors that contribute to demand, such as those shown in Table 13.1. Because of task demands, even single tasks can create workload overload, such as a task that demands rehearsal of 10 chunks of information in working memory.

4.   *Multiple resources.* Many aspects of task interference or task overload can be accounted for by the extent to which tasks demand common versus separate resources within the four dimensions of the multiple resource model (Wickens, 2002). For example, two visual tasks are likely to interfere more and create more performance-based workload, than are a visual and an auditory task. Some of the more advanced computational models of workload can ac-

TABLE 13.1   Demand Checklist

Legibility	Working-memory demand (number of
Visual search demand (parallel versus	chunks $\times$ number of seconds to retain)
serial; see Chapter 4)	Unprompted procedures
Display organization: Reduce scanning	S-R compatibility
Compatibility: Display compatible with	Delayed feedback of action
mental model	(intrinsic, tactile)
Consistency of format across displays	(extrinsic, visual)
Number of modes of operation	Precision of required action
Prediction requirements	Skill-rule-knowledge
Mental rotation	

count for overlapping of resources (Sarno & Wickens, 1995; Wickens 2002) although these are not described here.

Taken together, these four qualifications, particularly the latter three, indicate that some caution should be used in relying upon simple task timelines to quantify workload in the overload region without considering how to implement their impact. A pure timeline measure of workload is probably best suited for generating workload predictions within the spare capacity region of Figure 13.4.

Workload Overload Consequences.   Whether the result of pure time stress (TR/TA > 1.0) or from increases in task-resource demand, when task combinations enter the workload overload region, there are important consequences for human performance: *Something* is likely to suffer. Less predictable, however, is knowing how things will suffer. For example, Edland and Svenson (1993) report any of the following effects to have been found in making decisions under time pressure (decreasing TA/TR): more selectivity of input, more important sources of information given more weight, decrease in accuracy, decreasing use of strategies that involve heavy mental computation, and locking onto a single strategy.

The study of task management strategies discussed in Chapter 6, can begin to provide some evidence as to the nature of which tasks are more likely to suffer under overload conditions. Most critical is the operator's continuing awareness of the objective importance of all tasks that may compete for attention, such that those of lesser importance will be shed first (Chao, Madhavan and, Funk, 1996; Wickens et al, 2003, in press.).

Remediations

On the basis of these observed trends in behavior, certain remediations are suggested. Most obviously these include task redesign by trying to assign certain time-loading tasks to other operators or to automation (see Chapter 16). They also include developing a display design such that information for the most objectively important tasks are available, interpretable, and salient. Training for high time-stress workload can focus on either of two approaches. One is training on the component tasks to try to speed or automate their performance (Schneider, 1985; see Chapter 6). This means that tasks will either occupy less time in the timeline or will require little attention so that they can be overlapped with others without imposing workload. The other approach is to focus on training of *task management skills* (Chao et al., 1996) and to ensure that operators are properly *calibrated* regarding the relative importance of tasks and information sources (Raby & Wickens, 1994). Dismukes and colleagues (2003) have developed specific training packages regarding task management for pilots, and some are embedded in the FAA rules for Cockpit Resource Management (see Chapter 19). As another example, the nuclear regulatory agency has explicitly stated the policy that in the case of emergency, the operator's first task priority should be to try to stabilize the plant (to keep the situation from growing worse), the second is to take steps to ensure safety, and the third is to try to diagnose the cause of the emergency (Chapter 16).

Mental Workload Measurement

We discussed the manner in which workload can be defined in terms of TR/TA, and indeed time *is* a major driver of workload (Hendy et al., 1997). However, mental workload can be defined more generally by the ratio of the resources required to the resources available, where time is one of those resources but not the only one. This is shown by relabeling the *x* axis of Figure 13.4 to encompass the more general definition of resource demands. For example, we know that some tasks are time consuming but not particularly demanding of cognitive resources or effort (e.g., a repetitive action on an assembly line), whereas others may be very effortful but occupy only a short time (e.g., answering a difficult logic question on a test). As noted, predictive workload techniques based purely on timelines have limits, and so workload researchers must turn to various forms of assessing or **measuring** the resource demands of tasks as humans actually perform them (O'Donnell & Eggemeier, 1986; Tsang & Wilson, 1997).

The assessment of workload can serve three useful functions. First, we have already seen how assessing the workload of component tasks can contribute to predictive models of workload. Second, workload assessment after a system has been built (or put in use) can provide a very important contribution to usability analysis (Chapters 3 and 15) because, even though performance with the system in question may be satisfactory, if the workload experienced while using it is excessive, the system may require improvement. Third, workload may be assessed online to make inferences about an operator's capability to perform (e.g., blocking out cellular phone calls in vehicles when workload is inferred to be high). Traditionally, workload has been assessed by one of four different techniques.

Primary Task Measures. Primary task measures are measures of system performance on the task of interest. For example, in assessing an interface for an ATM, the primary task measure may be the speed and accuracy with which a user can carry out a transaction. The primary task measure is not really a workload measure per se, but it is often *influenced* by mental workload and hence assumed to reflect workload (i.e., higher workload will make performance worse). However, this may not always be the case. For example, a car driver can perform equally well, in terms of lane keeping (the primary task measure), on a crowded, rainy freeway at night as on an empty, dry freeway in the daytime, despite the higher workload associated with the former condition. As this example suggests, there are many circumstances in which very good primary task performance is attained but only at a cost of high workload. This means that there will be no margin of reserve capacity if unexpected increases in load occur, close to the breakpoint in the spare capacity region of Figure 13.4. It may also mean that users will choose not to use the high-workload device in question when given an option. The ATM customer may simply choose to go inside the bank to the teller.

Secondary Task Methods. Performance on a secondary or concurrent task provides a method of measuring *reserve capacity,* roughly the distance to the left of the breakpoint in Figure 13.4. The assumption is that performance of the primary task takes a certain amount of cognitive resources. A secondary task will use whatever residual resources are left. To the extent that fewer resources are left over from the primary task, performance on the secondary task will suffer. Most researchers using secondary tasks to assess workload have used external secondary tasks or tasks that are not usually part of the job (Tsang & Wilson, 1997; Kantowitz & Simsek, 2001). In this method, people are asked to perform the primary task as well as possible and then to allocate whatever effort or resources are still available to the secondary task. Increasing levels of difficulty on the primary task will then yield diminishing levels of performance on the secondary task. Examples of common secondary tasks are time estimation, tracking tasks, memory tasks, mental arithmetic, and reaction time tasks (Tsang & Wilson, 1997).

The use of a secondary task for measuring workload is good because it has high face validity in that it seems like a reasonable measure of demands imposed by the primary task. However, the secondary task is problematic because, it often seems artificial, intrusive, or both to operators performing the tasks. Several researchers therefore have suggested the use of *embedded* secondary tasks, which are secondary tasks that are normally part of the job but have a lower priority (Raby & Wickens, 1994). An example might be using the frequency of glances to the rearview mirror as an embedded secondary task measure of driving workload, or monitoring for the appearance of a call sign of your own aircraft.

Physiological Measures. Because of problems with intrusiveness and multiple resources, some researchers favor using physiological measures of workload (Tsang & Wilson, 1997; Kramer, 1991). In particular, measures of *heart rate variability* have proven to be relatively consistent and reliable measures of mental workload (just as mean heart rate has proven to be a good measure of physical workload and stress; see Chapter 12). At higher levels of workload, the heart rate (interbeat interval) tends to be more constant over time, whereas at lower workload levels it waxes and wanes at frequencies of around 0.1 Hz and those driven by respiration rate (Tattersall & Hockey, 1995).

Measures of visual scanning are also useful in understanding the qualitative nature of workload changes. For example, in driving we can measure fixations on the dashboard as a measure of the workload demands (head-down time) associated with in-vehicle instrumentations (Landsdown, 2001; see Chapter 17). Many other physiological workload measures are associated with variables such as blink rate, pupil diameter, and electroencepholographic (EEG) recording, which are not described here (see Tsang & Wilson, 1997, and Kramer, 1991, for a fuller discussion). Generally speaking, physiological measures correlate with other measures of workload and hence are valid. The equipment and instrumentation required for many of these, however, may sometimes limit their usefulness.

Subjective Measures. The most intuitive measure of mental workload, and that which is often easiest to obtain, is to simply ask the operator to rate workload on a subjective scale. The best scales are often anchored by explicit descriptions of the high and low endpoints of the scale. Sometimes they may be associated with a structured decision tree of questions that guide the rater to a particular number (Wierwille & Casali, 1983). Researchers have argued that subjective workload should be rated on more than just a single scale because workload is a complex multidimensional construct (e.g., Derrick, 1988). For example, the NASA Task Load Index (TLX; Hart & Staveland, 1988) imposes five different subscales with seven levels (Wickens & Hollands, 2000).

While subjective ratings are easy to obtain, they also have the limitation that they are, by definition, subjective, and it is a fact of life that people's subjective reports do not always coincide with their performance (Andre & Wickens, 1995). It is also possible to envision raters intentionally biasing their reports to be low (or high) under certain circumstances for motivational reasons. However, to the extent that subjective effort sometimes guides the choice of actions, strategies, and tasks (favoring those that involve lower effort; see Chapter 6), then collection of such data can be extremely helpful in understanding such choices.

Workload Dissociations. Workload measures will not always agree (Yeh and Wickens, 1988). For example, if operators were more motivated to "try harder" with one system than another, they will perform better on the first system (better primary task performance → lower workload), but their subjective rating of the effort invested would also be higher for the first system (more effort → higher workload). Because of these, and other forms of dissociation (Yeh and Wickens, 1988), it is important that multiple measures of workload be collected.

FATIGUE AND SLEEP DISRUPTION

High mental workload can have two effects. While performing a task, performance may degrade. But the effects of high and even moderate mental workload are also cumulative in terms of the buildup of *fatigue* in a way that can adversely affect performance on subsequent tasks or on the same tasks after a prolonged period of performance without rest (Orasanu & Backer, 1996; Desmond & Hancock, 2001; Gawron et al., 2001). Fatigue may be defined as "a transition state between alertness and somnolence" (Desmond & Hancock, 2001), or more elaborately, "a state of muscles and the central nervous system in which prolonged physical activity or mental processing, in the absence of sufficient rest, leads to insufficient capacity or energy to maintain the original level of activity and/or processing" (Soames-Job & Dalziel, 2001).

Fatigue, as a stressor, clearly degrades performance and creates problems in maintaining attention. Mental as well as physical fatigue becomes relevant in scheduling rest breaks or maximum duty cycles in high-workload tasks. For example, the Army establishes limits on the amount of helicopter flight time based on the level of workload imposed during flight. Night flying imposes higher workload (and hence shorter duty) than day flight; flight low to the ground imposes higher workload than that at higher altitudes.

The role of fatigue also becomes relevant in predicting the consequences of long-duration, sustained operations, or continuous performance, such as that which might be observed on a military combat mission (Orasanu & Backer, 1996). Major negative influences of fatigue were documented in operation Desert Storm, in 1991–92 (Bisson et al., 1992), as well as with long-haul truck drivers (Hamelin, 1987) and represents a potential source of many of the medical errors that plague workers of long hours in hospitals (Kohn et al., 2000). In these examples, of course, the effects of fatigue from continuous work are often confounded with those of sleep loss, although their influences are not identical. We return to the issue of sleep loss at the end of this chapter. We note here that fatigue may result not only from the accumulated effects of doing too much work, but also from prolonged periods of doing very little (Desmond & Hancock 2001), the issue of *vigilance.*

Vigilance and Underarousal

At first glance, circumstances in which the operator is "doing little" might seem like less of a human factors problem than circumstances in which the operator is overloaded. Yet a long history of research, as well as accident and incident analysis, reveals that maintaining *sustained attention* to vigilance tasks in low-arousal environments can be just as fatiguing and just as prone to human vulnerabilities as the high-workload situation, and can indeed be a source of high mental effort, as reflected in subjective ratings (Hancock & Warm, 1989). For example, several studies have found that some quality-control inspectors on the assembly line, whose only job is to look for defects, show an alarmingly high miss rate.

Causes of the Vigilance Decrement.   The stage for the vigilance problem was set in our discussion of signal detection theory in Chapter 4. We outlined how signal detection problems are analyzed in terms of the four classes of joint events: hits, correction rejections, misses, and false alarms. The main problem in vigilance appears to be the increased number of misses that occur as the vigil progresses. Years of research (Warm, 1984; Warm & Parasuraman, 1987; Davies & Parasuraman, 1982; Wickens & Hollands, 2000) have identified certain key characteristics of the environment that lead to the loss of performance in detecting signals or events of relevance. The characteristics include

1. *Time.* The longer duration an operator is required to maintain vigilance, the greater is the likelihood that misses will occur.

2. *Event salience.* Bright, loud, intermittent, and other salient events are easily detected. The event that is subtle, like a typesetting error in the middle of a word, a small gap in the wiring of a circuit board, or the offset of a light, will show a larger loss in detection over time.

3. *Signal rate.* When the signal events themselves occur at a relatively low rate, monitoring for their presence is more effortful, and the likelihood of their detection is reduced, partly because low signal expectancy causes the operator to adopt a more conservative response criterion (producing more misses and fewer false alarms) and partly because the presence (and detection) of events appear to act as stimulants that better sustain arousal. When these events are fewer in number, arousal falls.

4. *Arousal level.* A problem with vigilance situations is that there is generally little intrinsic task-related activity to maintain the information-processing system in the state of alertness or arousal to optimize perception. The operator is often at the far left end of the inverted U curve shown in Figure 13.2, and attentional resources are diminished (Young & Stanton, 2001). As might be expected, anything that further decreases arousal, like sleep deprivation, has particularly profound effects on vigilance performance.

Vigilance Remediations. The four primary factors identified above suggest some appropriate solutions to the vigilance problem (Wickens & Hollands, 2000). First, watches or vigils should not be made too long, and operators should be given fairly frequent rest breaks. Second, where possible, signals should be made more salient. This is not always easy to achieve, but there are certain techniques of **signal enhancement** that can be cleverly employed in areas such as quality control inspection (Drury, 1982; Wickens & Hollands, 2000).

Third, if miss rates are high, it is possible to alter the operator's criterion for detecting signals through payoffs (large rewards for detecting signals) or changing the signal expectancy. However, in a situation in which the signals (or events) to be detected occur only rarely, the only way to change signal expectancy effectively (and credibly) is by *introducing false signals* (e.g., put a few known defective parts on the assembly line or intentionally concealed weapons in luggage for inspection). Of course, as discussed in Chapter 4, designers and practitioners should always remember that such alterations in the response criterion will invariably produce more false alarms and should therefore assume that the costs of a false alarm to total system performance are less than the benefits of reducing the miss rate.

Fourth, efforts should be made to create or sustain a higher level of arousal. Frequent rest breaks will do this, as will intake of appropriate levels of stimulants such as caffeine. Other forms of external stimulation may be effective (e.g., music, noise, or conversation), but caution should be taken that these do not form sources of *distraction* from the inspected product (or monitored environment). Finally, every effort should be made to ensure that operators are not sleep deprived because of the particular vulnerability of vigilance tasks to fatigue from sleep loss.

Before we turn to the discussion of sleep disruption, however, it is important to preview an issue that we will revisit in Chapter 16 when we discuss automation. Increasingly, automated systems are removing both physical and cognitive activity from the human, as such activity is now carried out by computers. Such a trend often leaves humans in a purely monitoring role, which makes sustained vigilance for the rare computer failure a very challenging task (Parasuraman, 1987).

Sleep Disruption

Sleep disruption is a major, although not the only, contributor to fatigue. Sleep disruption incorporates the influence of three separate factors: (1) sleep deprivation or sleep loss, referring to less than the 7 to 9 hours of sleep per night that the average adult receives; (2) performance at the low point of the *circadian rhythms* in the early hours of the morning; (3) disruption of those circadian rhythms from jet lag or shift work.

There is no doubt that sleep disruption is a major stressor that has a negative impact on both safety and productivity. We are, for better or for worse, becoming a 24 hour a day society, with obligations to run transportation systems, generate energy, deliver products, staff medical facilities, and maintain security around the clock. The sleep disruption that results can take its toll. For example, 60 percent of class A aircraft mishaps in the Air Force were attributed to fatigue (Palmer et al., 1996); four of the largest nuclear power plant disasters, attributed to human error, occurred in the early morning shifts (Harrison & Horne, 2000); and the tragic explosion of the space shuttle Challenger was attributed, in large part, to the poor decision making of the launch team, who had received very little sleep prior to their early morning decision to launch the rocket in excessively cold temperatures (President's Commission, 1986). It is estimated that over 200,000 auto accidents per year are attributed in part to sleep disruption and fatigue. Impairment on many other sorts of tasks, such as medical treatment in the hospital (Asken & Raham, 1983; Rosa, 1995) or performance on the battlefield (Ainsworth & Bishop, 1971), have been shown to suffer substantially from sleep loss (Huey & Wickens, 1993).

Sleep Deprivation and Performance Effects

As we all know, losing sleep, the "all nighter" before an exam or paper is due, can hinder performance. To some extent, almost all aspects of performance suffer when a person is sufficiently sleepy. After all, when we fall asleep, little performance of any kind can be expected! However, short of this, some aspects of performance are more susceptible to sleep deprivation than others (Huey & Wickens, 1993). Given that sleepiness causes increased blinks, eye closures, and brief durations of "microsleep" (nodding off), it is understandable that tasks depending on visual input are particularly sensitive to sleep disruption. Furthermore, tasks that are not themselves highly arousing will also be unable to compensate for sleepiness by sustaining operator attention. As we saw in the previous section, this is particularly true of vigilance or monitoring tasks, which seem to be the first to go when operators are sleep deprived (Horne et al., 1983; Hockey et al., 1998).

In addition, researchers have reported that tasks particularly sensitive to sleep disruption are those involving higher level cognition, such as decision making (Harrison & Horne, 2000), innovation and creativity (Harrison & Horne, 2000), learning or storing new material (Williams et al., 1966), as well as those tasks involving self-initiated cognitive activity, like maintaining situation awareness and planning. Hockey and colleagues (1998) report that in a multi-task situation, central tasks are more resistant to the negative effects of sleep loss than are peripheral or secondary tasks. Not surprisingly, the tasks that are relatively less susceptible to sleepiness are those with a great deal of intrinsic arousal, such as those involving a lot of motor activity or highly interesting material. For example, Haslem (1982) reports that sleep deprivation of soldiers has little effect on their rifelry performance but has a substantial effect on their cognitive activity.

Sleep loss has particular implications for performance in *long-duration missions,* defined as intense periods of job-related activity, away from home, lasting more than a day. This might include military combat missions or long-haul truck driving, or an airline pilot's trip (which typically is a series of flights over 3–4 days). Two factors combine in these situations to create sleep deprivation. First, the quality of sleep "on the road" is typically less, and so a *sleep debt* is typically built up as the mission progresses (Graeber, 1988). Second, there is usually a less than adequate amount of sleep the night prior to the mission, a period often involved with preparations, an early morning departure, and so on. Thus, the mission typically *begins* with a sleep debt, which only grows during subsequent days, a finding documented with both aircrews and long-haul truck drivers (Graeber, 1988; Feyer & Williamson, 2001).

Circadian Rhythms

In addition to sleep loss, a second cause of sleepiness is related to the time of the day-night cycle, our phase in the natural *circadian rhythms* (Horne, 1988). These rhythms have a clear physiological base. As shown in Figure 13.5, our body temperature undergoes a natural fluctuation, reaching a minimum in the early hours of the morning and climbing progressively during the day to reach a maximum in the late afternoon/early evening hours before declining again. This rhythm of arousal is correlated with and "entrained by" the natural day-night cycle on Earth.

There are at least three important variables correlated with body temperature, as also shown in the figure. These include *sleepiness* (which can be mea-

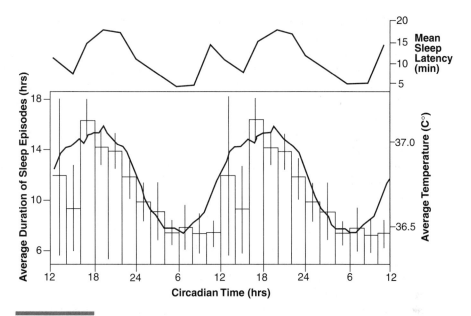

FIGURE 13.5

Graph plotting mean sleep latency (top), circadian rhythms (body temperature), and sleep duration (bottom) against time for two day-night cycles. The bars around sleep duration represent the variability. (*Source:* Czeisler, C. A., Weitzman, E. D., Moore-Ede, M. C., Zimmerman, J. C., & Knauer, R. S., 1980. Human sleep: Its duration and organization depend on its circadian phase. *Science,* 210, pp. 1264–1267. Reprinted with permission. Copyright 1980, American Association for the Advancement of Science.)

sured by the *sleep latency test*—how long it takes a volunteer to go to sleep in a dark room on a comfortable bed); *sleep duration,* which measures how long we can sleep (greater at night); and measures of *performance.* Shown in Figure 13.6 are the performance fluctuations observed with four different kinds of tasks; all four show the same consistent drop in performance in the early morning hours, a drop that is mirrored in the real-world observations such as the greater frequency of errors by air traffic controllers (Stager et al., 1989) or accidents by truck drivers (Czeisler et al., 1986; Harris, 1977). It is not surprising that the effects of sleep loss and circadian cycle essentially add, so that the early morning lows are substantially lower for the sleep-deprived worker (Gawron et al., 2001). The sleep deprived person may be able to compensate the following day, after one night's deprivation, but when this deprivation is experienced during the following early morning hours, compensation becomes exceedingly difficult.

Circadian rhythms also influence intentional sleep. Just as the low point in Figure 13.5 is a period during which it is hard to stay awake, so the high point is one during which it is hard to sleep. As a consequence, sleep cycles in which the sleep must be undertaken during the day or early evening will reduce the quality of sleep and further contribute to a sleep debt.

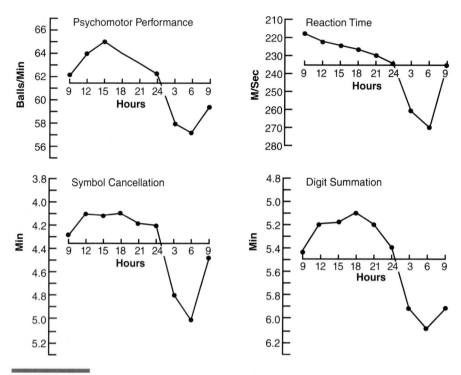

FIGURE 13.6

Graph showing how performance on four kinds of tasks varies as a function of circadian rhythms, shown for a one day cycle. (*Source:* Klein, K. E., and Wegmann, H. M., 1980. *Significance of Circadian Rhythms in Aerospace Operations* [*NATO AGARDograph #247*]. Neuilly sur Seine, France: NATO AGARD.)

Circadian Disruption

Circadian disruption, or *desynchronization,* characterizes the circumstances in which a person is trying to sustain a level of activity that is out of synchrony with the internal circadian rhythm and its associated level of arousal. It has implications for both long distance east-west travel (jet lag) and shift work.

Jet Lag. Jet lag is caused after crossing several time zones, when the ongoing circadian rhythm becomes out of synchrony with the day-night cycle at the destination, in which case it may take as much as 3-5 days to adjust, or adapt. For a variety of reasons the adjustment period is considerably longer following eastbound flights (e.g., U.S. to Europe) than westbound flights (U.S. to Asia).

The most successful ways to reduce the disruptive effects of jet lag are to try to bring the body into the local cycle of the destination as rapidly as possible. One way to do this is by waiting until the local bedtime after one has landed rather than napping during the day (Graeber, 1988). A second way to "hurry" the adaptation process along is by exposure to intense light prior to departure at a time that approximates daylight at the destination (Czeisler et al., 1989). Similar effects on biochemically adapting the circadian rhythms can be achieved by taking the drug melatonin (Comperatore et al., 1996).

Shift Work.   Given that certain jobs must be performed round the clock, some workers must be active in the early morning hours when the circadian rhythms are at their lowest level of arousal. Three strategies can help deal with the resulting problem. They vary considerably in their effectiveness.

One strategy is simply to assign workers permanently to different shifts, under the assumption that the circadian rhythms of the "night shift" worker will eventually adapt. The problem with this approach is that full adaptation never entirely takes place as long as the worker is exposed to *some* evidence of Earth's natural day-night cycle, such as sunlight and the natural daytime activity of most of the rest of the population. Such evidence will be desynchronized from the intended circadian cycle. The quality of daytime sleep will, as a result, continue to be less than adequate (Rosa, 2001). Another problem with this strategy is the smaller pool of people who are willing to work the night shift because of personal preference and a need to retain an activity cycle more compatible with other family members.

A second strategy, employed, for example, in shipboard watches, is to maintain a fairly continuous rotation of shifts; a worker might have an 8-hour night watch one "day," a morning watch the next, an afternoon watch the next, and so forth. Here the problem is that desynchronization remains in a continuous state of flux. The circadian rhythms never have a chance to catch up to the levels of alertness that the person is trying to obtain via the scheduled shift. Hence, the worker's arousal will never be optimal during the work time (particularly in the early morning hours), nor, for the same reasons, will his or her sleep be optimal during the off time (Carskadon & Dement, 1975; Rosa, 2001). The third, and more successful strategy is to alter the shift periods but to do so relatively infrequently (e.g., following 14 to 21 days on a given cycle; Wilkinson, 1992). This strategy has the advantage of allowing the circadian rhythm to synchronize with (adapt to) the desired schedule, an adaptation which takes 4 to 5 days to occur and yet still allows all workers to share in the same inconveniences of night and early morning shifts (Czeisler et al., 1982; Rosa, 2001). However, when such slow rotation shifts are used, *workers are particularly vulnerable on the first shift after the change;* naturally, they are even more vulnerable on the first *night shift* after a change, a period of time that should raise a red flag of danger in safety-critical jobs.

Whether schedules are rotated rapidly or slowly, a second finding is that shift changes that are clockwise or *delayed* are more effective than those that are counterclockwise or *advanced* (Barton & Folkard, 1993; Rosa, 2001).

There are other shift work issues besides the particular time of day that impact fatigue and human performance. One of the most important of these is the *longer shift,* (i.e., 10 to 14 hours) that might be created for a number of reasons: overtime, a desire to create longer weekends by working four consecutive 10-hour days, or, with small crews in remote cites like oil rigs, the need to sustain a 12-on 12-off cycle. In all cases, the data are fairly conclusive (Rosa, 2001; Gawron et al., 2001): The longer shifts produce greater fatigue and more errors. For example, truck driver shifts of 14 hours were found to produce three times the accident rate as those shifts of less than 10 hours (Hamelin, 1987); and in hospitals, extensive hours were found to be associated with workers skipping important procedures to assure hygiene (Rosa, 2001).

Remediation to Sleep Disruption

We have described a host of problems that can result from all three forms of sleep disruption. The solutions or remediations we propose can, to an extent, be applied to all of them. Some of the remediations that can be suggested to combat sleepiness and fatigue are as obvious as the source of the problem itself: *Get more sleep.* In fact, even small amounts of sleep, such as 3 to 4 hours per night, can quite beneficial in sustaining performance through several days even though such an amount will still not come close to sustaining the performance level of a well-rested individual (Orasanu & Backer, 1996; Gawron et al., 2001).

Napping has by now been well documented as an effective countermeasure (Rosa, 2001). For example, Dinges et al. (1987) found that a single strategically placed 2-hour nap could significantly improve the level of performance of people after 54 hours of sustained wakefulness. Rosekind and colleagues (1994) documented the benefits of controlled naps in the cockpit of aircraft on long transoceanic flights. Such naps improve the level of vigilance performance and still allow pilots to get just as good sleep after the flight as if they had not napped at all. In general, a nap should be at least 15 minutes in duration to be effective (Naitoh, 1981). In the workplace, it is also important to provide good conditions for napping. This may sometimes involve the creation of an explicit "sleep room".

The one possible drawback with naps (or any other sleep in operational environments) is the presence of *sleep inertia*. This is the tendency of the mind not to function with full efficiency for the first 8 to 10 minutes following awakening (Dinges et al., 1985). Hence, any controlled napping strategy must be implemented with allowance made for full recovery of mental functions following the nap. For example, watchkeepers should be awakened 10 minutes prior to assuming their watch.

A third remediation is to build up *sleep credits*, that is, trying to gain extra sleep prior to a mission or period in which sleep deprivation is anticipated (Huey & Wickens, 1993). Unfortunately, this procedure is very often the opposite of reality. For example, Graeber (1988) noted that pilots typically sleep *less* than an average amount on the night before a 3 to 4 day series of flights is initiated.

Perhaps the best way of implementing all three remediations is through implementation of a careful program of *sleep management* (deSwart, 1989), that is, endorsed and supported by the organizational management. This option may be particularly feasible in relatively controlled units, such as those found in the military. While less controllable in other circumstances, such as the medical facility or industrial factory, it is still feasible for organizations to emphasize the importance of adequate sleep for operational safety, and, for example, to disapprove of rather than admire the individual who may brag of "not sleeping for two nights to get the job done." Clearly, it should be the role of organizations to avoid conditions in which operators must work long hours in life-critical jobs with little sleep (the pattern often reported by medical students, interns, and residents; Asken & Raham, 1983; Friedman et al., 1971).

There are, finally, two remediations that have far less consistent records of success for quite different reasons. First, stimulant drugs like caffeine can be used to combat sleepiness in the short run, and these as well as other motivators can be used to sustain performance through and after one night's sleep deprivation (Gawron et al, 2001; Lipschutz et al., 1988). However, after two nights, the compensatory ability of such drugs is limited (Horne, 1988). Furthermore, while excessive consumption of caffeine may be adequate in the short run, in the long run it disrupts the ability to sleep soundly when sleep time *is* available and hence may be counterproductive in reducing overall fatigue. A caffeine-induced sleep resistance is particularly disruptive when trying to sleep during the daytime. Other stimulant drugs, such as dexamphetamine (Caldwell et al., 1995) may be effective in sustaining arousal over a longer, multiday duration, and also may be less likely to disrupt sleep after their termination, although their long-term effects have been not well studied (Gawron et al., 2001)

A second remediation that has only limited success is simply to not require (or to prohibit) work during the late night-early morning hours at the low arousal point of the circadian rhythm. If this is done, then the periods of lowest performance will be avoided, and workers will not be required to sleep during the day when adequate sleep is more difficult to attain. The problem with this remediation is simply that many organizations *must* function round the clock: Ships must sail all night, trucks must drive, and many factories and industrial plants must keep running 24 hours a day to provide services or products, often on a just-in-time basis, hence requiring management to address the issues of shift work.

CONCLUSION

Stress comes in a variety of forms from a variety of causes, and exhibits a variety of symptoms. The underlying concern for human factors is the potential risk to health and degradation in performance on tasks that may be otherwise well human factored. Whether the underlying cause is overarousal and overload or underarousal and underload stress reveals the clear vulnerabilities of the human operator. Such vulnerabilities can be a source of accident or error, as we describe in the next chapter. Issues of workload overload have always confronted the worker in society. However, two trends appear to make the issue of underload one of growing concern. First, the continued push for productivity in all domains appears to be increasing the frequency of round-the-clock operations, thereby inviting concerns about night work and sleep disruption (Desmond &

Hancock, 2001). Second, increasing capabilities of automation are now placing the human more frequently in the role of the passive monitor—the underarousing task that is most vulnerable to conditions of fatigue. In this role, the human's only other responsibility may be to make sudden creative decisions in response to the rare but critical circumstances when the automation *does* fail, a task that we have also seen as vulnerable to sleep disruption (Harrison & Horne, 2000).

Chapter 14

Safety and Accident Prevention

Marta loved her new job at the convenience store. One morning, as she was busy restocking shelves, she turned a corner to go down an aisle on the far side of the store. A glare came in through the large window, which is probably why she did not see the liquid that had spilled on the floor. She slipped on the substance and fell, impaling her arm on a blunt metal spike meant to hold chips. Her arm never healed properly, and she had back problems for the remainder of her life.

John walked across a bare agricultural field to where a 6-inch-diameter irrigation pipe came out of the ground. The opening was filled by a large chunk of ice, so John began using a steel pry bar to dislodge the chunk. As the ice chunk broke free, air pressure that had built up in the pipe suddenly drove the ice up against the pry bar. The force sent the bar through John's neck and impaled him backward to the ground. Amazingly, John was taken to the hospital and lived.

Steve and Pete were fighting a canyon forest fire along with several other relatively new firefighters. Suddenly, a high wind drove the fire toward them, and all of the men began running to escape the oncoming blaze. Realizing that they would be overtaken at any moment, Steve and Pete quickly set up their survival tents and crawled inside. In the meantime, two other men (who had thrown aside their heavy survival tents in order to run faster) were forced to try to escape by running up a steep hill. The men in the survival tent died, and the men who had to run out made it to safety.

A 4-year-old boy in California climbed up on a new concrete fountain in his backyard to retrieve a ball from the basin area. As he pulled himself up, the fountain toppled over and crushed him to death. His parents successfully sued the manufacturer and landscape company who installed it.

As we saw in Chapter 1, a major goal of human factors is to increase the *health* and *safety* of people in a variety of environments, such as work, home, transport systems, and so on. Health and safety are related but can be distinguished in at least two ways. First, in general, safety concerns itself with injury-causing situations, whereas health is concerned with disease-causing situations. Also, safety focuses on *accidents* resulting from acute (sudden or severe) conditions or events, while health focuses on less intense but more prolonged conditions, such as poor design of a data-entry keyboard (DiBerardinis, 1998; Goetsch, 2001; Manuele, 1997). Hazards in the workplace can lead to health problems, safety problems, or both (noise is one example). We focused on hazards that affect health in Chapters 10, 11, and 12, presenting information on the design of physical work environments so as to reduce hazards and decrease long-term ergonomic-based health problems, such as cumulative trauma disorders.

In this chapter, we focus on hazardous conditions that may result in more sudden and severe events, causing injury or death. This includes such things as human performance failures, mechanical failures, falls, fires, explosions, and so forth. While the majority of our discussion centers on occupational safety, many of the factors that cause accidents in the workplace are applicable to other more general tasks, such as driving. More specifically, we review safety and accident prevention by discussing (1) general factors that contribute to, or directly lead to, accidents, (2) methods for systematically identifying hazards in equipment and the workplace, (3) methods for hazard control, and (4) factors that affect human behavior in hazardous environments.

INTRODUCTION TO SAFETY AND ACCIDENT PREVENTION

All of the scenarios at the beginning of this chapter are based on true stories. They represent just a few of the thousands of ways in which people are injured or killed in accidents every year. Safety and accident prevention is a major concern in the field of human factors. In a typical year in the United States, 47,000 people die in motor vehicle accidents, 13,000 die in falls, and 7,000 people die from poisoning. In 1993, there were 10,000 deaths in the workplace alone; Table 14.1

TABLE 14.1   Most Frequent Causes of Workplace Deaths and Injuries

Injury	*Deaths*
Overexertion: Working beyond physical limitations	Motor-vehicle related
Impact accidents: Being struck by or against an object	Falls
Falls	Electrical current
Bodily reaction to chemicals	Drowning
Compression	Fire related
Motor vehicle accidents	Air transport related
Exposure to radiation or caustics	Poison
Rubbing or abrasions	Water transport related
Exposure to extreme temperatures	

shows the major causes of workplace injury and death as reported by the National Safety Council (1993a). The major causes of injuries are overexertion, impact accidents, and falls. The major causes of death are accidents related to motor vehicles and falls; however, other causes are common as well, such as fire, drowning, explosion, poison, and electrical hazards. Finally, NIOSH estimates that over 10 million men and women are exposed annually to hazardous substances that could eventually cause illness (Goetsch, 2001). In addition to the human tragedy of injury and death, accidents carry a high monetary cost. Workplace deaths and injuries alone typically cost at least $50 billion per year. This reflects factors such as property damage, lost wages, medical expenses, insurance administration, and indirect costs. According to Kohn, Friend, and Winterberger (1996), each workplace fatality costs U.S. society $780,000 per victim. Statistics such as these show that workplace health and safety is not only a moral concern, but now also an economic one. However, businesses have not always viewed safety as a high priority issue, which becomes most evident by reviewing the history of safety legislation in the United States.

SAFETY LEGISLATION

Safety in the workplace has been strongly impacted by legislation over the last 100 years. It is generally recognized that during the 1800s, workers performed their duties under unsafe and unhealthful conditions. The philosophy of businesses was that of laissez-faire, which means to let things be—letting natural laws operate without restriction. Although technically, under common law, employers were expected to provide a safe place to work and safe tools with which to work, in reality the public accepted accidents as inevitable. When an accident occurred, the only means for the employee to obtain compensation was to prove the employer's *negligence,* which was defined as "failure to exercise a reasonable amount of care, or to carry out a legal duty so that injury or property damage occurs to another." The problem was that *reasonable amount of care* was ill-defined. Companies argued that hazardous conditions were normal. In addition, companies could defend themselves by claiming that either (1) there had been *contributory negligence*—meaning that an injured person's behavior contributed to the accident; (2) a fellow employee had been negligent; or (3) the injured worker had been aware of the hazards of the job and had knowingly assumed the risks (Hammer, 2000). For example, if a fellow employee contributed in any way to an accident, the employer could not be held responsible. As a result of these loopholes favoring businesses, until the early 1900s, working conditions were poor and injury rates continued to climb.

Workers' Compensation and Liability

Between 1909 and 1910, various states began to draft workers' compensation laws. These early laws were based on the concept of providing compensation to workers for on-the-job injuries regardless of who was at fault. The first two such laws were passed in Montana for miners and in New York for eight highly hazardous occupations. Both laws were thrown out as unconstitutional. Shortly after that, a tragic and highly publicized fire in a shirt factory in New York killed 146 workers and seriously injured 70 more. This increased public demand for some type of legislative protection, and by 1917, the Supreme Court declared that state workers' compensation laws were constitutional. Today there are different workers' compensation laws in each state, with approximately 80 percent of all workers covered by the laws (Hammer, 2000). Overall, the goals of workers' compensation include

- Provide sure, prompt, and reasonable income and medical benefits to work-accident victims or income benefits to their dependents, regardless of fault.
- Provide a single remedy to reduce court delays, costs, and workloads arising out of personal-injury litigation.
- Eliminate payment of fees to lawyers and witnesses as well as time-consuming trials and appeals.
- Encourage maximum employer interest in safety and rehabilitation through an experience-rating mechanism.
- Promote the study of causes of accidents.

Workers' compensation is a type of insurance that requires companies to pay premiums just like any other type of insurance. The workers' compensation insurance then pays set rates for benefits, depending on the job and type of injury. To be covered under workers' compensation insurance, an injury must meet three conditions: (1) it arose from an accident, (2) it arose out of the worker's employment, and (3) it occurred during the course of employment.

Under workers' compensation law, workers are not allowed to sue their employer for negligence; however, they are allowed to sue a third party. This can include the manufacturer of the equipment that caused the injury, the driver or company of other involved vehicles, the architect that designed the building, or the safety inspector. Many of the large product liability suits are claims for injuries to industrial workers because it is a way to get benefits beyond the relatively small workers' compensation benefits. As an example, a man in California lost eight fingers in a press that had a defective safety switch. He received $40,000 plus a life-time disability pension from workers' compensation, but was also awarded $1.1 million in a product liability suit. While claims of negligence are common, claims of *strict liability* are increasing also. Strict liability means that a manufacturer of a product is liable for injuries due to defects without a necessity for the injured party to show negligence or fault.

Establishment of OSHA and NIOSH Agencies

In the 1960s, many people felt that the state legislated laws were still inadequate; many industries still had poor safety and health standards, and injury and death rates were still too high. As a result, in 1970, the federal government acted to impose certain safety standards on industry by signing into effect the Occupational Safety and Health Act. This act established the administrative arm, Occupational

Safety and Health Administration (OSHA), under the U.S. Department of Labor. OSHA implements safety programs, sets and revokes health and safety standards, conducts inspections, investigates problems, monitors illnesses and injuries, issues citations, assesses penalties, petitions the courts to take appropriate action against unsafe employers, provides safety training, provides injury prevention consultation, and maintains a database of health and safety statistics (see Goetsch, 2001). OSHA publishes standards for general industry (Department of Labor, 1993) and also for specific industries such as construction, agriculture, and maritime. Employers must comply with OSHA regulations through activities such as complying with standards for injury avoidance, keeping records of work-related injuries and death, keeping records of exposure of employees to toxic materials or other hazards, and keeping employees informed on matters of safety and health.

One other federal organization is also important to the human factors profession, the National Institute for Occupational Safety and Health (NIOSH). NIOSH performs research and educational functions. It conducts or reviews research to identify hazardous types of conditions in the workplace. It prepares recommendations that often become provisions of the OSHA standards. Human factors specialists working in the area of workplace design or safety often use NIOSH standards or recommendations.

Product Liability

While OSHA has resulted in greater industrial safety, there are still numerous problems. As with all large bureaucracies, the agency is cumbersome and slow. OSHA is also heavily influenced by political lobbying, has fines that are ineffectively small, and has too few inspectors. For this and other reasons, safety in both industry and product manufacturing is increasingly influenced by civil and criminal suits.

Whether an injury or death occurs in the workplace or elsewhere, people are increasingly bringing suit against businesses. Most of these suits are *product liability* claims, alleging that a product was somehow defective, and the defect caused the injury or death. Product liability cases usually assume one of three types of defect: a *design defect* (inherently unsafe), a *manufacturing defect,* or a *warning defect.* Also, an increasing number of suits allege improper instruction as well as warning. For example, the suit described earlier for the backyard fountain alleged that the manufacturer failed to properly instruct the retailer on installation of the 500-pound fountain (using adhesive between the fountain tiers) and that both manufacturer and retailer failed to warn the consumer of hazards. The case was tried in California, and a settlement of $835,000 made to the mother of the 4-year-old who was killed. The number and size of product liability cases is growing so alarmingly that in 2003, Congress attempted to enact a bill limiting the scope and award value of product liability cases.

A critical question that must be answered for each product liability case is whether the product is defective or simply inherently "dangerous." For example, a carving knife is dangerous but would not be considered defective. An important precedent was set by the California Supreme Court in the 1970s. It specified that a product is defective when it "failed to perform safely as an ordinary user would expect when it was used in an intended or reasonably foreseeable manner, or if the risks inherent in the design outweighed the benefits of that design." There are two important implications of this judgment for human factors:

1. The concept of *reasonably foreseeable.* Human factors specialists are often asked to act as expert witnesses to testify concerning what could be considered "reasonably foreseeable." For example, is it reasonably foreseeable that a child would climb on a fountain? Most people would say yes, and this was the verdict in the fountain suit. In another notorious case, a person was injured in the act of using a lawnmower as a hedge trimmer. Is this a reasonably foreseeable use of the equipment?

2. *The tradeoff between risk and benefit.* Human factors specialists act as expert witnesses by providing information and analyses relevant to tradeoff questions. For a given design, the original designer should have weighed the positive effects of the hazard control against the negative effects such as cost or other disadvantages. Factors considered in assessing the tradeoff include the likelihood of injury, the likely severity of injury, possible alternative designs, costs or feasibility of a given design versus alternative designs, the effectiveness of alternative designs, and so forth. A knife can be made safer by making it dull, but the tradeoff is that it loses most of its functionality.

A final area where human factors specialists are central to product liability is in helping manufacturers design safer products to avoid litigation in the first place. Professionals trained in hazard and safety analysis work with design teams to ensure that the product is safe for reasonably foreseeable uses. Some of the methods used for such safety analyses are presented later in this chapter.

FACTORS THAT CAUSE OR CONTRIBUTE TO ACCIDENTS

A variety of theories and models have been proposed to explain and predict accidents. Most of these only consider some of the factors that contribute to accidents, for example, the social environment. Probably the most comprehensive model, the *systems approach,* is also one that is compatible with the human factors approach. The systems approach assumes that accidents occur because of the interaction between system components (Firenzie, 1978; Slappendel et al., 1993). It is assumed that some factors are closely or directly involved in task performance and therefore are direct causal factors in safety. These factors include characteristics of (a) the employee performing a task, (b) the task itself, and (c) any equipment directly or indirectly used in the task. Other factors also significantly impact safety. These can be categorized as social/psychological factors and environmental factors. Figure 14.1 shows one particular view of the systems approach proposed by Slappendel et al. (1993).

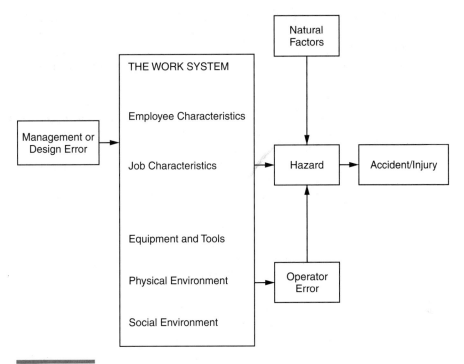

FIGURE 14.1

Model of causal factors in occupational injuries. (*Source:* Slappendel, C., Laird, I., Kawachi, I., Marshall, S., & Cryer, C., 1993. Factors affecting work-related injury among forestry workers: A review. *Journal of Safety Research, 24,* 19–32. Reprinted with permission.)

Some factors affect performance of the worker more indirectly. For example, one social/psychological factor is the existence of social norms in the workplace. Social norms may support unsafe behavior, such as taking off protective gear, using unsafe lifting practices, or walking into unsafe work areas. Construction workers more often than not install roofing without being tied off, as they are supposed to. The predominant reason is that the *social norm* is to not bother with this protective equipment. Table 14.2 shows some of the more important causal and contributing factors. Notice that many of the causal factors include topics discussed in other chapters of this text. For example, illumination is covered in Chapter 4, and Job Characteristics are covered in Chapters 10–13.

Safety concerns permeate much if not most of the field of human factors. In the remainder of this section, we review contributing and causal factors not covered elsewhere; we first discuss the five "work system" factors shown in Figure 14.1 and then briefly discuss operator error.

Personnel Characteristics

A number of factors associated with industry personnel increase the likelihood of accidents; see Figure 14.2. Generally, the factors fall into clusters that affect hazard recognition, decisions to act appropriately, and ability to act appropriately.

TABLE 14.2   Causal and Contributing Factors for Accidents

Task Components

Employees	*Job*	*Equipment and Tools*
Age	Arousal, fatigue	Controls, displays
Ability	Physical workload	Electrical hazards
Experience	Mental workload	Mechanical hazards
Drugs, alcohol	Work-rest cycles	Thermal hazards
Gender	Shifts, shift rotation	Pressure hazards
Stress	Pacing	Toxic substance hazards
Alertness, fatigue	Ergonomic hazards	Explosive hazards
Motivation	Procedures	Other component failures
Accident proneness		

Surrounding Environment

Physical Environment	*Social/Psychological Environment*
Illumination	Management practices
Noise	Social norms
Vibration	Morale
Temperature	Training
Humidity	Incentives
Airborne pollutants	
Fire hazards	
Radiation hazards	
Falls	

In this section we review only some of the more important factors that affect safe behavior.

Age and Gender.   One of the most highly predictive factors for accident rates is age. Research has shown that overall, younger people have more accidents, with accident rates being highest for people between the ages of 15 and 24 (Bell et al., 1990). Industrial accident rates peak at around age 25. Since this is correlational data, it is difficult to determine why age affects accident rates. Some people speculate that the primary reason is that as people get older, they become more conservative, and their estimations of risk become more conservative; that is, younger people think there is less likelihood of accidents and injury occurring to themselves than do older workers (Leonard et al., 1990). In addition, young males perceive themselves as less at risk and therefore have a greater number of accidents (e.g., Alexander et al., 1990; Lyng, 1990).

However, there are certain exceptions to the general relationship between age and the accident rates; that is, when accidents are tied to the physical and cognitive abilities of the employee, accident rates go up for the elderly (Slappendel et al., 1993). For physically intensive occupations, such as logging, performance may decline at an age as early as 35. For perceptual and cognitive abilities, people approaching 50 to 60 years of age show a decreased "useful field of vi-

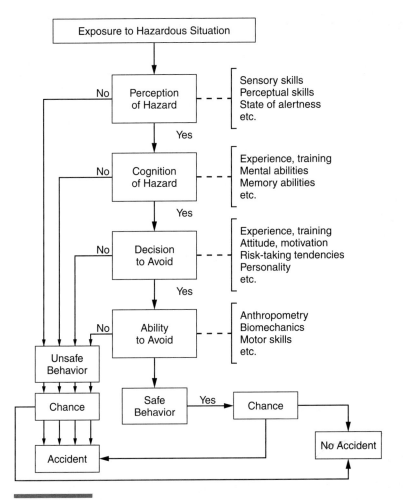

FIGURE 14.2

Operator characteristics that affect various steps in the accident sequence. (Adapted from Ramsey, I., 1985. Ergonomic factors in task analysis for consumer product safety. *Journal of Occupational Accidents*, 7, 113–123.)

sion," a slowing in information processing, and more difficulties in encoding ambiguous stimuli. If a job, such as driving, requires information-processing capabilities, accident rates tend to rise.

Job Experience.   A second characteristic of employees that predicts accident rate is time on the job, or work experience. A high percentage of accidents (approximately 70 percent occur within a person's first 3 years on the job, with the peak at about 2 to 3 months. This point represents a transition stage: The person has finished training and is no longer supervised but still does not have the experience necessary for hazard recognition and appropriate response.

Stress, Fatigue, Drugs, and Alcohol. Other, more temporary characteristics of the employee affect performance and therefore accident rates. For example, stress and fatigue are both factors found to be related to accidents (see Chapter 13). Performance decrements sometimes also result from life stressors outside of work, such as death of a loved one or divorce (e.g., Hartley & Hassani, 1994). These factors can make people more likely to be preoccupied with nonwork-related thoughts.

Employees under the influence of drugs or alcohol are shown to have a higher accident rate (Holcom et al., 1993). Field studies demonstrate a relationship between drug use and job performance indicators such as injury rates, turnover, and workers' compensation claims (e.g., Lehman & Simpson, 1992).

Many employers now drug-test employees for this reason. Data show that organizations adopting drug-testing programs show a reduction in personal injury rates (Taggart, 1989). While these data imply that drug use directly affects accident rate, this is not necessarily the case. Some theorists believe that drug use simply indicates a general characteristic of employees. It is this characteristic, a sort of "social deviancy," that is the operating mechanism responsible for work-related accidents (Holcom et al., 1993). According to this view, drug screening simply reduces the numbers of such people being employed, which results in a lower accident rate.

Holcom and colleagues (1993) suggest that there are several personality factors that seem to predict accident rates in high-risk jobs, including general deviance, job dissatisfaction, drug use, and depression. This finding is consistent with descriptive research indicating that some people seem to have a greater likelihood of incurring numerous accidents than others (e.g., Mayer et al., 1987). Although these employees might be termed *accident prone*, the term is not particularly diagnostic, and we must continue to work toward determining exactly what characteristics make such people more likely to have accidents. Thus, employee assistance programs need to deal with an entire range of psychosocial problems rather than just targeting drug use.

Job Characteristics

Many characteristics of the job or task can cause difficulties for the operator. Some of these include high physical workload, high mental workload, and other stress-inducing factors such as vigilance tasks that lower physiological arousal levels. Other characteristics associated with an increase in industrial hazards include long work cycles and shift rotation—factors that increase fatigue levels (see Chapter 13 for a review of these factors).

Equipment

Many of the hazards associated with the workplace are localized in the tools or equipment used by the employee, and as a consequence, much of the safety analysis performed in an industrial environment focuses on hazards inherent in the equipment itself. Additional hazards may be created by a combination of equipment and environmental conditions.

Controls and Displays. As we have seen throughout the text, controls and displays can be poorly designed so as to increase the likelihood of operator error (e.g., see Chapter 8 and 9). While good design of controls and displays is always desirable, any time there are hazards present in the equipment and/or environment, it is especially critical.

Electrical Hazards. Electric shock is a sudden and accidental stimulation of the body's nervous system by an electric current. The most common hazards are electrical currents through the body from standard household or business currents and being struck by lightning. Electricity varies in current, volts, and frequency. Some levels of these variables are more dangerous than others. The lowest currents, from 0 to 10 milliamperes, are relatively safe because it is possible to let go of the physical contact. However, at a point known as "let-go" current, people lose the ability to let go of the contact. The let-go point for 60-Hertz circuits for males is about 9 milliamperes, and for females it is about 6 milliamperes. Above this point, prolonged contact makes the electrical current extremely dangerous due to paralysis of the respiratory muscles. Paralysis lasting over three minutes usually causes death. As the current reaches 200 milliamperes, it becomes more likely to throw the person from the source. This is good, because at this level, any current lasting over 1/4 second is essentially fatal. Thus, we can say that prolonged exposure due to contact generally makes the 10 to 200 milliampere current range the most dangerous. Higher currents stop the heart and cause respiratory paralysis, but the person can often be resuscitated if done immediately.

In general, AC, or alternating current, is more dangerous than DC, direct current, because alternating current causes heart fibrillation. In addition, currents with frequencies of 20 to 200 Hertz are the most dangerous. Note that the standard household current is AC, with a 60-Hertz current, which is in the most dangerous range. Exposure to such electrical current is damaging after only 25 msec. Home and industrial accidents frequently occur when one person turns off a circuit to make repairs and another person unknowingly turns it back on. Circuits turned off for repairs should be locked out or at least marked with warning tags. Accidents also occur from the degradation of insulating materials. Recent methods to reduce electrical hazards include regulations regarding wiring and insulation; requirements for grounded outlets; insulation of parts with human contact; rubber gloves and rubber mats; and the use of fuses, breakers, and ground-fault circuit interrupts (GFCI). GFCIs monitor current levels, and if a change of more than a few mAmps is noted, the circuit is broken. These mechanisms are now required in most household bathrooms (and are visually distinct).

Mechanical Hazards. Equipment and tools used in both industrial and home settings often have an incredibly large number of mechanical hazards. At one time, most injuries in industrial plants arose from mechanical hazards (Hammer, 1989). Machines had hazardous components such as rotating equipment, open-geared power presses, and power hammers. More recently, such equipment has been outfitted with safeguards of various types. However, mechanical hazards are still common, and can result in injuries induced by actual physical contact with a part or component. Examples include the following hazards:

- *Cutting* or *tearing* of skin, muscle, or bone. Typical sources are sharp edges, saw blades, and rough finishes. Tearing can occur when a sharp object pierces the flesh and then pulls away rapidly.
- *Shearing* is most commonly a problem where two sharp objects pass close together. An example is power cutters or metal shears. In industrial plants, workers often position materials in shears and then, realizing at the last moment that the material is not correctly in position, reach in to perform a readjustment. This results in loss of fingers and hands.
- *Crushing* is a problem when some body part is caught between two solid objects when the two objects are coming closer together. These are referred to by OSHA as *pinch points*—any point other than the point of operation at which it is possible for any part of the body to be caught between moving parts.
- *Breaking,* which occurs when crushing is so extreme that bones are broken.
- *Straining* refers to muscle strains, usually caused by workers overexerting themselves, for example, trying to lift more than they are capable. Many workers strain their arms or back by relying too much on those body parts and not enough on the legs. Other common sources of strain are when employees are lifting objects and slip on a wet floor because the attempt to maintain an upright position puts an undue strain on muscles (Hammer, 1989). Chapter 10 discusses these problems in more detail.

Guards are commonly used to reduce mechanical hazards, although sometimes people remove them, which defeats the purpose (Hammer, 1989). Various types of guards include total enclosures, enclosures with interlocks (if guard is removed, the machine is stopped), and movable barriers such as gates (see extensive review in National Safety Council, 1993b). Other common safety devices are systems that interrupt machine operation if parts of the body are in the hazardous area. This can be accomplished by mechanisms such as optical sensors, electrical fields using wrist wires, two hand controls, and arms that sweep the front of the hazardous area.

Pressure and Toxic Substance Hazards.   The most common problems associated with pressure are vessel ruptures. In many industrial settings, liquids and gases are contained in pressurized vessels. When the liquid or gas expands, the vessel, or some associated component, ruptures and employees may be injured. These can be considered hidden hazards because employees may not be aware of the inherent dangers. The factors that typically cause vessels to rupture are direct heat (such as fire), heat from the sun or nearby furnaces, overfilling, and altitude changes. When pressurized liquids or gases are released, injuries may be sustained from the contents themselves, fragments of the vessel, or even shock waves. An example of hazards associated with pressurized vessels is the use of compression paint sprayers. Paint sprayers aimed at a human have enough pressure to drive the paint molecules directly into the skin, causing toxic poisoning, a hazard of which many people are unaware. Steps that should be taken to deal with pressure hazards include safety valves, depressurizing vessels before maintenance activities, marking vessels with contents and warning labels, use of protective clothing, and so on (see Hammer, 2000).

Toxic substances tend to fall into classes depending on how they affect the body. *Asphyxiants* are gases that create an oxygen deficiency in the blood, causing asphyxiation. Examples include carbon dioxide, methane, and hydrogen. Natural gas is a hidden hazard, because it is normally odorless and colorless. Sometimes odorants are added to act as a warning mechanism. *Irritants* are chemicals that inflame tissues at the point of contact, causing redness, swelling, blisters, and pain. Obviously, these substances are particularly problematic if they are inhaled or ingested. *Systemic poisons* are substances that interfere with organ functioning. Examples include alcohol and other drugs. *Carcinogens* are substances that cause cancer after some period of exposure. Because of the length of time to see effects of carcinogens, they are particularly difficult to study in an industrial setting.

Hazardous substances have become a focus of federal concern, and since 1987, OSHA has required all employers to inform workers about hazardous materials. The purpose of the OSHA Hazard Communication Standard is to ensure that information about chemical hazards is communicated to employees by means of "comprehensive hazard communication programs, which are to include container labeling and other forms of warning, material safety data sheets and employee training" (OSHA Hazard Communication Standard 29 CFR 1910.1200). Because the category of toxic substances includes materials such as bleach, ammonia, and other cleaners, the OSHA standard applies to almost every business.

The Physical Environment

Illumination.   Lighting most directly affects safety by making it relatively easy or difficult to perform tasks. Other illumination factors that are important for safety include direct or indirect glare and light/dark adaptation. Another problem is the problem of *phototropism,* our tendency to move our eyes toward a brighter light. Not only does this take our attention away from the central task area but it may cause transient adaptation, making it more difficult to see once our attention does return to the task area. Large windows are especially problematic in this regard. In the case of the convenience store slip and fall case described earlier, phototropism may have been a contributing factor if the employee's visual attention was temporarily drawn toward the brighter window area.

Noise and Vibration.   Noise and vibration are two factors associated with equipment that can be hazardous to workers, as discussed in Chapters 5, 10, and 13.

Temperature and Humidity.   Working conditions that are either too hot or too cold pose serious safety hazards either directly by impacting body health or indirectly by impairing operator performance (see Chapter 13). Clothing is also a key factor in the body's ability to transfer or maintain heat. It is important to note that many types of *protective clothing* designed to guard the operator from other hazards may exacerbate the problems of thermal regulation by limiting airflow over the body, making the cooling mechanisms of vasodilation and sweating less effective.

Fire Hazards.   In order for a fire to start, there must be a combination of three elements: fuel, an oxidizer, and a source of ignition. Common fuels include paper products, cloth, rubber products, metals, plastics, process chemicals, coatings such as paint or lacquer, solvents and cleaning fluid, engine fuel, and insecticides. These materials are considered flammable under normal circumstances, meaning they will burn in normal air. Oxidizers are any substance that will cause the oxidation-reduction reaction of fire. Atmospheric oxygen is the most common oxidizer, but others include pure oxygen, fluorine, and chlorine. Some of these are powerful oxidizers and great care must be taken that they do not come in contact with fuels.

The activation energy for ignition is usually in the form of heat; however, light can sometimes also be an ignition source. Typical fire ignition sources include open flames, electric arcs or sparks (including static electricity), and hot surfaces (such as cigarettes, metals heated by friction, overheated wires, etc.). In spontaneous reaction or combustion, materials gradually absorb atmospheric gases such as oxygen and, due to decomposition processes, become warm. This is especially common for fibrous materials that have oils or fats on them. If materials are in an enclosed location, such as a garbage bin, the heat buildup from oxidization cannot be dissipated adequately. The heat accumulated from the numerous reactions in the materials eventually provides the ignition source. The length of time required for oily rags or papers to combust spontaneously can range from hours to days, depending on temperatures and the availability of oxygen. Preventing spontaneous combustion requires frequent disposal in airtight containers (thus eliminating the oxidizer). In industrial settings, there are numerous standard safety precautions to prevent hazardous combinations of fuel, oxidizers, and ignition sources (see Hammer, 1989).

Radiation Hazards.   Certain combinations of neutrons and protons result in unstable atoms, which then try to become stable by giving off excess energy in the form of particles or waves (radiation). These unstable atoms are said to be radioactive. *Radioactive material* is any material that contains radioactive (unstable) atoms.

The criticality of exposure to radiation depends on several factors, including the type of radiation (x-rays, gamma rays, thermal neutrons, etc.), the strength of the radiation (REM), and the length of exposure. These factors all affect the dose, which is the amount of radiation actually absorbed by human tissue. Biological effects of radiation can occur in a one-time acute exposure or from chronic long-term exposure. Chronic low levels of exposure can actually be safer than acute exposure because of the body's ability to repair itself. However, as chronic levels increase, long-term damage such as cancer will occur. Acute doses of radiation are extremely hazardous. The best defense against radioactivity is an appropriate shield (e.g., plastic or glass for beta particles, lead and steel for gamma rays).

Falls. Falls resulting in injury or death are relatively common. As noted in Table 14.1, these are the second most frequent source of workplace deaths. The most common type of injury is broken bones, and the most serious is head injury. Unfortunately, falls can be more serious than most people realize. According to one estimate, 50 percent of all persons impacting against a surface at a velocity of 18 mph will be killed (see Hammer, 2000). This represents a fall of only 11 feet. People can fall and sustain injuries in a number of ways, including slipping on wet flooring and falling, falling from one floor to another, falling from a natural elevation or building, falling from a ladder, and falling from a structural support or walkway. Falls from ladders are so common that there are now OSHA precautionary regulations for the design and use of various types of ladders.

Exits and Emergency Evacuation. Although evacuation is a critical mitigation measure for fire and other emergencies, until the tragic World Trade Center (WTC) events of September 11, 2001, this crucial safety issue has received little attention in human factors research and in building codes/standards development (Pauls & Groner, 2002). There is an urgent need for assessment and research on building codes and safety standards requirements for egress capacity, stair width, exit sign, and alarm design. Research on and design for emergency evacuation must consider the effects of crowd panic behavior, electric power failure, and potential presence of other concurrent hazards such as explosions and toxic materials. Other factors such as the height and the number of stories of a building, the total number of building occupants and their floor distributions, and the extent to which elevators can be used for egress must also be considered (Pauls, 1980, 1994; Proulx, 2001; Sime, 1993).

Emergency evacuation and exits pose special challenges to human factors research and design, and we must examine carefully how to apply human factors data and knowledge to this special environment. For example, to apply the anthropometric data and methods we learned in Chapter 10 to the design of exit stairs for a high-rise building, we must not assume building occupants would walk slowly side-by-side in an emergency evacuation. The design must deal with a possibly panicked crowd getting down a potentially dark and smoky stairway. Further, firefighters and rescue workers may be using the same stairs, but moving in the opposite direction than the crowd, carrying heavy and potentially large firefighting or rescuing equipment. Similarly, loss of power and lighting and the presence of loud sirens raise special questions about how to design displays and controls (Chapters 8 and 9) for emergency evacuation situations.

The Social Environment

A number of contextual factors indirectly affect accident rates. Researchers are realizing that hazard controls at the equipment level are not always successful because human behavior occurs within a social context. A ship captain may not see warning lights if he or she is in the next room having a drink. A construction worker will not wear safety equipment on the third story roof because his boss told him that none of the crew "bothers with that stuff." The social environment can provide extremely powerful influences on human behavior.

The list of social factors shown in Table 14.2 identified some of the major contributing factors to accidents, including management practices, social norms, morale, training, and incentives. Each factor affects the likelihood that an employee will behave in a safe manner. For example, management can implement incentive programs to reward safe behavior. Feedback concerning accident reduction has also been shown to reduce the rate of unsafe behaviors (e.g., Fellner & Sulzer-Azaroff, 1984). Training is also an important consideration, because this is one of the primary ways that people learn about hazards, what behaviors are appropriate or safe, and the consequences of unsafe behavior.

Finally, social norms refer to the attitudes and behavior of an employee's peers. People are extremely susceptible to social norms; they are likely to engage in safe or unsafe behaviors to the extent that others around them do so (e.g., Wogalter et al., 1989). For example, if no one else wears protective goggles on the shop floor, it is unlikely that a new employee will do so for very long. Later in this chapter we review some methods to facilitate safe behavior by affecting these social factors.

Human Error

Human error is a critical contributor to lapses in system safety. For example, medical error has been attributed as the cause of up to 98,000 preventable patient deaths per year, with a cost estimated to be as high as $29 billion annually (Kohn et al., 2000). A majority of the 40,000 deaths per year in auto accidents in this country have been attributed, in part, to driver error. We may define error as *inappropriate human behavior that lowers levels of system effectiveness or safety.* Much attention has been devoted to the role of human operator error in contributing to accidents. Woods and Colleagues (1994; 1999) often refer to this as a focus on the operator at the "sharp end" of the system. However, there are numerous other contributing causes within the system that lead a particular error by the operator to cause the accident. Before we discuss these other systemwide causes, however, we describe two particular efforts to classify human error.

Error Classification. Perhaps the simplest classification of human error distinguishes between errors of *commission* and errors of *omission.* The former describes an operator who does something that should not have been done—for example, hitting the delete key instead of the save key. The latter describes an operator who *fails* to do something that should have been done, such as a maintenance technician who fails to tighten a screw after completing a procedure.

The omission/commission classification can help to explain *what* was done, but does not contribute much to an understanding of *why.* Greater understanding of the why of human error is provided by a popular approach based, in part, on the distinction between whether the inappropriate action was intended or not (Norman, 1981; Reason, 1990). If the action, which turned out to be inappropriate was *intended,* this is labeled a *mistake.* (Note that the commission of an error is not intended, but the intended action turned out to be erroneous.) An example would be a lost traveler who intended to turn right at an intersection, but was not aware that it was a one-way street. Using terminology similar to the knowedge, rule, skill-based behavior taxonomy introduced in Chapter 7, Reason distinguishes between *knowledge-based mistakes* and *rule-based mistakes.* The former, describing the behavior of our driver, is committed when either knowledge in the head or in the world fails to be adequate to support the human's understanding of the situation. Included in these knowledge-based mistakes are both failures of understanding and *perceptual errors* (Wiegmann & Shappell, 2001). In contrast, the rule-based mistake results because the human is unaware of, or misapplies, the rules governing appropriate behavior. This might characterize the American driver who intentionally turns into the right lane of traffic on a British motorway, forgetting the rule that "if Britain, then drive left."

In contrast to mistakes (both rule-based and knowledge-based), if the incorrect act was *not* intended, but "slipped out" through the selection of action, this form of error is termed a *slip.* We often make "slips of the tongue" when we are talking. We hit the delete key when we intended to hit the save key. Another example is the cook who, working with the stove layout shown in Chapter 3, Figure 3.3b, grabs the wrong control and lights the wrong burner. Most slips can be thought of as commission errors of a nonintended action. When nonintentional errors are those of **omission,** they are called lapses. In the above example, the maintenance technician did not intend to leave the screw untightened. Reason (1997) highlights the role of omission errors as some of the most frequent in aircraft maintenance tasks.

The contrast between mistakes (rule and knowledge), slips, and lapses is useful because conditions that produce the different kinds of errors often have different remediations. For example, since most mistakes reflect a lack of knowledge, they can be addressed either by providing knowledge in the head (better training, as described in Chapter 18) or knowledge in the world (better displays, as described in Chapter 8). Furthermore, the lack of knowledge is more likely to be characteristic of the novice performer. In contrast, slips typically result from bad or confusing links between display or control; confusing, similar-appearing switches, or poor display-control compatibility are often responsible. Furthermore, unlike mistakes, slips are often shown by expert operators, who are performing their task without allocating close attention to it. Finally, lapses, which can often be represented as a failure of **prospective memory,** as described in Chapter 16, can be supported by checklists or explicit reminders. A nice example of such a lapse-fighting reminder is the prominent sign on the photocopier that says "**Remove the last page.**"

A final addition to this taxonomy of human error is the *violation.* In a sense, this is when the user intentionally does something inappropriate, as when we drive above the speed limit or a worker intentionally ignores a safety procedure. The accident at the Chernobyl nuclear power plant in the Soviet Union was caused, in part, by a violation (Nature, 1986). As we see below, violations are "caused" by the joint influences of an emphasis on productivity over safety and on an inadequate safety culture.

We may summarize this error categorization as follows, reflecting the organization of Reason (1997) and Wiegmann and Shappell (2001):

Intended
- knowledge-based mistake (failure of perception, of understanding)
- rule-based mistake (selection of the wrong if-then rule)
- violation (intentionally did the wrong thing)

Unintended
- slip
- lapse (the operator did not intend to not do the action)

These and other classifications of human error (Park, 1997) have sometimes been incorporated into models of *human reliability* (see Kirwan & Ainsworth, 1992 for a good review). Such models are designed to predict the overall reliability of high-risk systems, like nuclear power plants, that involve an interaction of humans and equipment. For example, they might be applied in an effort to prove that the design of a nuclear plant would lead to a catastrophic system failure with a probability of less than .0001. Unfortunately, such models have a large number of challenges to their effectiveness (see Dougherty, 1990; Wickens & Hollands, 2000), leading to suspicion of the meaningfulness of the actual reliability numbers that are produced.

Errors and System Safety.   When accidents occur, the human operator at the "sharp end" is often a contributing factor. But more often than not, this person can be seen as only the final "triggering" event at the end of a series of earlier events, or embedded in a set of preexisting conditions, all of which made the disastrous consequences nearly inevitable. To quote the familiar phrase, it was "an accident waiting to happen." Reason (1990, 1997) refers to these preexisting conditions as *resident pathogens,* and their potential list is long, including factors such as poor environmental conditions (Chapter 13), poor human factors of the interface (Chapters 8 and 9), inappropriate sleep schedules and fatigue (Chapter 13), poor training (Chapters 6 and 18) or job support (Chapter 18), poor maintenance, management attitudes that overemphasize productivity, poor workplace climate.

Many of these factors are embodied in what is called the *safety culture* of the organization, which may span a great range (Reason, 1997).

In addressing the problem that blame for accidents is often directed more at the operator at the sharp end than at the resident pathogens, it is also important to note the extent to which operator error is attributed in *bad decisions,* or decision errors that have only proven to be so in hindsight (Woods & Cook, 1999). That is, the accident investigator may reveal factors that in hindsight should have been obvious to the sharp-end operator, but, re-creating the actual conditions existing at the time of the error would not be seen at all as obvious. These examples of the *hindsight bias* or "Monday morning quarterbacking" were discussed in Chapter 7. Such findings suggest that great care should be taken to distinguish between establishment of the human operator behavior as partially *responsible* for an error, and pointing *blame* at that operator. Establishing responsibility can often lead to a better understanding of the cause of safety-compromising errors. However, directing blame is often unfair in hindsight and furthermore has an added detriment to safety investigation. To the extent that operators feel that they will be blamed for errors which, in hindsight, may lead to their punishment, this is likely to inhibit the free and useful self-reporting of incidents, which can otherwise provide valuable data about associated hazards and risks in the workplace.

Error Remediation.   Many approaches to reducing human error in the workplace can be directly associated with good human factors practices, as discussed throughout the book. The value of causal error taxonomies such as the slips-mistakes taxonomy, is that they can help reveal specific solutions, given the kinds of errors committed. In addition, however, it is important to highlight the role of **error containment** (Reason, 1997) embodied in the design of *error-tolerant systems* (Rouse, 1990, see also Chapters 3 and 15). Such systems are designed with the understanding that human operators are inherently fallible, but careful system design can often allow them to catch and recover their own errors, or "trap" the error so that it is not propagated to create an accident. As we discuss in Chapter 15, good feedback as well as some time-lag imposed between operator response and safety-critical system changes can often accomplish this goal (Rouse, 1990). Error tolerance can be achieved by methods such as feedback to the operator about current consequences, feedback about future consequences, and monitoring actions for possible errors. Design features can be included so that erroneous actions can be reversed (if they are noticed) before they have serious consequences on system performance. Computer systems now typically give the user a "second chance" before permanently deleting a file (e.g., by asking "Are you sure you want to delete?" or by providing an undo option; see Chapter 15).

HAZARD IDENTIFICATION AND CONTROL

System safety analysis and accident prevention consists of identifying potential hazards using accident frequency rates for the task in a particular environment. For example, a particular injury might occur in a plant at the rate of 5.0 per million man-hours. In a facility with multiple hazards, the most critical or high-risk hazards should receive top priority. If there are several methods for controlling hazards, then certain methods may be considered more optimal or reliable than others. In this section, we first address the meaning of a critical or high-risk hazard. We review a number of methods for identifying hazards in the design of a product or piece of equipment, and then we discuss the methods for hazard control.

Hazard Criticality and Risk

There have been many operational definitions of hazard *criticality*. It is often considered synonymous with risk, which is a combination of the *probability* and *severity* of the event or accident. Probability is the likelihood of an event taking place. Probability is measured in a number of ways and is often called *frequency*. Sometimes it is precisely quantified by using accident frequency rates for the task in a particular environment. Sometimes probability must be estimated because of the lack of adequate accident data. When probability is estimated, it is often categorized in a ranked scale of frequent, probable, occasional, remote, and improbable (Roland & Moriarity, 1990). Severity is usually scaled according to the severity of the injury. As an example, Military Standard MIL-STD-882B uses the following categories: catastrophic, critical, marginal, and negligible. These categories correspond to death or loss of a system, severe injury or major damage, minor injury or minor system damage, and no injury or system damage (Department of Defense, 1984).

One way of combining these two factors into a single *criticality* scale has been provided in MIL-STD-882B. A matrix combines the frequency and severity categories, and by using the hazard-assessment matrix (shown in Table 14.3), the hazard can be assigned a numerical value ranging from 1 to 20, with 1 representing the highest criticality and 20 the lowest. Using the language of expected-value decision making, discussed in Chapter 7, this scale roughly translates to "expected loss."

Hazard Identification

In designing equipment, one should ideally look for every possible hazard that could occur during each step in the operator's job. This must be done for all environmental conditions and for every possible foreseeable use of the equipment. In addition, the equipment must be analyzed as it exists in combination with other equipment and with other possible environmental hazards. Several complementary methods are used for identifying potential hazards.

TABLE 14.3 Hazard Matrix for Combining Frequency and Severity into a Single "Criticality" Variable

	Severity			
	Catastrophic	Critical	Marginal	Negligible
Frequency				
Frequent	1	3	7	13
Probable	2	5	9	16
Occasional	4	6	11	18
Remote	8	10	14	19
Improbably	12	15	17	20

Source: Adapted from Department of Defense *MIL-STD-882B*, 1984.

Preliminary Hazards Analysis.   The simplest method for hazard analysis, a preliminary hazards analysis, is often done before other more detailed methods, early in the conceptual design phase (Hammer, 2000). In a preliminary hazards analysis, the specialist evaluates the combinations of task actions, potential users, and environments to develop a list of the most obvious hazards that will be associated with a system (preliminary hazard, analyses are usually presented in a columnar table format). For example, if a power tool is being designed, the engineer will know that all standard electrical hazards must be considered. After each hazard is listed, columns are used to specify the cause of each hazard and the most likely effect on the system. The engineer then uses whatever data or knowledge is available to estimate the likelihood that an accident would occur as a result of the hazard and perhaps estimate the severity of the consequences. Potential corrective measures are then listed for each hazard. The problem with performing a preliminary hazards analysis is that the analyst may let it suffice and never complete the more thorough analyses.

Failure Modes and Effects Criticality Analysis (FMECA).   FMECA is an extension of a traditional method known as FMEA, which focused on the hazards associated with physical components of a system (Henley & Kumamoto, 1981). An FMEA first breaks down the physical system into subassemblies. For example, an automobile would be broken down into engine, cooling system, brake system, and so forth. Next, each subassembly is broken down into constituent components, and the analyst studies each component to identify the different ways that it could break down or function incorrectly, *the failure modes.* After this step, effects of the component failure on other components and subassemblies are estimated. For example, the component of an automobile fuel tank might be evaluated for the failure mode of "punctured," which would result in fuel leakage. The analyst would evaluate the effects of a fuel leak on other components in the fuel system, other subassemblies, and the entire system. This process is done for every system and environmental condition, including whether the automobile is running, outdoor temperature, and other factors such as potential surrounding heat sources. Many FMEAs also include a cause for each failure mode and corrective measures to control the failure or its effects (Kirwan & Ainsworth, 1992).

The FMECA is essentially an FMEA, but with an added factor. Once the component is analyzed for its effect on the system, the hazard is also given a score representing the hazard criticality of the effect.

While traditionally FMEAs have not focused on humans and human error, it is possible and desirable to extend the FMECA to analysis of the human system, that is, operator performance (Kirwan & Ainsworth, 1992). Instead of listing components and their failures, the analyst evaluates each step within the task analysis; that is, for each step, the engineer can list the types of errors that might occur (omission, incorrect performance, and so forth) and the possible effects of the error on the system. For example, if a person omitted the step of putting the gas cap back on a lawnmower, what would be the effects on system components and the system in general? How critical would those effects be? In this way, failures in human performance are analyzed for effects on the system in much the same way as failure of physical components. It is important to include foreseeable misuse in this analysis. An example of part of a FMECA focusing on human error is shown in Table 14.4.

Fault Tree Analysis. While FMECAs begin with a molecular view of the system and its components and work in a bottom-up fashion, other methods work in the opposite direction. One such analysis technique is *fault tree analysis,* which works from the top down from an incident or undesirable event down to possible causes (Green, 1983; Kirwan & Ainsworth, 1992). These causes could be conditions in the physical system, events, human error, or some combination. For each identified event or condition, the analyst works downward to identify all possible causes of that event. This is continued, and branches of the fault tree are added downward.

Fault trees show combinations of causal factors that result in the next level of event or condition through the use of Boolean AND/OR logic to represent the causal relationships. As an example, recall that a fire requires a fuel, oxidizer, and ignition source. All three must be present for a fire to occur. The fault tree would represent this as fuel *and* oxidizer *and* ignition source (see Figure 14.3.)

Fault trees are extremely powerful methods of hazard identification. One advantage of fault tree analysis is that it systematically identifies single causes and also multiple interacting causes of accidents. Single causes, known as *single-points failure,* are usually more likely to occur than combinations of conditions or events, and are therefore high in priority for controlling. Single-point failures are causes that pass upward or propagate through *or* gates rather than *and* gates. Because they are relatively difficult to build in isolation, fault trees are usually used in conjunction with other methods, such as FMECA.

Hazard Controls

After hazards are identified, how does an engineer or safety expert identify possible methods of hazard control reduction? Safety texts and articles are one source of information. For example, Hammer (2000) provides a fairly complete discussion of methods for reducing the various types of hazard listed earlier (fire, pressure, toxic, etc.). In addition, the National Safety Council publishes

TABLE 14.4   Example of "Human Error" Components for FMECA for Lawnmower

Human Error Component	Failure Mode	Effect on Component(s)	Effect on System/Subsystem	Criticality	Comments
Set blade torque	Torque set too high Torque set too low	Bolt experiences undue stress, breaks	Blade comes off mower	6	
Check mower blade	Fail to see blade cracks				

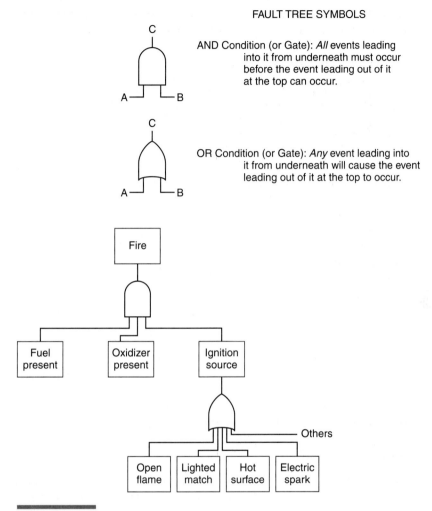

FAULT TREE SYMBOLS

C

A — B

AND Condition (or Gate): *All* events leading into it from underneath must occur before the event leading out of it at the top can occur.

C

A — B

OR Condition (or Gate): *Any* event leading into it from underneath will cause the event leading out of it at the top to occur.

Fire

Fuel present

Oxidizer present

Ignition source

Others

Open flame

Lighted match

Hot surface

Electric spark

FIGURE 14.3

Part of fault tree diagram that represents combinations of events that lead to a fire.

texts and documents (such as *Safeguarding Concepts Illustrated*, 6th ed., 1993), numerous publishers print texts specializing in health and safety (e.g., Mansdorf, 1995; Moran, 1996), and there are a number of journal and conference sources in the field of industrial safety, such as *Journal of Safety Research*.

A main step in safety analysis is to develop a list of hazard controls. Analyses such as FMECAs or fault trees yield a number of hazards, which can be listed in the first column of a *hazard controls* table. A second column can show the criticality of each hazard. The focus is then to generate all possible controls for each hazard, making sure first to generate controls that design the hazard out and then to generate ways to guard against the hazard. Different means of controlling each hazard should be generated if possible. Once the control methods are generated, they must be evaluated in terms of cost/benefit tradeoffs. Factors to consider include

- Other hazards that may be introduced by the various alternatives
- Effects of the control on the subsequent usefulness of the product
- Effect of the control on the ultimate cost of the product
- A comparison to similar products (What control methods do they use?)

If necessary, the designer may consult with others for information on factors such as manufacturing costs related to the hazard controls. Notes on the relative advantages and disadvantages of each alternative control should be made in the next column or in a separate document (for liability reasons). Finally, the designer should choose one control method and list it in a final "recommended control" column. Once a product or system is designed to include the hazard controls identified, the design team should do a final check to make sure the design does not have any defects that have historically led to litigation.

Hazards associated with a tool or piece of equipment can be thought of as originating at a source and moving along some path to a person. The reduction of hazards should be prioritized as follows: (1) source, (2) path, (3) person, (4) administrative controls.

The best hazard reduction is to eliminate it at the source. This is also called *designing out* a hazard. An example would be eliminating a sharp edge on a piece of equipment. Designing out hazards should always be attempted before other methods of hazard control. However, it is possible that the tool or equipment cannot function with the hazard designed out. An automobile can be designed to go only 2 miles per hour, eliminating the hazard of injuring a person on the inside and significantly reducing the likelihood of injury to someone on the outside. While a hazard has been designed out, the functionality has been designed out also.

After designing out, the next best solution is to provide a hazard control on the path between the hazard and the user. This usually means providing a barrier or *safeguard* of some sort. This method is considered less optimal because it is more likely to fail to control the hazard. For example barriers to unsafe acts could conceivably be removed by strong wind. *Personal protective equipment* can be removed by the person wearing it.

It is sometimes not possible to either design out or guard against a hazard. In this case, the hazard control must consist of trying to control the hazard at the point of the person: changing his or her behavior. This approach usually depends on *warning* or *training* and is considered even less reliable for hazard control than guarding. An example is training workers not to place their hands near a pinch point. The workers may be well intentioned, but human error could still result in an accident. Another example is the plastic bags from dry cleaners that may pose a serious suffocation hazard for children who may not understand the warning.

A final method of hazard control is through administrative procedures or *legislation*. In industry, administrative procedures might include shift rotation, mandatory rest breaks, sanctions for incorrect and risky behavior, and so forth. In addition to laws and regulations for industry, there are general public laws or regulations, such as requirements to use seat belts, requirements for motorcyclists to use helmets, and so on. The problem is that, like training or warning, these methods are meant to impact the behavior of a person. Since people ultimately do as they wish (including suffer the consequences), these methods are less reliable than design or even guarding. In addition, evidence suggests that legislative methods are generally less effective than warning or training methods of behavior change (e.g., Lusk et al., 1995).

SAFETY MANAGEMENT

Safety in industry is promoted in a number of ways: through proper design of equipment and facilities, safety management at specific facilities through activities such as assessing facility safety, taking remedial actions to enhance safety, and performing formal accident or incident investigations. In this section, we briefly summarize some methods for safety management in a company or facility.

Safety Programs

A person rarely has to go in and set up an entire safety program in a business from scratch, but occasionally it does happen. A safety program should involve the participation of both management and staff. Many studies have demonstrated that employee involvement makes a significant difference in the effectiveness of a safety program (e.g., Robertson et al., 1986). Manning (1996) suggests the following three stages:

1. *Identify risks to the company*
2. *Develop and implement safety programs*
3. *Measure program effectiveness*

Identifying Risks.   A full assessment should first be conducted to evaluate existing hazards, hazard controls, accident frequency, and company losses due to accident/incident claims. A safety officer usually begins by analyzing appropriate company documents, including accident/incident reports, safety records, training materials, and so on. Information from these documents should be tabulated for the different jobs or tasks, and according to OSHA injury categories:

Struck by	*Fall/slip/trip*
Body mechanics	*Caught-in-between*
Laceration/cut/tear/puncture	*Struck against*
Contact with temperature extremes	*Eye*
Miscellaneous	

After document analysis, the safety officer conducts interviews with supervisors and employees and performs observational analysis via walk-throughs. The purpose of this activity is to look for equipment or behavior-based hazards associated with task performance. A facility walk-through should also be conducted using a safety checklist based on OSHA General Industry Standard 1910 (Table 14.5 shows part of a typical checklist). Complete checklists can be found in Hammer (2000), Goetsch (2001), and Manning (1996).

From these activities, the safety officer or analyst can develop a list of hazards. In addition to this reactive approach, the analyst should use a proactive approach by using the system safety analysis methods described earlier and also by using the analysis methods described in Kohn, Friend, & Winterberger (1996). One particularly valuable method is *job safety analysis,* which relies on supervisors and employees to identify hazards associated with a particular job. The major advantages to this approach include (1) the heavy involvement of employees, a factor shown to have substantial effects of safety program effectiveness (Kohn et al., 1996; Ray et al., 1993), (2) the long-term benefits of having employees more knowledgeable about hazards, and (3) the efficiency of having employees working to identify hazards. Finally, the analyst should evaluate ergonomic factors that reflect potential hazards to long-term health, such as repetition and excessive force requirements (see Chapter 10).

The final result of this stage should be a table of hazards for each job, piece of equipment, and facility location, with hazard prioritization according to criticality scores. The analysis should also identify those hazards that result in large numbers of accidents and produce the greatest financial (or potential financial) loss.

Implementing Safety Programs. Safety programs should be developed with the assistance and buy-in of management and employees. Safety programs usually include the following elements:

> *Management involvement.* Involve executive management from the beginning, and have supervisors attend or be responsible for conducting monthly safety meetings. Develop procedures for management receiving and acting on labor suggestions. Develop and distribute a general safety policy signed by the chief officer.

TABLE 14.5 Example Checklist Items for Identifying Industrial Hazards

Fall-Related Hazards	*Electrical Hazards*
Are foreign objects present on the walking surface or in walking paths?	Are short circuits present anywhere in the facility?
Are there design flaws in the walking surface?	Are static electricity hazards present anywhere in the facility?
Are there slippery areas on the walking surface?	Are electrical conductors in close enough proximity to cause an arc?
Are there raised or lowered sections of the walking surface that might trip a worker?	Are explosive/combustible materials stored or used in proximity to electrical conductors?
Is good housekeeping being practiced?	Does the facility have adequate lightning protection?
Is the walking surface made of or covered with a nonskid material?	

Accident/incident investigation. Ensure that investigation procedures are in place, identify routing for investigation reports, and train personnel responsible for accident investigation.

Recommendations for equipment, environment, job changes. Develop recommendations for hazard control of high-priority hazards and make all facility changes necessary for OSHA compliance.

Safety rules. Develop general safety rules and job task rules; develop a plan for yearly evaluation of safety rules, and post safety rules in conspicuous places; cover safety rules in new employee orientation; and develop policies for safety rule violation.

Personal protective equipment (PPE). Write standards for use of PPE, compliance criteria, and policies for PPE violations. Develop and implement training on use of PPE.

Employee training. Develop training for job tasks, new employee orientation, hazard awareness, knowledge, and hazard avoidance behavior. Begin regular safety meetings, and develop employee manual to include safety rules and other safety information.

Safety promotion: Feedback and incentives. Display safety posters, notices, memos; display data on frequency of safe behavior and accidents and injury rates; and provide individual and group recognition or other incentives (incentive programs are effective over long periods as long as they are not dropped permanently at some point).

Suggestions and guidelines for implementing these components can be found in various sources. After changes have been implemented, safety checklists can be used for walk-throughs to check for OSHA compliance (e.g., see Davis et al., 1995; Keller & Nussbaum, 2000). Research to date suggests that the most effective means for increasing safety, after design and guarding methods, are to (1) use a participatory approach involving management and employees, (2) providing training for knowledge of hazards, safe behavior, and belief/attitude change, and (3) use behavior-change methods such as feedback and incentives (Ray et al., 1993).

Measuring Program Effectiveness.   After initial collection of baseline data (e.g., accidents, injury, monetary losses, etc.), it is important to continue to collect such data. Program effectiveness is usually evaluated by looking at changes in safe behaviors, accident/incident rates, number of injuries or death, and number of days off due to injury. OSHA logs (which are to be kept by the safety officer) are valuable for this purpose because they contain data on the type and number of injuries for each worker.

Accident and Incident Investigation

OSHA requires investigation of all accidents and for some industries, such as petrochemical plants, also requires investigation of *incidents* (OSHA Rule 29 CFR1910.119). An incident is the occurrence of some event that could have resulted in injury or death but did not. A near miss is considered an incident. The National Transportation Safety Board conducts corresponding investigations for accidents in air transport and ground vehicles. The Aviation Safety Reporting System (ASRS) run by NASA collects data on aviation incidents (see Chapter 2). There are some relatively standardized procedures for performing an accident or incident investigation. Like a police investigation, accident investigations often require careful securing of evidence, extensive interviewing, information collection, analyses of evidence, and drawing of conclusions. Some of the cognitive issues associated with testimony from accident witnesses are covered in Chapter 6. Training programs just for performing accident or incident investigations are becoming common.

Safety Regulators

Finally, the role of regulators in assuring safety compliance must be highlighted (Reason, 1997). OSHA can play a proactive role in assuring compliance with safety regulations through inspections and leveling fines when violations are found. Unfortunately, the small number of inspectors available compared to the vast number of industries where worker safety is of concern means that accidents will occur in unsafe workplaces, and the regulator's role will become **reactive,** leveling penalties only after the damage to a worker has been done. Unfortunately too, some company's tendency to "behave safely" in a proactive fashion may be viewed in the context of the *framing* bias discussed in Chapter 7: When a decision is framed as a choice between a sure loss and a risky loss, decision makers tend to choose the risky option. In the case of an industry manager's choice to implement a safety program, which may cost money and slow productivity, this option can be represented as a sure loss. Too often, the bias is to select the risky option of allowing unsafe practices to continue, gambling that the serious accident will not occur. Such a choice, however, can be counterproductive, given that the **expected costs** of unsafe operation (penalties, workman's compensation, bad publicity) generally outweigh the actual smaller costs of behaving safely. This tendency amplifies the role of regulators to insure that safety choices are made.

RISK-TAKING AND WARNINGS

Risk-Taking as a Decision Process

When hazards are not designed out or guarded, people are ultimately responsible for safe behavior. Examples include proper use of ladders, following correct job procedures, cautious driving behavior, and use of seat belts. Even when safeguards are employed, people frequently have the option of overriding them, such as in the choice not to use personal protective equipment. The choice between safe and unsafe behavior is initially a knowledge-based decision process; eventually, it may become rule-based behavior or simply automatic (see Chapter 7 for discussion of knowledge- and rule-based behavior). One area of research in human factors considers the factors that affect the decision to act safely. The decision to act safely is a function of the factors that affect this decision process: People must know a hazard exists (diagnosis), know what actions are available (generation of alternative actions), and know the consequences of the safe behavior versus alternative behaviors in order to make a wise decision (evaluate alternative actions).

The view of choosing to act safely as an analytical knowledge-based decision suggests that people might sometimes use simplifying heuristics, such as satisficing, and other times use more extensive decision analysis. In the first case, satisficing the individual would consider an action and then evaluate the consequence of that one action. If the consequence is seen as positive to some criterion level, the action will be carried out. For example, a person wants to cut a piece of wood with a circular saw. The cord does not reach an outlet, so he connects an extension cord to the tool. He might briefly consider the positive and negative consequences associated with the action. On the positive side, the tool is now operable, and he does not think of any likely negative consequences. Thus, based on satisficing, the person goes ahead and uses the equipment. Taking this view, decision making relative to use of hazardous tools or equipment would depend heavily on the processes of "generation of an action" and "evaluation of the action." If the person performs the evaluation via running a mental model, the quality of evaluation depends on the quality and completeness of the person's knowledge base plus the availability of different types of information in memory.

We might also assume that in some cases, people perform a decision analysis to evaluate alternative choices. If this were the case, we would expect subjective expected-utility theory to be applicable to behavioral data (DeJoy, 1991), and in fact, several researchers have demonstrated that both expected frequency of consequences and severity of consequences affect decisions or intentions to act safely (e.g., Wogalter et al., 1987). However, it appears that *severity* of injury has a greater effect than *likelihood* on risk perception (Young et al., 1992) and that other variables impact the decision process as well. For example, Young and Laughery (1994) and Schacherer (1993) found that intentions to behave in a safe manner were affected by three psychological components: (1) variables related to perceived severity of the hazard/injury, (2) the *novelty* of the hazard and whether exposure was *voluntary*, and (3) *how familiar* the product or item was to the person.

In understanding the choice to act safely, it is helpful to think of the action-selection process as involving two closely related cognitive stages—risk perception and action choice (DeJoy, 1991). *Risk perception* is the process of determining the likelihood and severity of injury to one's self and may be closely determined by the *availability* of risk in memory. For example, if a vehicle driver has recently suffered a rear-end collision, this event will be available and hence judged as more likely. The perceived risk of tailgating will be greater. After this estimate, the person chooses between the safe and alternative actions by considering the subjective costs and benefits of each behavior outcome. For example, wearing safety goggles while mowing the yard would have the benefit of eliminating possible eye injury but might also have costs such as finding the goggles, wearing them with associated discomfort, not being able to see as well, and looking silly to the neighbors. We refer to these factors collectively as the *cost of compliance*. The alternative, not wearing goggles, has the cost of possible eye injury, but also benefits such as comfort and being able to see well.

A variety of studies have shown that people do, in fact, seem to weigh these types of consideration in making their decisions. For example, the costs of compliance associated with safe behavior, such as wearing personal protective equipment, have an extremely strong, negative effect on the frequency of safe behavior (Wogalter et al., 1989). Greater costs are tolerated for behaviors only where probability and particularly the severity of injury are perceived to be relatively high. However, in the context of the framing bias, the cost of compliance may viewed as a *certain* negative cost, which is balanced against the uncertain, probabilistic negative cost of an accident or injury (if compliance is not undertaken). As we might infer from the framing bias, individual people have a tendency to choose the risky, unsafe behavior, just as we described the tendency of some management to make the same choice (Reason, 1997).

Written Warnings and Warning Labels

We saw that hazard control often relies on instruction or warning about hazards. Especially in the area of consumer products, warnings are becoming increasingly common. One of the reasons for this is that manufacturers have found that warnings are the easiest and cheapest means of protecting themselves against product liability suits. Unfortunately, to be fully defensible, warnings must be targeted for every foreseeable use of a tool or piece of equipment, which is not usually feasible. As a result, there is often disagreement, even among human factors experts, about the number and type of warning labels that should be placed on products.

Written warnings are meant to convey the hazards of a product or piece of equipment. Their goal is to affect people's intentions and behavior so that their actions do not bring about an accident, injury, or death. As we noted earlier, warnings and warning labels are third on the priority list of hazard reduction techniques and thus should only be used when design and safeguard hazard controls are not feasible. Most guidelines suggest that a warning should include a signal word plus information pertaining to the hazard, consequences, and necessary behavior (Wogalter et al., 1987):

- *Signal word* conveying the seriousness, such as Danger, Warning, or Caution
- *Description* of the hazard
- *Consequences* associated with the hazard
- *Behavior needed* to avoid the hazard

An example including these elements is given by Strawbridge (1986):

DANGER:
Contains Acid
To avoid severe burns, shake well before opening.

Another example using both the standard caution icon and a pictograph is shown in Figure 14.4.

In designing warning labels, one must remember several factors. First, people may not see or read a warning label. Therefore, designers should attempt to make such labels as noticeable as possible, for example, by using bright orange

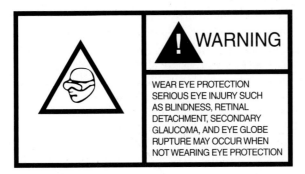

FIGURE 14.4

Warning label with pictograph, caution icon, and hazard information. (*Source:* Dingus, T. A., Hathaway, J. A., & Hunn, B. P., 1991. A most critical warning variable: Two demonstrations of the powerful effects of cost on warning compliance. Proceedings of the Human Factors Society 35th Annual Meeting [pp. 1034–1038]. Santa Monica, CA: Human Factors Society.)

in all or part of the warning or placing the warning next to a part of the equipment that the user *must* look at to operate (e.g., the power switch). Gaining a person's attention is the first goal. Second, people must actually read the words and interpret any pictures or icons. This means the warning must use legible font size and contrast (see Chapter 4), short and relatively simple text (Chapter 6), and easily interpreted pictures or icons (Chapter 8). Traditionally, designers use different *signal words* to convey different degrees of hazard severity:

- *Danger:* An immediate hazard that would likely result in severe injury or death.
- *Warning:* Hazards that could result in personal injury or death.
- *Caution:* Hazards or unsafe practices that could result in minor personal injury or property damage.

However, research indicates that the public is not particularly good at interpreting the difference between the three signal words (e.g., Wogalter et al., 1992), and people especially seem to have difficulty recognizing differences in meaning for *warning* and *caution* (Kalsher et al., 1995). When in doubt, designers are usually encouraged to provide more rather than less information on warnings and warning labels. The problem is that a hazardous tool such as a table saw could end up with hundreds of warning labels, each with a considerable amount of information. At some point, the labels are ignored and become ineffective. Furthermore, when warnings must be printed in a small area, as in a label on a medicine bottle, more warnings requires finer print, and this reduces legibility, a major problem particularly for the older adult.

Third, people must comply with the warning. Compliance is encouraged by clear articulation of the consequences and the behavior needed, but in the workplace, compliance can also be supported by administrative controls and enforcement, as we discuss in Chapter 17. But of course, compliance can never be

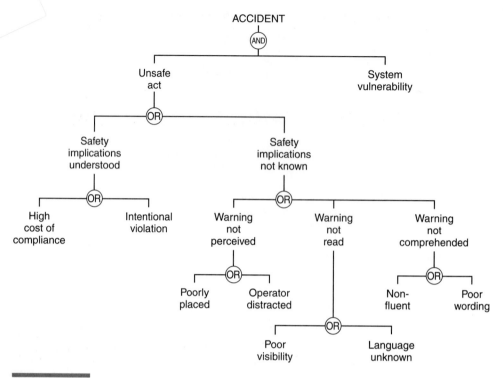

FIGURE 14.5

Fault tree analysis showing the causes of an accident. The unsafe act must be committed at a time when the system is vulnerable (thus, the *and* gate). The unsafe act might be committed when its safety implications are understood but dismissed either because the cost of compliance is too high or for other intentional reasons. Alternatively, the safety implications may not be known, as a result of a series of possible breakdowns in the effectiveness of warnings, as described in the text.

assured to the extent that someone intentionally chooses to engage in hazardous behavior. Figure 14.5 summarizes, in terms of a fault tree, many of the human behavioral factors underlying hazardous behavior.

CONCLUSION

In conclusion, achieving safe behavior is a critical but complex goal of human factors. It depends on identifying and analyzing hazards, identifying the shortcomings of design (both inanimate components and human factors) that may induce those hazards, and proposing (and implementing) the various remediations that will reduce hazards and accidents. While the surest means is to eliminate the hazard itself, this is not always possible, given the hazards to which humans are inevitably exposed in certain tasks and environments. Thus, the most complex and challenging remediation is to address the human's choice to engage in safe versus unsafe behavior. Psychologists' knowledge of this and other choice processes still remains far from mature, but the contributions such knowledge can make to the human factors of safety are potentially quite large.

Ray Cox, a 33-year-old man, was visiting the East Texas Cancer Center for radiation treatment of a tumor in his shoulder. He had been in several times before and found that the sessions were pretty short and painless. He laid chest-side down on the metal table. The technician rotated the table to the proper position and went down the hall to the control room. She entered commands into a computer keyboard for the PDP-II that controlled the radiotherapy accelerator. There was a video camera in the treatment room with a television screen in the control room, but the monitor was not plugged in. The intercom was inoperative. However, Mary Beth viewed this as normal; she had used the controls for the radiation therapy dozens of times, and it was pretty simple.

The Therac-25 radiation therapy machine had two different modes of operation, a high-power x-ray mode using 25-million electron volt capacity and a relatively low-power "electron beam" mode that could deliver about 200 rads to a small spot in the body for cancer treatment. Ray Cox was to have treatment using the electron beam mode. Mary Beth pressed the x key (for the high-power x-ray mode) and then realized that she had meant to enter e for the electron beam mode. She quickly pressed the up arrow key to select the edit function. She then pressed the e key. The screen indicated that she was in the electron beam mode. She pressed the return key to move the cursor to the bottom of the screen. All actions occurred within 8 seconds. When she pressed the b to fire the electron beam, Ray Cox felt an incredible pain as he received 25 million volts in his shoulder. In the control room, the computer screen displayed the message "Malfunction 54." Mary Beth reset the machine and pressed b. Screaming in pain, Ray Cox received a second high-powered proton beam. He died 4 months later of massive radiation poisoning. It turned out that similar accidents had happened at other treatment centers because of a flaw in the software. When the edit function was used very quickly to change *the x-ray mode to electron beam mode, the machine displayed the correct mode but incorrectly delivered a proton beam of 25,000 rads with 25-million electron volts. (A true story adapted from S. Casey,* Set phasers on stun and other true tales of design, technology, and human error, *1993).*

Computers profoundly impact all aspects of life, whether at work or in the home. They have revolutionized the way people perform office tasks such as writing, communicating with coworkers, analyzing data, keeping databases, and searching for documents. Computers are increasingly being used to control manufacturing processes, medical devices, and a variety of other industrial equipment, as well as to promote individual and group creative activity (Fischer, 1999).

Computers are becoming so small that they can be implanted in the human body to sense and transmit vital body statistics for medical monitoring. Because the application of computers is spreading so rapidly, we must assume that much, if not most, of human factors work in the future will deal with the design of complex computer software and hardware.

Human factors work related to computers can roughly be divided into topics related to hardware design, functionality of the software, and design of the software interface. Functionality refers to what the user can do with the software and how it supports or replaces human activities. Chapter 16 addresses functionality in describing how software should be designed when it is used to automate tasks once performed by people. Software interface refers to the information provided by the computer that we see or hear and the control mechanisms for inputting information to the computer. Currently, for most computers, this means the screen, keyboard, and mouse. Software that increases productivity must be useful (provide the appropriate functionality) and usable (have an interface that can be used easily). A well-designed interface does not guarantee a useful product.

On the hardware side, computer workstations should be designed to maximize task performance and minimize ergonomic problems or hazards, such as cumulative trauma disorders. Chapter 10 discussed some of the more well-known design methods for computer workstations and specific hardware components such as keyboards and video display terminals. Chapter 9 discussed various methods for system control with common input devices for computers.

Good software interface design must take into account the cognitive and perceptual abilities of humans, as outlined in Chapters 4, 5, and 6. Interface design also requires the application of display principles, described in Chapter 8, and control principles, described in Chapter 9. Finally, the human–computer interaction (HCI) process will affect and/or be affected by other factors such as fatigue, mental workload, stress, and anxiety. Clearly, most of the material in this text is relevant to the design of the software interface to one extent or another. While we can successfully apply general human factors principles and guidelines to interface design, there is also a solid line of research and methodology that is unique to HCI (Olson & Olson, 2002). A variety of books and journals are written exclusively on this topic (e.g., *Human–Computer Interaction* and *International Journal of Human–Computer Interaction*), and annual meetings result in proceedings reflecting the cutting-edge views and work, such as

Computer–Human Interaction (CHI). Some of this research has been compiled in a recent handbook for HCI (Jacko & Sears, 2002). Given the expanding role of HCI in the field of human factors, we present some of the basic concepts and principles from the subspecialty of HCI.

THE TROUBLE WITH COMPUTERS AND SOFTWARE DESIGN

Computers are relatively new tools; because they change rapidly and tend to be complex, they are high on most peoples' list of "things that are difficult to use." The fact that computer software is sometimes poorly designed and therefore difficult to use causes a variety of negative consequences. First, user performance suffers; researchers have found the magnitude of errors to be as high as 46 percent for commands, tasks, and transactions in some applications. Other consequences follow, such as confusion, panic, boredom, frustration, incomplete use of the system, system abandonment altogether, modification of the task, compensatory actions, and misuse of the system (Galitz, 1993). A comprehensive analysis of how computers influence productivity demonstrates that computers have failed to deliver the promised improvements (Landauer, 1995). Between 1980 and 1989, investment in computer technology in the service sector increased by 116 percent per worker, but productivity increased only 2.2 percent (Tenner, 1996). No systematic relationship exists between investment in information technology and worker productivity. In fact, some industries that spend the most on information technology see the smallest gains in productivity. This relationship has been changing as more emphasis is put on designing to meet user needs (Norman, 1998), but increased computer technology does not guarantee increased productivity. In fact, poor software design has been implicated in disasters and accidents, such as the software design error in the radiation therapy machine mentioned at the start of the chapter (Leveson, 1995).

Human factors designers strive to maximize the ease, efficiency, and safety of products and environments. These goals all apply to software interface design. As Shneiderman (1992) notes, the well-designed software interface can have a sizable impact on learning time, performance speed, error rates, and user satisfaction. In industry this often translates into large monetary savings, and in consumer products these factors can mean success or failure. When the software controls life-critical systems, such as air traffic control systems, power utilities, ship navigation, and medical instruments (such as a device for delivering radiation treatment), the usability of the software can easily become a matter of life and death (Leveson, 1995). Usability is thus one of the greatest concerns for those designing software.

Design Criteria for Usable Software

A number of researchers have specified factors that define or at least suggest high system usability. The concept of usability is discussed in Chapter 3, along with its five criteria of efficiency, accuracy, learnability, memorability, and satisfaction. While designers should evaluate all five criteria, it is important to note that sometimes certain criteria will have either greater or lower priority than others depending on the characteristics of users and the task. For a medical device, such as the one mentioned at the beginning of the chapter, or almost any device with safety-critical implications, errors would be the most important criterion, and satisfaction would be less important.

SOFTWARE DESIGN CYCLE: UNDERSTAND, DESIGN, AND EVALUATE

In Chapter 3, we outlined a basic method for system design used by human factors specialists to enhance system effectiveness and safety. In HCI, a similar design method is used. In the design sequence, the critical components include (1) involvement of typical users throughout the design lifecycle to be sure their needs are understood, (2) use of guidelines and principles in design, and (3) iterative usability testing beginning early in the design process. While there are many models for software interface design, most include steps such as those suggested by Mayhew (1992). One important aspect of the design process is that users should be heavily involved.

Incorporating users as actual members of the design team from beginning to end, an approach termed *participatory design,* has been very successful (Dayton, McFarland, & White, 1994). However, as Nielson (1993) cautions, users working with design teams become steeped in the designers' ways of thinking and familiar with the software system. A different set of users must be brought in for system usability testing.

The design cycle can be simplified into three major phases: *understand* the user, *design,* and *evaluate* (Woods, Patterson, Corban & Watts, 1996). The task analysis provides the initial data to understand the user (see Chapter 3). Designers combine this understanding with a theoretical understanding of the user, interface guidelines and principles of human behavior to create initial design concepts. Soon after these initial concepts are developed, designers conduct *heuristic evaluations* and *usability tests* with low-fidelity mock-ups or prototypes (Carroll, 1995). Usability evaluations are particularly useful because they often help designers better understand the users and their needs. This enhanced understanding can then guide new design concepts. Many iterations of design should be expected, so it is not efficient to worry about details of screen design or making the screens look elegant at the beginning. Rather, the emphasis should be on identifying useful functions, and how the user responds to those functions. When the system becomes more final it may be placed in an operational environmental and a comprehensive test and evaluation may be performed (see Chapter 2). This final evaluation can be considered to be the final step of product development. It can also be considered as the first step in developing a better understanding of the user for the next version of the product. Usability tests are conducted multiple times as the interface design goes through modifications. Each repetition of the testing and modification cycle can produce significant improvements. This process is so valuable that even after 60 cycles testing can provide benefits that outweigh the costs (Landauer, 1995). The balance of this chapter describes some

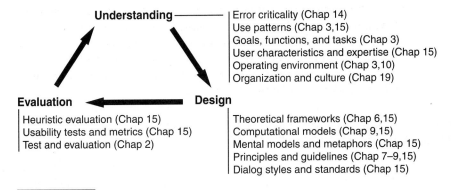

FIGURE 15.1

An iterative cycle of system development.

of the more critical elements of each of these three phases. Other elements are discussed elsewhere in this book as noted in Figure 15.1

UNDERSTAND SYSTEM AND USER CHARACTERISTICS

Software varies from performing very simple functions such as basic arithmetic to extremely complex functions such as control of a chemical-processing plant. The *functionality* of a system generally refers to the number and complexity of things the computer system can do. Software designers usually strive to build in as much functionality as is feasible. However, as a rule of thumb, the greater the functionality, the more difficult it is to design the interface to be usable or user-friendly. If the product is complex, the interface will likely have numerous displays, menus, display formats, control systems, and many levels of interface functions. The trend toward a greater number of functions, called *creeping featurism,* is an important problem because the additional functions make the interface more complex and increase the number of choices a user must make. Microsoft Word has over 1000 commands, up from 311 in 1992. Any one user will find a small number of these commands useful, and the rest simply complicate the system (Norman, 1998).

Complex products designed with no attention to the user often leave a large gulf between the demands they place on the user and the user's capabilities. The goal of the human factors specialist is to help create a product that narrows this gulf by focusing on the needs of the user rather than on the capability of the technology. Imagine designing an interface to the Internet so that any literate person could sit down and successfully search for whatever item he or she happens to need at the moment.

The gulf between user capabilities and product demands depends on more than the product characteristics alone. The product is often part of a system composed of other products, and the complexity that faces the user will depend on all the products in the users' environment. Think about the added complexity of working with a word processor compared to working with a word processor, a spreadsheet, and a database program. The demands facing the user include the overall work responsibilities and not only those associated with the specific product being developed. Likewise, the demands also depend on the organizational and cultural situation, discussed in Chapter 19. Narrowing the gulf between user capabilities and system demands often requires that designers carefully consider the overall environment in which their product will be used. Chapter 3 describes task analysis techniques to address this issue.

Complex software requires a complex interface with many functions. This will, almost by definition, mean some learning time for the user. The reality is that each designer must strive to find the correct balance between making the system usable and expecting the user to expend some effort on learning to use the software. As we describe below, three considerations central to this balancing act between functionality and ease of use are (1) the frequency of task performance using the particular software, (2) mandatory versus discretionary use, and (3) the knowledge level of the user. These influence the relative importance of different usability criteria.

Some computer-based tasks, such as word processing, might be done by a user 8 hours a day, every day. Other tasks, such as making a will, might be done only once or twice in a lifetime. *Frequency of use* has important implications for the software interface design for several reasons. For example, people who will be using a software system frequently are more willing to invest initial time in learning; therefore, performance and functionality can take precedence (to some degree) over initial ease of learning (Mayhew, 1992). In addition, users who perform tasks frequently will have less trouble remembering interactive methods such as commands from one use to the next. This means that designers can place efficiency of operation over memorability (Mayhew, 1992).

There is also a difference between mandatory use of software and *discretionary* use, where people use a system because they want to, not because they are required to. Discretionary users are people who use a particular software program somewhat frequently but are not broadly knowledgeable, as in the case of an expert. Santhanam and Wiedenbeck (1993) describe discretionary users as having expertlike characteristics on a small number of routine tasks, but they may know little regarding anything beyond those tasks. Mayhew (1992) suggests that for high frequency of use or mandatory use, designers should emphasize ease of use. However, for low or intermittent frequency of use or for discretionary users, ease of learning and remembering should have priority over ease of use.

Finally, users may range from *novice* to *expert.* Shneiderman (1992) describes three common classes of users along this experience scale:

> *Novice users:* People who know the task but have little or no knowledge of the system.
> *Knowledgeable intermittent users:* People who know the task but because of infrequent use may have difficulty remembering the syntactic knowledge of how to carry out their goals.
> *Expert frequent users:* Users who have deep knowledge of tasks and related goals, and the actions required to accomplish the goals.

Design of software for novice users tends to focus on ease of learning and low reliance on memory. Vocabulary is highly restricted, tasks are easy to carry out, and error messages are constructive and specific. Systems that are built for first-time users and are extremely easy to use are called "walk up and use" systems typical of an electronic check-in system at an airport. Currently, the technologies predominantly being used for novice users rely heavily on icons, menus, short written instructions, and a *graphical user interface* (GUI). A GUI consists of buttons, menus, windows, and graphics that enable people to *recognize* what needs to be done and then do it through intuitive actions. Users select items from menus or groups of icons (recognition memory) rather than recalling text commands, thus reducing the load on long-term memory ("knowledge in the head") or the need to look things up. Rather than typing commands, users directly manipulate objects on the screen with a mouse, touch screen, or thumb pad. In contrast, a command-line interface requires users to *recall* commands and then type them on a keyboard. Because memory for recognition is more reliable than recall, a GUI is often more effective than command-line interaction, particularly for novice users (Chapter 6). For example, a portion of text can be marked and then moved on the screen from one section of the document to another. In addition to reducing memory load, the GUI makes the task easier because it maps onto how the task might be done without a computer (e.g., cut a section out and move it to a different section of the document).

Reducing the load on memory is especially critical for *intermittent* users, whether they are expert or not. Such users may have a good idea of how the software works but be unable to recall the specific actions necessary to complete a task. However, typing in commands is often preferred by experts, especially if they are frequent users, giving them a feeling of control and quick performance (Shneiderman, 1992). This point demonstrates the difficulty of designing one software interface to meet the needs of multiple types of users. To deal with this, a software interface might have features that accommodate several types of user, as in the case of software that has input *either* from clicking on buttons *or* from typed-command entry. However, once people use a GUI such as menus, even when they become experienced, they will not be prone to switching to the more efficient command-entry format. For this reason, adaptive interfaces are often desirable, automatically monitoring performance and prompting the user to switch entry styles as particular tasks become familiar (e.g., Gong & Salvendy, 1994). In Chapter 16 we discuss adaptive automation, which takes this idea a step further by intervening with automatic control when human control performance declines.

Initial ease of learning and memorability are often less important for systems that will be primarily used by *experts*. For a nuclear power control panel, the designer strives to develop an interface that provides information and input mechanisms that map onto the task. If the task is complex, then learning the software interface will probably take a period of time. In addition, for life-critical systems or hazardous equipment, designers may perceive that error rates are by far the most important of the five criteria listed above; that is, longer training periods are acceptable but should result in fast, efficient, and error-free performance. However, while designers may occasionally lower the priority for ease of learning, it is still generally the case that software interface design strives to maximize all five of the usability criteria listed above. Although the expert or mandatory user differs from the novice or discretionary user in terms of which criteria are considered most important (efficiency, accuracy for the expert, learnability and memorability for the novice), an important challenge is to have a single product satisfy all five criteria for *both* populations.

Although the classes of novice, intermittent, and expert users provide clear distinctions that can help guide designs, reality is often more complex. Frequently, people may use certain parts of a program frequently and other parts infrequently. This might mean a person is an expert user of the drawing tools of a word processor, an intermittent user of the automatic table of contents function, and a novice user of the mail merge function. In addition, expertise may refer to experience with the software or with a particular domain. A secretary with 20 years of experience may be an expert in document production, but a novice with a particular word processor. These distinctions demonstrate the potential danger in using the simple categories of expert, intermittent, and novice users to guide software design. A more sophisticated approach requires a deep understanding of the specific types of expertise of the likely users. This understanding can be summarized with the concept of personas described by Cooper (1997) and in Chapter 3.

DESIGN USING THEORIES AND MODELS

Contemporary researchers strive to provide guidance to software designers so that design can be something more than sheer intuition. This guidance for designers falls into several categories: high-level theories and models, basic principles and guidelines, and methods for evaluation and testing. In this section, we review a few of the more commonly used theories and models. Such theories provide a general framework for designers to conceptualize their problem and discuss issues, using a language that is application independent. Models can provide more specific answers regarding how people might respond to the system. A theory and a model described below, can help designers develop an overall idea of user capabilities, including a description of the kinds of cognitive activity taking place during software use.

Seven Stages of Action

One theory that has been useful in guiding user-oriented interface design is Norman's (1986) *seven stages of action*. It consists of two "bridges" and seven steps (Figure 15.2). A user starts with goals, needs to understand *what* to do to accomplish those goals, *how* to do it. These steps bridge the gulf of execution, which is the mismatch between the user's intentions and the actions supported by the software. This gulf can be narrowed by good, well-human factored controls designed according to control principles discussed in Chapter 9. Next, the user then processes, and evaluates feedback on **whether**

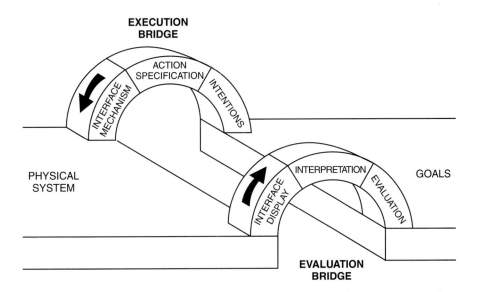

EXECUTION
BRIDGE

ACTION
SPECIFICATION

INTERFACE MECHANISM

INTENTIONS

PHYSICAL
SYSTEM

INTERPRETATION

INTERFACE DISPLAY

EVALUATION

GOALS

EVALUATION
BRIDGE

FIGURE 15.2

Bridging the gulf of execution and gulf of evaluation. (*Source:* Norman, D., 1986. Cognitive engineering. In D. A. Norman & S. W. Draper [eds.], *User-Centered System Design.* Hillsdale, NJ: Lawrence Erlbaum. Copyright ©1986. Reprinted by permission of Lawrence Erlbaum Associates.)

and **how well** those goals are achieved. These steps bridge the gulf of evaluation, which is the mismatch between the user's expectations and the system state. This gulf can be narrowed by providing good, dynamic information, in interpretable displays, following the principles of display design discussed in Chapter 8.

The user first establishes a goal, such as sending an email to a friend. If the person feels that this goal is something that he or she might be able to accomplish using the system, the user forms an intention to carry out actions required to accomplish the goal. Next, the user identifies the action sequence necessary to carry out the goal. It is at this point that a user may first encounter difficulties. Users must translate their goals and intentions into the desired system events and states and then determine what input actions or physical manipulations are required. The discrepancy between psychological variables and system variables and states may be difficult to bridge. Closing this gap is particularly important for novices who use a system infrequently. For situations where people "walk up and use" the system, it must be very clear how they should begin the interaction. Supporting the first step of the interaction is critical because these users are likely to walk away and use another system. This is particularly true of Web sites and Web-based applications.

Even if the user successfully identifies needed input actions, the input device may make them difficult to carry out physically. For example, the "hot" portion of a small square to be clicked using a mouse might be so small that it is difficult to be accurate. Norman notes that the entire sequence must move the user over the gulf of execution (see Figure 15.2). A well-designed interface makes that translation easy or apparent to the user, allowing him or her to bridge the gulf. A poorly designed interface results in the user not having adequate knowledge and/or the physical ability to make the translation and therefore be unsuccessful in task performance.

Once the actions have been executed, users must compare the system events and states with the original goals and intentions. This means perceiving system display components, interpreting their meaning with respect to system events and current state, and comparing this interpretation with the goals. The process moves the user over the gulf of evaluation. If the system displays have been designed well (Chapter 8), it will be relatively easy for the user to identify the system events and states and compare them with original goals. As a simple example, consider a user who is trying to write a friend via email. This user has composed a letter and is now ready to send it. The goal is to "send letter," and the user clicks on the button marked "send." This is a relatively straightforward mapping, allowing easy translation of goal into action. However, after the button is pressed, the button comes up and the screen looks like it did before the user clicked on it. This makes evaluation difficult because the user does not know what system events occurred (i.e., did the letter get sent?). Viewed in terms of this theory, system design will support the user by making two things clear—what actions are needed to carry out user goals and what events and states resulted from user input. The seven steps needed to bridge the gulfs of execution and evaluation provide a useful way of organizing the large number of more specific design guidelines and principles.

Models of User Performance for Design: GOMS

A model of user performance that also centers around users goals and actions is the *goals, operators, methods, and selection rules* (GOMS) model developed by Card, Moran, and Newell (1983) and extended by Kieras (1988a). Like the seven stages of action theory, GOMS helps designers understand the challenges users might face in bridging the gulfs of evaluation and execution. Although the seven-stage model of action provides a very useful integrating framework for thinking about HCI guidelines, it is not very useful in predicting the specific response of users to particular design alternatives. GOMS provides a detailed description of user tasks and can even be used to make specific quantitative predictions of how users will respond to a particular system.

GOMS assumes that users formulate goals (such as write email) and subgoals (make blank page to write on) that they achieve through methods and selection rules. A method is a sequence of steps that are perceptual, cognitive, or motor operators. Since several methods often can be used to accomplish a goal or subgoal, selection rules must be postulated to identify the conditions under which a user will use one method or another. As an example, consider the goal of printing a document using a typical Windows type of word processor. The person could use the *method* of

1. Using the mouse to move the cursor over the button with the printer symbol.
2. Quickly depressing and releasing the upper left area of the mouse one time.

Alternatively, the user could use the *method* of

1. Using the mouse to move the cursor over the word *File* at the top of the screen.
2. Quickly depressing and releasing the upper left area of the button.
3. Using the mouse to move the cursor down to the word *Print.*
4. Quickly depressing and releasing the upper left area of the button, and so forth.

There are also other methods for printing the document, such as using the keyboard instead of the mouse. Selection rules would specify the conditions under which the user would choose each method. Note that different users might have varying selection rules, and these might be different from what the software designers would consider to be the "best" selection rules.

The GOMS model has been useful to designers in a number of ways. Probably the most common is use of the GOMS language for describing software functionality and interface characteristics (e.g., Irving et al., 1994). This supports a systematic analysis of potential usability problems. Designers generally do the following: (1) explicitly identify and list users' goals and subgoals; (2) identify all of the alternative methods (sequences of operators) that could be used for achieving each goal/subgoal; and (3) write selection rules, specifying the conditions under which each method should be used. Evaluation of the GOMS structure reveals problems such as too many methods for accomplishing a goal, similar goals supported by inconsistent methods, and methods that rely too heavily on long-term memory (e.g., see Gong & Kieras, 1994). When there are multiple methods to accomplish one goal, designers may realize that one method is so clearly preferable that there will be no conditions under which a person would need to choose the alternative. This alternative can then be altered or dropped altogether. Designers may also realize that users will not ever notice that an alternative method exists or be able to infer the correct selection rules to discriminate between different methods. One recent solution to both of these problems is the idea of "helpful hints." For example, a word-processing program might open with a different Helpful Hint box each day, suggesting new and easier methods for accomplishing a task or conditions under which the person might choose one method over another.

Other researchers have developed computer models of software systems using the GOMS notation. For example, Kieras and Polson (1985) used production rules to specify the conditions and actions in an interactive text editor. They found that the number and complexity of production rules predicted actual user performance with respect to both learning time and performance time. Thus, the GOMS model provides a language and structure for modeling interactive software. This allows designers to make modifications to the software interface and predict the impact of such changes on performance time of importance for a cost-benefit analysis (see Chapter 3). Finally, some experts have studied the weaknesses of online help systems and developed design principles based on the use of specific GOMS elements for presenting information to users (Elkerton, 1988).

A CASE STUDY OF THE APPLICATION OF GOMS TO EVALUATE A NEW COMPUTER WORKSTATION

Gray, John, and Atwood (1993) used a GOMS model to evaluate a new telephone operator workstation for NYNEX. Their model predicted and explained why a new and supposedly improved workstation would actually lower productivity by $2.4 million a year if it were implemented. Instead of making operators faster, the new GUI made operators slower than the old command-line workstation.

To arrive at this conclusion, they created a GOMS model for each of 20 call types. The model included detailed keystroke information that described how the operator used the workstation to handle each type of call. The model provided extremely precise time estimates to describe how the new system affected the time to process each call.

Surprisingly, the model predicted that the new system would increase the time to process calls by an average of 3 percent. It also showed that this difference was much greater for some calls compared to others, with an increase of between 0.2 seconds and 3.4 seconds. These data were then compared to actual calls with the old and new system. Based on 78,240 calls, the new system took operators 4 percent more time to process each call on average. This result is particularly surprising because the new system required fewer keystrokes.

More useful than simply predicting the poor performance of the new system, the model also explained this surprising result. The detailed task sequence information of the model showed that several sequences of activity occur in parallel as an operator handles a call. When several sequences of activities occur in parallel, one sequence will take longer than the others. The *critical path* is the sequence of activities that takes the longest to complete. The critical path determines how long the overall set of activities will take to complete. Even though the new system had fewer keystrokes overall, it had more keystrokes on the critical path. One reason for this was the spacing and configuration of the function keys. The keyboard of the new system forced operators to use only their right hand in making selections. Other contributors to the slower response depended on the complex interaction between the caller, operator, and computer. These causes for the increased response time would have been difficult to identify without the GOMS model.

GOMS and the seven stages of action theory demonstrate how theories and models can support interface design in very different ways. The seven stages of action theory describes, in a very general way, how people must bridge the gulf of evaluation (understanding the state of the system and establishing goals) and the gulf of execution (understand what to do to accomplish those goals and how to do it). The GOMS model is another way of describing this process, but it focuses on generating a response to achieve a goal. In particular, GOMS is useful in understanding the problems that arise when there are multiple methods for accomplishing a single goal—a situation that can sometimes be confusing for the user. GOMS also has the advantage of being developed to make quantitative predictions, which the seven stage theory cannot.

DESIGN TO SUPPORT MENTAL MODELS
WITH CONCEPTUAL MODELS AND METAPHORS

Bridging the gulfs of execution and evaluation often depends on the mental model of the user, which can best be described as a set of expectancies regarding what human actions are necessary to accomplish certain steps and what computer actions will result. As described in Chapter 6, an effective mental model is one that is relatively complete and accurate, and supports the required tasks and subtasks. It allows the user to correctly predict the results of various actions or system inputs. As a consequence, a good mental model will help prevent errors and improve performance, particularly in situations that the user has not encountered before. The development of effective mental models can be facilitated by system designers.

One way to promote an accurate mental model is by developing a clearly defined conceptual model. A *conceptual model* is "the general conceptual framework through which the functionality is presented" (Mayhew, 1992). Often the success of a system hinges on the quality of the original conceptual model. For example, the success of the cut-and-paste feature in many programs is due to the simple but functional conceptual model of this component (cut and paste). Mayhew (1992) suggests several specific ways a conceptual model can be made clear to the user:

Making invisible parts and processes visible to the user. For example, clicking on an icon that depicts a file and dragging it to a trash can makes an invisible action (getting rid of a file) visible to the user.

Providing feedback. When an input command is given, the system can report to the user what is happening (e.g., loading application, opening file, searching, etc.).

Building in consistency. People are used to organizing their knowledge according to patterns and rules. If a small number of patterns or rules are built into the interface, it will convey a simple yet powerful conceptual model of the system.

Presenting functionality through a familiar metaphor. Designers can make the interface look and act similar to a system with which the user is familiar. This approach uses a *metaphor* from the manual real world with which the user is supposedly familiar.

Metaphors are a particularly useful approach for helping users develop an effective mental model. A metaphor is the relationship between objects and events in a software system and those taken from a noncomputer domain (Wozny, 1989), which supports the transfer of knowledge. The use of a metaphor provides knowledge about what actions are possible, how to accomplish tasks, and so forth. Many of the GUI interfaces currently in use are strongly based on well-known metaphors.

An example of a powerful metaphor is that of "rooms." The Internet has different types of rooms, including chat rooms, where people can "go" to "talk." Obviously, none of these actions are literal, but the use of the concepts provides some immediate understanding of the system (Carroll et al., 1988). People then only need to refine their mental model or add a few specific rules.

Using a metaphor can have adverse consequences as well as positive benefits (Halasz & Moran, 1982; Mayhew, 1992). For example, overreliance on a physical metaphor can cause users to overlook powerful capabilities available in the computer because they simply do not exist in the real world. In addition, there are always differences between the metaphorical world and the software system. If these differences are not made explicit, they can cause errors or gaps in users' mental models of the software system (Halasz & Moran, 1982). For example, anywhere between 20 percent and 60 percent of novice errors on a computer keyboard could be attributed to differences between the typewriter metaphor and actual editor functions (Douglas & Moran, 1983; Alwood, 1986).

In summary, users will invariably develop a mental model of the software system. Designers must try to make this mental model as accurate as possible. This can be done by making the conceptual model of the system as explicit as possible and can sometimes be aided by the use of real-world metaphors.

DESIGN USING PRINCIPLES AND GUIDELINES

Theories are helpful tools for generating the conceptual design or *semantic* level of the software system. Theories are relatively widely used because there really are no specific guidelines for how to design an effective software system at the conceptual stage. However, when designers are ready to translate the conceptual model into the *syntactic* components, which are the actual interface elements, a large number of design principles and guidelines are available to enhance the effectiveness and usability of software interfaces (e.g., Lynch & Horton, 1999; Mayhew, 1992; Nielson, 1993; NASA, 1996; Park & Hannafin, 1993; Shneiderman, 1992).

General Usability Guidelines

Nielson (1994a) recognized that some usability guidelines might be more predictive of common user difficulties than others. In order to assess this possibility, he conducted a study evaluating how well each of 101 different usability guidelines explained usability problems in a sample of 11 projects. Besides generating the predictive ability of each individual heuristic, Nielson performed a factor analysis and successfully identified a small number of usability factors that "structured" or clustered the individual guidelines and that accounted for most of the usability problems. Table 15.1 summarizes the general usability principles identified by Nielson (1994a), which provide direction to a design team developing software and can also be used as the basis for heuristic evaluation of prototypes.

The first principle, matching the system to the real world, should sound familiar to readers. This is the idea that the software interface should use concepts, ideas, and metaphors that are well known to the user and map naturally onto the user's tasks and mental goals. Familiar objects, characteristics, and actions cannot be used unless the designer has a sound knowledge of what these things are in the user's existing world. Such information is gained through performing a task analysis, as discussed in Chapter 3. This is not to say that the interface should only reflect the user's task *as the user currently performs it.* Computers can provide new and powerful tools for task performance that move beyond previous methods (Nielson, 1994c). The challenge is to map the new interface onto general tasks required of the user while still creating new computer-based tools to support performing the tasks more effectively and efficiently.

The second principle is to make the interface consistent, both internally and with respect to any existing standards. *Internal consistency* means that design elements are repeated in a consistent manner throughout the interface: The same type of information is located in the same place on different screens, the same actions always accomplish the same task, and so forth (Nielson, 1989). An application should also be consistent with any platform standards on which it will run. For example, a Windows application must be designed to be consistent with standardized Windows icons, groupings, colors, dialog methods, and so on. This consistency acts like a mental model—the user's mental model of "Windows" allows the user to interact with the new application in an easier and quicker fashion than if the application's interface components were entirely new.

The third principle, visibility of system status, should also sound familiar. The goal is to support the user's development of an explicit model of the system—making its functioning transparent. Features important in this category are showing that input has been received, showing what the system is doing, and indicating progress in task performance. Sometimes this can be accomplished fairly easily with the right metaphor. For example, showing the cursor dragging a file from one file folder to another provides feedback regarding what the user/system is doing. However, showing the cursor dragging a short dotted line from one file folder to another makes system functioning slightly less visible.

The principle of *user control and freedom* centers around the idea that users need to be able to move freely and easily around in an interface, undo actions that may have been incorrect, get out of somewhere they accidentally "entered," cancel tasks midpoint, go to a different point in their task hierarchy, put away a subtask momentarily, and so forth. Exits can be provided in the form of "undo" commands that put the user back to the previous system state (Abowd & Dix, 1992; Nielson, 1993). The interface should provide alternative ways of navigating through screens and information and alternative paths for accomplishing tasks.

TABLE 15.1   General Interface Design Principles

Match between system and real world

Speak the user's language.
Use familiar conceptual models and/or metaphors.
Follow real-world conventions.
Map cues onto user's goals.

Consistency and standards

Express the same thing the same way throughout the interface.
Use color coding uniformly.
Use a uniform input syntax (e.g., require the same actions to perform the same functions).
Functions should be logically grouped and consistent from screen to screen.
Conform to platform interface conventions.

Visibility of system status

Keep user informed about what goes on (status information).
Show that input has been received.
Provide timely feedback for all actions.
Indicate progress in task performance.
Use direct manipulation: visible objects, visible results.

User control and freedom

Forgiveness: Obvious way to undo, cancel, and redo actions.
Clearly marked exits.
Allow user to initiate/control actions.
Avoid modes when possible.

Error prevention, recognition, and recovery

Prevent errors from occurring in the first place.
Help users recognize, diagnose, and recover from errors.
Use clear, explicit error messages.

Memory

Use see-and-point instead of remember-and-type.
Make the repertoire of available actions salient.
Provide lists of choices and picking from lists.
Direct manipulation: visible objects, visible choices.

Flexibility and efficiency of use

Provide shortcuts and accelerators.
User has options to speed up frequent actions.
System should be efficient to use (also, ability to initiate, reorder, or cancel tasks).

Simplicity and aesthetic integrity

Things should look good with a simple graphic design.
Use simple and natural dialog; eliminate extraneous words or graphics.
All information should appear in a natural and logical order.

Source: Nielson, J. *Enhancing the explanatory power of visibility heuristics.* Chi '94 Proceedings. New York: Association for Computing Machinery.

This brings us to a closely related category, errors and error recovery. Errors can be described in terms of slips and mistakes. Slips are unintentional, inappropriate actions, such as hitting the wrong key. Mistakes are intentional acts that reflect a misunderstanding of the system, such as entering an incorrect command. Chapter 14 provides a more detailed discussion of these error types. It is a basic fact of life that computer users will make errors, even minor ones such as hitting the wrong key, and even with a well-designed system.

Because all errors cannot be prevented, the design should minimize the negative consequences of errors or to help users recover from their errors (Nielson, 1993). Such *error-tolerant* systems rely on a number of methods. First, systems can provide "undo" facilities as discussed previously. Second, the system can monitor inputs (such as "delete file") and verify that the user actually understands the consequence of the command. Third, a clear and precise error message can be provided, prompting the user to (1) recognize that he or she has made an error, (2) successfully diagnose the nature of the error, and (3) determine what must be done to correct the error. Shneiderman (1992) suggests that error messages should be clearly worded and avoid obscure codes, be specific rather than vague or general, should constructively help the user solve the problem, and should be polite so as not to intimidate the user (e.g., "ILLEGAL USER ACTION").

The accident described in the beginning of this chapter occurred because (1) the software system had a bug that went undetected, (2) there was not good error prevention, and (3) there was not good error recognition and recovery. As an example, when the operator saw the message "Malfunction 54," she assumed the system had failed to deliver the electron beam, so she reset the machine and tried again.

While the previous guidelines are aimed mostly at the novice or infrequent user, expert users need to be accommodated with respect to efficiency and error-free performance. Principle seven, *flexibility and efficiency of use,* refers to the goal of having software match the needs of the user. For example, software can provide shortcuts or accelerators for frequently performed tasks. These include facilities such as function or command keys that capture a command directly from screens where they are likely to be most needed, using system defaults (Greenberg, 1993; Nielson, 1993). In other words, they are any technique that can be used to shorten or automate tasks that users perform frequently or repeatedly in the same fashion.

One common tendency among designers is to provide users with a vast assortment of functions, meeting every possible need in every possible circumstance. While this creeping featurism may seem to be providing a service to users, it may be doing more harm than good. Facing a complex interface or complex choice of multiple options is often overwhelming and confusing to users. Designers must remember that users do not bring rich knowledge and understanding to their perception of the system in the way that designers do. Principle eight concerns the need to create simple and aesthetically pleasing software. An interface that presents lots of information and lots of options will simply seem difficult. The ultimate design goal is to provide a broad functionality through a simple interface (Mayhew, 1992). One common way to accomplish this is to layer the interface so that much of the functionality is not immediately apparent to the novice. System defaults are an example of this approach. Once users become more familiar with the system, they can go in and change defaults to settings they prefer. An example is the typical graphical word-processing software with "invisible" defaults for page layout, font, style, alignment, and so on. Design goals of simplicity and consistency will pay off in software that users find easy to learn and easy to use. This will make them more likely to appreciate and use its unique functionality.

Basic Screen Design

Most interaction with computers depend on various *manual* input methods (as opposed to voice or other means) and viewing text or graphic displays on a monitor. Although there is a great deal of dynamic interaction, designers still must focus heavily on the components and arrangement of *static* screen design, that is, what each screen looks like as a display panel (Galitz, 1985). Most current screen layout and design focuses on two types of elements, output displays (information given by computer) and input displays (dialog styles, buttons, slider switches, or other input mechanisms that may be displayed directly on the screen). For information related to output displays, see Chapter 8.

One general design consideration of output displays is the use of color. Novice designers tend to overuse color, most professional designers suggest that, because of factors such as the prevalence of color-blindness, one should always design the interface so that it can be understood in black and white (e.g., Mayhew, 1992; Shneiderman, 1992). Nielson (1993) also recommends using light gray or light colors for background. Color should then be used conservatively and only as redundant coding. More specific guidelines have been developed specifically within the field of HCI. For example, Mayhew (1992) divides screen layout and design principles into five categories: general layout, text, numbers, coding techniques, and color. By reviewing research and published applications, she identified a number of design principles relevant to each of these categories. For more information related to output displays, see Chapter 8.

Dialog Styles

Given that computers are information-processing systems, people engage in a *dialog* with computers, which consists of iteratively giving and receiving information. Computers are not yet technologically sophisticated enough to use unrestricted human natural language, so the interface must be restricted to a dialog that both computer and user can understand. There are currently several basic dialog styles that are used for most software interfaces:

> *Menus:* Provides users with a list of items from which to choose one of many.
> *Fill-in forms:* Provides blank spaces for users to enter alpha or numeric information.

Question/answer: Provides one question at a time, and user types answer in field.

Command languages: At prompt, user types in commands with limited, specific syntax.

Function keys: Commands are given by pressing special keys or combinations of keys.

Direct manipulation: Users perform actions directly on visible objects.

Restricted natural language: Computer understands a restricted set of spoken messages.

While it is sometimes difficult to distinguish perfectly between these dialog styles, it is still convenient to categorize them as such for design purposes. Some dialog styles are suited to specific types of application or task, and a number of dialog styles are frequently combined in one application. Mayhew (1992) describes such guidelines in great depth, a few of which are included in the following discussion. For further information, see Mayhew (1992), Nielson (1993), Helander (1988), and Shneiderman (1992).

Menus. Menus have become very familiar to anyone who uses the Macintosh or Windows software environment. Menus provide a list of actions to choose from, and they vary from menus that are permanently displayed to pull-down or multiple hierarchical menus.

Menus should be used as a dialog style when users have one or more of the following negative attitudes, low motivation, poor typing skills, little computer or task experience.

One approach to menu design is to rely on simple guidelines. For example, a series of studies have found that each menu should be limited to between four and six items to reduce search time (see Chapter 4) (Lee & MacGregor, 1985). This number can be increased by grouping menu items into categories and separating them with a simple dividing line.

Menus that have a large number of options can be designed to have few levels with many items per level ("broad and shallow") or to have many levels with few items per level ("narrow & deep"). In general, usability is higher with broad & shallow menus. Mayhew (1992) provides the following guidelines (among others):

- Use graying out of inactive menu items.
- Create logical, distinctive, and mutually exclusive semantic categories.
- Menu choice labels should be brief and consistent in grammatical style.
- Order menu choices by convention, frequency of use, order of use, functional groups, or alphabetically, depending on the particular task and user preference.
- Use existing standards, such as: File, Edit, View that are common on most Windows applications.

Unfortunately, menu design is more complex than these simple guidelines suggest. Even with a relatively simple set of menu items, the number of possible ways to organize the menu design options explodes combinatorially as the number of menu items increase (Fisher et al., 1990). For example, a system with eight menu items at each of three levels in a menu hierarchy generate 6.55×10^{13} possible menu designs, prompting the need for computational approaches for menu design (Francis, 2000). A model of the menu selection will minimize the time to select menu option. One important limit of these models is that they do not tend to consider the meaning of the menu items and how people remember the menu structure. A comprehensible menu structure enables users to select correct menu options more quickly. One way to make menus comprehensible is to use a variant of the cluster analysis described in Chapter 3. With this approach, cluster analysis tools identify groups of related menu item that might not be otherwise obvious.

Fill-in Forms. Fill-in forms are like paper forms: They have labeled spaces, termed *fields,* for users to fill in alphabetical or numeric information. Like menus, they are good for users who have a negative to neutral attitude, low motivation, and little system experience. However, they should be reasonably good typists and be familiar with the task. Otherwise, very strong guidance is needed for filling out the form spaces. Fill-in forms are useful because they are easy to use, and a "form" is a familiar concept to most people.

Like menus, fill-in forms should be designed to reflect the content and structure of the task itself. An example is a form filled out by patients visiting a doctor's office. The form could look very similar to the traditional paper forms, asking for information about the patient's name, address, medical history, insurance, and reason for the visit. Having the patient type this information on a computer in the waiting room would alleviate the need for a receptionist to type the information for the patient's file. Fill-in forms should be designed according to the following basic principles:

- Organize groups of items according to the task structure.
- Use white space and separate logical groups.
- Support forward and backward movement.
- Keep related and interdependent items on the same screen.
- Indicate whether fields are optional.
- Prompts should be brief and unambiguous.
- Provide direct manipulation for navigation through fields.

Question-Answer. In this dialog style, the computer displays one question at a time, and the user types an answer in the field provided. The method is good for users who have a negative attitude toward computer technology, low motivation, little system experience, and relatively good typing skills. It is appropriate for tasks that have low frequency of use, discretionary use, and low importance. Question-answer methods must be designed so that the intent of the question and the required response is clear: (1) Use visual cues and white space to clearly distinguish prompts, questions, input area, and instructions; (2) state questions in clear and simple language; (3) provide flexible navigation; and (4) minimize typing requirements.

Command Languages.   At a prompt, such as >, the user types in commands that require use of a very specific and limited syntax (such as C++ or Basic), and unlike menus, they do not require much screen space. Command languages are appropriate for users who have a positive attitude toward computer use, high motivation, medium- to high-level typing skills, high computer literacy, and high task-application experience. Designers who are creating a command language should strive to make the syntax as natural and easy as possible; make the syntax consistent; avoid arbitrary use of punctuation; and use simple, consistent abbreviations (see Mayhew, 1992, for additional guidelines).

Function Keys.   In this dialog style, users press special keys or combinations of keys to provide a particular command. An example is pressing/holding the control button and then pressing the 'B' key to change a highlighted section of text to boldface type. The use of function keys as input mechanisms in computer dialog is declining, probably because they are arbitrary and taxing on human memory. However, for users who perform a task frequently, want application speed, and have low-level typing skills, function keys are extremely useful.

Because of their arbitrary nature and demands on memory, design of function key commands is tricky. Designers should use the following guidelines, among others:

- Reserve the use of function keys for generic, high-frequency, important functions.
- Arrange in groups of three to four and base arrangement on semantic relationships or task flow.
- Label keys clearly and distinctly.
- Place high-use keys within easy reach of home row keys.
- Place keys with serious consequences in hard to reach positions and not next to other function keys.
- Minimize the use of "qualifier" keys (alt, ctrl, command, etc.) which must be pressed on the keyboard in conjunction with another key.

Direct Manipulation.   Direct manipulation means performing actions directly "on visible objects" on the screen. An example is using a mouse to position the cursor to a file title or icon, clicking and holding the mouse button down, dragging the file to a trash can icon by moving the mouse, and dropping the file in the trash can by letting up on the mouse key. Direct manipulation dialog styles are becoming extremely popular because they map well onto a user's mental model of the task, are easy to remember, and do not require typing skills. Direct manipulation is a good choice for users who have a negative to moderate attitude toward computers, low motivation, low-level typing skills, and moderate to high task experience. Mayhew (1992) provides the following design guidelines, among others:

- Minimize semantic distance between user goals and required input actions.
- Choose a consistent icon design scheme.
- Design icons to be concrete, familiar, and conceptually distinct.
- Accompany the icons with names if possible.

Direct manipulation interface design requires a strong understanding of the task being performed and high creativity to generate ideas for metaphors or other means of making the direct manipulation interface make "sense" to the user.

Natural Language.   Finally, natural language is an interface dialog style that currently has some limited applications. In this method, users speak or write a constrained set of their natural language. Because it is a natural rather than artificial style for human operators, natural language can be thought of as the "interface of choice." As technology improves, this dialog style will become more common. However, the technology required to enable computers to understand human language is quite formidable. Recognizing spoken commands is not the same as understanding the meaning of a spoken sentence, and natural language interfaces may never be the ultimate means of bridging the gulfs of execution and evaluation (see Chapter 9 for a more detailed discussion).

Conclusion.   No dialog style is best for all applications. The choice depends on matching the characteristics of the dialog style to those of the user and the tasks being performed. For example, certain tasks are better performed through direct manipulation than natural language. Consider the frustration of guiding a computer to tie your shoe compared to simply tying the shoe manually. More realistically, it is feasible to control the volume of a car stereo by voice, but the continuous control involved in volume adjustment makes the standard knob more appropriate.

DESIGN OF USER SUPPORT

It is a worthwhile goal to make computer software so "intuitive" that people require no training or help to be able to use it. Unfortunately, much of the time this goal cannot be entirely achieved. Like other complex equipment, many software systems have features that are ultimately useful for task performance but require time and learning from the user. This means that as people begin to use a software system, they will need assistance or user support from time to time. User support refers to a variety of assistance mechanisms, which may include software manuals, online help, tutorials, software wizards, and help lines.

There is also variety even within this array of user support services. For example, online help methods may include keyword help (Houghton, 1984), command prompting (Mason, 1986), context-sensitive help (Fenchel & Estrin, 1982; Magers, 1983), task-oriented help (Magers, 1983), and intelligent online help (Aaronson & Carroll, 1987; Dix et al., 1993; Elkerton, 1988). All of these methods provide users with information concerning how to use the interface to accomplish tasks.

We now briefly consider the two most commonly used support mechanisms: manuals and online help. Also refer to Chapter 18, which reviews material relevant to manuals and tutorials.

Software Manuals

Most software systems are sufficiently complex to require a manual and possibly online help systems. The difficulty in writing manuals is that users do not read them as instructional texts ahead of time as much as they use them as reference manuals when they need immediate help (Nielson, 1993; Rettig, 1991). Because of this type of use, manuals should have well-designed, task-oriented search tools. Writers should keep in mind that users use search words based on their *goals* and *tasks,* not on system components or names.

The software manual is a system in and of itself, so it should be designed using standard human factors principles and guidelines to maximize efficiency and effectiveness. Software manuals can be subjected to usability testing just like any other system. Table 15.2 gives some general guidelines for designing software manuals; refer also to Adams and Halasz (1983), Gordon (1994), and Weiss (1991).

Online Help Systems

Hardcopy manuals are the default help system for software, but many designers are realizing the advantages of offering online help systems. Among other things, online help offers the following:

- Easy to find information (where hardcopy manuals maybe lost or loaned out).
- Can be context-sensitive.
- Users do not have to find available workspace to open manuals.
- Information can be electronically updated.
- On-line help systems can include powerful search mechanisms such as string search, multiple indices, electronic bookmarks, hypertext navigation, and backward tracing.

TABLE 15.2   General Guidelines for Writing Software Manuals

Guideline	Example
Make information easy to find	
Let the user's tasks guide organization: Base the labeling, entry names, and sequencing on user goals and tasks rather than system components.	Poor: "Upload file to server," "Access server data," and similar system labels.
	Better: Base labeling and entries on terms such as "Send letter" and "Open mail."
Include entry points that are easy to locate by browsing.	Poor: Uniform blocks of text with no indication of the content.
	Better: Show each text section name in bold-face, arrange sections alphabetically, and put the terms at the top of each page as headers.
Use both table of contents and index.	Use system and user task terms such as *format, callout, button,* and *create a callout.*
Include entries based on both the system components and user goals, and index both types of entries along with extensive synonyms at back of manual.	

Because of these advantages, most commercial interfaces now come packaged with some type of online help system in addition to the paper manual. Search effectiveness and efficiency is a general difficulty for online help systems. For example, in one study, Egan and colleagues (1989) found that it took almost 50 percent longer for users to find information in an online help system than it took to find the information in a hardcopy manual (7.6 min versus 5.6 min). This longer search time was reduced to a smaller amount than the hardcopy manual only after extensive usability testing and iterative design. Other problems that may be associated with online help systems include the following:

- Text may be more difficult to read on a computer screen.
- Pages may overlap with task information and therefore interfere with the task.
- Pages on the computer may contain less information than hardcopy manuals.
- People are used to finding information in manuals, whereas navigation may be more difficult in online help systems.

Careful design can reduce many of these problems. For example, powerful browsing systems can make access to online manuals successful, and a well-designed table of contents can help users locate information. Shneiderman (1992) suggests a properly designed table of contents that stays on the screen when text is displayed and the use of an expanding/shrinking table of contents. As with other software systems, design of online help should follow principles of good design and be subjected to extensive usability testing and iterative design.

EVALUATE WITH USABILITY HEURISTICS

Even systems created with careful attention to design guidelines require evaluation. A useful preliminary evaluation is *heuristic evaluation,* discussed in Chapter 3 in which several HCI specialists evaluate how well the candidate design adheres to interface guidelines, principles, or heuristics (rules for interface design that generally, but not always, work). The heuristic evaluation can be quite effective because it can be less expensive and less time consuming than a usability test.

The heuristic evaluation first identifies the most relevant interface design principles that address the tasks the product is meant to support, such as the display and control principles in Chapters 8 and 9, and the guidelines in Table 15.1. The second step is to apply the selected guidelines, principles, and heuristics to the product. Potential usability problems are identified when the interface violates one or more of the heuristics. Different HCI experts will be likely to discover a *different* set of problems. For this reason, it is important that two to four experts evaluate the system independently.

EVALUATE WITH USABILITY TESTS AND METRICS

Even a carefully designed system that uses the best theories must be evaluated in usability tests. Usability tests involve typical users using the system in realistic situations. Difficulties and frustrations they encounter are recorded to identify opportunities to enhance the software.

Prototypes

An important issue for designers is the kind of prototypes to use for usability testing. While it may seem that all usability testing would be done with screens that look much like the final software, this is not the case. Most usability specialists use a variety of prototypes, which may range in fidelity, with low-fidelity methods, used early in the design process including index cards, stickies, paper and pen drawings, and storyboards. *Storyboards* are a graphical depiction of the outward appearance of the software system, without any actual system functioning. High-fidelity methods include fully interactive screens with the look and feel of the final software. Use of low fidelity methods early has several advantages. (1) They are often faster and easier, and can be modified more easily *during* usability testing; (2) since designers are less invested in work, they are more willing to change or discard ideas; (3) users don't focus on interface details, such as the type of font, and therefore give more substantive feedback to the functionality of prototypes that are obviously low fidelity (Carroll, 1995). As Salasoo and colleagues (1994) write when describing the use of paper and pen technique, "Several people can work at once (collaboratively) and users have as much access to the prototype medium as product team members." The goal is to move through different design ideas until one is identified that works well for users. Once the interface has gone through several loops of iterative design, the prototype is then moved to computer screens for more advanced usability testing.

Usability Metrics

When designers are conducting usability testing, whether early in the low-fidelity prototyping stages or late in the design lifecycle, they must identify what they are going to measure, often called *usability metrics*. Usability metrics tend to change in nature and scope as the project moves forward. In early conceptual design phases, usability can be done with a few users and focuses on qualitative assessment of general usability (whether the task can even be accomplished using the system) and user satisfaction. Low-fidelity prototypes are given to users who then imagine performing a very limited subset of tasks with the materials or screens (Carroll, 1995). At this point, there is usually little to no quantitative data collection; simply talking with a small number of users can yield a large amount of valuable information.

As the design takes on more specific form, usability testing becomes more formalized and often quantitative. Table 15.3 shows some of the more common usability categories of effectiveness, efficiency, and subjective satisfaction (from Mayhew, 1992; Nielson, 1993). To collect data on these measurements, a fully functioning prototype is built, and users are given a set of task scenarios to perform as they would under normal circumstances (e.g., see Carroll, 1995). In addition to collecting quantitative measures such as those shown in Table 15.3, designers ask users to think aloud during task performance and answer questions. The goal is to find a prototype design that users like, learn easily, and can use to successfully perform tasks. Observation of users gives the designers insight into difficulties with controls, navigation, general conceptual models, and so on.

TABLE 15.3 **Examples of Software Usability Metrics (with usability defined in a broad sense)**

Effectiveness	Efficiency	User Satisfaction
Percent of tasks completed	Time to complete a task	Rating scale for usefulness of the software
Ratio of successes to failures	Time to learn	
Number of features or commands used	Time spent on errors	Rating scale for satisfaction with functions/features
	Percent or number of errors	
	Frequency of help or documentation use	Number of times user expresses frustration or dissatisfaction
	Number of repetition of failed commands	
		Rating scale for user versus computer control of task
		Perception that the software supports tasks as needed by user

Users are asked to use the software to perform the task. They may be video-taped, observed, and/or timed on task performance. The task times and number of errors are then compared to goals originally identified by the design team. After task performance, users can indicate their reaction to the software on a rating scale ranging, for example, from 1 = extremely difficult to use to 9 = extremely easy to use. However, it is important to note that sometimes user satisfaction does not predict which software is best at supporting task performance (Andre & Wickens 1995; Bailey, 1993). Finally, in response to reports of the usability test, the interface design is modified until time, error rates, and subjective reaction are all within the acceptable limits set by the team and project managers.

When writing reports of usability testing, it is important that evaluators carefully articulate **what** was observed (e.g., user errors, confusions, recovery steps), as well as their inferences as to **why** such behavior was observed (e.g., in terms of psychological mechanisms, or violations of principles and guidelines). But the "what" and the "why" should be kept very distinct. An effective usability test need not, and perhaps should not contain prescriptions of how to fix usability problems, because, in the general cycle of testing, it should be up to a team of individuals to generate solutions.

Number of Users and Data Interpretation

A usability test is not a research experiment. Its purpose is to identify specific problems with the software design. In contrast, human factors research experiments evaluate theories and compare alternate designs to understand how people respond to technology (see Chapter 2). As a consequence of this difference in perspective, usability testing is less concerned with large sample sizes to provide adequate statistical power for hypothesis testing (i.e., the "condition A is better than condition B"), because identifying problems in a single system, is a qualitatively different process, than identifying significant differences between two systems.

Increasing the number of participants will identify more problems, but after 5–6 people, the benefit of using additional people diminishes. Developers are better off running a usability test with six people, making changes, and then running another test on the revised software than running a single usability test with another six participants. Each cycle of evaluation and redesign tends to enhance performance by approximately 50 percent (Landauer, 1995), so even a large number of evaluation and redesign cycles are cost effective. Figure 15.3 shows the cost/benefit ratio for different numbers of usability cycles (Landauer, 1995). Although the maximum cost/benefit ratio occurs with approximately five evaluation and redesign cycles for the type of system shown in Figure 15.3, even as many as 60 cycles produce a cost/benefit ratio greater than one. Another important consideration for usability tests is that the consumer of the resulting data is a programmer creating a product, not an academic creating a new theory, so it is helpful to have programmers observe the test directly. Programmers who observe usability tests invariably respond more favorably to suggested changes that arise from those tests. Seeing real people struggling with what the programmers might consider a simple, easy-to-use interface is often more effective than an elaborate report of the data collected during the usability test.

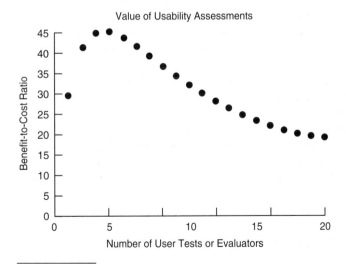

FIGURE 15.3

For a typical software system that will be used by 1,000 people, the cost/benefit ratio is highest with three to five evaluation-redesign cycles. (*Source:* From T. K. Landauer (1995). *The Trouble with Computers. Usefulness, Usability and Productivity.* Cambridge MA.: MIT Press.)

Pitfalls of Usability Testing

Usability testing is a necessary part of software development. Almost every feature of every Microsoft product now undergoes usability testing. However, usability testing is not sufficient to guarantee effective software. In fact, it may even hinder good design. Usability testing conducted with prototypes that are not grounded in an understanding of the users and their tasks, or prototypes that are developed without consideration of basic human factors design principles and theories, are likely to produce refinements of a prototype that are fundamentally flawed. Like rearranging the deck chairs on the Titanic, many iterations of usability testing may not address the underlying problems. Understanding the users and their tasks is required as a foundation for good design.

A second problem with usability testing is a fixation on the laboratory environment. Usability testing is often done in a dedicated laboratory setting that includes a one-way mirror, a specially instrumented computer, and video cameras. Like the experimental laboratory of the human factors researcher, the carefully controlled environment limits extraneous factors that could influence performance. This controlled environment is also artificial. For certain products, critical problems may surface only in more realistic environments. For example, usability problems with a personal digital assistant (PDA) may surface only when the PDA is taken outside and the glare of the sun on the screen makes it unusable. It is often a mistake to limit usability testing to the laboratory.

INFORMATION TECHNOLOGY

Although the computer was originally designed as a traditional computational device, one of its greatest emerging potentials is in information handling. Currently, computers can make vast amounts of information available to users far more efficiently than was previously possible. Large library databases can be accessed on a computer, eliminating trips to the library and long searches through card catalogs. As suggested in Chapter 7, diagnostic tests and troubleshooting steps can be called up in a few key presses instead of requiring the maintenance technician to page through large and cumbersome maintenance manuals containing out-of-date information or missing pages. Physicians can rapidly access the records of hundreds of cases of a rare disease to assess the success of various treatments. All of these uses require people to be able to interact with the computer in order to search for information. Supporting information search and retrieval is a critical emphasis in human factors and software interface design. We briefly consider some of the issues involved in designing information systems.

Hypertext, Hypermedia, and the Internet

Many computer-based documents now make use of computational power by linking one information "chunk" to another, changing the traditional linear text format to a nonlinear one. This technique of linking chunks of information is termed *hypertext*. The most common example of hypertext occurs when certain words in text are highlighted and the user clicks on the highlighted area to see additional information. Sometimes, clicking on the highlighted material brings up a pop-up screen, and other times it simply "moves" the user to a different part of the document.

Hypertext essentially stores all text and graphics as chunks of information, called nodes, in a network. The chunks are electronically linked in whatever manner the designer chooses. Although hypertext was originally designed for use on static text and graphics, it has now been extended to linking chunks of any information, including text, audio clips, video clips, animation, and so forth. Because this information is *multimedia,* the technique of linking these types of material is called *hypermedia.* Hypermedia is the basic principle behind a variety of information networks, such as the Internet.

The "browsing" metaphor for the Web was appropriate when it was primarily an information repository. Recently, the nature of the Internet has changed, particularly with the increase in Web-based applications. "Weblications"—a term coined by Bruce Tognazzini (Norman, 1998)—are software delivered as a service over the Web. Examples of weblications range from computer-based training and travel planning to Internet banking. Such services involve more complex interactions that are quite different than information retrieval. The browsing metaphor, therefore, is not well suited for weblications and often may undermine weblication usability. Although weblications have become an important part of the Internet, no effective guidelines for their design and implementation exist. Wroblewski and Rantanen (2001) argue that typical desktop application guidelines do not cover the spectrum of possibilities available on the Web. Likewise, Web usability guidelines do not address the new levels of interaction needed within weblications. In addition, Web guidelines may be inappropriate because a weblication user's motivation differs from a Web site users' goals.

Information Database Access

As computers become more technologically sophisticated, we are using them to access increasingly large and complex databases. However, while the computer holds the potential to allow users to access, search, and manipulate extremely large information databases, it is less apparent how computers should be designed to support these processes. This question represents an important aspect of HCI (Wickens & Seidler, 1995).

As with any other domain of human factors, a fundamental first step is to perform a task analysis (Chapter 3). What are the tasks, or user needs, in interacting with an information database? We can list at least four general needs that vary along the degree to which the user can specify the information needed from the database in advance:

1. The user knows a precise label for a piece of information that needs to be retrieved; for example, a telephone operator needs to retrieve the phone number of a person whose name is known.
2. The user knows some general characteristics of the desired item but can identify it positively only when he or she sees it; for example, you know the general topic of the particular book you want, but you can remember neither the specific author nor title.

3. The user wants to learn what exists in the database regarding a general topic but wishes initially to browse the topic, searching opportunistically for items that may be of interest and does not know in advance if those items are there. For example, you may be searching an accident database for particular cases and are not aware of the existence of what turns out to be a very relevant class of accidents until you encounter them in the database.

4. The user simply wants to understand the overall structure of the database: what cases exist, what cases do not, and how certain classes of cases relate to others. For example, the epidemiologist may wish to examine the occurrence of a disease over space and time to gain some understanding of its transmission.

While specific applications for document retrieval interfaces will require more elaborate task analysis than this simple categorization scheme (e.g., Belkin et al., 1993), the four types of search illustrate the important tradeoffs between different database interaction techniques. Across the four classes, there are different dialog interface methods that may be better suited for one than another, as we discuss next.

Mediated Retrieval. When the nature of information required can be precisely specified in advance, a command language or keyword search is often an adequate technique for directly retrieving the information. Much as one uses an index to look up a precise information in a book, one formulates a list of keywords to specify, as uniquely as possible, the attributes of the desired information (fact, book, passage, etc.). In designing interfaces for such direct retrieval systems, it is important for designers to label the index or keyword terms according to standard conventions within the domain, using semantic labels that users will be most likely to generate, rather than using their own intuition.

This principle may be somewhat difficult to carry out if one interface must be designed for multiple classes of users. People use an extremely diverse set of terms when looking for the same object; some estimates suggest that the chances of two people choosing the same term for a familiar object are less than 15 percent (Furnas et al., 1987). For example, if you are an engineering student, access to topics in human factors may be most likely through more familiar engineering terminology (e.g., displays, controls, manual control). However, for psychologists, similar topics may be more familiarly accessed through psychological terms like *perception, response,* or *tracking.* The key to good keyword access whether for electronic data bases or paper indexes is to provide *multiple routes* to access the same entities. Such a design may not produce the shortest, most parsimonious index or keyword list, but it will greatly reduce the frustration of users.

Even with well-designed multiple access routes, keyword searches are not always satisfactory from a user's point of view for two primary reasons. First, it is sometimes difficult for users to specify precisely the queries or combinations of keywords that identify their needs. In particular, people are not very good at using the Boolean logic that forms the backbone of most query systems (Mackinlay et al., 1995). An example of such a query might be, "All the information on people with incomes above a certain level, who live in a certain region, *and* incomes above a different level, who live in a different region and are female."

The second problem, somewhat related to the first, is that users are not always fully satisfied with the results of such keyword searches (Muckler, 1987). For example, in searching a library database, large numbers of documents that are irrelevant to the target search may be retrieved (false alarms) and large numbers of relevant documents may well remain unretrieved (misses). To compound this possibility, users may have more confidence in the exhaustiveness of the search than is warranted, as they have no way of assessing the rate of misses—they don't know what they don't know (Blair & Maron, 1985).

Intelligent Agents. Because people often have difficulties finding the information they want or need, a new approach under development is the concept of a computer-based helper to act as an interface agent between the user and the information database (Maes, 1994). These intelligent agents take input in the form of general needs and goals of the user. A detailed knowledge about the organization of the information database as well as access mechanisms allows the intelligent agent to search for the most relevant and useful information. The agent "goes to get" information, and then displays either the information itself (as in decision aids) or a list of the information it has obtained. Intelligent agent interfaces provide expert assistant to users, saving users the time and effort of doing their own search.

Spatially Organized Databases. An alternative approach to computer-mediated retrieval is to rely on a spatial representation of the information space to support search processes (e.g., Fowler et al., 1991). Good arguments can be made for dialog interfaces that support *navigation* or *travel* through the database or information space rather than direct retrieval. Because navigation is a spatially relevant term, we describe these as *spatially organized databases* (Lund, 1994). Such organization is logical, in part because the items in many databases bear analog *similarity relations* to each other; that is, certain items are more similar to each other than to others, perhaps because they share more common keywords or are more related to the same task, and hence are "closer" to each other. Spatial organization also makes sense because space and navigation are such natural metaphors, coming from interaction with the everyday environment (Hutchins et al., 1985).

Different kinds of spatially organized databases or information spaces have different ways of defining *proximity,* or "near" and "far" (Durding et al., 1977). For example, in a menu structure (Figure15.4a), proximity is typically defined by the lowest common ancestor in the hierarchy. Thus, in the figure, X is "closer" to Y than to Z. In a network information space, like a communications network defining "who talks to whom" on the Internet, proximity may be defined in terms of the number of links joining a pair of models (Figure 15.4b; see also Figure 10.8). In a matrix database (Figure 15.4c), like a spreadsheet, proximity may be defined simply in terms of the nearness of cells (rows and columns) to each other. Euclidean spaces, like those in Figure 15.4d, define distance or proximity

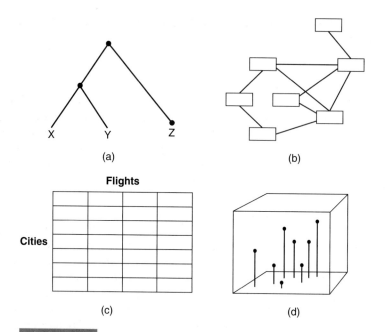

FIGURE 15.4

Four examples of spatially organized databases. (a) hierarchical (e.g., menu structure) (b) information network (c) spreadsheet (d) Euclidean space. In each of these, the concept of "distance" has an important spatial representation.

in terms that are directly equivalent to those used to describe natural 3-D space. Such spaces may often depict scientific data which itself is located at different Euclidean coordinates (e.g., a severe thunderstorm, geological strata, or the spread of pollution).

Spatially represented databases have both benefits and costs. The benefits are first that they generally position task-related elements close together and hence allow the user to consider these elements in a single glance or at least with little time spent traveling from one to the other (e.g., comparing two related entries to see which is more relevant or tagging a set of related entries for later retrieval). Hence, like the guidelines for good display layout discussed in Chapter 8, a spatially defined database can adhere to the layout principles of relatedness and sequence of use (Seidler & Wickens, 1992; Wickens & Baker, 1995). The second benefit is that such databases can allow the user to better understand the full structure of the database by examining a broad "map" of its elements. The design of this map can be a challenging exercise when the information space is large and multidimensional. It may well capitalize on familiar metaphors like rooms and walls (Robertson et al., 1993).

Third, users should be allowed an option to "recover" when they are lost, allowing them to backtrack to the previous item (Nielson, 1987), perhaps by making readily available the same pop-up option to the top of the menu (or any other well-known landmark within the space). Fourth, it is often useful to provide a historical record of where one has recently been within the space. Many systems offer "bookmarks" that can label entries the user may want to rapidly revisit (Bernstein, 1988).

Against these benefits, we can identify at least two potential costs, which careful design should strive to minimize:

1.   *Getting lost.* As databases become more complex and populated, getting lost in the information space, can be a problem. Several specific solutions can remedy this problem. First, the database should be spatially organized in a manner consistent with the user's mental model (Roske-Hofstrand & Paap, 1986; Seidler & Wickens, 1992). Second, users can be provided with an overall "map" of the space (Vicente & Williges, 1988; Beard & Walker, 1990) that is either continuously viewable on the screen or, if screen space is at a premium, can be rapidly and easily called up on a window.

2.   *Update rate.* Spatially organized databases are vulnerable to frustrating delays. The screen graphics of such systems are often very complex and the time required to redraw the depicted space as a user moves through it can be long enough to interfere with the perception of motion. Also, if a direct manipulation interface is provided to allow the user to "travel" through the space, the delays encountered in this tracking task can be devastating to control stability, as we learned in Chapter 9. The user can easily overshoot targets, turn too far, and so forth, thus defeating the very purpose of interactivity. What is important to realize about the update-rate problem is that users may quickly become frustrated with such a system and simply choose not to use it at all.

Virtual and Augmented Reality

It is but a short leap from our discussions of interactive spatial databases to those involved in virtual reality (VR) interfaces. The latter are certainly examples of HCI in which the computer is expected to be extremely powerful and render the user's experience as closely as possible to direct interaction with the natural world; that is, the computer interface should be entirely "transparent" to the user. We have discussed other issues of the human factors of virtual reality elsewhere in this book (see also Durlach & Mavor, 1995; Barfield & Furness, 1995; Sherman & Craig, 2003). Here we wish only to reemphasize two points made above: (1) Designers should ensure that the *task* for which a VR interface is designed is, in fact, one that is best suited for full immersion; (2) designers should be extremely sensitive to the negative effects of delayed updates. Such lags become progressively more disruptive (and likely) the more immersed is the environment, and the more richly the designer tries to create a visual reality. It is important to understand that simpler images, updated more frequently, are usually far more effective in an interactive VR environment than are complex images, updated with longer lags (Wickens & Baker, 1995).

Affective Computing

Considering the emotional or affective elements of software design may become increasingly important as computers become more complex and ubiquitous. Many studies show that humans respond socially to technology and react to computers similarly to how they might respond to human collaborators (Reeves & Nass, 1996). For example, the similarity attraction hypothesis in social psychology predicts people with similar personality characteristics will be attracted to each other. This finding also predicts user acceptance of software (Nass et al., 1995). Software that displays personality characteristics similar to that of the user tend to be more readily accepted. Similarly, the concept of affective computing suggests that computers that can sense and respond to user's emotional states may be more readily accepted by users (Picard, 1997). One potential outcome of affective computing is that future computers will sense your emotional state and change the way they respond when they sense you are becoming frustrated. Considering affect in system design is not just about designing for pleasure. Designers should consider how to create unpleasant emotional responses to signal dangerous situations (Liu, in press).

Emotion is important to making appropriate decisions and not just in reducing frustration and increasing the pleasure of computer users. Norman, Ortony, and Russell (2003) argue that affect complements cognition in guiding effective decisions. A specific example is the role of trust in Internet-based interactions. People who do not trust an Internet service are unlikely to purchase items or provide personal information. In many cases, trust depends on surface features of the interface that have no obvious link to the true capabilities (Tseng & Fogg, 1999). Credibility depends heavily on "real-world feel," which is defined by factors such as speed of response, listing a physical address, and including photos of the organization. Visual design factors of the interface, such as cool colors and a balanced layout, can also induce trust (Kim & Moon, 1998). Similarly, trusted Web sites tend to be text-based, use empty space as a structural element, have strictly structured grouping, and use real photographs (Karvonen & Parkkinen, 2001). These results show that trust tends to increase when information is displayed in a way that provides concrete details in a way that is consistent and clearly organized. Chapter 16 describes the role of trust in guiding reliance on automation.

Information Appliances

Technological advances continue to make computers increasingly powerful, portable, and inexpensive. This trend will move computers from the desktop to many other places in our lives. Rather than multifunction devices tied to a desk, computers of the future may be more specific information appliances that serve specific function (Norman, 1998). Rather than using a computer to locate weather or traffic information on the Internet, this information might be shown and continuously updated on a display hung on the wall by your front door. Already, this trend has introduced important new challenges to software design. One challenge is the reduced screen size and need for alternate interaction devices. Specifically, cellular telephones and PDAs cannot rely on the typical keyboard and mouse. Designers must develop creative ways of displaying complex graphics on small screens. Increasingly, computers are embedded in common devices such as cars. These computers combine wireless connections and Global Positioning System (GPS) data to provide drivers a wide array of functions such as route guidance and congestion information (Lee, 1997). With desktop computers, users are typically focused on one task at a time. In contrast, users of in-car computers must also control their car. Poor software design that might only frustrate a desktop user might kill a driver. As computers go beyond the desktop and become information appliances, software designers must consider human factors design practices developed in other safety critical applications such as aviation. Chapter 16 and 17 discuss some of the challenges that emerge when computers move beyond the desktop.

CONCLUSION

Creating effective and satisfying software requires a process that begins by developing an understanding of the user as described in the front-end analysis techniques, such as task analysis, in Chapter 3. Software design must build on this understanding with theories, principles, and design knowledge from the fields of HCI and human factors. Usability testing and multiple design iterations are required, even for designs with good theoretical justification, because users respond to new technology in unpredictable ways. Designers must accept the concept that the purpose of the software is to support the user in some task, not to provide all kinds of great features that are fun, interesting, handy, useful once in a lifetime, or that might be used by 1 percent of the population. User-centered design requires a concerted effort to make the software fit the user, not count on the user adapting to the software. Having users highly involved or even on the design team can make it easier to stay focused on the true purpose of the project.

Computers are being used to support an ever-widening array of tasks, including complex cognitive functioning (see Chapter 7), group work such as problem solving and decision making (see Chapter 7 and 19), scientific visualization, database management, and so on. One just has to look at the growing use of the Internet and the emergence of information appliances to understand the complexities and challenges we face in designing software. It is important for human factors specialists to design and evaluate the system so that it works effectively for the user in the sense of "deep" task support, as well as usability at the superficial interface level. This requires full evaluation of cognitive, physical, and social functioning of users and their environment during task performance (see Chapter 19).

Chapter **16**

Automation

The pilots of the commercial airlines transport were flying high over the Pacific, allowing their autopilot to direct the aircraft on the long, routine flight. Gradually, one of the engines began to lose power, causing the plane to tend to veer toward the right. As it did, however, the autopilot appropriately steered the plane back to the left, thereby continuing to direct a straight flightpath. Eventually, as the engine continued to lose power, the autopilot could no longer apply the necessary countercorrection. As in a tug-of-war when one side finally loses its resistance and is rapidly pulled across the line, so the autopilot eventually "failed." The plane suddenly rolled, dipped, and lost its airworthiness, falling over 30,000 feet out of the sky before the pilots finally regained control just a few thousand heart-stopping feet above the ocean (National Transportation Safety Board, 1986; Billings, 1996). Why did this happen? In analyzing this incident, investigators concluded that the autopilot had so perfectly handled its chores during the long routine flights that the flight crew had been lulled into a sense of complacency, not monitoring and supervising its operations as closely as they should have. Had they done so, they would have noted early on the gradual loss of engine power (and the resulting need for greater autopilot compensation), an event they clearly would have detected had they been steering the plane themselves.

Automation characterizes the circumstances when a machine (nowadays often a computer) assumes a task that is otherwise performed by the human operator. As the aircraft example illustrates, automation is somewhat of a mixed blessing and hence is characterized by a number of ironies (Bainbridge, 1983). When it works well, it usually works *very* well indeed—so well that we sometimes trust it more than we should. Yet on the rare occasions when it does fail, those failures may often be more catastrophic, less forgiving, or at least more frustrating than would have been the corresponding failures of a human in the same circumstance. Sometimes, of course, these failures are relatively trivial and benign—like my copier, which keeps insisting that I have placed the book in an orientation that I do not want (when that's exactly what I *do* want). At other times, however, as with the aircraft incident and a host of recent aircraft crashes that have been attributed to automation problems, the consequences are severe (Billings, 1996; Dornheim, 1995; Sarter & Woods, 2000).

If the serious consequences of automation resulted merely from failures of software or hardware components, then this would not be a topic in the study of human factors. However, the system problems with automation are distinctly and inexorably linked to human issues of attention, perception and cognition in *managing* the automated system in its normally operating state, when the system that the automation is serving has failed or has been disrupted, or when the automated component itself has failed (Parasuraman & Riley, 1997). The performance of most automation depends on the interaction of people with the technology. Before addressing these problems, we first consider why we automate and describe some of the different kinds of automation. After discussing the various human-performance problems with automation and suggesting their solution, we discuss automation issues in industrial process control and manufacturing, as well as an emerging area of agent-based automation and hortatory control.

WHY AUTOMATE?

The reason designers develop machines to replace or aid human performance are varied but can be roughly placed into four categories.

1. *Impossible or hazardous.* Some processes are automated because it is either *dangerous or impossible* for humans to perform the equivalent tasks. In Chapter 9, we learned that teleoperation, or robotic handling of hazardous material (or material in hazardous environments), is a clear example. Also, there are many circumstances in which automation can serve the particular needs of special populations whose disabilities may leave them unable to carry out certain skills without assistance. Examples include automatic guidance systems for the quadriplegic or automatic readers for the visually impaired. In many situations, automation enables people to do what would otherwise be impossible.

2. *Difficult or unpleasant.* Other processes, while not impossible, may be *very challenging* for the unaided human operator, such that humans carry out the functions poorly. (Of course, the border between "impossible" in category 1 and "difficult" is somewhat fuzzy). For example, a calculator "automatically" multiplies digits that can be multiplied in the head. But the latter is generally more effortful and error producing. Robotic assembly cells automate highly repetitive and fatiguing human operations. Workers can do these things but often at a cost to fatigue, morale, and sometimes safety. Autopilots on aircraft provide more precise flight control and can also unburden the fatiguing task of continuous control over long-haul flights. Chapter 7 describes expert systems that can replace humans in routine situations where it is important to generate very consistent decisions. As another example, we learned in Chapter 4 and 13 that humans are not very good at *vigilant monitoring*. Hence, automation is effective in monitoring for relatively rare events, and the general class of warning and alert systems, like the "idiot light" that appears when your oil pressure or fuel level is low in the car. Of course, sometimes automation can impose more vigilant monitoring tasks on the human, as we saw in the airplane incident (Parasuraman, 1987). This is one of the many "ironies of automation" (Bainbridge, 1983). Ideally, automation makes difficult and unpleasant tasks easier.

3. *Extend human capability.* Sometimes automated functions may not replace but may simply *aid humans* in doing things in otherwise difficult circumstances. For example, we saw in Chapter 6 that human working memory is vulnerable to forgetting. Automated aids that can supplement memory are useful. Consider an automated telephone operator that can directly print the desired phone number on a small display on your telephone or directly dial it for you (with a $.17 service charge). The decision aids discussed in Chapter 7 and the predictive displays discussed in Chapters 6 and 8, are examples of automation that relieve the human operator of some cognitively demanding mental operations. Automated planning aids have a similar status (Layton et al., 1994). Automation is particularly useful in extending human's *multitasking* capabilities. For example, pilots report that autopilots can be quite useful in temporarily relieving them from duties of aircraft control when other tasks demands temporarily make their workload extremely high. In many situations automation should extend rather than replace the human role in a system.

4. *Technically possible.* Finally, sometimes functions are automated *simply because the technology is there* and inexpensive, even though it may provide little or no value to the human user. Many of us have gone through painfully long negotiations with automated "phone menus" to get answers that would have taken us only a few seconds with a human operator on the other end of the line. But it is probable that the company has found that a computer operator is quite a bit cheaper. Many household appliances and vehicles have a number of automated features that provide only minimal advantages that may even present costs and, because of their increased complexity and dependence on electrical power, are considerably more vulnerable to failure than are the manually operated systems they replaced. It is unfortunate when the purported "technological sophistication" of these features are marketed, because they often have no real usability advantages. Automation should focus on supporting system performance and humans' tasks rather than showcasing technical sophistication.

STAGES AND LEVELS OF AUTOMATION

One way of representing what automation does is in terms of the *stages* of human information processing that automation replaces (or augments), and the amount of cognitive or motor *work* that automation replaces, which we define by the *level* of automation. A taxonomy of automation offered by Parasuraman et al (2000) defines 4 stages, with different levels within each stage.

1. *Information acquisition, selection, and filtering.* Automation replaces many of the cognitive processes of human selective attention, discussed in Chapter 6. Examples include warning systems and alerts that guide attention to inspect parts of the environment that automation deems to be worthy of further scrutiny (Woods, 1995). Automatic highlighting tools, such as the spell-checker that redlines my misspelled words, is another example of attention-directing automation. So also are automatic target-cueing devices (Dzindolet et al., 2002; Yeh & Wickens, 2001b). Finally, more "aggressive." examples of stage 1 automation may filter or delete altogether information assumed to be unworthy of operator attention.

2. *Information integration.* Automation replaces (or assists) many of the cognitive processes of perception and working memory, described in Chapters 6 and 7, in order to provide the operator with a situation assessment, inference, diagnosis, or easy-to-interpret "picture" of the task-relevant information. Examples at lower levels may configure visual graphics in a way that makes perceptual data easier to integrate (Chapter 8). Examples at higher levels are automatic pattern recognizers, predictor displays, diagnostic expert systems (Chapter 7). Many intelligent warning systems (Pritchett, 2002) that guide attention (stage 1) also include sophisticated integration logic necessary to infer the existence of a problem or dangerous condition (Mosier et al., 1998; stage 2).

3. *Action selection and choice.* As described in Chapter 7, diagnosis is quite distinct from choice, and in Chapter 4, sensitivity is quite different from the response criterion. In both cases, the latter entity explicitly considers the *value* of potential outcomes. In the same manner, automated aids that diagnose a situation at stage 2 are quite distinct from those that recommend a particular course of action. In doing the latter, the automated agent must explicitly or implicitly assume a certain set of values for the operator who depends on its advice. An example of stage 3 automation is the airborne traffic alert and collision avoidance system (TCAS), which explicitly (and strongly) advises the pilot of a vertical maneuver to take in order to avoid colliding with another aircraft. In this case, the values are shared between pilot and automation (avoid collision); but there are other circumstances where value sharing might not be so obvious, as when an automation medical decision aid recommends one form of treatment over another for a terminally ill patient.

4. *Control and action execution.* Automation may replace different levels of the human's action or control functions. As we learned in Chapter 9, control usually depends on the perception of desired input information, and therefore control automation also includes the automation of certain perceptual functions. (These functions usually involve *sensing* position and trend rather than *categorizing* information). Autopilots in aircraft, cruise control in driving, and robots in industrial processing are examples of control automation. More

mundane examples of stage-4 automation include electric can openers and automatic car windows.

We noted that levels of automation characterized the amount of "work" done by the automation (and therefore, workload relieved from the human). It turns out that it is at stages 3 and 4, where the levels of automation take on critical importance. Table 16.1, adapted from Sheridan (2002), summarizes eight levels of automation that apply particularly to stage-3 and stage-4 automation characterizing the relative distribution of authority between human and automation in choosing a course of action.

The importance of both stages and levels emerges under circumstances when automation may be *imperfect* or *unreliable*. Here automation at different stages and levels may have different costs to human and system performance, issues we address in the following section.

PROBLEMS IN AUTOMATION

Whatever the reason for choosing automation, and no matter which kind of function (or combination of human functions) are being "replaced," the history of human interaction with such systems has revealed certain shortcomings (Sheridan, 2002, Parasuraman & Riley, 1997; Billings, 1996). In discussing these shortcomings, however, it is important to stress that they must be balanced against the number of very real *benefits* of automation. There is little doubt that the ground proximity warning system in aircraft, for example, has helped save many lives by alerting pilots to possible crashes they might otherwise have failed to note (Diehl, 1991). Autopilots have contributed substantially to fuel savings; robots have allowed workers to be removed from unsafe and hazardous jobs; and computers have radically improved the efficiency of

---

TABLE 16.1   Levels of Automation Ranging from Complete Manual Control to Complete Automatic Control

1. Automation offers no aid; **human in complete control.**
2. Automation **suggests multiple alternatives;** filters and highlights what it considers to be the best alternatives.
3. Automation **selects an alternative,** one set of information, or a way to do the task and suggests it to the person.
4. Automation **carries out the action if the person approves.**
5. Automation provides the person with **limited time to veto the action** before it carries out the action.
6. Automation **carries out an action and then informs** the person.
7. Automation carries out an action and **informs the person only if asked.**
8. Automation selects method, executes task, and **ignores the human** (i.e., the human has no veto power and is not informed).

(Adapted from Sheridan (2002))

many human communications, computations, and information-retrieval processes. Still, there is room for improvement, and the direction of those improvements can be best formulated by understanding the nature of the remaining or emerging problems that result when humans interact with automated systems.

Automation Reliability

To the extent that automation can be said to be reliable, it does what the human operator expects it to do. Cruise control holds the car at a set speed, a copier faithfully reproduces the number of pages requested, and so forth. However, what is important for human interaction is not the reliability per se but the *perceived reliability*. There are at least four reasons why automation may be perceived as unreliable.

First, it *may be unreliable*: A component may fail or may contain design flaws. In this regard, it is noteworthy that automated systems typically are more complex and have more components than their manually operated counterparts and therefore contain more components that could go wrong at any given time (Leveson, 1995), as well as working components that are incorrectly signaled to have failed. The nature of these "alarm false alarms" in warning systems was addressed in Chapter 5.

Second, there may be *certain situations* in which the automation is not designed to operate or may not perform well. All automation has a limited operating range within which designers assume it will be used. Using automation for purposes not anticipated by designers leads to lower reliability. For example, cruise control is designed to maintain a constant speed on a level highway. It does not use the brakes to slow the car, so cruise control will fail to maintain the set speed when traveling down a steep hill.

Third, the human operator may *incorrectly "set up"* the automation. Nurses sometimes make errors when they program systems that allow patients to administer periodic doses of painkillers intravenously. If the nurses enter the wrong drug concentration, the system will faithfully do what it was told to do and give the patient an overdose (Lin et al., 2001). Thus, automation is often described as "dumb and dutiful."

Fourth, there are circumstances when the automated system does exactly what it is supposed to do, but the logic behind the system is sufficiently complex and poorly understood by the human operator—a poor mental model—that it *appears to be acting erroneously* to the operator. Sarter and Woods (2000; Sarter et al., 1997;) observed that these *automation induced surprises* appear relatively frequently with the complex flight management systems in modern aircraft. The automation triggers certain actions, like an abrupt change in air speed or altitude, for reasons that may not be readily apparent to the pilot. If pilots perceive these events to be failures and try to intervene inappropriately, disaster can result (Strauch, 1997; Dornheim, 1995).

The term *unreliable automation* has a certain negative connotation. However, it is important to realize that automation is often asked to do tasks, such as weather forecasting or prediction of aircraft trajectory or enemy intent, that are simply impossible to do perfectly given the uncertain nature of the dynamic world in which we exist. (Wickens et al., 2000). Hence, it may be better to label such automation as "imperfect" rather than "unreliable." To the extent that such imperfections are well known and understood by the operator, even automation as low as 70 percent reliable can still be of value, particularly under high workload situations (Lee & See, 2002; Wickens & Xu, 2003). The value that can be realized from imperfect automation relates directly to the concept of trust.

Trust: Calibration and Mistrust

The concept of *perceived* automation reliability is critical to understanding the human performance issues because of the relation between reliability and *trust*. As we know, trust in another human is related to the extent to which we can believe that he or she will carry out actions that are expected. Trust has a similar function in a human's belief in the actions of an automated component (Muir, 1987; Lee & Moray, 1992). Ideally, when dealing with any entity, whether a friend, a salesperson, a witness in a court proceeding, or an automated device, trust should be well *calibrated*. This means *our trust in the agent, whether human or computer, should be in direct proportion to its reliability.* Mistrust occurs when trust is not directly related to reliability. As reliability decreases, our trust should go down, and we should be prepared to act ourselves and be receptive to sources of advice or information other than those provided by the unreliable agent.

While this relation between reliability, trust and human cognition holds true to some extent (Kantowitz et al., 1997; Lee & Moray, 1992; Lewandowsky et al., 2000), there is also some evidence that human trust in automation is not entirely well calibrated: Sometimes it is too low (distrust), sometimes too high (overtrust) (Parasuraman & Riley, 1997). *Distrust is a type of mistrust where the person fails to trust the automation as much as is appropriate.* For example, in some circumstances humans prefer manual control to automatic control of a computer, even when both are performing at precisely the same level of accuracy (Liu et al., 1993). A similar effect is seen with automation that enhances perception, where people are biased to rely on themselves rather than the automation (Dzindolet et al., 2002). Distrust of alarm systems with high false alarm rates is a common syndrome across many applications as discussed in Chapter 5. Distrust in automation may also result from a failure to understand the nature of the automated algorithms that function to produce an output, whether that output is a perceptual categorization, diagnostic recommendation, a decision, or a controlled action. This can be a particularly important problem for decision-making aids, as described in Chapter 7.

The consequences of distrust are not necessarily severe, but they may lead to inefficiency when distrust leads people to reject the good assistance that automation can offer. For example, a pilot who mistrusts a flight management system and prefers to fly the plane by hand may become more fatigued and may fly routes that are less efficient in terms of fuel economy. Many times "doing things by hand" rather than, say, using a computer can lead to slower performance that may be less accurate, when the computer-based automation is of high reliability.

As noted in Chapter 5, distrust of faulty automated warning systems can lead to the real danger of ignoring legitimate alarms (Sorkin, 1988).

Overtrust and Complacency

In contrast to distrust, *overtrust* of automation, sometimes referred to as *complacency, occurs when people trust the automation more than is warranted* and can have severe negative consequences if the automation is less than fully reliable (Parasuraman et al., 1993; Parasuraman & Riley, 1997). We saw at the beginning of the chapter the incident involving the airline pilot who trusted his automation too much, became complacent in monitoring its activity, and nearly met disaster. The cause of complacency is probably an inevitable consequence of the human tendency to let *experience* guide our expectancies, as discussed in Chapters 4, 5, and 6. Most automated systems *are* quite reliable. (They would not last long in the marketplace if they were not.) It is likely that many people using a particular system may not ever encounter failures, and hence their perception of the reliability of the automation is that it is perfect (rather than the high, but still less than 100 percent, that characterize all operations of the system in question). Perceiving the device to be of perfect reliability, a natural tendency would be for the operator to cease monitoring its operation or at to least monitor it far less vigilantly than is appropriate (Bainbridge, 1983; Moray, 2003). This situation is exacerbated by the fact that, as we learned in Chapter 3, people make pretty poor monitors in the first place, when they are doing nothing but monitoring (Parasuraman, 1986; Warm et al., 1996).

Of course, the real problem with complacency, the failure to monitor adequately, only surfaces in the infrequent circumstances when something does fail (or is perceived to fail) and the human must (or feels a need to) intervene. Automation then has three distinct implications for human intervention related to detection, situation awareness, and skill loss.

1. *Detection.* The complacent operator will likely be slower to detect a real failure (Parasuraman et al., 1994; Parasuraman et al., 1992). As noted in chapters 4 and 13, detection in circumstances in which events are rare (the automation is reliable) is generally poor, since this imposes a vigilance monitoring task. Indeed, the more reliable the automation, the rarer the "signal events" become, and the poorer is their detection (Parasuraman et al., 1996).

2. *Situation awareness.* People are better aware of the dynamic state of processes in which they are active participants, selecting and executing its actions, than when they are passive monitors of someone (or something) else carrying out those processes, a phenomenon known as the generation effect (Slamecka & Graf, 1978; Endsley & Kiris, 1995; Hopkin & Wise, 1996). Hence, independent of their ability to detect a failure in an automated system, they are less likely to intervene correctly and appropriately if they are out of the loop and do not fully understand the system's momentary state (Sarter & Woods, 2000). With cruise control, the driver may remove her foot from the accelerator and become less aware of how the accelerator pedal moves to maintain a constant speed. Thus, she may be slower to put her foot on the brake when the car begins to accelerate down a hill. The issue of situation awareness can be particularly problematic if the system is designed with poor feedback regarding the ongoing state of the automated process.

3.   *Skill loss.* A final implication of being out of the loop has less to do with failure response than with the long-term consequences. Wiener (1988) described *deskilling* as the gradual loss of skills an operator may experience by virtue of not having been an active perceiver, decision maker, or controller during the time that automation assumed responsibility for the task. Such a forgetting of skill may have two implications. First, it may make the operator less self-confident in his or her own performance and hence *more* likely to continue to use automation (Lee & Moray, 1994). Second, it may degrade still more the operator's ability to intervene appropriately should the system fail. Imagine your calculator failing in the middle of a math or engineering exam, when you have not done unaided arithmetic for several years. The relation between trust and these features of automation is shown in Figure 16.1.

Another irony of automation is that the circumstances in which some automated devices fail are the same circumstances that are most challenging to human: automation tends to fail when it is most needed by the human operator. Such was the case with the failed engine in our opening story. These circumstances may also occur with decision aids that are programmed to handle ordinary problems but must "throw up their hands" at very complex ones. It is, of course, in these very circumstances that the automated system may hand off the problem to its human counterpart (Hopkin & Wise, 1996). But now, the human,

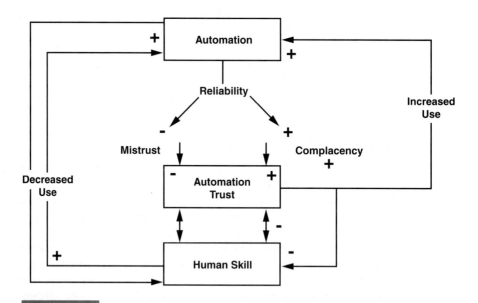

FIGURE 16.1

Elements of automation reliability and human trust. The + and − indicate the direction of effects. For example, increased (+) automation reliability leads to increased (+) trust in automation, which in turn leads to increased (+) use and a decrease (−) in human skill.

who is out of the loop and may have lost skill, will be suddenly asked to handle the most difficult, challenging problems, hardly a fair request for one who may have been complacent and whose skill might have been degraded.

Workload and Situation Awareness

Automation is often introduced with the goal of reducing operator workload (see Chapter 13). For example, an automated device for lane keeping or headway maintenance in driving may be assumed to reduce driving workload (Hancock et al., 1996; Walker et al., 2001; see Chapter 17) and hence allow mental resources to drive more safely. However, in practice, sometimes the workload is reduced by automation in environments when workload is already too low and *loss of arousal* rather than high workload is the most important problem (e.g., driving at night). In fact, as we saw in Chapter 13 it is probably incorrect to think that vigilance tasks are low workload at all if attention is adequately allocated to them so that event detection will be timely (Warm et al., 1996). In addition, sometimes the reduced workload achieved via automation can also directly lead to a loss in situation awareness, as the operator is not actively involved in choosing the actions recommended or executed by the automation. There is a correlation between situation awareness and workload; as automation level moves up the scale in Table 16.1 both workload and situation awareness tends to go down (Endsley and Kiris, 1995).

Sometimes automation has the undesirable effect of both reducing workload during already low-workload periods and increasing it during high-workload periods. This problem of *clumsy automation* is where automation *makes easy tasks easier and hard tasks harder*. For example, a flight management system tends to make the low-workload phases of flight, such as straight and level flight or a routine climb, easier, but it tends to make high-workload phases, such as the maneuvers in preparation for landing, more difficult, as pilots have to share their time between landing procedures, communication, and programming the flight management system.

Training and Certification

Errors can occur when people lack the training to understand the automation. As increasingly sophisticated automation eliminates many physical tasks, complex tasks may appear to become easy, leading to less emphasis on training. On ships, the misunderstanding of new radar and collision avoidance systems has contributed to accidents (NTSB, 1990). One contribution to these accidents is training and certification that fails to reflect the demands of the automation. An analysis of the exam used the by the U.S. Coast Guard to certify radar operators indicated that 75 percent of the items assess skills that have been automated and are not required by the new technology (Lee & Sanquist, 2000). Paradoxically, the new technology makes it possible to monitor a greater number of ships, enhancing the need for interpretive skills such as understanding the rules of the road and the automation. These very skills are underrepresented on the test. Further, the knowledge and skills may degrade because they are used only in rare but *critical* instances. Automation design should carefully assess the effect of automation on the training and certification requirements (Lee & Sanquist, 2000). Chapter 18 describes how to identify and provide appropriate training.

Loss of Human Cooperation

In nonautomated, multiperson systems, there are many circumstances in which subtle communications, achieved by nonverbal means or voice inflection, provide valuable sources of information (Chapter 19; Bowers et al., 1996). The air traffic controller can often tell if a pilot is in trouble by the sound of the voice, for example. Sometimes automation may *eliminate valuable information channels*. For example, in the digital datalink system (Kerns, 1999), which is proposed to replace air-to-ground radio communications with digital messages that are typed in and appear on a display panel, such information will be gone. Furthermore, there may be circumstances in which negotiation between humans, necessary to solve nonroutine problems, may be eliminated by automation. Many of us have undoubtedly been frustrated when trying to interact with an uncaring, automated phone menu in order to get a question answered that was not foreseen by those who developed the automated logic.

Job Satisfaction

We have primarily addressed performance problems associated with automated systems, but the issue of job satisfaction goes well beyond performance (and beyond the scope of this book) to consider the morale implications of the worker who is replaced by automation. In reconsidering the reasons to automate, we can imagine that automation that improves safety or unburdens the human operator will be well received. But automation introduced merely because the technology is available or that increases job efficiency may not necessarily be well received. Many operators are highly skilled and proud of their craft. Replacement by robot or computer will not be well received. If the unhappy, demoralized operator then is asked to remain in a potential position of resuming control, should automation fail, an unpleasant situation could result.

FUNCTION ALLOCATION BETWEEN THE PERSON AND AUTOMATION

How can automation be designed to avoid these problems? One approach is a systematic allocation of functions to the human and to the automation based on the relative capabilities of each. We can allocate functions depending on whether the automation or the human generally performs a function better. This process begins with a task and function analysis, described in Chapter 3. Functions are then considered in terms of the demands they place on the human and automation. A list of human and automation capabilities guides the decision to automate each function. Table 16.2 lists the relative capabilities originally developed by Fitts (1951) and adapted from Sheridan (2002) and Fuld (2000).

As an example, an important function in maritime navigation involves tracking the position and velocity of surrounding ships using the radar signals.

TABLE 16.2 "Fitts's List" Showing the Relative Benefits of Automation and Humans

Humans are better at	*Automation is better at*
Detecting small amounts of visual, auditory, or chemical signals (e.g., evaluating wine or perfume)	Monitoring processes (e.g., warnings)
Detecting a wide range of stimuli (e.g., integrating visual, auditory, and olfactory cues in cooking)	Detecting signals beyond human capability (e.g., measuring high temperatures, sensing infrared light and x-rays)
Perceiving patterns and making generalizations (e.g., "seeing the big picture")	Ignoring extraneous factors (e.g., a calculator doesn't get nervous during an exam)
Detecting signals in high levels of background noise (e.g., detecting a ship on a cluttered radar display)	Responding quickly and applying great force smoothly and precisely (e.g., autopilots, automatic torque application)
Improvising and using flexible procedures (e.g., engineering problem solving, such as on the Apollo 13 moon mission as described in Chapter 7)	Repeating the same procedure in precisely the same manner many times (e.g., robots on assembly lines)
Storing information for long periods and recalling appropriate parts (e.g., recognizing a friend after many years)	Storing large amounts of information briefly and erasing it completely (e.g., updating predictions in a dynamic environment)
Reasoning inductively (e.g., extracting meaningful relationships from data)	Reasoning deductively (e.g., analyzing probable causes from fault trees)
Exercising judgment (e.g., choosing between a job and graduate school)	Performing many complex operations at once (e.g., data integration for complex displays, such as in vessel tracking)

This "vessel tracking" function involves many complex operations to determine the relative velocity and location of the ships and to estimate their future locations. According to Table 16.2, automation is better at performing many complex operations at once, so this function might best be allocated to automation. In contrast, the course selection function involves considerable judgment regarding how to interpret the rules of the road. According to Table 16.2, humans are better at exercising judgment, so this task should be allocated to the human. A similar analysis of each navigation function can be done to identify whether it should be automated or left to the person.

Applying the information in Table 16.2 to determine an appropriate allocation of function is a starting point rather than a simple procedure that can completely guide a design. One reason is that there are many *interconnections between functions*. In the maritime navigation example, the function of vessel tracking interacts with the function of course selection. Course selection involves substantial judgment, so Table 16.2 suggests that it should not be automated, but the mariner's ability to choose an appropriate course depends on the vessel-tracking function, which is performed by the automation. Although vessel tracking and course selection can be described as separate functions, the automation must be designed to support them as an integrated whole. In general, you should not fractionate functions between human and automation but strive to give the human a coherent job.

Any cookbook approach that uses comparisons like those in Table 16.2 will be only partially successful at best; however, Table 16.2 contains some general considerations that can improve design. Human memory tends to organize large amounts of related information in a network of associations that can support effective judgments requiring the consideration of many factors. People tend to be effective with complete patterns and less effective with details. For these reasons it is important to leave the "big picture" to the human and the details to the automation (Sheridan, 2002).

HUMAN-CENTERED AUTOMATION

Perhaps the most important limit of the function allocation approach is that the *design of automation is not an either/or decision* between the automation or the human. It is often more productive to think of how automation can support and complement the human in adapting to the demands of the system. Ideally, the automation design should focus on creating a human–automation partnership by incorporating the principles of *human-centered automation* (Billings, 1996). Of course, human-centered automation might mean keeping the human more closely in touch with the process being automated; giving the human more authority over the automation; choosing a level of human involvement that leads to the best performance; or enhancing the worker's satisfaction with the workplace. In fact, all of these characteristics are important human factors considerations, despite that they may not always be totally compatible with each other. We present our own list of six human-centered automation features that we believe will achieve the goal of maximum harmony between human, system, and automation.

1. *Keeping the human informed.* However much authority automation assumes in a task, it is important for the operator to be informed of what the automation is doing and why, via good displays. As noted, humans should have the "big picture." What better way is there to provide this than through a well-designed display? As a positive example, the pilot should be able to see the amount of thrust delivered by an engine as well as the amount of compensation that the autopilot might have to make to keep the plane flying straight. A negative example is a small feature that contributed to the catastrophe at the Three Mile Island nuclear power plant (Rubinstein & Mason, 1979). Among the myriad displays, one in particular signaled to the crew that an automatic valve had closed. In fact, the display only reflected that the valve had received a signal to close; the valve had become stuck and did *not* actually close, hence continuing to pass liquid and drain coolant from the reactor core. Because the operator only saw the signal "closed" and was not informed of the processes underlying the au-

tomation, the status of the plant coolant was misdiagnosed, and the radioactive core was eventually exposed.

Of course, merely presenting information is not sufficient to guarantee that it will be understood. Coherent and integrated displays as discussed in Chapter 8 are also necessary to attain that goal.

2. *Keeping the human trained.* Automation can make complex tasks seem simple when manual interactions are automated. At the same time, automation often changes the task so that operators must perform more abstract reasoning and judgment in addition to understanding the limits and capabilities of the automation (Zuboff, 1988). These factors strongly argue that *training for the automation-related demands* is needed so that the operator uses the automation appropriately and benefits from its potential (Lee & Sanquist, 2000). The operator should have lots of training in exploring the automation's various functions and features in an interactive fashion (Sherry & Polson, 1999). In addition, as long as any automated system might conceivably fail or require rapid human intervention, it is essential that the human's skill in carrying out the otherwise automated function be maintained at as high a level as possible to avoid the problems of skill loss described above.

3. *Keeping the operator in the loop.* This is one of the most challenging goals of human-centered automation. How does one keep the operator sufficiently in the control loop so that awareness of the automated state is maintained without reverting fully to manual control so that the valuable aspects of automation (e.g., to reduce workload when needed) are defeated? Endsley and Kiris (1995) compared different levels of automation (similar to those shown in Table 16.1) in an automated vehicle navigational task, the authors found that the highest levels of automation degraded the drivers' situation awareness and their ability to jump back into the control loop if the system failed. There was, however, some evidence that performance was equivalent across the situations of moderate level of automation; that is, as long as the human *maintained some involvement* in decision making regarding whether to accept the automation suggestions (by vetoing unacceptable solutions at level 5), then adequate levels of situation awareness were maintained even as workload was reduced. Stated in other terms, the tradeoff of situation awareness and workload is not inevitable.

4. *Selecting appropriate stages and levels when automation is imperfect.* Designers may often have to choose the stage and level of automation to incorporate into a system. For example, should the decision aid designed for a physician be one that highlights important symptoms (stage 1), makes a diagnosis through an expert system (stage 2), recommends a form of treatment (stage 3), or actually carries out the treatment (stage 4). An emerging pattern of data suggests that, to the extent that automation *is* imperfect, the negative consequences of late stage automation imperfection (3 and 4) are more harmful than early stage imperfection (Sarter & Schroeder, 2001; Lee et al., 1999).

Such findings have three explanations. First, the lower stages (and levels) force the operator to stay more in the loop, making active choices (and increasing situation awareness as a result). Second, at the higher stages, action may be more or less completed by the time the error has been realized, and therefore it will be harder to reverse its consequences. (Consider the drug infuser pump example.) Third, stage-3 automation explicitly considers values, whereas earlier stages do not. This adds another aspect in which the automation at high levels of stage 3-automation may "fail" by using values in its decision that are different from those of the human user.

In implementing the recommendation for levels and stages in automation for high-risk decision aiding, it is important to realize the tempering effect of time pressure. There is no doubt that if a decision must be made in a time-critical situation, later stages of automation (choice recommendation or execution) can usually be done faster by automation than by human operators. Hence the need for time-critical responses may temper the desirability for low levels of stage 3 automation.

5. *Making the automation flexible and adaptive.* A conclusion that can be clearly drawn from studies of automation is that the *amount of automation needed for any task is likely to vary* from person to person and within a person to vary over time. Hence, a flexible automation system in which the level could vary is often preferable over one that is fixed and rigid. Flexible automation simply means that different levels are possible. One driver may choose to use cruise control, the other may not. The importance of flexible automation parallels the flexible and adaptive decision-making process of experts. As discussed in Chapter 7, decision aids that support that flexibility tend to succeed, and those that do not tend to fail. This is particularly true in situations that are not completely predictable. Flexibility seems to be a wise goal to seek.

Adaptive automation, goes one step further than flexible automation by implementing the level of automation based on some particular characteristics of the environment, user, and task (Rouse, 1988; Scerbo, 1996; Wickens & Hollands, 2000). For example, an adaptive automation system would be one in which the level of automation increases as either the workload imposed on the operator increases or the operator's capacity decreases (e.g., because of fatigue). For example, when psychophysiological (e.g., heart rate) measures indicate a high workload, the degree of automation can be increased (Prinzel et al., 2000). While such systems have proven effective (Rouse, 1988; Parasuraman et al.,

Level of Automation		Roles	
		Human	System
None	1	Decide, Act	———
Decision Support	2	Decide, Act	Suggest
Consensual AI	3	Concur	Decide, Act
Monitored AI	4	Veto	Decide, Act
Full Automation	5	———	Decide, Act

FIGURE 16.2

Continuum of shared responsibility between human and computer. (*Source:* Endsley, M. R., and Kiris, E. O., 1995. The out-of-the-loop performance problem and level of control in automation. *Human Factors,* 37(2), pp. 381–394. Reprinted with permission. Copyright 1993 by the Human Factors and Ergonomics Society. All rights reserved.)

1993), for example, in environments like the aircraft flight deck in which there are wide variations in workload over time, they should be implemented only with great caution because of their potential pitfalls (Wickens & Hollands, 2000). First, because such systems are adaptive closed-loop systems, they may fall prey to problems of negative feedback, closed-loop instability, as discussed in Chapter 9. Second, humans do not always easily deal with rapidly changing system configurations. Remember that *consistency* is an important feature in design (Chapter 8). Finally, as Rouse (1988) has noted, computers may be good at taking control (e.g., on the basis of measuring degraded performance by the human in the loop) but are not always good at giving back control to the human.

   6. *Maintaining a positive management philosophy.* A worker's acceptance and appreciation of automation can be greatly influenced by the management's philosophy (McClumpha & James, 1994). If, on the one hand, workers view that automation is being "imposed" because it can do the job better than they can, their attitudes toward it will probably be poor. On the other hand, if automation is introduced as an aid to improve human–system performance and a philosophy can be imparted in which the human remains the master and automation the servant, then the attitude will be likely to remain more accepting (Billings, 1996). This can be accompanied by good training of what the automation does and how it does its task. Under such circumstances, a more favorable attitude will also probably lead to better understanding of automation, better appreciation of its strengths, and more effective utilization of its features. Indeed, studies of the introduction of automation into organizations show that *management is often responsible for making automation successful* (Bessant et al., 1992).

SUPERVISORY CONTROL AND AUTOMATION-BASED COMPLEX SYSTEMS

Process Control

Automation plays a particularly critical role in situations when a small number of operators must control and supervise a very complex set of remote processes, whose remoteness, complexity, or high level of hazard prevents much "hands on" control, of the sort described in Chapter 9. Automation here is not optional, it is a necessity (Sheridan, 2002). Examples of such systems include the production of continuous quantities, such as energy, in the area of chemical process control, the production of discrete quantities, in the area of manufacturing control (Karwowski et al, 1997, Sanderson, 1989), and the control of remotely operated vehicles and robots, in the area of robotics control. In all of these cases, the human supervisor/controller is challenged by some or all of several factors with major human factors implications: the remoteness of the entity controlled from the operator, the complexity (multiple-interacting elements) of the system, the sluggishness of the system, following operator inputs, and the high level of risk involved, should there be a system failure. More detailed treatments of the human factors of these systems are available elsewhere (Moray, 1997, Sheridan, 1997, 2002; Wickens and Hollands, 2000 Chapter 13), and so, below, we only highlight a few key trends with human factors relevance.

For process control, such as involved in the manufacturing of petro-chemicals, nuclear or conventional energy, or other continuous commodities, the systems are so complex that high levels of automation must be implemented. Perhaps the key human factors question is how to support the supervisor in times of failures and fault management, so that disasters such as Three Mile Island (Rubinstein and Mason, 1979) and Chernobyl (Read, 1993) do not occur as a result of poor diagnosis and decision making. Tools for such support were suggested in Chapter 7, where the importance of decision support for knowledge-based behavior was emphasized, and in chapter 8, where the concepts of predictor displays and of *ecological interfaces* were introduced (Woods & Roth, 1988; Vicente, 2002; Wickens & Hollands, 2000). Such interfaces have two important features: (1) they are highly graphical, often using configural displays to represent the constraints on the system, in ways that these constraints can be easily perceived, without requiring heavy cognitive computations. (2) they allow the supervisor to think flexibly at different levels of abstraction (Burns, 2000, Vicente, 2002), ranging from physical concerns like a broken pipe or pump, to abstract concerns, like the loss of energy, or the balance between safety and productivity. In some regards, many aspects of air traffic control mimic those of process control (Wickens et al, 1998; Hopkin, 1995).

The automation served by robotics control in manufacturing is desirable because of the repetitive, fatiguing, and often hazardous mechanical operations involved, and is sometimes a necessity because of the heavy forces often required. Here, as discussed in chapter 9, a critical emerging issue is that of *agile manufacturing*, in which manufacturers are able to respond quickly to the need for high-quality customized products (Gunasekaran, 1999). In this situation, decision authority is often transferred from the traditional role of management to that of operators empowered to make important decisions. In this situation, automation needs to support an integrated process of design, planning and manufacturing, and integrate information so that employees can make decisions that consider a broad range of process considerations (See Chapter 7 for a description of how decision aids might support this process).

A second use of robots is in navigating unmanned air and ground vehicles, such as the air vehicles (UAVs) that provide surveillance for military operations, or ground vehicles that can operate in cleaning up hazardous waste sites. Here a major challenge is the control-display relationships with remote operators. How can the resolution of a visual display, necessary to understand a complex environment, be provided with a short enough delay, so that control can be continuous and relatively stable. If this is not possible, because of bandwidth limitations on the remote communications channels, what form and level of automation of control is best.

Finally, remote operators must sometimes supervise the behavior of a group, or collection of agents, not through direct (and delayed) control, but rather, by encouraging or "exorting" the desired behavior of the group. This concept of *hortatory control* describes such systems where the systems being controlled retains a high degree of autonomy (Lee, 2001, Murray & Liu, 1997). An example might be road traffic controllers, trying to influence the flow of traffic in a congested area around a city by informing travelers of current and expected road conditions and encouraging them to take certain actions; however, they have limited authority or technical ability to actually direct the behavior of the individual travelers. They also routinely cope with unexpected events. Other examples of hortatory operations environments include educational systems, safety management programs, and certain financial management programs, in which administrators can encourage or attempt to penalize certain behaviors, but they often lack reliable and complete knowledge of the system or the means of direct and strict supervisory control. In such circumstances, the greatest challenge to human factors is to identify ways of providing advisory information that is most effective in attracting the users to adopt certain behavior, and ways of gathering and integrating information so as to achieve situational awareness in an ill-structured environment.

CONCLUSION

Automation has greatly improved safety, comfort, and job satisfaction in many applications; however, it has also led to many problems. Careful design that considers the role of the person can help avoid these problems. In this chapter, we described automation classes and levels and used them to show how function allocation and human-centered approaches can improve human–automation performance. In many situations automation supports human decision making, and Chapter 7 discusses these issues in more detail. Although the domains of process control, manufacturing, and hortatory control already depend on automation, the challenges of creating useful automation will become more important in other domains as automation becomes more capable and pervasive—entering our home and even our cars, as described in Chapter 17.

Automation is sometimes introduced to replace the human and avoid the difficulties of human-centered design. This chapter identified several ironies of automation that show that *as systems become more automated, the need for careful consideration of human factors in design becomes more important, not less.* In particular, requirements for decision support (Chapter 7), good displays (Chapter 8), and training (Chapter 18) become more critical as automation becomes more common.

Chapter 17

Transportation Human Factors

$\mathbf{A}$*fter a fun-filled and sleep-deprived spring break in Daytona Beach, a group of four college students begin their trip back to university. To return in time for Monday morning classes, they decide to drive through the night. After eight hours of driving, Joe finds himself fighting to stay awake on a boring stretch of highway at 3:00 A.M. Deciding to make up some time, he increases the setting of the cruise control to 80 mph. After setting the cruise control, Joe begins to fall asleep, and the car slowly drifts towards the shoulder of the highway. Fortunately, a system monitoring Joe's alertness detects the onset of sleep and the drift of the vehicle. Based on this information, the system generates an alert that vibrates the seat by means of a "virtual rumble strip." This warning wakes Joe and enables him to quickly steer back onto the road and avoid an otherwise fatal crash.*

Every day, millions of people travel by land, water, and air. Several features of vehicles make them stand apart from other systems with which human factors is concerned and hence call for a separate chapter. Tracking and continuous manual control are normally a critical part of any human–vehicle interaction. Navigational issues also become important when traveling in unfamiliar environments. Furthermore, those environments may change dramatically across the course of a journey from night to day, rain to sunshine, or sparse to crowded conditions. Such changes have major implications for human interaction with the transportation system. The advent of new technologies, such as the satellite-based global positioning system, and the increased power of computers are revolutionizing many aspects of ground and air transportation. Because aircraft and ground vehicles often move at high speeds, the safety implications of transportation systems are tremendously important. Indeed, in no other system than the car do so many people have access to such a high-risk system and particularly a system in which their own lives are critically at risk. Every year, 500,000 people worldwide lose their lives in auto accidents, and around 40,000 lives per year are lost in the United States alone (Evans, 1996), while the cost to the U.S. economy of traffic accident-related injuries is typically over $200 billion per year.

In this chapter, we place greatest emphasis on the two most popular means of transportation: the automobile (or truck) and the airplane; these have received the greatest amount of study from a human factors perspective. However, we also consider briefly some of the human factors implications of maritime and public ground transportation, both with regard to the operators of such vehicles and to potential consumers, who may choose to use public transportation rather than drive their own car.

AUTOMOTIVE HUMAN FACTORS

The incredibly high rate of accidents on the highways and their resulting cost to insurance companies and to personal well-being of the accident victims (through death, injuries, and congestion-related delays) make driving safety an issue of national importance; that the human is a participant in most accidents and that a great majority of these (as high as 90 percent) are attributable to human error bring these issues directly into the domain of human factors.

Many of the human factors issues relevant to driving safety are dealt with elsewhere in the book. Here, we integrate them all as they pertain to driving a vehicle, often at high speeds along a roadway. We first present a *task analysis* of the vehicle roadway system and then treat critical issues related to *visibility, hazards and collisions, impaired drivers, driving safety improvements,* and *automation.*

It is important to note that driving typically involves two somewhat competing goals, both of which have human factors concerns. *Productivity* involves reaching one's destination in a timely fashion, which may lead to speeding. *Safety* involves avoiding accidents (to oneself and others), which is sometimes compromised by speeding. Our emphasis in this chapter is predominantly on the safety aspects. Safety itself can be characterized by a wide range of statistics (Evans, 1991, 1996), including fatalities, injuries, accidents, citations, and measures of speeding and other violations.

Two aspects of interpreting these statistics are important to remember. First, figures like fatality rates can be greatly skewed by the choice of *baseline*. For example, a comparison of fatalities per year may provide very different results from a comparison of fatalities per passenger mile. In the United States, the former figure has increased or remained steady over the past decade, while the latter has declined (Evans, 1996). Second, statistics can be heavily biased by certain segments of the population, so it is important to choose the appropriate baseline. For example, accident statistics that use the full population as a baseline may be very different from those that do not include young drivers, male drivers, or young male drivers.

Task Analysis of the Vehicle Roadway System

Strategic, Tactical, and Control Aspects of Driving.   Three levels of activity describe the complex set of tasks that comprise driving—*strategic, tactical,* and *control* (Michon, 1989). *Strategic* tasks focus on the purpose of the trip and the driver's overall goals; many of these tasks occur before we even get into the car. Strategic tasks include the general process of deciding where to go, when to go, and how to get there. In the opening vignette, a strategic task was the decision to drive through the night to return in time for classes Monday morning. *Tactical* tasks focus on the choice of maneuvers and immediate goals in getting to a destination. They include speed selection, the decision to pass another vehicle, and the choice of lanes. In the opening vignette, Joe's decision to increase the car's speed is an example of a tactical driving task. *Control* tasks focus on the moment-to-moment operation of the vehicle. These tasks include maintaining a desired speed, keeping the desired distance from the car ahead, and keeping the car in the lane. Joe's failure to keep the car in the lane as he falls asleep reflects a failure in performing a critical control task. Describing driving in terms of strategic, tactical, and control tasks identifies different types of driving performance measures, ways to improve driving safety, and types of training and assessment.

The challenge of measuring driver performance shows the importance of considering the different levels of driving activity. At the strategic level, driving performance can be measured by the drivers' ability to select the shortest route or to choose a safe time of day to drive (e.g., fatigue makes driving during the early morning hours dangerous). At the tactical level, driving performance can be measured in terms of drivers' ability to make the appropriate speed and lane choice as well as respond to emerging hazards, such as an upcoming construction zone. For example, drivers talking on a cell phone sometimes fail to adjust their speed when the road changes from dry to slippery (Cooper & Zheng, 2002). At the control level, driver performance can be measured in terms of drivers' ability to stay in the lane, control their speed, and maintain a safe distance from the vehicle ahead. The three levels of driving activity provide complementary descriptions of driving performance that all combine to reflect driving safety.

Control Tasks.   The control tasks are a particularly important aspect of driving. As shown in Figure 17.1, the driver performs in a multitask environment. At the core of this environment, shown at the top, is the two-dimensional tracking task of vehicle control. Using the terminology introduced in Chapter 9, the lateral task of maintaining lane position can be thought of as a second-order control task with preview (the roadway ahead) and a predictor (the heading of the vehicle). The "longitudinal" task can be thought of as a first-order tracking task of speed keeping, with a command input given by either the internal goals (travel fast but do not lose control or get caught for speeding) or the behavior of the vehicles, hazards, or traffic control signals in front. Thus, the tracking "display" presents three channels of visual information to be tracked along the two axes: Lateral tracking is commanded by the roadway curvature; longitudinal tracking is commanded by the flow of motion along the roadway and the location or distance of hazards and traffic control devices. The quality of this visual input may

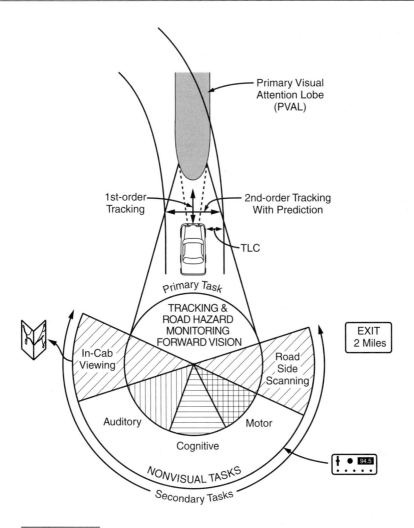

FIGURE 17.1

Representation of the driver's information-processing tasks. The top of the figure depicts the tracking or vehicle control tasks involved with lane keeping and hazard avoidance. The bottom of the figure presents the various sources of competition for resources away from vehicle tracking. These may be thought of as secondary tasks.

be degraded by poor visibility conditions (night, fog) or by momentary glances away from the roadway.

Gibson and Crooks (1938) described the control task in Figure 17.1 in terms of a *field of safe travel* in which drivers adjust their speed and direction to avoid hazards and move themselves towards their goal. Importantly, this perspective leads to a time-based description of the distance to hazards, such as the car ahead or the roadside. The time-to-line-crossing (TLC) measure is an example. Likewise, time-to-contact is often used as a time-based measure of drivers' performance in maintaining a safe distance to the vehicle ahead and is calculated by dividing the distance between the vehicles by the relative velocities of the vehicles (Lee, 1976).

Multitask Demands.   Driving frequently includes a variety of driving and non-driving tasks that vary in their degree of importance (see Chapters 6 and 13). We can define the primary control task as that of lane keeping and roadway hazard monitoring. Both of these depend critically on the *primary visual attention lobe* (PVAL) of information, a shaded region shown in Figure 17.1 that extends from a few meters to a few hundred meters directly ahead (Mourant & Rockwell, 1972). Figure 17.2 shows side and forward views of this area. Most critical for driving safety (and a force for human factors design concerns) are any tasks that draw visual attention away from the PVAL. Figure 17.1 shows that tactical driving tasks, such as roadside scanning for an exit sign, can also draw visual attention from the PVAL. Likewise, in-cab viewing of a map can compete for visual attention. Other tasks, such as the motor control involved in tuning the radio, talking to a passenger, and eating a hamburger also have visual components that compete for visual attention (Dingus et al., 1988; Table 17.1).

While the visual channel is the most important channel for the driver, there are nontrivial concerns with secondary motor activity related to adjusting controls, dialing cell phones, and reaching and pulling, which can compete with manual resources for effective steering. Similarly, as noted in our discussion of multiple resources in Chapter 6, intense cognitive activity or auditory information processing can compete for perceptual/cognitive resources with the visual channels necessary for efficient scanning. Any concurrent task (auditory, cognitive, motor) can create some conflict with monitoring and processing and visual information in the PVAL (Chapter 6). For example, Recarte and Nunes (2000) found that cognitive tasks reduce drivers' scanning of the roadway.

Cabin Environment.   From the standpoint of vehicle safety, one of the best ways to minimize the dangerous distracting effects of "eyes-in" time is to create the simplest, most user-friendly design of the internal displays and controls that is

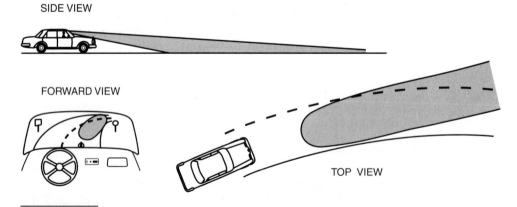

FIGURE 17.2
Representation of the PVAL from the forward view, top view, and side view.

TABLE 17.1   Single Glance Time (Seconds) and Number of Glances for Each Task

	Glance Duration	Number of Glances
Speed	0.62 (0.48)	1.26 (0.40)
Destination Direction	*1.20 (0.73)*	*1.31 (0.62)*
Balance	0.86 (0.35)	2.59 (1.18)
Temperature	1.10 (0.52)	3.18 (1.66)
Cassette Tape	0.80 (0.29)	2.06 (1.29)
Heading	*1.30 (0.56)*	*2.76 (1.81)*
Cruise Control	0.82 (0.36)	5.88 (2.81)
Power Mirror	0.86 (0.34)	6.64 (2.56)
Tune Radio	1.10 (0.47)	6.91 (2.39)
Cross Street	*1.66 (0.82)*	*5.21 (3.20)*
Roadway Distance	*1.53 (0.65)*	*5.78 (2.85)*
Roadway Name	*1.63 (0.80)*	*6.52 (3.15)*

Standard deviations in parentheses; tasks in italic are associated with a route guidance system.

Adapted from T. A. Dingus, J. F. Antin, M. C. Hulse, and W. Wierwille. 1988. Human factors associated with in-car navigation system use. *Proceedings of the 32nd Annual Meeting of the Human Factors Society.* Santa Monica, CA: Human Factors Society, pp. 1448–1453. Copyright 1988 by the Human Factors and Ergonomics Society. All rights reserved.

possible, using the many principles described earlier in the book. Displays should be of high contrast (Chapter 4), interpretable, and easy to read (Chapters 6 and 8); and design of the task environment within the vehicle should strive toward simplicity by avoiding unnecessary features and gizmos. Controls should be consistently located, adequately separated, and compatibly linked to their associated displays (Chapter 9). Simply put, the *vehicle controls should be designed so they would be easy to use by blind person.* Such a design philosophy can help minimize the demands on visual attention.

An important way to reduce the eyes-in time is to make any text display or label large enough that it can be read easily. The visual angle should be at least 16 arcminutes, but 24 arcminutes is ideal (Campbell et al., 1988). A simple rule ensures that the text will be large enough: The height *(H)* of the text divided by the distance *(D)* should be greater than .007 ($H/D > 0.007$). Because of the constant, this guideline is known as the "James Bond" rule (Green, 1999). In applying this guideline, it is critical that the units for H and D are the same.

Visibility

Great care in the design of motor vehicles must be given to the visibility of the critical PVAL for vehicle control and hazard detection. Four main categories of visibility issues can be identified: anthropometry, illumination, signage, and resource competition.

Anthropometry.   First, the application of good human factors requires that attention be given to the anthropometric factors of *seating,* discussed in Chapter 10 and in (Peacock & Karwowski, 1993). Can seat adjustments easily allow a wide range of body statures to be positioned so that the eye point provides adequate visibility down the roadway, or will drivers with the smallest seated eye height be unable to see hazards directly in front of their cars? (Such anthropometric concerns must also address the *reachability* of different controls.) Vehicles provide a clear example of where the philosophy of *design for the mean is not appropriate*. However, in creating various flexible options for seating adjustment to accommodate individual differences in body stature, great care must be given to making adjustment controls both accessible (so they can and will be used) and interpretable (i.e., compatible) so that they will be used correctly (e.g., a forward movement of the control will move the seat forward).

Illumination. Second, as we noted in Chapter 4, putting information within the line of sight does not guarantee that it will be sensed and perceived. In vehicle control, night driving presents one of the greatest safety concerns, because darkness may obscure both the roadway and the presence of hazards like pedestrians, parked cars, or potholes. Schwing and Kamerud (1988) provide statistics that suggest the relative fatality risk is nearly ten times greater for night than for day driving (although this higher risk is only partly related to visibility). *Adequate highway lighting can greatly improve visibility and therefore highway safety.* For example, an analysis of 31 thoroughfare locations in Cleveland, Ohio, revealed that placing overhead lights reduced the number of fatalities from 556 during the year before illumination to 202 the year after (Sanders & McCormick, 1993). Using adequate reflectors to mark both the center of the lane and the lane's edges can enhance safety.

Signage. A third visibility issue pertains to signage (Dewar, 1993; Lunenfeld & Alexander, 1990). As we noted, both searching for and reading critical highways signs can be a source of visual distraction. Hence, there is an important need for highway designers to (1) *minimize visual clutter* from unnecessary signs, (2) *locate signs consistently*, (3) *identify sign classes distinctly* (a useful feature of the redundant color, shape, and verbal coding seen, for example, in the stop sign), and (4) *allow signs to be read efficiently* by giving attention to issues of contrast sensitivity and glare, as discussed in Chapter 4. An important issue in roadway design is the potential for a large number of road guidance signs to create a high level of visual workload for the driver. When several guidance and exit signs are bunched together along the highway, they can create a dangerous overload situation (Lunenfeld & Alexander, 1990). Signage should be positioned along the road so that visual workload is evenly distributed.

We note below how all of the above visibility issues can become amplified by deficiencies in the eyesight of the older driver (Klein, 1991; Shinar & Schieber, 1991). Contrast will be lost, accommodation may be less optimal, and visual search may be less effective.

Resource Competition. The fourth visibility issue pertains to the serious distraction of in-cab viewing due to radios, switches, maps (Dingus & Hulse, 1993, Dingus et al., 1988; Table 17.1), or auxiliary devices such as cell phones (Violanti & Marshall, 1996).

The in-cab viewing that competes for visual resources can be described in terms of the number and duration of glances. The duration of any given glance should be relatively short. Drivers feel safe when glances are shorter than 0.8 seconds, provided they have about 3 seconds between glances (Green, 1999). On this basis, the last three tasks in Table 17.1 might not be acceptable. Not only are the individual glance times longer than 1.5 seconds, but each task requires more than five glances. In addition, because these tasks are critical in determining when to make a turn, the driver may be pressured into making these glances in rapid succession, with much less than 3 seconds between each glance.

The number of glances and duration of glances has been used to estimate the crash risk posed by a particular task. Several studies have combined crash reports, glance data, and frequency of use data to estimate the contribution of visual demand to the number of fatalities per year (Green, 1999; Wierwille; 1995; Wierwille & Tijerina, 1996). This analysis provides a rough estimate of the number of fatalities as a function of glance duration, number of glances, and frequency of use (Green, 1999). The equation predicts the number of fatalities per year *(Fatalities)* in the United States associated with in-vehicle tasks based on the market penetration *(MP)*, mean glance time in seconds *(GT)*, number of glances *(G)*, and frequency of use per week *(FU)*. The first term in the equation accounts for the 1.9 percent increase in the number of vehicle miles driven each year. The 1.5 power associated with mean glance time reflects the increasingly greater danger posed by longer glances. This equation shows that *we must balance the benefits of any in-vehicle device with the potential fatalities that it may cause.*

$$Fatalities = (1.019^{(Currentyear-1989)})(MP)[-0.33 + 0.0477(GT)^{1.5}(G)(FU)]$$

Resource competition issues can be dealt with in a number of ways. In addition to simplifying in-cab controls and displays, more technology-oriented solutions include using *auditory displays* to replace (or augment) critical navigational guidance information (i.e., maps), a design feature that improves vehicle control (Dingus & Hulse, 1993; Parkes & Coleman, 1990; Srinivasan & Jovanis, 1997). *Speech recognition* can also reduce resource competition by allowing people to speak a command rather than press a series of buttons. However, even auditory information and verbal commands are not interference-free. These tasks still compete for perceptual resources with visual ones, leading to some degree of interference (Lee et al., 2001). Voice-activated, hands-free cell phones can reduce lane-keeping errors and glances away from the road (Jenness et al., 2002), but they *do not eliminate distraction* (Strayer Drews & Johnston, 2003). For example, using a hands-free phone while driving is associated with impaired gap judgment (Brown et al., 1969), slower response to the braking of a lead vehicle (Lamble et al., 1999), and increased crash risk (Redelmeier & Tibshirani, 1997).

Automotive designers have also proposed using *head-up displays* to allow information such as speedometers to be viewed without requiring a glance to the dashboard (see Chapter 8, Figure 8.10; Kaptein, 1994; Kiefer, 1995; Kiefer & Gellatly, 1996). In driving, as in aviation, HUDs have generally proven valuable by keeping the eyes on the road and supporting faster responses to highway events (Srinivasan and Jovanis, 1997; Horrey & Wickens, 2002). If images are simple, such as a digital speedometer, HUD masking does not appear to present a problem (Kiefer & Gellatly, 1996). However, this masking may be more serious when more complex imagery is considered for head-up display location (Ward & Parkes, 1994). Horrey and Wickens (2002) show that displaying automotive HUDs slightly downward so that they are visible against the hood rather than against the roadway appears to reduce the problems of hazard masking.

Although technological advances, such as auditory route guidance, may sometimes reduce the competition for the critical forward visual channel, compared to reading a paper map, it is also the case that many of these advances, designed to provide more information to the driver, may also induce a substantial distraction. For example, the negative safety implications of cellular phones are by now well established (Redelmeier & Tibshirani, 1997). By one estimate, eliminating the use of all cellular phones while driving would save 2,600 lives and prevent 330,000 injuries anually (Cohen & Graham, 2003). Emerging applications, such as Internet content and email that is made possible by speech recognition technology may also distract drivers (Lee et al., 2001; Walker et al., 2001). It may also be the case that poorly designed electronic maps or other navigational aides can be just as distracting as paper maps (Dingus & Hulse, 1993). When such electronic aids are introduced, it becomes critical to incorporate human factors features that support easy, immediate interpretation (Lee et al., 1997). These include such properties as a "track-up" map rotation, designs that minimize clutter (see Chapter 8).

Hazards and Collisions

Nearly all serious accidents that result in injury or death result from one of two sources: loss of control and roadway departure at high speed (a failure of lateral tracking) or collision with a roadway hazard (a failure of longitudinal tracking or speed control). The latter in turn can result from a failure to detect the hazard (pedestrian, parked vehicle, turning vehicle) or from an inappropriate judgment of the time to contact a road obstacle or intersection. In the United States, rear-end collisions are the most frequent type of crash, accounting for nearly 30 percent of all crashes; however, roadway departure crashes cause the greatest number of fatalities, accounting for over 40 percent of driving-related fatalities (National Safety Council, 1996).

Control Loss. Loss of control can result from several factors: Obviously slick or icy road conditions are major culprits, but so are narrow lanes and momentary lapses in attention, which may contribute to a roadway departure. A major contributor to this type of crash is fatigue, as described in the opening vignette of this chapter. Another cause is a minor lane departure followed by a rapid overcorrection (a high-gain response), which can lead to unstable oscillation resulting in the vehicle rolling over or a roadway departure, as we saw in Chapter 9. In all of these cases, the likelihood of loss of control is directly related to the *bandwidth* of correction, which in turn is related to vehicle speed. *The faster one travels, the less forgiving is a given error* and the more immediate is the need for correction; but the more rapid the correction at higher speed, the greater is the tendency to overcorrection, instability, and possible loss of control (e.g., rollover).

Human factors solutions to the problems of control loss come in several varieties. Naturally, any feature that keeps vision directed outward is useful, as is anything that prevents lapses of attention (e.g., caused by fatigue, see the section

"Driving Safety Improvements"). Wider lanes lessen the likelihood of control loss. Two-lane rural roads are eight times more likely to produce fatalities than are interstate highways (Evans, 1996). Most critical are any feedback devices that provide the driver with natural feedback of high speed. Visible marking of lane edges (particularly at night) are useful, as are "passive alerts" such as the "turtles" dividing lanes or "rumblestrips" on the lane edge that warn the driver via the auditory and tactile sense of an impending lane departure and loss of control (Godley et al., 1997). As described at the start of this chapter, new technology makes it possible to generate "virtual rumble strips" in which a potential lane departure is detected by sensors, and the system alerts the driver with vibrations through the seat (Raby et al., 2000).

Hazard Response. A breakdown of visual monitoring because of either poor visibility or inattention can cause a failure to detect hazards. In understanding hazard response, a key parameter is the *time to react* to unexpected objects, which is sometimes called the perception-reaction time, or in the case of the time to initiate a braking response, *brake reaction time.* Brake reaction time includes the time to detect a threat, release the accelerator, and move from the accelerator to the brake. Moving from the accelerator to the brake takes approximately 0.2 to 0.3 seconds. On the basis of actual on-the-road measurements, brake reaction time has been estimated to be around 1.0 to 2.0 seconds for the average driver, with a mean of around 1.5 seconds (Summala, 1981; Dewar, 1993; American Association of State Highway & Transportation Officials, 1990; Henderson, 1987; Green, 2000; Sohn & Stepleman, 1998). This is well above the reaction-time values typically found in psychology laboratory experiments. It is important to note the 1.5 second value is a mean and that driver characteristics and *the driving situation can dramatically affect the time taken to react.* For example, drivers respond to unexpected events relatively slowly (see Chapter 9), but respond to severe situations more quickly (Summala, 2000). The relatively slow response to unexpected events is the greatest contribution to hazard situations (Evans, 1991). In the specific situation in which a driver is 10 meters behind another vehicle and the driver is unaware that the lead vehicle is going to brake, the 85th percentile estimate of reaction time is 1.92 s, and the 99th percentile estimate is 2.52 s (Sohn & Stepleman, 1998). Driver characteristics (e.g., age, alcohol intoxication, and distraction) also increase reaction time, so designers may need to assume longer reaction times to accommodate those drivers (or conditions; Dewar, 1993; Triggs & Harris, 1982).

Speeding. High vehicle speed poses a quadruple threat to driver safety (Evans, 1996): (1) It increases the likelihood of control loss; (2) it decreases the probability that a hazard will be detected in time; (3) it increases the distance traveled before a successful avoidance maneuver can be implemented; and (4) it increases the damage at impact. These factors are illustrated in Figure 17.3, which shows how the time to contact a hazard declines with higher speeds. Drivers should maintain speeds so that this time is less than the time available to respond, creating a positive safety margin.

Why, then, do people speed? Obviously, this tendency is sometimes the result of consciously formed goals—for example, the rush to get to a destination on time

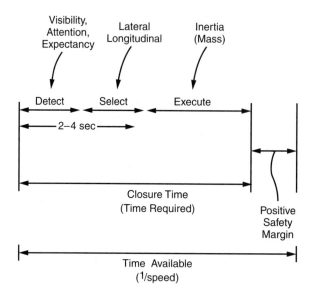

FIGURE 17.3

The components of the hazard response time, which is the time required to stop before contacting a hazard, the influences on these components, and the need to maintain a positive safety margin between the time required and the time available. Time available will be *inversely* proportional to speed.

after starting late, which is a tactical or strategic decision. There are also reasons at the control level of driving behavior that explain why drivers tend to *overspeed* relative to their braking capabilities (Wasielewski, 1984; Evans, 1991, 1996; Summala, 1988). For example, Wasielewski (1984) found that the average separation between cars on a busy freeway is 1.32 seconds, despite that the minimum separation value recommended for safe stopping, based upon total braking time estimates, is 2 seconds! The sources of such a bias may be perceptual (i.e., underestimating true speed) or cognitive (i.e., overestimating the ability to stop in time).

Perceptual biases were seen in the study by Eberts and MacMillan (1985), discussed in Chapter 4, in which small cars were found to be more likely to be hit from behind because their size biased distance judgments (the small cars were perceived as farther away than they really were). Any factors that reduce the *apparent* sense of speed (quieter engines, higher seating position above the ground, less visible ground texture) will lead to a bias to overspeed (Evans, 1991). Adaptation also plays a role. Drivers who drive for a long period of time at a high speed will perceive their speed to be lower than it really is and hence may overspeed, for example, when reaching the off-ramp of an interstate highway or motorway. Chapter 4 describes a technique of roadmarking to increase the *perceived* speed, and therefore induce appropriate braking, when approaching a traffic circle. (Denton, 1980, Godley et al. 1997).

Risky Behavior. Equally as important as the perceptual biases, but less easily quantifiable, are the *cognitive biases* that can lead to overspeeding and other risky behaviors. Such biases are induced by the driver's feeling of *overconfidence* that hazards will not suddenly appear, or if they do, that he or she will be able to stop in time; that is, overconfidence yields an underestimation of *risk* (Brown et al., 1988; Summala, 1988; see Chapters 7 and 14). This underestimation of risk is revealed in the belief of drivers that they are less likely to be involved in an accident than "the average driver" (Svenson, 1981). We may ascribe some aspect of this bias in risk perception to the simple effects of *expectancy,* discussed in Chapters 4 and 8; that is, most drivers have not experienced a collision, and so their mental model of the world portrays this as a highly improbable or perhaps "impossible" alternative (Summala, 1988; Evans, 1991). For example, the normal driver simply does not entertain the possibility that the vehicle driver ahead will suddenly slam on the brakes or that a vehicle will be stationary in an active driving lane. Unfortunately, these are exactly the circumstances in which rear-end collisions occur.

Interestingly, this expectancy persists even after a driver experiences a severe crash—a fatal crash has little effect on the behavior of survivors. Any change is typically limited to the circumstances of the accident, does not last more than a few months, and does not generalize to other driving situations (Rajalin & Summala, 1997). Risky choices in driving are a good example of decisions that provide poor feedback (see Chapter 7)—a poor decision can be made for years without a negative consequence and then a crash situation can emerge in which the same poor decision results in death.

The Impaired Driver

Vehicle drivers who are fatigued, drunk, angry (Simon & Corbett, 1996), or otherwise impaired present a hazard to themselves as well as to others on the highway.

Fatigue. Along with poor roadway and hazard visibility, fatigue is the other major danger in night driving (Summala & Mikkola, 1994). The late-night driver may be in the lower portion of the arousal curve driven by circadian rhythms (see Chapter 13) and may also be fatigued by a very long and tiring stretch of driving, which was initiated during the previous daylight period. Fatigue accounts for over 50 percent of the accidents leading to the death of a truck driver and over 10 percent of all fatal car accidents. A survey of long-haul truck drivers reported that 47.1 percent had fallen asleep behind the wheel, 25.4 percent of whom had fallen asleep at the wheel within the past year (McCartt et al., 2000). Some of the factors associated with the tendency of truck drivers to fall asleep while driving include arduous schedules, long hours of work, few hours off-duty, poorer sleep on road, and symptoms of sleep disorders (e.g., sleep apnea) (McCartt et al., 2000). Like truck drivers, college students, such as those mentioned at the start of the chapter, are particularly at risk because they commonly suffer from disrupted sleep patterns. Sleeping less than 6.5 hours a day can be disastrous for driving performance (Bonnet & Arand, 1995). As we noted in Chapter 13, the kind of task that is most impaired under such circumstances is that of *vigilance:* monitoring for low-frequency (and hence unexpected) events. In driving, these events might involve a low-visibility hazard in the roadway or even the nonsalient "drift" of the car toward the edge of the roadway.

Alcohol. Historically, alcohol has contributed to approximately 50 percent of fatal highway accidents in this country—in 1999 alone, there were 15,786 alcohol-related deaths (Shults et al., 2001). The effects of alcohol on driving performance are well known: With blood alcohol content as low as 0.05 percent, drivers react more slowly, are poorer at tracking, are less effective at time-sharing, and show impaired information processing (Evans, 1991). All of these changes create a lethal combination for a driver who may be overspeeding at night, who will be less able to detect hazards when they occur, and who are far slower in responding to those hazards appropriately. Exhortations and safety programs appear to be only partially successful in limiting the number of drunk drivers, although there is good evidence that states and countries that are least tolerant have a lower incidence of driving under the influence (DUI) accidents (Evans, 1991). A dramatic illustration of the effect of strict DUI laws on traffic fatalities in England is provided by Ross (1988), who observed that the frequency of serious injuries on weekend nights was reduced from 1,200 per month to approximately 600 per month in the months shortly after a strict DUI law was implemented. Beyond consistent enforcement of DUI laws, Evans notes that the most effective interventions may be *social norming*, in which society changes its view of drinking and driving, in the same manner that such societal pressures have successfully influenced the perceived glamour and rate of smoking. The organization Mothers Against Drunk Driving (MADD) provides a good example of such an influence.

Age. Although age is not in itself an impairment, it does have a pronounced influence on driving safety. As shown in Figure 17.4, safety increases until the mid-20s and then decreases above the mid-50s. The reasons for the higher acci-

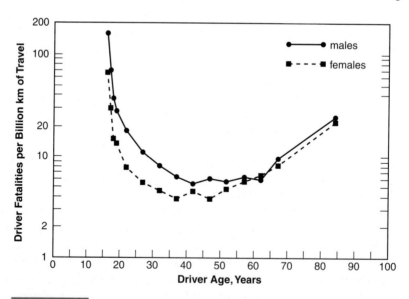

FIGURE 17.4

Fatality rate as a function of age and gender. (*Source:* Evans, L., 1988. Older driver involvement in fatal and severe traffic crashes. *Journal of Gerontology: Social Sciences, 43*(5), 186–193. Copyright, Gerontological Society of America.)

dent rates at the younger and the older ends of the scale are very different and so shall be treated separately. Younger drivers may be less skilled and knowledgeable simply because of their lack of training and experience. Furthermore, younger drivers tend to have a greater sense of overconfidence (or a greater underestimation of dangers and risks; Brown et al., 1988). For example, younger drivers tend to drive faster and are more likely to drive at night (Waller, 1991) and while under the influence of alcohol.

Statistics show that the brand new driver of age 16 is particularly at risk, a characteristic that is probably heavily related to the lack of driving skill and increased likelihood of driving errors (Status Report, 1994). For example, such drivers have a greater proportion of fatalities from rollover loss-of-control accidents, suggesting driving-skill deficiency. The 16-year-old is also much more likely to suffer a fatality from speeding (Status Report, 1994). After age 17, however, the still-inflated risk (particularly of males) is due to other factors. The driver at this age (1) is more exposed to risky conditions (e.g., driving fast, at night, while fatigued, or under the influence; Summala & Mikkola, 1994; Brown, 1994); (2) is more likely to experience risk as intrinsically rewarding (Fuller, 1988); (3) has greater overconfidence (Brown et al., 1988); and (4) has not sufficiently acquired more subtle safe-driving strategies (as opposed to the pure perceptual motor control skills; Evans, 1991). For example, when interacting with a radio, cassette, or cell phone, no experienced drivers took glances longer than 3 seconds, but 29 percent of the inexperienced drivers did (Wikman, Nieminen, & Summala, 1998). After the first year of driving, *young drivers have acquired the basic control skills of driving but not the tactical and strategic judgment needed for safe driving* (Ferguson, 2003; Tränkle et al., 1990).

In contrast to the skill deficiencies and the risk-taking tendencies of younger drivers, *information-processing impairments* lead to a relatively high crash rate with older drivers (Barr & Eberhard, 1991; Evans, 1988). Increasing age leads to slower response times; to a more restricted field of attention (Owsley et al., 1998) and reduced time-sharing abilities (Brouwer et al., 1991; Kortelling, 1994); and of course, to reduced visual capabilities, particularly at night due to glare (Shinar & Schieber, 1991; see Chapter 4). Many older drivers are able to compensate fully for these impairments during normal conditions simply by driving more slowly and cautiously or by avoiding certain driving environments, such as freeways or bad weather (Waller, 1991). Considered in terms of strategic, tactical, and control elements of driving, *older drivers drive less safely at the control level but can compensate with appropriate choices at the tactical and strategic levels* (e.g., choosing not to drive at night) (De Raedt & Ponjaert-Kristoffersen, 2000).

Impairment Interactions. Fatigue, alcohol, and age can combine to degrade driving performance in a way that might not be predicted by each alone. This is particularly true with younger drivers. For example, young college students tend to have varied sleep patterns that can induce high levels of fatigue. They are also more likely to drive after drinking, and the combined effect of alcohol and fatigue can be particularly impairing. The presence of passengers can further compromise the situation by distracting the driver or inducing risky behavior. A young person driving with friends at night and after drinking is an *extremely* dangerous combination (Williams, 2003).

Driving Safety Improvements

There is no single answer to the driving safety problem. Instead, enhancing driving safety requires a systematic approach that considers drivers, the vehicles, and the roadway environment. Haddon's matrix (Table 17.2) addresses each of these considerations and shows how safety improvements can prevent crashes, minimize injury during crashes, and influence the treatment of injuries after the crash (Noy, 1997). Each cell in the matrix represents a different set of ways to improve driving safety.

Table 17.2 shows several examples of driver, vehicle, and roadway characteristics that *help avoid crashes* at the "pre-crash" level of the Haddon matrix. Earlier in this chapter we discussed the role of roadway design and sign visibility on crash prevention; now we describe other driver, vehicle, and roadway characteristics that can also enhance driving safety by avoiding crashes.

Driver Characteristics: Training and Selection. We saw that reasons for higher accident rates are related to both limited skills (for the very young driver) and limited information-processing abilities (for the elderly), which can be addressed by training and selection (see Chapter 18). In driver education programs the two solutions are carried out to some extent in parallel. However, despite its mandatory nature, there is little evidence that driver training programs actually serve to improve driver safety (Evans, 1991; Mayhew et al., 1997), and these might actually undermine safety if they allow drivers to be licensed at a younger age.

Changing when young drivers are granted a full license can compensate for some limits of driver training programs. For example, several states have raised the minimum driving age. New Jersey and 16 other states have a minimum driving age of 18 for an unrestricted license, and they receive a corresponding benefit to traffic safety (as do most European countries in which the age minimum is 18). Increases in the minimum drinking age in this country have also been associated with a 13 percent reduction in driving fatalities (National Highway Traffic Safety Administration, 1989). *Graduated licensing* is another promising legislative solution, which capitalizes on an understanding of the factors that in-

TABLE 17.2 **Haddon's Matrix Showing Examples of How Driver, Vehicle, and Roadway Characteristics Contribute to Crash Avoidance, Crash Mitigation, and Post-Crash Injury Treatment**

	Drivers	*Vehicle*	*Roadway*
Pre-crash	Training and selection, compliance with laws, risk calibration, fitness to drive measures.	Collision warnings, distractions, and antilock brake system (ABS).	Roadway design, consistency, and sign visibility. Availability of public transport.
Crash	Seat belt use.	Airbag and relative vehicle size.	Barriers and lane separation.
Post-crash	Manual emergency call with cell telephones. Knowledge of first aid.	Automatic emergency call.	Traffic congestion, availability of rescue vehicles, and emergency room resources.

crease the crash risk for younger drivers and *restricts the driving privileges of young drivers during their first years of driving.* For example, crash risk is low during the learner period (when drivers are accompanied by an adult) and particularly high immediately after licensure, at night, with passengers, and after consuming alcohol. As a consequence of this risk assessment, graduated licensing restrictions include daytime-only driving, driving only to and from school or work, no young passengers, and driving only with an adult (Williams, 2003). This strategy effectively extends the behind-the-wheel training of the young driver for several years and reduces the crash rate for 16-year-old drivers by approximately 25 percent (McCartt, 2001; Shope & Molnar, 2003).

We learn in Chapter 18 that actual behind-the-wheel training in a vehicle may not be the best environment for all forms of learning. Such an environment may be stressful, performance assessment is generally subjective, and the ability of the instructor to create (and therefore teach a response to) emergency conditions is very limited. For these reasons, increasing attention is being given to driving simulators (Green, 1995; Kaptein et al., 1996).

How to address older driver safety is a more difficult issue (Nicolle, 1995). Clearly, the requirement for more frequent driving tests above a certain age can effectively screen the age-impaired driver. One problem with this approach is that there is no consensus on what cognitive impairments degrade driving safety and how these impairments can be measured (Lundberg et al., 1997). Because older drivers adopt a compensating conservative behavior (drive more slowly, avoid driving at night), many older drivers who might fail performance tests would not show a higher rate of crashes on the road (De Raedt & Ponjaert-Kristoffersen, 2000). At the same time, depriving older drivers of their privilege to drive can severely degrade the quality of life of many older people, and so any move to restrict driving privileges must be carefully considered (Waller, 1991).

For both younger and older drivers, certain aspects of training and selection call for improvement. In terms of selection, for example, research has found that the standard *visual acuity test,* an assessment of 20/40 corrected vision, has very little relevance for driving (Wood & Troutbeck, 1994). More appropriate screening tests might evaluate *dynamic visual acuity* (Burg & Hulbert, 1961). Furthermore, driver selection tests fail to examine critical abilities related to *visual attention* (Ball & Owsley, 1991), as described in Chapter 6. More generally, it is important to emphasize that the *pure perceptual-motor control components of driving skill are but a small component of the skills that lead to safe driving compared to strategic and tactical aspects.* For example, professional race car drivers, who are surely the most skilled in the perceptual-motor aspects, have an accident and moving violation rate in normal highway driving that is well above the average for a control group of similar age (Williams & O'Neill, 1974; Evans, 1996). For this reason, training and selection should address issues associated with not just the control but also the tactical and strategic aspects of driving.

Driver Characteristics: Driver Adaptation and Risk Calibration. Risk-based solutions must address ways of leading people to better appreciate the probability of these low-frequency events and hence better *calibrating* their perceived risk level to the actual risk values (e.g., publishing cumulative likelihood of fatality over a lifetime of not wearing a seat belt; Fischhoff & MacGregor, 1982). Drivers should be encouraged to adopt an attitude of awareness, to "expect the unexpected" (Evans, 1991).

The concept of driving risk has been incorporated into a model explaining why innovations designed to improve traffic safety do not always produce the expected full benefits. According to the *risk homeostasis* model (Wilde, 1988), drivers seek to maintain their risk at a constant level. They negate the safety benefits of any safety measure (e.g. antilock brakes) by driving faster and less cautiously. In fact, highway safety data appear to be only partially consistent with this viewpoint (Evans, 1991; Summala, 1988). Evans argues that drivers are rarely conscious of any perceived risk of an accident (in such a way that they might use perceived risk to adjust their driving speed). Instead, driving speed is dictated by either the direct motives for driving faster (e.g., rush to get to the destination) or simply force of habit. Evans points out that different safety-enhancing features can actually have quite different effects on safety. Some of those that actually improve vehicle performance (e.g., antilock brakes system ABS) may indeed have a less than expected benefit (Farmer et al., 1997; Wilde, 1988). Specifically, a study showed that taxi drivers with ABS had significantly shorter time headways compared to taxis without ABS thereby possibly offsetting their safety benefits (Sagberg et al., 1997). But others, such as widening highways from two to four lanes, have clear and unambiguous safety benefits (Evans, 1996), as do those features like protection devices (seat belts and air bags) that have no effect on driving performance but address safety issues of crashworthiness. *Any safety intervention must consider the tendency for people to adapt to the new situation* (Tenner, 1996).

Driver Characteristics: Regulatory Compliance.   Intuitively, an ideal solution for many driving safety problems is to have everyone drive more slowly. Indeed, the safety benefits of lower speed limits have been clearly established (Summala, 1988; McKenna, 1988; Evans, 1996). Yet, despite these benefits, public pressure in the United States led to a decision first to increase the national limit to 65 miles per hour, causing an increase in fatalities of 10 to 16 percent, and then to remove the national speed limit altogether. *Effective enforcement of speed limits can make a difference.* While "scare" campaigns about the dangers of high speeds are less effective than actual compliance enforcement (Summala, 1988), a more positive behavior-modification technique that proved effective was based on posting signs that portrayed the percentage of drivers complying with speed limits (Van Houten & Nau, 1983). Automatic speed management systems may offer the most effective tool to enforce speed limit compliance. These systems automatically limit the speed of the car to the posted limit by using GPS and electronic maps to identify the speed limit and then adjust throttle and brakes to bring the car into compliance with the speed limit; however, driver acceptance may be a substantial challenge (Varhelyi, 2002). Similarly, automated systems for issuing tickets for those who run red lights can promote better compliance, but are controversial.

Driver and Vehicle Characteristics: Fitness to Drive.   To counteract the effects of fatigue, drivers of personal vehicles have few solutions other than the obvious ones discussed in Chapter 13, designed to foster a higher level of arousal (adequate sleep, concurrent stimulation from radio, caffeine, etc.). For long-haul truck drivers, administrative procedures are being imposed to limit driving time during a given 24-hour period and to enforce rest breaks. Highway safety researchers have also examined the feasibility of *fitness for duty* tests that can be required of long-haul drivers at inspection stations or geographical borders (Miller, 1996; Gilliland & Schlegel, 1995). Such tests, perhaps involving a "video game" that requires simultaneous tracking and event detection, can be used to infer that a particular driver needs sleep before continuing the trip.

A possible future solution to fatigue and alcohol problems is a *driver monitoring system* (Brookhuis & de Waard, 1993) that can monitor the vehicle (e.g., steering behavior) and the driver (e.g., blink rate, EEG; Stern et al., 1994) and can then infer a pending loss of arousal or control. Following such an inference, the system could alert the driver via an auditory warning or haptic warning, as described at the start of the chapter (Bittner et al., 2000; Eriksson & Papanikolopoulos, 2001).

Vehicle Characteristics: Sensing and Warnings.   Speed limit enforcement may have little influence on the behavior of "tailgaters" who follow too closely (since safe separation can be violated at speeds well below the limit) nor on those drivers whose inattention or lack of relevant visual skills prevents them from perceiving the closure with the vehicle in front. Since rear-end collisions account for almost 30 percent of all motor vehicle accidents (National Safety Council, 1996), the potential human factors payoffs in this area are evident. *Sensors and alerts can enhance drivers' perception of vehicles ahead and enable them to follow more safely.* Here, human factors research has revealed some modest success of the high mounted brake lights that can make the sudden appearance of a braking vehicle more perceptually evident (Kahane, 1989; McKnight & Shinar, 1992; Mortimer, 1993). Other *passive* systems can make it easier for drivers to perceive the braking of leading vehicle. For example, a "trilight" system illuminates amber tail lights of the leading vehicle if the accelerator is released, which makes it possible for the following vehicle to anticipate emergency braking of the lead vehicle (Shinar, 1996).

Active solutions integrate information from radar or laser sensors of the rate of closure with the vehicle ahead. If this rate is high, it can then be relayed directly to the following driver by either visual, auditory, or kinesthetic signals (Dingus et al., 1997). In the latter case, the resistance of the accelerator to depression is increased as the driver gets closer to the car ahead. A system that provided continuous feedback regarding time headway through a visual display and an auditory alert reduced the amount of time drivers spent at headways below 1 second (Fairclough et al., 1997). Similarly, an auditory and visual warning for rear-end crash situations helped drivers respond more quickly and avoid crashes (Lee et al., 2002).

Roadway Characteristics: Expectancy.   A point that we have made previously is that people perceive and respond rapidly to things that they expect but *respond slowly when faced with the unexpected.* The role of *expectancy* is critical in driver perception (Theeuwes & Hagenzieker, 1993). Hence, design should capitalize on expectancy. For example, standardization of roadway layouts and sign placements by traffic engineers lead drivers to expect certain traffic behaviors and information sources (Theeuwes & Godthelp, 1995). However, roadway design and traffic control devices should also try to communicate the unexpected situation to the driver well in advance. Using this philosophy, a series of solutions can help drivers anticipate needed decision points through effective and visible signage in a technique known as *positive guidance* (Alexander & Lunenfeld, 1975; Dewar, 1993). While these points (i.e., turnoffs, traffic lights, intersections) are not themselves hazards, a driver who fails to prepare for their arrival may well engage in hazardous maneuvers—sudden lane changes, overspeeding turns, or running a red light. As one example, a shorter-than-expected green light will lead the driver to fail to anticipate the change, say, from green to yellow to red, and hence increase the possibility of delayed braking and running through the red light (Van der Horst, 1988). Light cycles should be standardized according to the speed with which the typical driver approaches the intersection in question.

Expectancy and standardization also apply to sign location and intersection design. For example, left exits off a freeway (in all countries outside of Great Britain, India, and Japan) are so unexpected that they represent accident invitations. So too are sharper-than-average curves or curves whose radius of curvature decreases during the turn (i.e., spiral inward).

Another approach to driving safety is to *reduce the consequence of an accident* (rather than reducing accident frequency per se). The "crash" level of Haddon's matrix in Table 17.2 shows several examples of how driver, vehicle, and roadway features can contribute to surviving a crash. Drivers can dramatically enhance their chance of survival by wearing a seat belt. Vehicle designs that include airbags and other occupant protection mechanisms (e.g., crumple zones) can also play an important role, as can roadway design. In particular, guardrail design and lane separation can make important contributions to crash survival. For example, current guardrail designs are based on the center of gravity of a passenger vehicle, so SUVs that collide with these barriers may suffer more severe consequences than typical cars because they are more prone to roll on impact (Bradsher, 2002).

Driver and Vehicle Characteristics: Use of Protective Devices. The mandatory requirement for collision restraints (seat belt laws, airbags) is a good example of how vehicle design can reduce the potential injuries associated with a crash. The effectiveness of such devices is now well established (Evans, 1996). For example, the failure to use lap/shoulder belts is associated with a 40 percent increase in fatality risk (Evans, 1991), and airbags have a corresponding protective value (Status Report, 1995). Of course, the mere availability of protective devices like seat belts does not guarantee that they will be used. As a consequence, mandatory seat belt laws have been introduced in several states, and their enforcement clearly increases in both compliance and safety (Campbell et al., 1988). Correspondingly, safety gains are associated with passage and enforcement of motorcycle helmet laws (Evans, 1991). One study showed that the *combined effect of seat belt laws and enforcement* served to increase compliance in North Carolina from 25 percent to nearly 64 percent and was estimated to reduce fatalities by 11.6 percent and serious injuries by 14.6 percent (Reinfurt et al., 1990). An interesting study that used the "carrot" rather than the "stick" approach found that if police officers randomly rewarded drivers with cash or coupons when seat belts were being worn increased the proportion of people using seat belts and provided more enduring behavioral changes than pure enforcement (Mortimer et al., 1990).

Table 17.2 shows several examples of driver, vehicle, and roadway features that *enhance post-crash response* at the post-crash level of Haddon's matrix. These interventions try to keep the driver alive after the crash. The most critical factor contributing to driver survival after the crash is the time it takes to get the driver to an emergency room. Cell phones make it possible for the driver to call for help quite easily, reducing the time emergency vehicles take to arrive on the scene. However, in severe crashes the driver may be disabled. For this reason, new systems automatically call for aid if the airbag is deployed. The roadway and traffic infrastructure also plays an important role in the post-crash response. Traffic congestion might prevent ambulances from reaching the driver in a timely manner, and appropriate emergency room resources may not be available. Navigation systems in ambulances that indicate low-congestion routes to the victim could also enhance post-crash response.

AUTOMOTIVE AUTOMATION

The previous sections have described several roadway safety concerns that automation might address, including collision warning systems, automated navigation systems, driver monitors, and so forth (Lee, 1997). Collectively, many of these are being developed under the title of *Intelligent Transportation System (ITS)*. The development of the various functions within ITS depends on several recent technological developments. For example, automated navigation aids depend on knowing the vehicle's momentary location, using satellite-based *global positioning system (GPS)*. Collision warning devices require accurate *traffic sensing devices* to detect rate-of-closure with vehicles ahead, and route planning aids must have an accurate and up-to-date description of the road network *(a digital map database)* and the associated traffic (a *wireless connection* for real-time traffic data). Once a GPS, traffic sensors, a map database, and wireless connections become standard equipment, many types of in-vehicle automation become possible. For example, many General Motors vehicles already have a system that automatically uses sensor data (e.g., airbag deployment) to detect a crash, calls for emergency aid, and then transmits the crash location using the car's GPS. This system could save many lives by substantially reducing emergency response times. Table 17.3 shows more examples of possible in-vehicle automation that is being developed for cars and trucks.

The introduction of automated devices such as these raises three general issues. First, as we discussed in Chapter 16, the introduction of automation must be accompanied by considerations of user *trust and complacency* (Stanton & Marsden, 1996; Walker, et al., 2001). Suppose, for example, that an automated collision warning device becomes so heavily trusted that the driver ceases to carefully monitor the vehicle ahead and removes his or her eyes from the roadway for longer periods of time (Lee et al., 2002). In one study of automated braking, Young and Stanton (1997) found that many drivers intervened too slowly to prevent a rear-end collision should the automated brake fail to

TABLE 17.3   Examples of General Capabilities, Functions and Specific Outputs of In-Vehicle Automation

General Capabilities	Functions	Example of Specific Outputs
Routing and Navigation	1.1 Trip planning	Estimate of trip length
	1.2 Multi-mode travel coordination and planning	Bus schedule information
	1.3 Pre-drive route and destination selection	Route to destination
Motorist Services	2.1 Broadcast services/attractions	Lodging location
	2.2 Services/attractions directory	Electronic yellow pages
	2.3 Destination coordination	Location of and distance to restaurant
Augmented Signage	3.1 Guidance sign information	Scenic route indicators
	3.2 Notification sign information	Sharp curve ahead warning
	3.3 Regulatory sign information	Speed limit information
Safety and Warning	4.1 Immediate hazard warning	Emergency vehicle stopped ahead
	4.2 Road condition information	Traffic congestion ahead
	4.3 Automatic aid request	Notification of aid request
Collision Avoidance and Vehicle Control	5.1 Forward collision avoidance	Auditory warning
	5.2 Road departure collision avoidance	Virtual rumble strips through seat
	5.3 Lane change and merge collision avoidance	Graded alert based on lane position
Driver Comfort, Communication, and Convenience	6.1 Real-time communication	Cellular phone call
	6.2 Contact search and history	Name of last person called
	6.3 Entertainment and general information	Track number of current CD
	6.4 Heating, ventilation, air conditioning, and noise	Heater setting

function. Will the net effect be a compromise rather than an enhancement of safety? If the automated systems become so good that the reliability is extremely high, might not this lead to still more complacency and risk adaptation?

Second, for the kinds of automation that provide secondary information, such as navigation or trip planning aids, there is a danger that *attention may be drawn more into the vehicle,* away from the critical visual attention lobe, as the potentially rich automated information source is processed (Dingus et al., 1988; Lee et al., 1997). The wireless connection that makes it possible to automatically send your location to the ambulance dispatcher in the event of a crash also makes the content of the Internet available while you drive. The pressure for greater productivity may lead people to work on their email messages as they drive. New in-vehicle automation may tempt drivers with distractions far worse than the acknowledged hazards of the cell telephone (Walker et al., 2001).

Third, these devices *introduce a new type of productivity and safety tradeoff in driving.* At the start of the chapter, productivity and safety were described in terms of speed. Cell phones and Internet access make it possible to conduct work while driving, so productivity can also be considered in terms of how much work can be done while driving. The productivity benefit of cell phones may be substantial. One estimate of increased productivity enabled cell phones totals of $43 billion per year (Cohen & Graham, 2003). Interestingly, this estimate does not consider the decreased productivity and impaired work-related decision making that the distraction driving poses to the business transaction. Parkes (1993) showed that business negotiations conducted while driving led to poorer decisions. The cost to driving safety posed by cell phones is also substantial: $43 billion per year according to one estimate (Cohen & Graham, 2003). This recent analysis suggests the safety costs may outweigh the productivity gains.

These cautions do not mean that automobile automation is a bad idea. As we have seen, many of the safety-enhancing possibilities are clearly evident. But as we pointed out in Chapter 16, automation must be carefully introduced within the context of a *human-centered philosophy.* One promising approach described in Chapter 16 is *adaptive automation* in which sensors would monitor the driver for signs of fatigue, distraction, and hazardous driving conditions and adapt vehicle control, infotainment, and warning systems accordingly, perhaps locking out certain of the more distracting devices.

Conclusion

Driving is a very hazardous undertaking compared to most other activities both in and outside of the workplace. Following a comprehensive review of the state of roadway-safety enhancement programs, accident statistics, and safety interventions, Evans (1991) identified where greatest safety benefits to this serious problem can be realized. He argues that interventions in the human infrastructure will be more effective than those in the engineering infrastructure. The most effective solutions are those that address social norms—emphasizing the "noncost" dangers of driving with alcohol and the fact that fast driving and driving while distracted have the potential of killing many innocent victims. Legislation can help in this direction, but society's pressure can exert a more gradual but enduring change. As an example, a growing number of people refuse to hold telephone conversations with a person who is driving. This social pressure might change people's attitude and behavior regarding the use of cell phones while driving.

Finally, it can be argued that American society should be investing more research dollars into ways to improve this glaring safety deficiency. As Table 17.4

TABLE 17.4 Relation Between Research Expenditure and Fatalities

	Research Expenditures (Million $)	Years of Preretirement Life Lost (Millions)	Research Expenditure/ Years Lost ($/Year)
Cause:			
Traffic Injuries	112	4.1	27.3
Cancer	998	1.7	587.0
Heart Disease & Stroke	624	2.1	297.1

Adapted from L. Evans, 1991. *Traffic safety and the driver.* New York: Van Nostrand Reinhold.

shows, the ratio of research dollars expended to preretirement life lost is vastly lower for driving compared to cancer and heart/stroke ailments (Evans, 1991).

PUBLIC GROUND TRANSPORTATION

Statistically, it is far safer to take the bus (30 times), plane (30–50 times), train (7 times), or subway than it is to drive one's own vehicle (National Safety Council, 1989). For this reason, making public transportation more accessible is one of the ways to decrease the crash risk of drivers shown in Table 17.2. Bus drivers and airline pilots are more carefully selected and trained than automobile drivers, and rail-based carriers are, of course, removed from the hazardous roadways. Their added mass makes them considerably more "survivable" in high-speed crashes. As an added benefit, the increased use of public ground transportation is much kinder to the environment because the amount of pollution per passenger mile is much less than it is with personal vehicles. Finally, as any city commuter will acknowledge, it is sometimes much more efficient to take public transportation than to sit immobile in traffic jams during rush hour.

As a consequence of these differences in safety, efficiency, and environmental pollution, one of the important human factors issues in public ground transportation lies in the efforts to induce *behavioral changes* of the traveling and commuting public—making this segment of the population more aware of the lower risks, lower costs, and greater efficiency of public transportation (Nickerson & Moray, 1995; Leibowitz et al., 1995). Equally important are systemwide efforts to improve the *accessibility* of public transportation by designing schedules and routings in accordance with people's travel needs and so on.

Because the vehicles in public transportation are larger, the control inertia characteristics for hazard avoidance discussed in the section on hazards and collisions also become more critical. A long train, for example, may travel as far as a mile before it can come to a full stop following emergency braking, and elaborate energy-based displays can help the train engineer compute optimal speed management on hilly tracks (Sheridan, 2002). Trucks are much more susceptible to closed-loop instability (Chapter 9) than are cars.

Unlike buses and other road vehicles, subways and trains depend much more on a fully operating infrastructure. Tracks and roadbeds must be maintained, and railway safety is critically dependent on track switch and signal management. Recent major train accidents have resulted because of possible failures of ground personnel to keep tracks in the right alignment or to signal switches in appropriate settings. Fatigue, circadian rhythms, and shift work, discussed in Chapter 13, remain a major concern for many railroad workers.

Finally, air travel has a tremendous need for infrastructure support, both in maintaining the physical surfaces of airports, but in particular, providing the safety critical services of air traffic control. The human factors of air traffic control, having many features in common with industrial process control (sluggish, high risk, and complex), has many facets that we will not cover in the current text, reserving our discussion for the human factors of the pilot. Readers are referred to books by Hopkin (1997) and Wickens et al. (1997, 1998) for coverage of ATC human factors.

Maritime Human Factors

Maritime transportation operates in a particularly harsh and demanding environment, which presents several human factors challenges that make human error the predominant cause of maritime accidents (Wagenaar & Groeneweg, 1987). Maritime transportation is a 24-hour, 7-day a week operation, where people generally live on the ships for 30 to 60 days or more. This can make it difficult to get adequate sleep, particularly in rough seas where a rolling ship makes it difficult or impossible to stay in a bunk (Raby & Lee, 2001). New automation technology and economic pressures have led crew sizes on oil tankers and other large ships to shrink from over 30 to as few as 10 or 12 (Grabowski & Hendrick, 1993). These reductions can make the problems of fatigue worse because the system has less slack if someone gets sick or unexpected repairs need to be made. Fatigue and crew reductions contribute to many maritime accidents, including the grounding of the *Exxon Valdez* (NTSB, 1990). The social considerations, discussed in Chapter 19, and the factors affecting fatigue, discussed in Chapter 13, are particularly important for maritime transportation.

In addition to human performance issues of fatigue, large ships share many features with industrial process control, being extremely sluggish in their handling qualities, benefiting from predictive displays (von Breda, 1999; see Chapters 8 and 9), and also encouraging high levels of automation. Automation in ships includes not only autopilots, but also automatic radar plotting aids that help mariners plot courses to avoid collisions and electronic charts that automatically update the ship's position and warn mariners if they approach shallow water (Lee & Sanquist, 1996). These systems have great potential to enhance safety but can also cause problems. For example, mariners onboard the *Royal Majesty* only realized that the electronic chart had plotted the ship's position incorrectly when they hit ground (NTSB, 1997). One contribution to this accident is that the mariners believed the system to be infallible and failed to consult other information sources, a classic case of overtrust and automation-induced complacency. Another reason this automation can cause problems is that the training and certification for mariners does not always keep pace with the rapidly changing technology (Lee & Sanquist, 2000). The problems of trust in automation, discussed in Chapter 16, and training, discussed in Chapter 18, must be addressed if these systems are to enhance rather than diminish maritime safety.

Aviation Human Factors

The number of pilots is far smaller than the number of drivers, and aircraft crashes are much less frequent than auto accidents. However, the number of people who fly as passengers in aircraft is large enough, and the cost of a single air crash is sufficiently greater than that of a single car crash, that the human factors issues of airline safety are as important as those involved with ground transportation. In the following section, we discuss the aircraft pilot's task, the social context in which the pilot works, and the implications of stress and automation on aviation human factors (Tsang & Vidulich, 2003; Wickens, 2002a; Orlady & Orlady, 1999; Garland et al., 1999; O'Hare & Roscoe, 1990).

The Tasks

The task of the aircraft pilot, like that of the vehicle driver, can be described as a primary multiaxis tracking task embedded within a multitask context in which resources must be shared with other tasks. As compared with car driving, more of these tasks require communications with others. Furthermore, compared with driving, the pilot's tracking task is in most respects more difficult, involving higher order systems (see Chapter 9), more axes to control, and more interactions between axes. However, in some respects it is less difficult, involving a lower bandwidth (more slowly changing) input and a somewhat greater tolerance for deviations than the car driver experiences on a narrow roadway. The most important task is *aviating*—keeping the flow of air over the wings such as to maintain lift. The competing tasks involve maintaining *situation awareness* for hazards in the surrounding airspace, *navigating* to 3-D points in the sky, following *procedures* related to aircraft and airspace operations, *communicating* with air traffic control and other personnel on the flight deck, and *monitoring* system status. Much of the competition for resources is visual (see Chapter 6), but a great deal more involves more general competition for perceptual, cognitive, and response-related resources. Depending on the nature of the aircraft, the mission, and the conditions of flight, pilot workload ranges the extreme gamut from underload conditions (transoceanic flight) to conditions of extreme overload (e.g., military combat missions, helicopter rescue missions, single pilots in general aviation aircraft flying in bad weather).

Tracking and Flight Control.   To understand the considerable tracking demands of flying, review the material covered in Chapter 9. As Figure 17.5 shows at the top, the aircraft has six degrees of freedom of motion. It can rotate around three axes of rotation (curved white arrows), and it can translate along three axes of displacement (straight black arrows). Conventionally, rotational axes are described by *pitch, roll (or bank),* and *yaw.* Translational axes are described by lateral, vertical, and longitudinal (airspeed or "along track") displacement. (Actually, lateral displacement is accomplished, as in driving, by controlling the heading of the vehicle.) All six axes are normally characterized by some target or command input, such as a heading toward a runway, and tracking is perturbed away from these inputs by disturbance inputs, usually winds and turbulence. In controlling these degrees of freedom (a six-axes tracking task), the pilot has two primary goals. One is aviating—keeping the plane from *stalling* by maintaining adequate air flow over the wings, which produces lift. This is accomplished through careful control of the airspeed and the *attitude* of the aircraft (pitch and roll). The other goal is to navigate the aircraft to points in the 3-D airspace. If these points must be reached at precise times, as is often the case in commercial

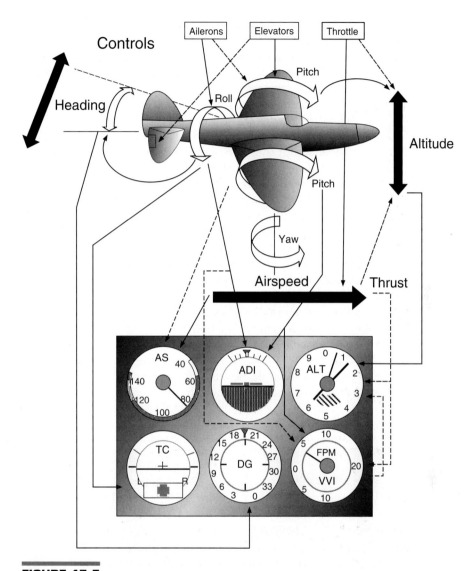

FIGURE 17.5

Flight control dynamics; controls (top) and primary flight displays (bottom). The thin solid lines represent direct causal influences. For example, adjustment of the throttle directly influences airspeed. The thin dashed lines represent axes interactions. For example, adjustment of the throttle, intended to influence airspeed, may also lead to an undesired loss of altitude.

aviation, then the task can be described as 4-D navigation, with time representing the fourth dimension.

To accomplish these tasks, the pilot manipulates three controls, shown at the top of the figure: The *yoke* controls the elevators and ailerons, which control the pitch and bank respectively, each via first-order dynamics (i.e., yoke position determines the rate of change of bank and of pitch). The *throttle* controls air-speed, and the *rudder pedals* are used to help coordinate turning and heading changes. Those direct control links are shown by the solid, thin arrows at the top of Figure 17.5.

Three facets make this multielement tracking task much more difficult than driving: the displays, the control dynamics, and interaction between the six axes. First, in contrast to the driver's windshield, the pilot receives the most reliable information from the set of "steam guage" displays shown at the bottom of the figure, and these do not show a good, integrated, pictoral representation of the aircraft that directly supports the dual tasks of aviating and navigating. Second, the dynamics of several aspects of flight control are of higher order, which we saw in Chapter 9, are challenging because of lags and instability, imposing needs for anticipation. For example the control of altitude is a 2nd order task, and that of the lateral deviation from a desired flight path is a 3rd order task. Third, the axes often have *cross-couplings*, signaled by the dashed lines in the figure, such that if a pilot makes a change in one axis of flight (e.g., decreasing speed), it will produce unwanted changes in another (e.g., lowering altitude).

With the development of more computer-based displays to replace old electromechanical "round dial" instruments in the cockpit (Figure 17.5, see also Figure 8.8), aircraft designers have been moving toward incorporating human factors display principles of *proximity compatibility, the moving part,* and *pictorial realism* (Roscoe, 1968, 2002; see Chapter 8) to design more user-friendly displays. Compare, for example, the standard instrument display shown in Figure 17.5 with the current display in many advanced transport aircraft (Figure 17.6a)

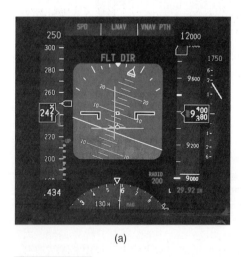

(a)

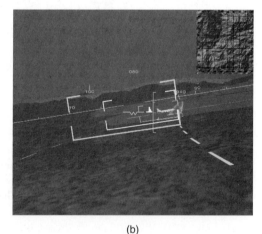

(b)

FIGURE 17.6

(a) Flight displays for modern commercial aircraft. (*Source:* Courtesy of the Boeing Corporation); (b) Flight display envisioned for future aircraft. Note the preview "tunnel in the sky" and the digital representation of terrain beyond.

and with even more integrated displays proposed for future design (Figure 17.6b).

Given the sluggish nature of aircraft dynamics, a valuable feature on almost every advanced display is the availability of *prediction* (of future aircraft position) and *preview* (of future command input) (Jensen, 1982; Haskell & Wickens, 1993). Somewhat further in the future is the implementation of 3-D displays, such as that shown in Figure 17.6b (Fadden et al., 2001). In spite of their promise, the advantages of such displays, in terms of their ability to integrate three axes of space, may sometimes be offset by their costs, in terms of the ambiguity with which 3-D displays depict the precise spatial location of aircraft relative to ground and air hazards (Wickens, 2003; see Chapter 8).

Maintaining Situation Awareness.   Piloting takes place in a dynamic environment. To a far greater extent than in driving, much of the safety-relevant information is not directly visible in its intuitive "spatial" form. Rather, the pilot must depend on an understanding or *situation awareness* of the location and future implications of hazards (traffic, terrain, weather) relative to the current state of the aircraft (see Chapter 6) as well as awareness of the state of automated systems within the aircraft itself (Sarter & Woods, 2000).

Solutions to situation awareness problems include head-up displays which allow parallel integrated electronic displays, such as that shown in Figure 17.6b, can allow the pilot to visualize a much broader view of the world in front of the aircraft than can the more restricted conventional flight instruments, supporting awareness of terrain and traffic hazards. However, as with head-up displays, potential human factors costs are associated with more panoramic 3-D displays. They may need to occupy more display "real estate" (a requirement that imposes a nontrivial engineering cost), and as noted above and in Chapter 8, when such displays are portrayed in 3-D perspective, they can make the precise judgments of where things are in space difficult because of the ambiguity of projecting a 3-D view onto a 2-D (planar) viewing surface (McGreevy & Ellis, 1986; Wickens, 2003).

Following Procedures.   As described in Chapter 6, a pilot's prospective memory for carrying out particular procedures at particular times, is supported both by vast amounts of declarative and procedural knowledge, but, in particular by *knowledge in the world* in the form of several checklists, devoted to different phases of flight (e.g., preflight, taxi, takeoff, etc.) and to different operating conditions (e.g., normal, engine-failure, fire).

As noted elsewhere, however, following checklists (e.g., set switch A to "on") can be susceptible to two kinds of errors. First, as discussed in Chapter 8, top-down processing (coupled with time pressure) may lead the pilot to "see" the checklist item in its appropriate (expected) state, even if it is not. Second, as discussed in Chapters 6 and 13, the distractions and high workload of a multitask environment can lead the pilot to skip a step in the checklist; that is, the distraction may divert the pilot's attention away from the checklist task, and attention may return to it at a later step than the one pending when the distraction occurred. This might have been the case in the Detroit Airport crash. The pilot's attention was diverted from the checklist by a procedural change called for by air traffic control, and attention apparently returned to the checklist at a point just after the critical "set flaps" item. These problems may be addressed by designing redundancy into checklist procedures (Degani & Wiener, 1993) and by automation (Bresley, 1995).

The Social Context

Since the 1980s, the aviation community has placed great emphasis on understanding the causes of breakdowns in pilot team performance. Often these breakdowns have been attributed to oral communications problems of the sorts discussed in Chapter 5; and these may interact with personality, as when a domineering senior pilot refuses to listen to an intimidated junior co-pilot who is trying to suggest that "there may be a problem" on board (Helmreich, 1997; Wiener, Kanki, & Helmreich, 1993). Sometimes the breakdowns may relate to poor leadership, as the captain fails to use all the resources at his or her disposal to resolve a crisis. Training programs called *Crew Resource Management*, designed to teach pilots to avoid these pitfails have been developed over the past two decades (see Chapter 19). Some have shown success (Diehl, 1991), but the record of these programs in consistently producing safer behavior in flight has been mixed (Salas et al., 2002).

Supporting the Pilot

Finally, we return to the importance of the three critical agents that support the pilots. First, maintenance technicians, and their inspection and trouble shooting skills, described in Chapters 3, 4, and 6, play a safety-critical role. Second, pilots are increasingly dependent upon automation. Paralleling our discussion of general issues of automation in Chapter 16, aircraft automation can take on several forms: Autopilots can assist in the tracking task, route planners can assist in navigation, collision avoidance monitors can assist in monitoring the dangers of terrain and other aircraft (Pritchett, 2001), and more elaborate *flight management systems* can assist in optimizing flight paths (Sarter & Woods, 2000). Some automated devices have been introduced because they reduce workload (autopilots), others because they replace monitoring tasks that humans did not always do well (collision alerts), and still others like the flight management system were introduced for economic reasons: They allowed the aircraft to fly shorter, more fuel-conserving routes.

As we noted in Chapter 16, many of the human factors issues in automation were directly derived from accident and incident analysis in the aviation domain, coupled with laboratory and simulator research. From this research has evolved many of the guidelines for introducing *human-centered automation* (Billings, 1996; Parasuraman & Byrne, 2003) that were discussed in that chapter.

Third, pilots are supported by air traffic control. On the whole, the air traffic control system may be viewed as remarkably safe, given what it has been asked to do—move millions of passengers per year through the crowded skies at speeds of several hundred miles per hour. On September 11, 2001, air traffic controllers safely landed 4,546 aircraft within 3 hours following the terrorist hijackings in New York, Washington, DC, and Pennsylvania (Bond, 2001). This safety can be attributed to the considerable redundancy built into the system and the high level of the professionalism of the ATC workforce. Yet arguments have been made that the high record of safety, achieved with a system that is primarily based on *human* control, has sacrificed efficiency, leading to longer-than-necessary delays on the ground and wider-than-necessary (to preserve safety) separations in the air. A consequence has been a considerable amount of pressure exerted by the air carriers to automate many of the functions traditionally carried out by the human controller under the assumption that intelligent computers can do this more accurately and efficiently than their human counterparts (Wickens et al., 1998).

CONCLUSION

The human factors of transportation systems is a complex and global issue. An individual's choice to fly, drive, or take public ground transportation is influenced by complex forces related to risk perception, cost perception, and expediency. The consumer's choice for one influences human factors issues in the others. For example, if more people choose to fly, fewer automobiles will be on the road, and unless people drive faster as a result, highways will become safer. However, the airspace will become more congested and its safety will be compromised. Air traffic controllers will be more challenged in their jobs. In the continuing quest for more expediency, demands will appear (as they have now appeared) for either greater levels of air traffic control automation or for more responsibility to be shifted from air traffic control to the pilots themselves for route selection and for maintaining separation from other traffic, in a concept known as "free flight" (Wickens et al., 1998; Metzger & Parasuraman, 2001). The technology to do so becomes more feasible with the availability of the global positioning system. Collectively, if these factors are not well managed, all of them may create a more hazardous airspace, inviting the disastrous accident that can shift risk perceptions (and airline costs) once again.

Such global economic issues related to risk perception and consumer choice (itself a legitimate topic for human factors investigation) will impact the conditions in which vehicles travel and the very nature of those vehicles (levels of automation, etc.) in a manner that has direct human factors relevance to design.

Chapter

18

Selection and Training

In 2002, the new Transportation Security Agency was tasked with creating a large workforce of airport inspectors who could reliably discriminate the large number of regular passengers from the tiny fraction of those who might board an aircraft with hostile intent. Various approaches can be proposed to support this effort. Good displays, workstations, and job design could help. So could automatic screening devices and intelligent decision aids, as discussed in Chapter 16. But a key component in supporting effective performance is the selection of workers who have the good skills in visual search and decision making, along with the high degree of motivation and the interpersonal skills necessary to avoid giving passengers a negative experience during the screening process. Are there tests to predict such skills and personality traits? Suppose there are not enough people who possess those skills to fill the necessary positions? In this case, skill deficiency may be supported by online job aids that assist the person in carrying out the task: a set of instructions on how to carry out a personal search, for example, or a picture of what a typical weapon's image might look like. Finally, it is inevitable that even those who do possess the skills will benefit from some training regarding what to look for, characteristics of people who might be suspicious, and the best scan pattern to find weapons in the shortest period of time.

Throughout this book, we have emphasized the importance of high-quality human performance. At the most general level, there are three routes to achieving this goal: design, selection, and training. Most of the book so far has focused on design of the task, of the environment, and of the interface. In this chapter, we address the second two routes to effective performance: *selection* and *training*. Selection involves choosing the right person for the job, a choice that, ideally, should be made via assessment before hiring the person or before the person is assigned to the job where the necessary job skills will be acquired. That is, much of selection involves **prediction,** on the basis of an assessment, of who will do well or poorly in a particular job. Such prediction can be made, given that we have faith that certain enduring *abilities* and *personality* traits can be measured in advance, before hiring or job assignment, and these attributes will carry over to the workplace to support effective performance.

Training assumes the necessity of putting knowledge in the head (Norman, 1988) to support effective performance. The question is, How can we support workers in rapidly acquiring this knowledge so that it can be used effectively in the workplace and so that it will endure, not being forgotten? Clearly, both selection and training work hand in hand. For example, not everyone has the abilities to be an effective combat pilot, teacher, or leader, and it would be nice to select those in advance who have the potential to succeed, without "wasting" training time on those who will fail. But all of these professions, and many more, require vast amounts of declarative and procedural knowledge, which must be acquired on the job or in separate specialized training programs. This chapter discusses both topics.

In addition to selection and training, which provide different complementary approaches to supporting job skills, we consider a third element closely allied with training: *performance support*. Performance supports can be thought of as training tools that are present at the time the job is performed in the workplace. They provide knowledge in the world to support effective performance, but at the same time, support the acquisition of knowledge in the head regarding how to do the job. The importance of performance support for people with disabilities is addressed here.

PERSONNEL SELECTION

Personnel selection is chronologically the first approach taken to maximize the skills and knowledge needed by an employee to perform a job. Selection has been a critical concern for government agencies such as the armed forces, and a long tradition of research in areas such as personnel psychology has grown out of this concern (Borman et al., 1997). The major focus of selection research is to identify reliable means of predicting future job performance. A second focus is to categorize accepted applicants into the job type for which they may be most suited.

A number of methods are used today to select employees for a particular job; such methods include interviews, work histories, background checks, tests, references, and work samples. Some use techniques that have been scientifically developed and validated; others use methods that are informal and depend heavily on intuition. A long line of research has demonstrated that, in general, the best techniques for selection include tests of skills and abilities and job-related work samples. The poorest methods (although they are still widely used) are interviews and references from previous employers (Osburn, 1987; Smither, 1994; Ulrich & Trumbo, 1965).

Selection can be conceptualized in terms of signal detection theory (see Chapter 4); where

hit = hiring a person who will be good at the job
miss = not hiring someone who would do a good job
false alarm = hiring someone who ends up being unacceptable or doing a poor job
correct rejection = not hiring someone who in fact would not do a good job if he or she had been hired

Framed this way, selection is usually performed using any means possible to maximize the number of employee hits (successes) and minimize the number of false alarms. Employers have traditionally been less concerned with the people that they do not hire. However, recent Equal Employment Opportunity (EEO) laws require that all individuals have equal opportunity with regard to employment. While no employer is required to hire individuals who cannot do the work, neither can they arbitrarily refuse to hire those who can. Obviously, this means that employers must be careful to use selection procedures that are valid and fair; that is, the selection criteria are *directly related* to job skills and abilities. Selection using irrelevant criteria is considered employment discrimination. As an example, firefighters cannot be selected on the basis of gender alone. However, a selection test could require applicants to lift and move 100 pounds of coiled fire hose if that task is considered part of the normal job.

Basics of Selection

Identifying people who will successfully perform a job first requires a thorough analysis of the duties or behaviors that define a job, a process termed *job analysis*. Job analysis (which is closely related to task analysis) is the basis of many related activities, such as selection, training, performance appraisal, and setting salary levels. Job analysis typically includes specifying the tasks normally accomplished, the environments in which the tasks are performed, and the related knowledge, skills, and abilities required for successful task performance (Smither, 1994).

Once the job knowledge, skills, and abilities have been identified, employers must prioritize them with respect to which knowledge and skills are essential for job entry and which are desirable but not essential. Employers then look for applicants who either already have the task-specific knowledge and skills required for a job or show evidence of having basic knowledge and abilities (such as mathematical ability or psychomotor skills) that would eventually lead to successful job performance. Many businesses and government agencies face high numbers of cases in the second category. This is because students directly out of high school or college rarely have enough specific job skills to allow selection on the basis of job skills alone. Instead, employers must select people based on criteria that are not measures of job skills but of basic abilities that are fundamental to eventual job performance.

A measure that is highly correlated with ultimate job performance is said to have high *criterion-related validity*. A measure with high validity is extremely useful for selection because employers can assume that applicants receiving a high score on the test will probably perform well on the job. Obviously, the higher the correlation coefficient, the more confidence the employer can have that high scores are predictive of high job performance. No test scores are perfectly related to job performance, and thus employers must deal with uncertainty. Figure 18.1 shows this uncertainty problem in the context of a signal detection analysis. The employer must select a score cutoff for the predictive measure that will maximize selection success (hits). This is relatively easy if there are enough applicants with high scores to eliminate the people falling in the lower right quadrant (false alarms). However, when the applicant pool is relatively small, setting the cutoff level so high may not be possible. This gives us some insight into why the armed forces seem to recruit so vigorously and offer big dividends for enlistment, thereby raising their applicant pool and assuring more people to the right of the criterion cutoff in Figure 18.1.

Selection Tests and Procedures

Not all selection procedures are equally effective, and the unsuccessful false alarms in Figure 18.1 can translate into thousands or millions of dollars lost for an organization (e.g., it costs over $1 million to train a competent fighter pilot).

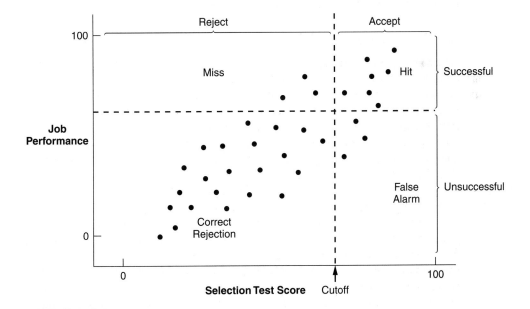

FIGURE 18.1

Hypothetical relationship between selection test and eventual job performance. The criterion related validity of the test can be expressed as the correlation between the test scores (*x* axis) and the measure of job performance (*y* axis).

Selection errors result in problems such as more training time and expense than necessary, supervisors or other staff having to compensate for inadequate performance, and supervisors having to spend time on reviews, feedback, and documentation of problems. In addition, poor selection can result in low employee morale, resentment, and complaints. This section describes some of the commonly used selection tests and procedures and notes those that seem to be most predictive of job performance.

Measures of Cognitive Ability.　Many commonly used selection tests are standardized tests of cognitive or information-processing abilities. People have numerous abilities, which are used in various combinations for task performance. Typical categories of cognitive ability measured for selection include general ability or intelligence, verbal ability, perceptual ability, numerical ability, reasoning or analytical ability, perceptual speed, memory, and spatial-mechanical abilities (Osburn, 1987; Ackerman & Cianiallo, 2000). Schmidt and Hunter (1981) presented evidence that cognitive ability tests are valid predictors of job performance, usually more valid than other assessment procedures. For complex jobs, measures of general intelligence are often very effective predictors (Borman et al., 1997). General intelligence is closely related to working-memory capacity, and we saw in Chapter 6 the importance of working memory in a variety of complex mental operations. In addition, Hunter and Hunter (1984) found that tests of verbal and numerical ability were better predictors for jobs with high complexity, while tests of motor coordination and manual dexterity were better predictors for jobs with low complexity.

Some jobs may have additional or more restricted requirements for specific information-processing capabilities. For example, some researchers suggest that driving and flying tasks rely heavily on abilities in the areas related to attention breadth and flexibility (e.g., Gopher & Kahneman, 1971; Kahneman et al., 1973; Gopher et al., 1994; Ball et al., 1993). Measures of selective attention could therefore be used for employment decisions (e.g., Gopher, 1982). Finally, certain jobs require a complex combination of skills, and selection methods should reflect this complexity. For example, in the aerospace domain, Hunter and Burke (1994) performed an analysis using 68 published studies of methods for pilot selection. They found that a *battery* of several measures of cognitive ability was best able to predict pilot success, including tests of verbal and numerical ability, mechanical knowledge, spatial ability, perceptual speed, and reaction time. On the whole, specific ability tests appear to be better equipped to make specific job classification assignments to one who is already accepted than to make overall selection decisions (Borman et al., 1997).

Measures of Physical Ability and Psychomotor Skills.　Some jobs require physical strength in particular muscle groups, physical endurance, manual dexterity, and/or psychomotor skills. It is therefore common and legally acceptable to select employees on the basis of tests measuring these abilities. Physical ability measures often include static strength, dynamic strength, trunk strength, extent flexibility, gross body coordination, gross body equilibrium, stamina, and aerobic fitness characteristics, described in detail in Chapters 10, 11, and 12. Other tests focus on motor abilities such as manual dexterity, finger dexterity, and arm-hand steadiness (Osburn, 1987).

Personality Assessment. Personality assessment has become more popular for selection in recent years (Borman et al., 1997). There are generally two different types of standardized personality measures. The first is what might be termed "clinical" measures because they primarily identify people with mental illness or behavioral disorders. Examples include the well-known Minnesota Multiphasic Personality Inventory (MMPI). Such traditional personality tests are not particularly appropriate for employee selection; they have not proven to be valid for prediction of success (Newcomb & Jerome, 1995), and they are often troublesome from a legal point of view (Burke, 1995a).

The other type of personality test measures personality dimensions that are found in one degree or another in all people. Examples of tests that measure general personality characteristics include Cattell's 16PF (Cattell et al., 1970), and the Eysenck Personality Inventory (Eysenck & Eysenck, 1964). Recent work on using personality measures for selection has indicated that five basic personality factors or clusters are useful in predicting job performance (Barrick & Mount, 1991; Hogan et al., 1997):

Neuroticism: Cluster of traits such as anxiety, depression, impulsiveness, and vulnerability.
Extroversion: Cluster of traits such as warmth, gregariousness, activity, and positive emotions.
Openness: Includes feelings, actions, ideas, and values.
Agreeableness: Cluster of traits including trust, altruism, compliance, and straight-forwardness.
Conscientiousness: Includes competence, order, dutifulness, achievement striving, and self-discipline.

Barrick and Mount (1991) found that the *conscientiousness* factor was effective in predicting performance in a wide array of jobs, including police, managers, salespeople, and skilled or semiskilled workers. Consistent with this finding, researchers evaluating the potential of personality tests for pilot selection have found that conscientiousness is the most strongly predictive measure (Bartram, 1995). Recent research has also found some success in the predictive value of tests of honesty and conscientiousness (Borman et al., 1997).

Work Samples and Job Knowledge. Work sampling typically requires applicants to complete a sample of work they would normally be required to perform on the job. Examples include a driving course for forklift operators, a typing test for secretaries, and an "in-basket test" where management candidates must respond to memos frequently found in a manager's mailbox. While realistic samples are most valid (Burke, 1995b; Hunter & Hunter, 1984; Hunter & Burke, 1995), they are often expensive to assess. A less costly but still somewhat effective method is to provide a video assessment (Smither, 1994) in which job candidates view videotapes that portray workers in situations that require a decision. The applicants see a short scenario and then are asked how they would respond in the situation.

Work samples can of course extend for longer periods, in which case they may be described as *miniature job training,* a technique shown to have strong predictive validity (Reilly & Chao, 1982; Siegel, 1983). The demonstration of work samples on the part of the applicant of course requires some job knowledge. In this regard there is also good evidence that *job knowledge tests,* assessing knowledge about the domain in which the job is performed, often provide better predictive validity than do abilities tests (Borman et al., 1997). The advantage of such tests is probably twofold. First, those who possess high job knowledge should be able to transfer this knowledge to the job. Second, those who have acquired such knowledge are likely to be intrinsically interested in the job domain, reflecting a motivational factor that will also contribute to better job performance.

Structured Interviews.   As noted, interviews and "personal impressions" are relatively poor tools for selection compared to more objective tests (Osburn, 1987; Dawes et al., 1989). So too are reference letters. Smither (1994) describes several interesting reasons for the poor predictive ability but widespread use of interviews. Probably the strongest factor currently biasing references is past employers' fear of litigation (Liebler & Parkman, 1992). However, interviews can also be valuable as a recruitment tool for an applicant who is already inferred to exceed the acceptance criterion (Borman et al. 1997). While interviews have relatively poor predictive validity, they can be made more predictive by using certain **structuring** methods (Friedman & Mann, 1981; Borman et al., 1997). At a minimum, questions should be based on and related to knowledge and skills identified in the job analysis. Other methods for structuring the interview focus on asking applicants to describe previous work behaviors. For example, Hendrickson (1987) suggests using the "critical behavior interview" approach. With this method, applicants are asked to discuss recent occasions when they felt they were performing at their best. They are asked to describe the conditions, what they said or did, and so on. The interviewer looks for and scores behaviors that are consistent with job-related selection criteria. Interviews that culminate in scoring procedures are generally more valid than those that result in only a yes/no overall evaluation (Liebler & Parkman, 1992).

Conclusion.   In summarizing the collective evidence for the various forms of assessment to be used as job predictors, there is now substantial evidence that all assessments have something to offer (Schmidt & Hunter, 1998), and each can offer predictive power that is somewhat different from the others. However, the ability of assessment techniques to fully predict performance, particularly on complex jobs, will always be limited because of the great amount of knowledge that must be acquired through experience. How this knowledge is supported through job aids and training is the focus of the rest of the chapter.

PERFORMANCE SUPPORT AND JOB AIDS

Jobs have become increasingly complex, and the knowledge and skills needed for successful job performance are changing rapidly. It is difficult to provide enough training for employees to cope with the volume and rapid turnover of information and technology related to their tasks. As an example, imagine trying to provide training for the phone-in help service operators of a computer software company. These people need to know a vast amount of information or at least know where to find it within a matter of seconds. The amount of information required for many jobs is simply too large to impart through traditional training methods such as classroom instruction.

Because of the increasingly poor fit between job needs and standard training methods, such as seminars and instructional manuals, performance technology specialists are moving toward a direct *performance-support* approach. This philosophy assumes that information and training activities (such as practice) should be provided on an *as-needed* basis, shifting a "learn-and-apply" cycle to a "learning-while-applying" cycle (Rosow & Zager, 1990; Vazquez-Abad & Winer, 1992). It is considered more efficient to allow people to access information (and learn) while they are doing a task rather than to try to teach them a large body of knowledge and assume they will retrieve it from memory at some later time. Performance support is the process of providing a set of information and learning activities in a context-specific fashion *during* task performance. Performance support is frequently the preferred method (Geber, 1991; Gery, 1989; Vazquez-Abad & Winer, 1992); it is more efficient and often preferred by employees because it is less taxing on memory (training in one context does not have to be remembered and carried over to the job context).

This "efficiency" viewpoint is often applied to instruction of software users (e.g., Spool & Snyder, 1993). Figure 18.2 illustrates a continuum of methods used by software interface designers for helping users learn new software. The right side shows the most desirable circumstance, where system "affordances" make the software inherently easy to use. There is maximum knowledge in the

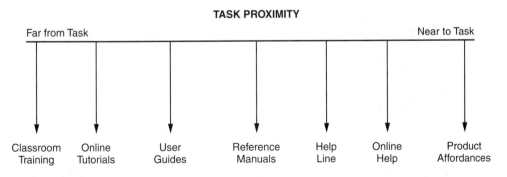

FIGURE 18.2

Continuum of computer interface training methods.

world. It wastes the least time for users and does not rely on user capabilities and motivation. The least desirable support is the traditional "learn-ahead-of-time" classroom instruction because it is so dependent on learner motivation, comprehension of the material, and retention of information. Consistent with this view, researchers in human factors are arguing more forcibly against traditional training that imparts a large body of declarative knowledge before people do the tasks in which the knowledge is used (e.g., Mumaw & Roth, 1995). As we noted in Chapter 6, such knowledge may often be inert and not easily transferable to the real world of the job environment.

Job Aids and Instructions.   A job aid is a device or document that guides the user in doing a task while the user is performing it (Swezey, 1987). In either paper or computer-based form, it should be available when and where the user needs it. Examples of job aids are the daily to-do list, a recipe, note cards for a speech, a computer keyboard template, instructions for assembling a product, or a procedural list for filling out a form (tax forms come with extensive job aids). A job aid can be a few words, a picture, a series of pictures, a procedural checklist, or an entire book. A well-designed job aid promotes accurate and efficient performance by taking into account the nature and complexity of the task as well as the capabilities of the user.

Traditionally, an important form of job aid for performance support is the instruction manual—often but not necessarily on paper. Psychologists know a fair amount about effective instructions, much of it drawn from material on comprehension (as discussed in Chapter 6), and effective display design (discussed in Chapter 8). Wright (1977) has outlined a particularly effective and compelling set of empirically based guidelines for printed technical instructions, which include the caution against using prose (or prose alone) to present very complex sets of relationships or procedures and the recommendation that such prose can often be replaced by well-designed flow charts. Wright's guidelines also highlight the effective use of pictures that are redundant with or related to words in conveying instructions, as illustrated in Figure 18.3. This is another illustration of the benefits of redundancy gain described in Chapter 8 (see Booher, 1975; Wickens & Hollands, 2000). Wright also notes the importance of locating pictures or diagrams in close proximity to relevant text, an example of the proximity-compatibility principle.

The phrasing of any text should of course be straightforward, as discussed in Chapter 6, and illustrations should be clear. In this regard it is important to emphasize that clarity does not necessarily mean photo realism (Spencer, 1988). In fact, in instructions such as emergency procedures in passenger aircraft evacuation, well-articulated line drawings may be better understood than photographs (Schmidt & Kysor, 1987). Finally, with voice synthesis becoming increasingly available as an option for multimedia instructions, it is important to note that research indicates an advantage for voice coupled with pictures when presenting instructions (Nugent, 1987; Tindall-Ford et al., 1997; Meyer, 1999). With this combination, words can be used to provide information related

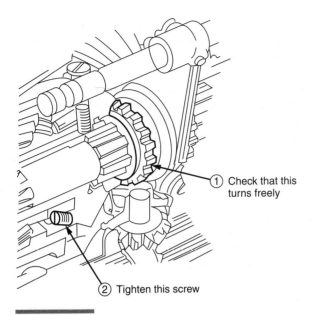

1 Check that this turns freely

2 Tighten this screw

FIGURE 18.3

Advantage of partially redundant combination of pictures and words. Imagine the difficulty of trying to convey this information entirely with words. (*Source:* Wright, P., 1977. Presenting technical information: A survey of research finding. *Instructional Science, 6,* 93–134. Reprinted by permission of Kluwer Academic Publishers.)

to pictures, but in contrast to print, the eyes do not have to leave the pictures as the words are being processed.

While job aids are often the right performance support solution, they are not without their shortcomings. Recent reviews have indicated that misuse of checklists was partially responsible for several major airline accidents (e.g., Degani & Wiener, 1993), as discussed in Chapter 17, and checklist problems have been identified in other industries as well (Swain & Guttmann, 1983). Degani and Wiener (1993) describe a number of human errors associated with the use of checklists, such as overlooking an item in a long checklist, thinking that a procedure on the checklist had been completed when it had not, and being temporarily distracted from checklist performance.

Embedded Computer Support. As so many tasks are now performed on computer workstations, it is quite feasible for intelligence within the computer system to infer the momentary information needed by the user and automatically provide access to additional information relevant to the inferred task at hand (Hammer, 1999). These on-line help systems were described in Chapter 15. Such an example of *adaptive automation,* discussed in Chapter 16, can certainly have its benefits, but may impose modest or even more serious problems in interrupting the ongoing task (Bailey et al., 2001; Czerwinski et al., 2000), as discussed in Chapter 6.

A final question involves knowing when to use performance support, training, or a combination of both. Most instructional design models have a step where this decision is made. Some guidelines also exist to help designers with this decision. Table 18.1 lists a number of guidelines provided by various researchers (e.g., Gordon, 1994). However, keep in mind that these suggestions assumed relatively basic performance support systems and may be less applicable for advanced displays or intelligent agents.

SUPPORTING PEOPLE WITH DISABILITIES

The issues of selection, individual differences, and job support are particularly critical in addressing the challenges of people with disabilities. Generally, these characterize broad classes of visual, hearing, cognitive, and physical impairment, the latter related either to injury or disease, such as multiple sclerosis. The 2000 U.S. census reveals that approximately 20 percent of the population possess formally defined disabilities. These disabilities increase in frequency for the older retirement-age population. But also for the younger population, disabled people represent a substantial portion of the workforce, and it is estimated that roughly one-third of those with disabilities who can and would like to work are unemployed (Vanderheiden, 1997). The issue of job support for people with disabilities has become particularly important, given the guidance of the Americans with Disabilities Act. However, the importance of such support extends well beyond the workplace to the schools, communities, and homes.

TABLE 18.1 **Factors Indicating Use of Performance Support Systems or Training**

Use Performance-Support Systems When

The job or tasks allow sufficient time for a person to look up the information.
The job requires use of large amounts of information and/or complex judgments and decisions.
Task performance won't suffer from the person reading instructions or looking at diagrams.
The job or task requires a great number of steps that are difficult to learn or remember.
Safety is a critical issue, and there are no negative repercussions of relying on a job aid.
The task is performed by a novice, or the person performs the job infrequently.
The job involves a large employee turnover rate.
The job is one where employees have difficulty obtaining training (due to distance, time, etc.).

Use Training Systems When

The task consists of steps performed quickly and/or in rapid succession.
The task is performed frequently.
The task must be learned in order to perform necessary higher level tasks (e.g., read sheet music in order to play an instrument).
The person wishes to perform the task unaided.
The person is expected to perform the task unaided.
Performance of the task would be hindered by attending to some type of aid.
The task is psychomotor or perceptual, and use of a job aid is not feasible.

The issues of selection and individual differences are relevant because of the need for formal classification of a "disability," in order to define those who are eligible for special services and accommodations that the particular impairment may require. For example, the formal definition of "legally blind" is a vision that is 20/200 *after* correction, or a functioning visual field of less than 20 degrees.

Vanderheiden (1997) identifies three general approaches can be adopted to support the disabled person on the job or elsewhere: (1) Change the individual through teaching and training strategies that may allow tasks to be done more easily. (2) Provide tools, such as hearing aids, wheelchairs, or prosthetic devices, that will restore some of the original functioning. In the design of such tools, several human factors principles of usability become evident. They should be functional, but they should also be employable without expending excessive mental or cognitive effort. Furthermore, where possible, designers should be sensitive to the possible embarrassment of using certain prosthetic devices in public places. (3) Change the design of "the world," in the workplace, school, community, or home, to better support effective performance of those with disabilities.

The third approach—changing design—might initially appear to be expensive and unnecessary, as it is intended to directly support a minority of the population. However, as Vanderheiden (1997) points out, in describing the concept of *universal design,* many of the design features that support the disabled make the world more usable for the rest of the population as well. For example, ramps for wheelchair users, rather than curbs, are less likely to lead to trips and falls for those who walk. Highly legible displays are more readable for all people in degraded reading conditions, and not just to those with visual impairments. In terms of cognitive impairment, making instructions simple, easy to read, and supported by graphics for the mentally retarded will greatly support those who are not native speakers of the language, may have low reading skills, or may need to follow the instructions in times of stress.

Many steps that can be taken toward universal design are those associated generally with "good design," as described elsewhere in this book. In addition, Vanderheiden (1997) provides an effective and exhaustive set of "design options and ideas to consider" for each general class of impairments.

TRAINING

Learning and Expertise

In Chapter 6, we described in detail the mental processes involved in storing information. Perceived information is given "deeper processing" via attention-demanding operations in working memory, and sufficient processing leads to long-term memory storage of facts—declarative knowledge—and the formation of connections and associations which are often characteristic of procedural knowledge. Also, practice and repetition of various perceptual and motor skills embody these more permanent representations in long-term memory.

Psychologists have sometimes associated the development of permanent memories with three different stages in the development of *expertise* (Anderson,

1995; Fitts & Posner, 1967). (1) Initially, knowledge about a job or a task is characterized primarily by *declarative knowledge*. Such knowledge is often not well organized, and it may be employed somewhat awkwardly in performance of the job. As discussed in Chapter 15, a novice computer user may be required to look up many needed steps to accomplish operations in order to support the more fragile declarative knowledge. (2) With greater familiarity and practice, *procedural knowledge* begins to develop, generally characterized by rules and if-then statements, which can be recalled and employed with greater efficiency. (3) Finally, there is a fine tuning of the skill, as *automaticity* develops after weeks, months, and sometimes years of practice. Automaticity was a concept discussed in Chapter 6, in which performance requires little attention and is carried out quite rapidly.

These three stages generally follow upon each other gradually, continuously and partially overlapping rather than representing sudden jumps. As a consequence, performance in the typical skill improves in a relatively continuous function. When, as shown in Figure 18.4, the performance measure is one measured by errors or time, such that high measures represent "poor" performance, the typical learning curve on a skill, proceeding through the three stages, follows an exponential decay function like the solid line in the graph (Newell & Rosenbloom, 1981). However, different performance aspects of the skill tend to emerge at different times, as shown by the three dotted lines. Error rates typically decrease initially, but after errors are eliminated, performance time becomes progressively shorter, and finally, continued practice reduces the attention demand until full automaticity is reached. For example if a skill is carried out with an inconsistent mapping of events to actions, full automaticity may never develop (Schneider, 1985).

Naturally, the representation shown in Figure 18.4 is schematic rather than exact. The rate of reduction of errors, time, and attention-demand varies from skill to skill. Furthermore, some complex skills may show temporary plateaus in the learning curve, as the limitations of one strategy in doing the task are en-

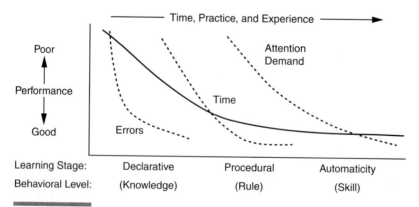

FIGURE 18.4

The development of skilled behavior.

countered and a new, more efficient strategy is suddenly discovered and practiced (Bryan & Harter, 1899). Finally, it is important to note that there is no absolute time scale on the x axis. Some skills may be fully mastered in as short a time as a few minutes of practice; others may take a lifetime to perfect (Fitts & Posner, 1967).

The three phases of skill learning, shown below the x axis of Figure 18.4, may strike you as somewhat analogous to the behavioral taxonomy proposed by Rasmussen (1983) and described in detail in Chapter 7: distinguishing knowledge-based from rule-based, from skill-based behavior, as these behaviors are shown at the bottom of the figure. Such a mapping between the two "trichotomies" is quite appropriate. They are not entirely identical, however, because in the context of Rasmussen's behavioral taxonomy, the *highly-skilled* operator, making decisions in a complex domain, will be able to move rapidly back and forth between the different behavioral levels. In contrast, what Figure 18.4 represents is the fact that any operator must benefit from extensive experience in the task domain to achieve the automaticity that is characteristic of skill-based behavior.

Figure 18.5 presents another way of looking at the acquisition of skills. The timeline shows the development from the novice (on the left) to the expert (on the right), a process that may take many years as various aspects of the task proceed from declarative to procedural knowledge and then to automaticity. In addition to automaticity, expertise in a domain also typically involves the possession of a vast amount of knowledge, understanding of different strategies, and often supports qualitatively different ways of looking at the world from those characterizing the novice (Ericsson 1996; Charness & Schultetus, 1999; Goldman et al., 1999). We noted in Chapter 6 that experts are capable of chunking material in a way that novices are not. The figure makes clear the obvious point that this progression from novice to expert requires practice. This practice may be supported by various job aids, which generally lead to retention of the skills; but as shown toward the bottom of the figure, the skill may be forgotten when it is not used.

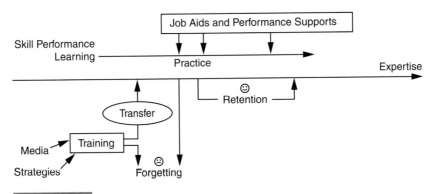

FIGURE 18.5

The contributing roles of practice, training, and transfer to the development of expertise.

Most importantly, as shown in the lower left portion of the figure, most job skills can be *explicitly trained* through various techniques designed to accelerate the development of expertise. Such training will certainly take place within the workplace, known as *on the job training,* or OJT, but this is not always effective or safe. Hence, a great premium is placed upon developing various training tools that may involve various media (classroom, computers, etc.) as well as various strategies that can shorten the trajectory to expertise. The effectiveness of such training tools must be evaluated by how well the knowledge and skills acquired during training *transfers* to the target job in the workplace. In the following, we discuss a number of features that can make training effective in its transfer of knowledge.

Methods for Enhancing Training

The human factors practitioner is usually concerned with four issues: identifying the training method that provides the (1) best training in the (2) shortest time, leads to the (3) longest retention of knowledge and skill, and is (4) the least expensive. Training programs that result in the best learning and job performance for the least time and expense are *efficient* and therefore desirable. In this section, we consider some of the important concepts and principles that influence training effectiveness and efficiency. More detailed discussions can be found in Bjork (1994) and Swezey and Llaneras (1997).

Practice and Overlearning. It is well understood that extensive practice has many benefits, as shown in Figure 18.4, leading to faster, more accurate, and less attention-demanding performance. As shown by the ordering of the three dashed lines of Figure 18.4, performance accuracy for some tasks may reach error-free levels well before time and attention demands have reached their minimum values. Thus, further practice, or *overlearning,* beyond error-free performance, does have training benefits in improving the *speed* of performance, whether involving cognitive or motor aspects. Overlearning would therefore be important in jobs where speed is critical. Because overlearning produces automaticity, it is particularly important in skills with high multitasking requirements, such as driving and flying. In addition, overlearning has been shown to decrease the *rate of forgetting* and increase the ease with which a task can be re-learned after some period of time (Anderson, 1990; Fisk & Hodge, 1992). As we discussed in Chapter 13, in some jobs, a skill that is critical in emergency or unusual situations might not be practiced on a routine basis. In these cases, overlearning is desirable so that when the emergency occurs, the operator is more likely to remember how to perform the task and to do so accurately, rapidly, and in a relatively automatic fashion.

Encouraging Deep, Active, and Meaningful Processing. In our discussion of working memory in Chapter 6, we described the role of active chunking in the formation of meaningful associations with material already in working memory in order to learn the new material. This mental activity is sometimes called "deep" processing (Craik & Lockhart, 1972), and there is by now plenty of evidence that encouragement of active processing of the material is important for effective learning (Goldman et al., 1999). Three techniques appear to be quite relevant here. First, we recall the "generation effect" from our discussion of automation, whereby those who actively generate actions are more likely to remember them (Slamecka & Graf, 1978). Thus, training techniques that encourage active participation lead to better recall. A relatively trivial example is that of note-taking during a lecture, which encourages understanding and re-expressing the material that was heard in written form. Second, both *active problem solving* and *group participation* encourage learners to apply or communicate material that is to be learned and hence think about it in a different way than that which is presented in more passive instructions (Goldman et al., 1999). Third, the meaning of some material, like procedural instructions, is better retained when the learner understands *why* something is to be done rather than just *what* is to be done, supporting the creation of a more effective mental model. However, when the theoretical material (the "why" of a process) is to be included in technical instructions, it should be embedded in the context of the procedural task to be learned, and not provided as a separate unit (Mumaw & Roth, 1995).

Offering Feedback.   It is well known that effective feedback is essential for effective learning (Holding, 1987). Such feedback can be of two types: *corrective* feedback informing the learner what was done wrong (and how to do it *right*) and *motivational* feedback or rewards for having done a job well. To be most effective, feedback should be offered immediately after the skill is performed. While it is therefore important that feedback should be delivered in a timely fashion, it should not be offered *while* attention is concurrently allocated to performing very difficult components of the skill. Under such circumstances, at best, the feedback may be ignored as the learner allocates resources to the skill being performed; and if the feedback is ignored, it will not be processed. At worst, if the skill is learned in a risky environment (e.g., driving behind the wheel), diversion of resources to feedback processing could compromise safety.

Consider Individual Differences.   Differences between learners exist both in terms of their preexisting level of knowledge of the task domain and their cognitive abilities, as discussed earlier. These differences matter. For example, those with greater preexisting levels of expertise will benefit from greater complexity of instructions for complex skills. For those with less knowledge, it is better if complex concepts are initially presented in a simplified fashion (Pollack et al., 2002). Presenting material in terms of spatial graphics assists both those of lower overall cognitive ability and those with high spatial ability (Meyer, 1999). In order to accommodate individual differences in cognitive ability with only a single version of training material, redundancy of graphics and words is most helpful (Meyer, 1999; Tindall-Ford et al., 1997).

Pay Attention to Attention.   Learning is information processing, and information processing, as discussed in Chapter 6, is generally resources limited. This fact has several implications for instructions, many embodied in the **cognitive load theory** offered by Sweller and his colleagues (Sweller 1994, 1999; Sweller & Chandler, 1994; Pollock et al., 2002) and supported by the work of Meyer

(1999). In particular, these researchers emphasize that instruction should not overload information-processing capabilities so that working memory will be unavailable for the creation of new associations in long-term memory. Negative things that might produce such an attentional overload are

- The concurrent processing of feedback and task performance, discussed above.
- Trying to work examples of problems that are so difficult that the relation between the problem-solving operations and the knowledge to be gained cannot be understood. (Sweller, 1993).
- Mentally integrating words of text and related pictorial diagrams that are removed physically from each other (Tindall-Ford et al., 1997), a violation that should remind you of the proximity-compatibility principle discussed as principle 9 in Chapter 8).
- Providing very difficult interacting concepts to the learner who has only basic knowledge (Pollack et al., 2002).
- Dealing with a poorly designed computer interface to learn material contained therein.
- Distracting "gee whiz" graphics and features that have little to do with the instructed content area, and divert attention from it.

To address these concerns, the following attentional principles of instruction can be identified:

1. Take care with the timing of feedback delivery so that concurrence is avoided.
2. Provide worked examples of problems to be solved (Sweller, 1993).
3. Use redundant or related pictures and words (Booher, 1975), placing the pictures close or connected to the related words (Tindall-Ford et al., 1997).
4. To avoid the resource competition of divided visual attention, consider capitalizing on multiple resources by using voice or synthetic speech concurrently with pictures (Meyer, 1999; Tindall-Ford et al., 1997; see principle 10 in Chapter 8).
5. Adapt the cognitive complexity of the material to the level of expertise of the learner (Pollock et al., 2002).
6. Take care in constructing the interface to instructional technology so that it is well human factored, or not distracting, and its use does not divert attention from the material to be learned.

Training in Parts.   The previous section suggests that some training programs for complex tasks can overwhelm the learner with their complexity For example, imagine the beginning flight student being asked to "fly the plane," having had no prior experience. As a consequence, in order to reduce cognitive load, human factors practitioners have argued that complex tasks be simplified or broken into parts, with each part trained in isolation, before recombining them (Fisk, 1987; Lesgold & Curtis, 1981). An example is learning a piece of piano music for each hand individually before combining the two hands. However, some reviews of the literature indicate that such *part-task training* is not always superior to *whole-task* training, a method where all subtasks are trained at once (Cream et al., 1978; Wightman & Lintern, 1985). In fact, some studies indicate a superiority for whole-task training in terms of the efficiency of transfer for a given amount of training. Wightman and Lintern (1985) suggest that one factor that affects the success of part-task training is how the task is broken down, which can be done by *segmentation* or *fractionation*.

The most successful use of part-task training is *segmentation*, where a task that has several components occurring in *sequence* is partitioned on the basis of nonoverlapping temporal components, which are then trained separately. This procedure makes sense if one or more (but not all) of the segments are very difficult. Then, by segmenting the whole task, relatively more time can be allocated to training the difficult segment(s), without spending time training the easier segment(s). For example, a particularly difficult musical piece can be more efficiently learned by practicing those difficult segments in isolation before combining them with the easier ones. In flying, the most difficult part of landing is the final "flare" phase during which the wheels touch down, (Benbassat & Abramson, 2002). Segmentation in a flight simulator can isolate the phase for extensive and repeated practice.

A less consistently successful use of part-task training is when a complex task is broken down into component tasks that are normally performed simultaneously or concurrently, termed *fractionation*. This would be like training on the left and right hand of a piano piece separately. Fractionation training consists of teaching only a subset of the components at first. Fractionated part-task training may or may not be successful depending on which subtasks are chosen for training (Wightman & Lintern, 1985). Anderson (1990) suggests that if subtasks are relatively independent of one another in total task performance, they are amenable to part-task training. An example in aviation might be radio communications and flying the plane. Fractionation may be particularly useful if the whole task is overwhelmingly complex in its information-processing demands, as in a task like flying (see Chapter 17). Here, at least in early phases of practice, part-task training can prove beneficial, particularly if parts can be trained to automaticity (Schneider, 1985). However, if the component parts are quite interdependent, the advantage of part-task training is eliminated. This interdependence occurs if performance on one of the part tasks depends on or affects performance of the other when combined. An example might be the tasks of using the clutch and manipulating the gear shift in a stick shift car.

Simplifying, Guiding, and Adapting Training. Another technique for reducing cognitive load on the learner is to simplify the task from its ultimate target level as it is performed in the real world (Wightman and Lintern, 1985). This simplification has the joint effect of reducing load and sometimes reducing errors of performance, thereby preventing learners from learning the task the wrong way. Actually, two different approaches can be taken here. *Simplification* involves making the actual task easier. For example, teaching a pilot how to fly could begin by using lower order flight dynamics, which we saw in Chapters 9 and 17 are easier to control. *Guiding* involves imposing means to prevent errors from occurring and assuring that only correct responses are given. For flying, it might involve an active guidance of the student pilots' controls along the correct trajectory to produce an accurate flight path. For teaching a computer skill, it might involve disabling or freezing keys that are not appropriate and highlighting, in sequence, the order of those keys that are. Guiding is sometimes described as a "training wheels" approach (Caroll and Carrithers, 1984; Cotrambone & Carroll, 1987), not unlike the training wheels on a child's bicycle that prevent the bike from falling as the child learns to ride. Using either mode, the level of difficulty can then be increased *adaptively* as the learner acquires the skill until the target level of difficulty of the final skill is reached (Mane et al., 1989; Druckman & Bjork, 1994). Researchers sometimes describe these techniques as *scaffolding* (Goldman et al., 1999; Brock, 1997), like the scaffolding on a building that can gradually be removed as its construction is completed.

Like part-task training, simplification and guidance can play a valuable role in supporting early phases of learning, both by reducing the distraction of unnecessary errors (Carroll & Carrithers, 1984) and by availing the learner of enough spare resources so that working memory can help to encode the necessary skills. However, both techniques have their potential dangers. Sometimes, learning a simplified version of a skill simply will not transfer to the complex version. For example, in the case of flight dynamics, learning a zero-order tracking task may not transfer at all to performing a second order-task. Sometimes, learners can become overly dependent on the guidance or scaffolding; using it as a "crutch," and suffer when it is removed (Lintern & Roscoe, 1980). This is particularly the case if, in the presence of the scaffolding, the learner does not acquire the necessary attention, perceptual, and cognitive skills to replace those that were provided by the scaffold. As an example, consider the child bike rider with training wheels, who never learns the skill necessary to balance the bike while it is in motion. Thus, with both simplification and guidance, care must be taken to adapt the training—remove the scaffold—in a way that fosters progressively more reliance upon performing the skill in its absence.

An aspect of scaffolding that deserves special mention is related to error reduction: Error prevention makes good sense only to the extent that errors are relatively *catastrophic* (e.g., a training session must be restarted from the beginning if an error is performed or equipment is damaged) or if the training regime allows errors to be *repeated* without corrective feedback. However, it is a myth to assume that error-free performance during training will produce error-free performance (or even, necessarily, effective performance) on transfer to the real skill (Druckman & Bjork, 1995). Indeed, there are many advantages for learners to be allowed to *commit* errors during training if part of the training focuses on *learning how to correct* those errors, since the error-correction skill will undoubtedly be important after the skill is transferred out of the training environment to the target job environment.

Media Matters?   The last 30 years have seen a wide interest among the training community in the exploitation of various forms of media in delivering training material (see Brock, 1997; Swezey & Llaneras, 1997; Meyer, 1999; Wetzel et al., 1994 for good overviews). These range from lecture, to video, to various forms of computer-based instruction, allowing interaction in either fairly artificial environments or the highly realistic ones characteristic of virtual reality (Sherman & Craig, 2003; Durlach & Mavor, 1995). A general conclusion seems to be that although there are some modest benefits of computer-based instruction over conventional instructions in terms of measures of knowledge acquisition (Brock, 1997), these gains are not large (Meyer, 1999) and are probably more related to how the particular aspects of the computer media are used than to any inherent advantage of the computer itself. Careful attention to how media are used can exploit the best properties of effective instruction. For examples,

- Use of interactive video can provide animation for skills in which processing of animation is important, like the prediction of motion by air traffic controllers.
- Use of concurrent sound (voice) and pictures can facilitate parallel processing and integration of the two media through multiple resources (Meyer, 1999).
- Use of computer-based instruction can provide immediate and timely feedback, provide support for active problem solving, and give performance-based adaptive training, whereby difficulty is increased or scaffolding removed as a skill progresses.
- Use of computer-based *intelligent tutoring systems* can provide individually tailored material based on a particular learner's needs (Farr & Psotka, 1992; Mark & Greer, 1995). (Note that this is a form of automation that simply replaces a human tutor with intelligent automation and may perform the task less effectively, but also less expensively, when a large number of learners are involved).
- Use of certain media can make the learning task more interesting and inherently motivating, particularly for children, so that they will *invest the cognitive effort* necessary for deep processing of the material so that it is well stored in long-term memory (Goldman et al., 1999).

Conclusion. There are a wide variety of techniques to influence training effectiveness. The list is much longer than that presented here. For example, Swezey and Llaneras (1997) actually present 156 guidelines! The training system designer should consider these in light of the skills to be trained, and their manifestations in the real job environment. However, in considering or evaluating these and other training strategies, one very important point that should be made is that *techniques that enhance performance during training, may not necessarily improve learning, as reflected by transfer to the job environment.* In an important article, Schmidt and Bjork (1992) provide several examples of cases where this dissociation between performance in training and skill after transfer is evident (see also Bjork, 1994, 1999). This caution is probably most relevant for some of the techniques designed to reduce cognitive demands or to improve performance during training via simplification or scaffolding.

Transfer of Training and Simulation

Transfer of training generally refers to how well the learning that has occurred in one environment, such as a training simulator or computer-based instruction, enhances performance in a new environment. As Holding (1987) words it, "When learning a first task improves the scores obtained in a second task (B), relative to the scores of a control group learning B alone, the transfer from A to B is positive" (pg. 955). The concept of *positive transfer of training* is important because it is a major goal of any training program, and measures of transfer of training are often used to evaluate training program effectiveness. While there are a variety of qualitative measures for transfer of training, a commonly used approach is to express the variable as a percentage of time saved in mastering a task in the target environment, using the training program compared to a no-training control group:

$$\% \text{ transfer} = \frac{(\text{control time} - \text{transfer time})}{\text{control time}} \times 100 = \frac{\text{savings}}{\text{control time}} \times 100$$

As applied to training programs, *control time* is the amount of performance time it takes for the untrained operators to come up to perform at some **criterion level** on Task B in the new environment, and *transfer time* is the amount of performance time it takes for the operators in that training group to reach the same performance criterion in the new environment. Thus, when put in the real job environment, it might take the control group an average of 10 hours to reach expected performance levels, and it might take the trained group 2 hours. This would be a transfer savings of $10 - 2 = 8$ hours, or a transfer of $8/10 = 80\%$. Notice, however, that this variable does not account for fact that the training itself takes time. If the training program required 8 hours, the savings would be nullified when considering the total time to master the skill. The ratio of savings/training time is called the *transfer effectiveness ratio* (Povenmire & Roscoe, 1973). Thus 8 hours of training that produced a 2 hour savings would have a transfer effectiveness ratio of $2/8 = 0.25$.

It is important to point out that training in environments other than the real world can be desirable for reasons other than transfer savings, including factors such as safety, greater variety of practice experiences, operational costs, and so forth. This is particularly the case for task simulators. For example, use of a high-fidelity flight simulator costs only a fraction of the operating cost for an F-16 airplane. For this reason, training systems may be quite desirable even if the transfer effectiveness ratio is less than 1.0. Both flight simulators and driving simulators are safer than training in air and ground vehicles. Another advantage of simulators is that they can sometimes optimize the conditions of learning better than can the real device. For example, a flight simulator can be programmed to fly rapid, repeated portions of the final flare segment of landing in segmentation part-task training. A real aircraft cannot. Simulators can also be paused to provide timely feedback without distraction.

An important issue in simulator training is the degree of *realism* or *fidelity* of the simulator to its counterpart in the real world. High fidelity simulations are usually quite expensive. Yet considerable research indicates that more realism does *not* necessarily produce more positive transfer (Swezey & Llaneras, 1997).

Sometimes, the expensive features of realism are irrelevant to the target task. Even worse, those expensive features may distract attention away from processing the critical information that underlies skill learning, particularly at the early stages.

Nearly all training devices produce *some* positive transfer. If they don't, they are worthless. Training devices should never produce *negative transfer* such that performance in the target tasks is worse than had training never been offered. However, other things do produce negative transfer when habits appropriate to one system are counterproductive in a new system. This may be the case in changing the layout of controls between the two systems. As we noted in Chapters 8 and 9, avoiding negative transfer can be achieved by standardization.

On the Job Training and Embedded Training

We described a series of training environments ranging from those that are quite different from the target job (the classroom) to those that may be quite similar (high fidelity simulation). Naturally, the maximum similarity can be obtained by training "on the job."

OJT is typically an informal procedure whereby an experienced employee shows a new employee how to perform a set of tasks. There are rarely specific guidelines for the training, and effective training depends highly on the ability of the person doing the training. OJT, as normally performed, has been shown to be much less effective than other training methods. However, if the training is done using Instructional System Design methods described below, with strong guidance to the trainer, this method can be very effective (Goldstein, 1986).

Finally, another type of instruction, *embedded training,* combines computer-based training with on-the-job performance. Evans (1988) defines embedded training as "training that is provided by capabilities built into or added into the operational system to enhance and maintain skill proficiency necessary to maintain or operate the equipment." Embedded training is most appropriate for jobs that rely at least partially on computers because the training is computer-based. This type of training is especially useful for people who just need occasional refresher training to keep up their skills. Embedded training should be considered for tasks when the task is critical with regard to safety concerns or when the task is moderate to high in cognitive complexity (Evans, 1988).

TRAINING PROGRAM DESIGN

There are many different ways to teach a person how to perform tasks. There are different types of media, such as lecture or text, and there are other considerations as well, such as how much and what type of practice is most efficient for learning skills. Like other topics in this book, training program design is really an entire course in itself. Here, we just skim the surface and describe some of the most prevalent concepts and issues in human factors. Before describing these concepts and issues, we first review a general design model for developing training programs and the major types of training media that specialists combine in designing a training program.

A Training Program Design Model

The majority of professionally designed business and government training programs are developed using a systematic design method termed *Instructional System Design* or ISD (Andrews & Goodson, 1980; Gordon, 1994; Reigeluth, 1989). ISD models are similar to human factors design models (see Chapter 3); they typically include a front-end analysis phase, design and development phase (or phases), implementation, and a final system evaluation phase. ISD models are also used to develop job aids and performance-support systems. Most professional instructional designers agree that the process used for designing the training program can be just as important as the type of program or the media chosen (e.g., video, computer-based training). A number of studies have demonstrated that use of systematic design methods can result in more effective training programs than less systematic methods, such as simply asking a subject matter expert to provide training (Goldstein, 1986).

An instructional program is a product or system and can therefore be designed using an "ergonomic" approach. Gordon (1994) modified a generic ISD model by incorporating methods derived from cognitive psychology and human factors. This model, carried out in three major phases described below, still has the traditional ISD phases of front-end analysis, design and development, and system evaluation. However, it also includes less traditional methods, such as early usability testing. The design model can be used for developing job aids, instructional manuals, and performance-support system in addition to more traditional training programs. The model contains four basic procedures or phases: front-end analysis, design and development, full-scale development, and final evaluation.

Phase 1: Front-End Analysis.   Like other types of design, training program design begins with an analysis of needs. In this model, front-end analysis is accomplished by performing an organizational analysis, task analysis, and trainee analysis. The information collected in the analyses is then used to determine whether training or some other intervention is needed and to define requirements and constraints for design of the training system.

The *organizational analysis* is an information-collection activity that looks at the broad context of the job or task; the goal is to identify any factors that would bear on the need for and success of a training program. Such factors include future company changes such as job redesign or acquisition or new technology, management attitude toward job duties, and so on. In this analysis, we answer questions related to the goals and priorities of the organization, management attitudes toward employees and toward training, and the performance levels expected of employees (see Gordon, 1994, for a complete discussion). The information can be collected through a combination of methods such as document analysis, interviews, questionnaires, job tests, and observation (Wexley & Latham, 1991). The answers to such questions determine whether training would be desirable and consistent with organizational and employee goals and values.

Task analysis is performed to identify the knowledge, skills, and behaviors required for successful task performance. Task analysis for front-end analysis can be performed using the same methods that are used for other types of human factors analysis (see Chapter 3). This will be followed by a brief *trainee analysis*. This process identifies (1) *prerequisite* knowledge and skills that should be possessed by trainees in order to begin the training program (e.g., eighth-grade English to take beginning course for auto mechanics); (2) *demographics* such as age, physical capabilities, primary language, and background; and (c) *attitudes* toward training methods if not done as part of organizational analysis.

Results from the organizational, task, and trainee analyses are used in a *training needs analysis* to determine whether the most appropriate performance improvement approach is task redesign, performance support, or develop a training program (if motivation is the problem, none of these would be used).

At this point, *functional specifications* are written that include the training program goal, training objectives, system performance requirements, and development constraints. Performance requirements are important because they include the characteristics to be possessed by the training program from an instructional design and human factors standpoint, such as desirable instructional strategies and interface requirements for ease of use or ease of learning (see Baird et al., 1983; Fisk & Gallini, 1989; Gordon, 1994; Holding, 1987; Jonassen, 1988).

Phase 2: Design and Development. The second phase, design and development, is where the analyst chooses a training program method or combination of methods and proceeds with further design and development while also performing formative evaluation. The steps for this phase are listed in a given sequence, but often there is iteration back through the steps many times. This is considered standard practice for most ISD models.

By considering the information contained in the functional specifications, the designer generates a number of *design concepts* that would work for the problem. If there is more than one possible solution, the alternatives can be compared by using a cost/benefit analysis in a matrix table format. By using such a table, the designer can choose the best overall design solution or, alternatively, complementary methods that can counteract the other's disadvantages. Once the design concept has been chosen, a project plan is written, including budget, equipment, personnel, and task timeline. In some cases, a cost/benefit analysis is performed to make sure that the proposed design solution will be adequately cost effective (Marrelli, 1993).

A prototype is used for *formative evaluation* of the design concept, to gain management approval and peer (human factors or instructional designer) approval, and to perform usability testing. In the latter case, representative trainees are asked to review the prototype and provide comments on its acceptability, perceived effectiveness, weaknesses, and so forth. As more fully functional prototypes are developed, trainees use the system prototype in the same way that standard usability evaluations are conducted, something now made possible by the use of *rapid prototyping* techniques.

After formative evaluation and usability testing has been accomplished, the *full-scale development* can proceed. Material is taken from the task analysis and translated into instructional units using instructional design guidelines such as those given by Clark (1989), Romiszowski (1984). As the system is developed, the design team should periodically perform additional formative evaluation. This prevents any unanticipated and unpleasant surprises at the end, when changes are more costly. Evaluation should focus on whether the training program appears to be acceptable to trainees and effective in meeting its objectives. If possible, the training program should be used with several naive trainees who have not been part of the design process. They should receive the training program and be tested on knowledge and skill acquisition both immediately after training and after a period of time similar to that expected to occur after training on the fielded system. Trainees should be asked questions via interview or questionnaire regarding their subjective reactions to the system (Gordon, 1994). This should be followed by a *final usability test*.

Phase 3: Program Evaluation.   The fielded training program or performance aid should be evaluated for system effectiveness and then periodically monitored. The evaluation process is carried out much like the evaluation processes described in Chapter 3. Goals of the evaluation process are to answer questions such as (Goldstein, 1986)

- Has a change occurred in trainee task performance?
- Is the change a result of the instructional program (as opposed to some other factor, such as a change in management or incentive programs)?
- Would the change occur with other trainees besides those in our sample?
- Would the change occur in other contexts or for other tasks?

To answer these questions, we design an evaluation plan by specifying *what* criteria (variables) to measure, *when* to measure the criteria, *who* (which trainees) to use in measuring the criteria, and *what context* to use. You can see that these are the same types of question involved in development of the research designs discussed in Chapter 2. While training programs are often not systematically evaluated, evaluation of a fielded training program should be performed by using either a pretest-posttest experimental design (with one group measured before and after training) or a control group design with one group of randomly selected trainees receiving the old training method (or none at all) and the other group receiving the training program being evaluated. Program evaluators strive to (1) conduct the evaluation in an environment as similar to the ultimate performance environment as possible; (2) conduct the knowledge and skill tests after a realistic period of time; and (3) base the evaluation on tasks and task conditions that are representative of the ultimate job (Gordon, 1994).

In addition to evaluation of trainee job performance, it may sometimes be desirable to evaluate the impact of a training program on an organization's productivity and performance levels. This is achieved by performing a longitudinal systematic evaluation incorporating multiple measures. Diehl (1991), for example, has described the impact of various decision and crew resource management training programs on overall flight safety of different flight organizations.

CONCLUSION

In conclusion, we have seen how selection, based upon valid tests, and training, can both be used in conjunction with job supports to complement and augment good design, in creating a well human factored system. In particular, two aspects of the synergy between these approaches must be emphasized. First, training and selection should never be considered as a satisfactory *alternative* to good human factors design, imposed to compensate for bad design. After all, a poorly human factored system may be used by an untrained operator, even if the intention is for this not to happen. Second, although the creation of training materials and job support may follow the completion of system design, it is imperative that these be given as much attention, in their clarity and usability, as the system that they are designed to support.

Chapter 19

Social Factors

George *entered the meeting room Monday morning, thinking that he could get a lot more accomplished without these 7:30 A.M. weekly meetings. His boss, Sharon, would already be there, ready and waiting with the two-page agenda. He could see the meeting play out already. By the second or third project item, the department critic, Martin Jones, would be going into a long lecture about all the problems associated with whatever they happened to be discussing. Last time, it was that the project had too many problems, they should not have ever taken it on, it was causing everyone to put in too much time, and on and on. Martin seemed to perpetually dominate the discussions, keeping anything from really getting accomplished. George wished they had some magic tool that could make the meetings a little more productive.*

Ergonomic interventions in business and industry usually focus on changing the workstation or equipment characteristics for the individual worker. For example, attempts to increase system safety might result in redesigning displays, adding alarms, or changing how a task is performed. However, there are many factors that can affect human performance that are larger than, or outside of, the envelope of the human–machine system. Most notably, *individual behavior is a function of the social context,* referring to the attitudes and behavior of coworkers and others in the work environment, and a *function of the organizational context,* which includes variables such as management structure, reward or incentive systems, and so forth.

In this chapter, we review some of the human factors topics that pertain to the larger social and organizational context. The organization structure describes the way individuals, technology, and the environment interact. We begin this chapter with a description of the general factors that govern system interaction at the organizational level—system *complexity* and *coupling.* The increasing complexity of many systems makes it necessary to decentralize management and increase the amount of time that people work together in either groups or teams. For this reason, we consider characteristics of *groups* and *teams* and how they interact with human performance. We also look at some of the concepts being applied in the emerging area of *team training*. Next, we consider how technology is being used to support work done by groups or teams who may be separated in time or space, an area termed *computer-supported cooperative work*. Finally, we briefly review some of the ways that *macroergonomic* intervention in industry is changing as a function of broader social and organizational ergonomic perspectives.

TYPES OF SYSTEMS

Two dimensions are particularly useful in describing organizations and the factors affecting their performance—complexity and coupling (Perrow, 1984). *Complexity* refers to the number of feedback loops, interconnected subsystems, and invisible, unexpected interactions. Nuclear power and petrochemical plants are complex because the behavior of one subsystem may affect many others, and these interactions can be perceived only indirectly. *Coupling* refers to the degree that there is little slack and a tight connection between subsystems. In a tightly coupled system, such as just-in-time supply chain, a disruption in part of the system quickly affects other parts of the system. The degree of complexity and coupling, examples of which are given in Table 19.1, has implications for the likelihood of catastrophic failures, with *highly complex, tightly coupled systems being vulnerable to catastrophic failure* (Perrow, 1984).

One reason that complex, tightly coupled systems are vulnerable to catastrophes is that the organizational requirements for their control are conflicting. High complexity generates unpredictable events that require the flexibility of a decentralized management structure (e.g., individual workers empowered to solve problems as they arise). However, the high degree of coupling leaves little room for error because there are few resources, little slack, or little time to recover from a mistake, so a centralized management structure would be most appropriate. In the case of the Three Mile Island and Chernobyl accidents, the tight coupling made it difficult to avoid catastrophic damage when operators made a mistake. At the same time, the unexpected behavior of these complex systems makes it impossible to use a centralized approach of procedures or

TABLE 19.1 System Characteristics of Complexity and Coupling

High Complexity and Low Coupling	High Complexity and High Coupling
Examples: Universities, government agencies	Examples: Nuclear power plant, airplane
Low Complexity and Low Coupling	Low Complexity and High Coupling
Examples: Traditional manufacturing	Examples: Marine transport, rail transport

Adapted from C. Perrow (1984) *Normal Accidents*, N.Y.: Basic Books.

automation to address every contingency. In general, *tightly coupled systems require centralization* to carefully coordinate resources, and *highly complex systems require decentralization* to cope with the unexpected. The degree of centralization has implications for organization structure (e.g., a centralized system has a hierarchical reporting structure and control comes from the top) and for the design of decision aids. A centralized approach might require operators to use an expert system or follow a procedure, whereas a decentralized approach might rely on good displays that help operators solve unexpected problems (see Chapters 7, 8, and 16).

GROUPS AND TEAMS

Because businesses must operate in an increasingly complex economic environment, recent trends in organizational design place a strong emphasis on "flattened" management structures, decentralized decision making (where workers at lower levels are making more important management decisions), and the use of work groups or teams for increased efficiency and flexibility (Hammer & Champy, 1993). Teams are also becoming more common as a way to respond to increasing job complexity and the associated cognitive demands placed on workers (Sundstrom et al., 1990). All indications suggest that the use of teams and work groups is a long-term trend in industry. Johnson (1993) reports that 27 out of the 35 companies surveyed responded that the use of work teams had resulted in favorable or strongly favorable results. In addition, 31 of the 35 companies said that work-team applications were likely to increase in their company. The role of teams in promoting the development of expertise as also been emphasized (Goldman et al., 1999).

Why would human factors specialists be concerned with the behavior of groups or teams of workers? One reason is that just as individuals vary with respect to performance and error, so do teams. In a growing number of industries, including the aviation industry, investigators have found that a large number of accidents have been caused primarily by a breakdown in team performance (Helmreich, 1997; Wiener et al., 1993). Human factors specialists are addressing this phenomenon as part of their traditional focus on safety and human error. They are identifying the skills responsible for successful teamwork and developing new methods that can efficiently and effectively train those skills. In this section, we briefly define and contrast the concepts of groups, teams, and crews. We also review a few of the basic concepts and findings concerning group performance and teamwork.

CHARACTERISTICS OF GROUPS, TEAMS, AND CREWS

Sociologists and social psychologists have studied group processes for 50 years but have only recently become seriously interested in teams (i.e., in the mid-1980s). Most of the groups and teams described in the literature are "small," with less than 12 members. However, teams can technically be quite large; for example, in the military, a combat team might have hundreds of members. As another example, the new business reengineering efforts are resulting in self-regulating work teams of all sizes. Peters (1988) suggested that organizations "organize every function into 10- to 30-person, largely self-managing teams."

Groups are aggregations of people who "have limited role differentiation, and their decision making or task performance depends primarily on individual contributions" (Hare, 1992). Examples include a jury, board of directors, or a college entrance committee. A team, however, is a small number of people with complementary skills and specific roles or functions (high role differentiation), who interact dynamically toward a common purpose or goal for which they hold themselves mutually accountable (Katzenbach & Smith, 1993). Teams tend to have the following characteristics (Sundstrom & Altman, 1989).

- Perception of the team as a work unit by members and nonmembers
- Interdependence among members with respect to shared outcomes and goals
- Role differentiation among members
- Production of a team-level output
- Interdependent relations with other teams and/or their representatives

There are numerous definitions of teams, but they all seem to center around the concepts of a common *goal* or output attained by *multiple people* working in an *interdependent* manner. *As compared to groups, teams have more role differentiation, and more coordination is required for their activities* (Hare, 1992). Group work is therefore not necessarily the same as teamwork.

If a team consists of interdependent members, how is this distinguished from the concept of a crew? The term *crew* is typically reserved for a team that manages some form of technology, usually some type of transportation system such as a ship, airplane, or spacecraft. Human factors specialists seem to be particularly interested in crew performance possibly because of the strong emphasis in the airline industry on aircrew performance and training (e.g., Helmreich, 1997; Helmreich & Wilhelm, 1991; Wiener et al., 1993).

Group Performance

In many group tasks, individuals often do some work (such as making a decision or performing problem solving), then have discussions and share information with the others. Comparing the productivity of the group with that of individuals shows that group productivity is *better than the average* of the individuals, *but not better than the best* individual (Hare, 1992; Hill, 1982). In terms of output or work productivity, a group will generally yield less than the sum of the individuals. This difference is increased to the extent that people feel their efforts are dispensable, their own contribution cannot be distinguished, there is shared responsibility for the outcome, and/or motivation is low. Even in the well-known method known as brainstorming, the number of ideas produced by a group is often less than the number produced by the members working individually (Street, 1974).

In some situations group interactions can generate substantially worse decisions compared to that of any individual. For example, *groupthink* occurs when group dynamics lead to *collective rationalization* in which members explain away contrary information (Janis & Mann, 1977). Groupthink also occurs when group dynamics produce a *pressure to conform* in which group members feel reluctant to voice concerns or contrary opinions. The Bay of Pigs fiasco is an example of how a group, under the leadership of John F. Kennedy, conceived and executed a highly flawed plan to invade Cuba. In retrospect, those involved quickly realized the problems with the plan, but at the time group dynamics and the pressure for consensus led group members to ignore critical considerations (Jones & Roelofsma, 2000). To combat groupthink, groups should emphasize the value of alternate perspectives, objections, and criticism. It is also useful to bring in outside experts to help evaluate decisions. In general, groups seem to have a poor awareness of group dynamics and fail to reflect on their behavior, an absence of what we might call "collective metacognition" (see Chapter 6). As seen in the opening vignette, many groups could benefit by clearly articulating their goals, planning meetings, and discussing ineffective behavior (Tipping et al., 1995).

Certain characteristics tend to make a group more productive. For example, if groups have members with personalities that allow them to take initiative, work independently, and act compatibly with others, productivity increases. Groups are also more productive if they have a high level of cohesiveness, appropriate or adequate communications, needed information, and adequate time and resources (Hare, 1992). Group attitude (e.g., "thinking we can") can improve group performance (Hecht et al., 2002). Group size can also have important implications for performance—for a job requiring discussion, the optimal size is five (Bales, 1954; Yetton & Bottger, 1983). The basis for group decisions is also important—a consensus model is often better for group productivity than a majority decision model (Hare, 1982).

Team Performance

Successful team performance begins with the selection of an appropriate combination of members: The leader should have a style that fits the project, individuals should have the necessary complementary task work skills and teamwork skills, and the team should not be so large that communication becomes difficult (Heenefrund, 1985). As projects become increasingly complex, such as in concurrent engineering design, constructing appropriate teams becomes more difficult. In these situations, it is important to decompose a large, interdependent team into smaller teams that can work relatively independently. One solution is to identify tasks of the team and then use cluster analysis techniques (as described in Chapter 3) to identify relatively independent sets of tasks that can be assigned to small teams (Chen & Li, 2003).

Several researchers have focused on the characteristics or preconditions that must exist for a team to be successful or effective. These include the following requirements (from Bassin, 1988; Patten, 1981; Katzenbach & Smith, 1993):

- A vision; a common, meaningful purpose; a natural reason for working together
- Specific performance goals and/or a well-defined team work-product
- A perceived, dependent need; members are mutually dependent on each others' experience and abilities
- Commitment from every member to the idea of working together as a team
- Leadership that embodies the vision and transfers responsibility to the team members
- Coordination; effective use of resources and the team members' skills
- Shared accountability; the team must feel and be accountable as a unit within the larger organization

While teams are usually developed with optimism, a number of problems may interfere with team performance, including problems centering around power and authority, lack of shared norms or values, poor cohesion or morale, poor differentiation or problems of team structure, lack of shared and well-defined goals and task objectives, poor or inadequate communication, and lack of necessary feedback or critique (Blake et al., 1989).

Some of these problems occur because of a poor choice of team members. However, many problems result from an organizational structure that is not aligned with the requirements of team success. For instance, the organization may lack a clear vision regarding the team goals, so the team has no shared values or objectives. In addition, the organization may reward individual performance rather than team performance, which can undermine commitment to team objectives. For example, a reward structure that reflects team effort and quality of communication led to improved team performance in concurrent engineering (Duffy & Salvendy, 1999).

Team Training

Another reason that teams often perform below the initial expectations is that they receive inadequate training and team-building in advance. Effective teams require that members have *task work* skills (see Chapter 18), which pertain to correct subtask performance, and also *teamwork* skills, which pertain to interpersonal skills such as communication. To illustrate, in one study, individuals practicing alone did not improve team performance, but the more the team practiced together, the more performance improved (Hollingshead, 1998). Morgan and colleagues (1986) suggest that teamwork skills include behaviors reflecting the following general categories of activity: cooperation, coordination, communication, adaptability, giving suggestions or criticisms, acceptance of suggestions or criticism, and showing team spirit. Responding to the need for teamwork skills, a number of researchers in the field of organizational development have created team-building workshops and seminars (e.g., George, 1987; Nanda, 1986). These team training programs generally are not uniformly beneficial (Salas et al., 1999).

The success of team development and the training depends on the type of team being assembled. Sundstrom and colleagues (1990) evaluated the concept of teams and determined that they can be placed in four categories. The categories are defined by factors such as whether the teams have high or low *role differentiation,* the work pace is externally or internally controlled, and the team process requires high or low *synchronization with outside units* (Sundstrom et al., 1990). According to the definition of teams and groups presented earlier, the teams with low role differentiation can sometimes include groups.

- *Advice/involvement teams.* Examples include review panels, boards, quality-control circles, employee involvement groups, and advisory councils. These teams are characterized by *low role differentiation* and *low external synchronization.* The work cycles for these teams may be brief and may not be repeated.
- *Production/service teams.* Examples include assembly teams, manufacturing teams, mining teams, flight attendant crews, data-processing groups, and maintenance crews. This category is characterized by *low role differentiation* and *high external synchronization* with other people or work units, and external pacing because of synchronization with other units. Work cycles are typically repeated continuously.
- *Project/development teams.* Examples include research and development groups, planning teams, architect teams, engineering teams, development teams, and task forces. The teams are typically characterized by *high role differentiation* and *low to medium external synchronization.* The pacing is not directly driven by outside units, although the work might require a large amount of communication with outside units.
- *Action/negotiation teams.* Examples include surgery teams, cockpit crews, production crews, negotiating teams, combat units, sports teams, and entertainment groups. The work requires *high role differentiation* (with frequently long team life spans) and *high synchronization* with outside units. The work is driven by externally imposed pacing and work cycles that are often brief and take place under new or changing conditions.

Each type of team needs different expertise and organizational support to be effective. For example, action/negotiation teams require a high degree of expertise among members and synchronization with outside people. This usually means that training and technology will play a major role in determining team effectiveness.

In general, team-building workshops that clarify the roles and responsibilities of team members have a greater effect on team performance than do workshops that focus on goal setting, problem solving, and interpersonal relations (Salas et al., 1999). The implication is that, at least for certain types of teams, team training must go beyond the usual organizational development team-building activities that focus on interpersonal relations and team cohesiveness. This is particularly true of teams in which there is a high demand for external synchronization. For example, training for production/service and action/negotiation teams must consider how work is synchronized and how to deal with time pressure. Training programs for flight crew resource management have attempted to enhance factors such as communication and stress management (e.g., Wiener et al., 1993).

For teams characterized by high levels of role differentiation, such as project/development teams, *job cross-training* can enhance knowledge of team members' information needs and increase the use of shared mental models (Volpe et al., 1996). Using exercises to give team members experience in other members' roles can increase team cohesiveness and improve knowledge about appropriate or necessary communications (Salas et al., 1997). In elite basketball and soccer teams, the degree of self-rated cohesiveness was an important factor in predicting the number of games the team would win (Carron et al., 2002). It takes more than the skill of the individual team members to win games or perform well.

When action/negotiation teams must perform tasks in a complex, dynamic environment with safety issues, such as an air traffic control room or hospital operating room, there is an even greater need to perform smoothly and effectively. In such environments, periods of stressful, fast-paced work activity lead to cognitive overload, and under most circumstances, the overall impact on teams appears to be a decline in communication and job performance (Bogner, 1994; Urban et al., 1996; Volpe et al., 1996; Williges et al., 1966; Xiao et al., 1996). Effective training must promote the development and use of shared mental models and development of strategies for effective communication, adaptation to stress, maintenance of situational awareness, and coordinated task performance (Orasanu & Salas, 1993; Stout, 1995; Volpe, 1993).

Team members' reduced ability to communicate during periods of high workload and stress can undermine team performance for a number of reasons. First, the members do not have the opportunity to build a *shared mental model* of the current problem and related environmental or equipment variables (Orasanu & Salas, 1993; Langon-Fox et al., 2000). This mental model makes it possible to communicate information between group members in an *anticipatory* mode, so that communication can occur before workload peaks. For example, a member can state relevant task information before or as it is needed by others rather than wait to be asked (Johannesen et al., 1994). Second, the members may not have the time and cognitive resources to communicate plans and strategies adequately. Third, members may not have the cognitive resources available to ask others for information they need.

Highly effective teams are able to overcome these challenges by making good use of the "downtime" between periods of high workload (Orasanu, 1990). That is, effective teams use low workload periods to share information regarding the situation, plans, emergency strategies, member roles, and so forth. Developing a shared mental model of the tasks also provides the team with a common understanding of who is responsible for what task and the information needs of others (Orasanu & Salas, 1993). This way, when teams encounter emergencies, they can use the shared mental model to support *implicit* coordination that does not require extensive communication. Highly effective teams also address the demands of high workload and time stress by distributing responsibilities beyond the allocation of formal responsibilities (Patel et al., 2000). The roles on some teams become fuzzy so that team members can cover for others in high demand situations.

COMPUTER-SUPPORTED COOPERATIVE WORK

The increasing use of groups and teams in the workplace, combined with rapid technological advances in the computer and communications industries, is resulting in a trend for group members to work separately and communicate via computer (Kies et al., 1998; Olson & Olson, 2003). As an example, control room displays are moving from large single-screen displays toward individual "cockpit" workstations for each operator or team member (Stubler & O'Hara, 1995). These people may be in the same room working at different stations or might even be in entirely separate locations. As organizations become increasingly global in their operations, teams may be dispersed across the world, so team members may be culturally distant as well (Bell & Kozlowski, 2002). The individual workstations use a computer-based graphical interface to combine and coordinate functions such as controls, displays, procedural checklists, communication support, decision aids, and so on (O'Hara & Brown, 1994).

The process of using computers to support group or team activity is termed *computer-supported cooperative work (CSCW)*, and the software that supports such activity is termed *groupware*. CSCW is a broad term that includes a number of different types of activities, including decision making, problem solving, design, procedural task performance, and so forth.

In the following, we discuss computer support first for groups and then for teams.

Decision Making Using Groupware

Kraemer and Pinsonneault (1990) distinguish between two types of support for group process: group communication support systems and group decision-support systems. *Group communication support systems* are information systems built primarily to support the communication among group members regardless of the task. Examples of communication support systems include teleconferencing, electronic mail, electronic boardrooms, and local group networks (Kraemer & King, 1988). *Group decision-support systems* are targeted mostly toward increasing the quality of a group decision by reducing noise in the decision process or by decreasing the level of communication barriers between group members (DeSanctis & Gallupe, 1987). Therefore, decision-support systems can be thought of as communication systems plus other aids to provide functions such as eliminating communication barriers, structuring the decision process, and systematically directing the pattern, timing, or content of discussion (DeSanctis & Gallupe, 1987). Chapter 7 describes the decision support aspects of this technology that support the performance of individuals. They support decision making or problem solving by:

- Providing anonymity
- Imposing structure on the process
- Providing word-processing functions for synthesis of writing
- Providing workspace for generating ideas, decisions, consequences, and so on

- Reducing counterproductive behavior such as disapproval
- Reducing authority and control problems exhibited by a minority of group members

This list demonstrates that much of the functionality of these systems resides in counteracting negative interpersonal dynamics of group meetings and decision processes, such as the problem described at the beginning of this chapter.

When looking at output, decisions are usually (although not always) of higher quality for groups using group decision-support systems (Sharda et al., 1988; Steeb & Johnson, 1981). It should be noted that the advantages of these systems could be caused by the promotion of more positive interaction among the group members or by the provision of specific decision aids such as computer-aided decision-tree analysis (e.g., Steeb & Johnson, 1981). Other benefits include the findings that use of a decision-support system increases the confidence of group members in the decision (Steeb & Johnson, 1981) and increases the satisfaction of group members with the decision (Steeb & Johnson, 1981).

Computer-Supported Team Performance

Some computer-supported groups are engaged in team performance activities such as cockpit management, maintenance tasks, or process control. Teams working via groupware are sometimes called "virtual teams" (Cano & Kleiner, 1996). Note that for groupware to support such collaborative task performance, the software functions must usually be much more elaborate than basic communication and decision-support systems for groups. This type of groupware is likely to support task performance via controls and displays, system status information, information concerning what other team members are doing, procedural checklists, and other types of support, such as those discussed in Chapters 7 and 16. Stubler and O'Hara (1995) evaluated some of the more critical display elements for groupware that support complex task performance, referring to the displays for these systems as *group-view displays* and proposed that group-view displays should provide the following categories of support:

1. *Provide a status overview.* The display should provide information that conveys a high-level status summary to inform all personnel about important status conditions and changes.

2. *Direct personnel to additional information.* The display should direct team members to other information that would be helpful or necessary but that is not currently displayed. The displays should generally follow human factors display design principles and guidelines, such as supporting easy manual retrieval of information (e.g., O'Hara et al., 1995; Woods, 1984; Woods et al., 1990).

3. *Support collaboration among crew members.* When crew members are sharing the same task, it is important that their collaboration is supported by activities such as recording actions of different personnel, providing whiteboard or other space for collaborative problem solving or brainstorming, displaying previous activity or highlights, and so forth. In face-to-face collaboration, the use of gestures, facial expressions, and body language is an important part of the communication process (Tang, 1991). If crew members are working remotely, their communication must be supported by some other means.

4. *Support coordination of crew activities.* Some team and crew members have highly differentiated roles and will therefore be doing different but related tasks. In this case, the groupware should support coordination of the work performed by the various crew members. Such support would facilitate a common understanding of each person's goals, activity, and information requirements. It would also support activities such as identifying and resolving errors, exchanging information, providing individual assistance to another, and monitoring the activity of others. These suggestions illustrate the need for groupware to support the numerous interpersonal activities critical for successful teamwork.

Difficulties in Remote Collaboration

Finally, researchers studying real-world, collaborative, computer-based work environments have focused on the disadvantages imposed by CSCW used by participants working remotely. As an example, there is evidence that people working in the same location use facial expressions and their bodies to communicate information implicitly and explicitly about factors such as task activity, system status, attention, mood, identity, and so forth (e.g., Benford et al., 1995; Tang, 1991). Inzana, Willis, and Kass (1994) found that collocated teams are more cohesive and outperformed distributed teams.

If we evaluate the difficulties of team performance under high workload or stress, we can assume that remote team performance would result in problems such as (1) increased difficulty in knowing who is doing what, (2) increased difficulty in communication because of the loss of subtle cues from eye contact and body language, and (3) increased difficulty in maintaining situation awareness because of a decrease in communication. Researchers have confirmed many of these assumptions. For example, in studying crew communication in the cockpit, Segal (1994) found that crew members watch each other during teamwork and rely heavily on nonverbal information for communication and coordination. Other field studies have shown the use of visual information for task coordination and communication (e.g., Burgoon et al., 1989; Hutchins, 1990) and have demonstrated that reducing available visual access significantly impacts group dynamics (e.g., Chapanis et al., 1972;). Supporting such visual information through video is therefore important in implementing remote collaboration.

CSCW and groupware supports the trend of many companies to adopt agile structures composed of self-directed teams that might be distributed across the world. These *virtual teams* enable rapid adaptation to change, but they also make life more complicated for team members (Bell & Kozlowski, 2002). Just as trust plays an important role in how well people deal with complex automation (as discussed in Chapter 16), trust helps people deal with the complexities of virtual teams (Grabowski & Roberts, 1999; Olson & Olson, 2003). High-performing teams tend to be better at developing and maintaining trust throughout the project (Kanawattanachai & Yoo, 2002). *Trust helps people to deal with the complexity of virtual teams* in several ways (Lee & See, in press). It supplants supervision when direct observation becomes impractical and facilitates choice under uncertainty by acting as a social decision heuristic (Kramer, 1999). It also reduces uncertainty in gauging the responses of others to guide appropriate reliance (Baba, 1999). In addition, trust facilitates decentralization and adaptive behavior by making it possible to replace fixed protocols, reporting structures, and procedures with goal-driven expectations regarding the capabilities of others. However, trust in virtual team members tends to be fragile and can be compromised by cultural and geographic distance between members (Jarvenpaa & Leidner, 1999). Trust between people tends to build rapidly with face-to-face communication but not with text-only communication (e.g., email). Interestingly, trust established in face-to-face communication transferred to subsequent text-only communication (Zheng et al., 2001). This suggests that an initial face-to-face meeting can greatly enhance trust and the performance of a virtual team.

It is clear that groupware methodologies are in their infancy, and as hardware technologies advance, the types of support provided by groupware will increase in power and sophistication. Currently, there is an important gap between the technical capabilities of groupware and the social requirements of the users (Ackerman, 2000)—computers are not able to convey the context of the users and support the subtleties of face-to-face communication. Whether the advances will be able to completely overcome the disadvantages of distance collaboration is not clear.

MACROERGONOMICS AND INDUSTRIAL INTERVENTION

Traditional ergonomic interventions in industry have focused on making changes in the workstation or physical environment for individual workers, an approach called *microergonomics* (Hendrick, 1986). Experience in industrial intervention has taught us that sometimes microergonomic changes are unsuccessful because they address performance and safety problems at the physical and cognitive levels but do not address problems at the social and organizational levels (Hendrick, 1986, 1994; Nagamachi & Imada, 1992). For this reason, recent decades have seen an emphasis on reengineering work systems whereby the analyst takes a larger perspective, addressing the social and organizational factors that impact performance as well as the more traditional human factors considerations (Alexander, 1991; Hendrick, 1986, 1995; Noro & Imada, 1991; Monk & Wagner, 1989). The *macroergonomic* approach addresses performance and safety problems, including analysis of the organization's personnel, social, technological, and economical subsystems (Brown, 1990; Hendrick, 1986, 1995); that is, it evaluates the larger system as well as the person-machine system for the individual worker.

The purpose of macroergonomics analysis is to combine jobs, technological systems, and worker abilities/expectations to harmonize with organizational goals and structure. After the initial analysis, macroergonomic solutions and interventions also focus on larger social and organizational factors, including actions such as increasing employee involvement, (changing communication patterns, restructuring reward systems, and integrating safety into a broad organizational culture) (Imada & Feiglstok, 1990). As Carroll (1994) notes when discussing accidents in high-hazard industries, "Decisions must be understood in context, a context of information, procedures, constraints, incentives, authority, status, and expectations that arise from human organizations" (p. 924). This approach mirrors Reason's (1990, 1997) approach to understanding organizational contributions to human error via differences in safety culture, as discussed in Chapter 14. Because human social factors are involved, they cannot necessarily be addressed with conventional engineering design solutions. The general goal of integrating technological systems with social systems is similar to goals of fields such as organizational development and industrial psychology. Therefore, human factors may begin to overlap with these fields more than it has done in the past.

One of the most commonly used methods for taking a macroergonomic approach is the use of *participatory ergonomics,* a method whereby employees are centrally involved from the beginning (e.g., Imada, 1991; King, 1994; Noro & Imada, 1991). They are asked to help with the front-end analysis, to do problem solving in identifying ergonomic or safety problems, to participate in generating solutions, and to help implement the program elements. Imada provides three reasons for using a participatory ergonomics approach: (1) Employees know a great deal about their job and job environment, (2) employee and management ownership enhances program implementation, and (3) end-user participation promotes flexible problem solving. Employee familiarity with the problems, what works and what does not, and the implicit social dynamics of the workplace allow them to see issues and think of design solutions that an outsider might not consider. It has also been widely noted that strong involvement and "buy in" of employees from the beginning of the intervention process tends to make the changes more successful and long-lasting (e.g., Dunn & Swierczek, 1977; Hendrick, 1995; Huse & Cummings, 1985). Participatory ergonomics does not mean that the end users are the primary or sole designers of an intervention, although they provide a particularly valuable perspective on the design. Their inputs must be guided by the knowledge of human factors professionals.

As noted in Chapter 3, management buy in to and acceptance of human factors can be gained by presenting clear cost/benefit analysis of the expected value realized by human factors applications (Hendrick, 1996).

These are all reasons for using the participatory approach that includes management involvement because strong management and employee participation are both needed to overcome these barriers. As in virtual teams, trust plays an important role when introducing innovations. Trust in senior management reduces the degree of cynicism towards change (Albrecht & Travaglione, 2003).

CONCLUSION

This chapter provided a brief overview of some of the social issues that can greatly influence system performance. The concepts of complexity and coupling that describe organizations have important implications, particularly for the degree of centralization. The increasing complexity of many systems has initiated a push towards decentralization. The trend towards decentralization and self-directed work groups makes it important to understand how to create, manage, and train effective groups and teams. Role differentiation and the degree of synchronization have important implications for training and the design of computers to support teamwork. To enact interventions that improve group and team performance discussed in the chapter, and ergonomic and human factors improvements discussed elsewhere in this book, requires a macroergonomic perspective. Critical to this perspective is the participation of the end users and management in defining a holistic strategy.

References

Aaronson, A., & Carroll, J. M. (1987). Intelligent help in a one-shot dialogue: A protocol study. In *Proceedings of CHI + GI 1987* (pp. 163–168). New York: Association for Computing Machinery.

Abowd, G. D., & Dix, A. J. (1992). Giving undo attention. *Interacting with Computers, 4*(3), 317–342.

Accorsi, R., Zio, E., & Apostolakis, G. E. (1999). Developing utility functions for environmental decision making. *Progress in Nuclear Energy, 34*(4), 387–411.

Ackerman, M. S. (2000). The intellectual challenge of CSCW: The gap between social requirements and technical feasibility. *Human-Computer Interaction, 15*(2–3), 179–203.

Ackerman, P. L., & Cianciolo, A. T. (2002). Ability and task constraint determinants of complex task performance. *Journal of Experimental Psychology: Applied, 8*(3), 194–208.

Adams, K. A., & Halasz, I. M. (1983). *Twenty-five ways to improve your software user manuals.* Worthington, OH: Technology Training Systems.

Adams, M. J., Tenney, Y. J., & Pew, R. W. (1995). Situation awareness and the cognitive management of complex systems. *Human Factors, 37,* 85–104.

Adelman, L., Bresnick, I., Black, P. K., Marvin, F. F., & Sak, S. G. (1996). Research with patriot air defense officers: Examining information order effects. *Human Factors, 38*(2) 250–261.

Ainsworth, L., & Bishop, H. P. (1971). *The effects of a 48-hour period of sustained field activity on tank crew performance.* Alexandria, VA: Human Resource Research Organization.

Albrecht, S., & Travaglione, A. (2003). Trust in pubic-sector senior management. *International Journal of Human Resource Management, 14*(1), 76–92.

Alexander, C. S., Kim, Y. J., Ensminger, M., & Johnson, K. E. (1990). A measure of risk taking for young adolescents: Reliability and validity assessments. *Journals of Youth & Adolescence, 19*(6), 559–569.

Alexander, D. C. (1991). Macro-ergonomics: A new tool for the ergonomist. In M. Pulat & D. Alexander (eds.), *Industrial ergonomics: Case studies* (pp. 275–285). Norcross, GA: Industrial Engineering & Management Press.

Alexander, D. C. (1995). The economics of ergonomics: Part II. *Proceedings of the 39th Annual Meeting of the Human Factors & Ergonomics Society* (pp. 1025–1027). Santa Monica, CA: HFES.

Alexander, D. C. (2002). Making the case for ergonomics. *Proceedings of the Association of Canadian Ergonomics.* Banff, Alberta, Canada.

Alexander, G., & Lunenfeld, H. (1975). *Positive guidance in traffic control.* Washington, DC: Federal Highway Administration.

Alkov, R. A., Borowsky, M. S., &Gaynor, M. S. (1982). Stress coping and the US Navy aircrew factor mishap. *Aviation, Space, & Environmental Medicine, 53,* 1112–1115.

Allen, P. A., Wallace, B., & Weber, T. A., (1995). Influence of case type, word frequency, and exposure duration on visual word recognition. *Journal of Experimental Psychology: Human Perception & Performance, 21*(4), 914–934.

Alty, J. L., & Coombs, M. J. (1980). Face-to-face guidance of university computer users-I: A study of advisory services. *International Journal of Man-Machine Studies, 12,* 390–406.

Alwood, C. M. (1986). Novices on the computer. A review of the literature. *International Journal of Man-Machine Studies, 25,* 633–658.

Aly, A. A., & Subramaniam, M. (1993). Design of an FMS decision-support system. *International Journal of Production Research, 31*(10), 2257–2273

American Association of State Highway & Transportation Officials. (1990). *A policy on geometric design of highways and streets.* Washington, DC: Author.

American Industrial Hygiene Association. (1975), *Industrial noise manual* (3rd ed.). Akron, OH.

American Psychological Association. (1992). *Ethical principles of psychologists and code of conduct.* Washington, DC: American Psychological Association.

Anderson, J. R. (1983). *The architecture of cognition.* Cambridge, MA: Harvard University Press.

Anderson, J. R. (1990). *Cognitive psychology and its implications* (3rd ed.). New York: W. H. Freeman.

Anderson, J. R. (1995). *Cognitive psychology* (4th ed.). New York: W. H. Freeman.

Andersson, G. B. T. (l981). Epidemiological aspects on low-back pain in industry. *Spine, 6*(1), 53–60.

Andersson, G. B. T., Ortengren, A., Nachemson, A., & Elfstrom, G. (1974). Lumbar disc pressure and myoelectric back muscle activity during sitting. I. Studies on an experimental chair. *Scandinavian Journal of Rehabilitation Medicine, 3,* 104–114.

Andre, A. D., & Wickens, C. D. (1992). Compatibility and consistency in display-control systems: Implications for aircraft decision aid design. *Human Factors, 34*(6), 639–653.

Andre, A. D., & Wickens, C. D. (1995, Oct.). When users want what's *not* best for them: A review of performance-preference dissociations. *Ergonomics in Design,* 10–13.

Andrews, D. H., & Goodson, L. A. (1980). A comparative analysis of models of instructional design. *Journal of Instructional Development, 3*(4), 2–16.

Annis, J. F. (1978). Variability in human body size, In *Anthropometric source book,* vol. 1, Chap. 2. NASA Reference Publication 1025, NASA Scientific and Technical Office. Houston, TX: NASA.

Arditi, A. (1986). Binocular vision. In K. Boff, L. Kaufman, & J. Thomas (eds.), *Handbook of perception and human performance, vol.* 1. New York: Wiley.

Aretz, A. J. (1991). The design of electronic map displays. *Human Factors, 33*(1), 85–101.

Arkes, H., & Harkness, R. R. (1980). The effect of making a diagnosis on subsequent recognition of symptoms. *Journal of Experimental Psychology: Human Learning & Memory, 6,* 568–575.

Arkes, H. R., & Hutzel, L. (2000). The role of probability of success estimates in the sunk cost effect. *Journal of Behavioral Decision Making, 13*(3), 295–306

Armstrong, T. T. (1983). *An ergonomics guide to carpal tunnel syndrome.* AIHA Ergonomics Guide Series, American Industrial Hygiene Association, Akron, Ohio.

Armstrong, T. T., & Silverstein, B. A. (1987). Upper-extremity pain in the workplace role of usage in causality. In N. Hadler (ed.), *Clinical concepts in regional musculoskeletal illness* (pp. 333–354). Orlando, FL: Grune & Stratton.

Armstrong, T. T., Buckle, P. D., Fine, L. T., Hagberg, M., Tonsson, B., Kilbom, A., Kuorinka, I., Silverstein, B. A., Sjogaard, G., Viikari-Tuntura, E. R. A. (1993). A conceptual model for work-related neck and upper limb musculoskeletal disorders. *Scandinavian Journal of Work, Environment & Health, 19,* 73–84.

Ashby, P. (1979). *Ergonomics handbook 1: Body size and strength.* Pretoria: SA Design Institute.

Asken, M. J., &Raham, D. C. (1983). Resident performance and sleep deprivation: A review. *Journal of Medical Education, 58,* 382–388.

Asmussen, E., & Heebol-Nielsen, K. (1961). Isometric muscle strength of adult men and women. Communications from the Testing & Observation Institute of the Danish National Association for Infantile Paralysis, NR-11, 1–41.

Association for the Advancement of Medical Instrumentation. (1988). *Human factors engineering guidelines and preferred practices for the design of medical devices* (AAMI HE-1988). Arlington, VA: AAMI.

Astrand, P. O., & Rodahl, L. (1986). *Textbook of work physiology* (3rd ed.). New York: McGraw-Hill.

Ayoub, M. M. (1973). Work place design and posture, *Human Factors, 15*(3), pp. 265–268.

Ayoub, M. M., & Lo Presti, P. (1971). The determination of an optimum size cylindrical handle by use of electromyography. *Ergonomics, 4*(4), 503–518.

Ayoub, M. M., Bethea, N., Bobo, M., Burford, C., Caddel, D., Intaranont, K., Morrissey, S., & Salan, J. (1982). *Mining in low coal, vol. 2: Anthropometry.* (OFR 162(2)-83). Pittsburgh: Bureau of Mines.

Baba, M. L. (1999). Dangerous liaisons: trust, distrust and information technology in American work organizations. *Human Organization, 58*(3), 331–346.

Baber, C. (1997). *Beyond the desktop.* San Diego: Academic Press.

Baddeley, A. D. (1986). *Working memory.* New York: Oxford University Press.

Baddeley, A. D. (1990). *Human memory. Theory and practice.* Boston: Allyn & Bacon.

Badler, N. I., Barsky, B. A., & Zelter, D. (eds.). (1990). *Making them move: Mechanics, control, and animation of articulated figures.* Palo Alto, CA: Morgan-Kaufmann.

Bailey, B. P., Konstan, J. A., & Carlis, J. V. (2001). The effects of interruptions on task performance, annoyance, and anxiety in the user interface. *Proceedings of the IFIP TC. 13 International Conference on Human-Computer Interaction,* Tokyo, Japan, pp. 593–601.

Bailey, G. D. (1993). Iterative methodology and designer training in human-computer interface design. *Proceedings of InterCHI '93, 198–205.* WHERE HELD?

Bailey, R. W. (1993). Performance vs. preference. *Proceedings of the 37th Annual meeting of the Human Factors & Ergonomics Society* (pp. 282–286). Santa Monica, CA: HFES.

Bailey, R. W. (1996). Human performance engineering using human factors/ergonomics to achieve computer system usability (3rd ed.). Englewood Cliffs, NJ: Prentice Hall.

Bainbridge, L. (1983). Ironies of automation. *Automatica, 19*(6), 775–779.

Bainbridge, L. (1988). Types of representation. In L. P. Goodstein, H. B. Andersen, & S. E. Olsen (eds.), *Tasks, errors, and mental models* (pp. 70–91). London: Taylor & Francis.

Baird, L. S., Schneier, C. E., & Laird, D. (eds.) (1983). *The training and development sourcebook.* Amherst, MA: Human Resource Development Press.

Bales, R. (1954). In conference. *Harvard Business Review, 32,* 44–50.

Ball, K. K., Beard, B. L., Roenker, D. L., Miller, R. L., & Griggs, D. S. (1988). Age and visual search: Expanding the useful field of view. *Journal of the Optical Society of America, 5*(12).

Ball, K., & Owsley, C. (1991). Identifying correlates of accident involvement for the older driver. *Human Factors, 33*(5), 583–596.

Ball, K., Owsley, C., Sloan, M., Roenker, D. L., & Bruni, J. R. (1993). Visual attention problems as a predictor of vehicle crashes among older drivers. *Investigate Ophthalmology & Visual Science, 34*(11), 3110–3123.

Balla, J. (1980). Logical thinking and the diagnostic process. *Methodology & Information in Medicine, 19,* 88–92.

Banbury, S. P., Macken, W. J., Tremblay, S., & Jones, D. M. (2001). Auditory distraction and short-term memory: phenomena and practical implications. *Human Factors, 43,* 12–29.

Barfield, W., & Furness, T. A. III (eds.) (1995). *Virtual environments and advanced interface design.* New York: Oxford University Press.

Barnes, R. M. (1963). *Motion and time study* (5th ed.). New York: Wiley.

Barnett, B. T., Arbak, C. T., Olson, T. L., & Walrath, L. C. (1992). A framework for design traceability. *Proceedings of the 36th Annual Meeting of the Human Factors Society* (pp. 2–6). Santa Monica, CA: HFS.

Barr, R. A., & Eberhard, J. W. (eds.) (1991). Safety and mobility of elderly drivers, Part 1. *Human Factors Special Issue, 33*(5).

Barrick, M. R., & Mount, M. K. (1991). The big five personality dimensions and job performance: A meta-analysis. *Personnel Psychology, 44,* 1–26.

Bartlett, F. C. (1932). *Remembering: An experimental and social study.* Cambridge: Cambridge University Press.

Barton, J., & Folkard, S. (1993). Advanced versus delaying shift systems. *Ergonomics, 36,* 59–64.

Barton, N. T., Hooper, G., Noble, T., & Steel, W. M. (1992). Occupational causes of disorders in the upper limb. *British Medical Journal,* 304, 309–311.

Barton, P. H. (1986). The development of a new keyboard for outward sorting foreign mail. *IMechE,* 57–63.

Bartram, D. (1995). The predictive validity of the EPI and 16PF for military flying training. *Journal of Occupational & Organizational Psychology, 68*(3), 219–236.

Bass, E. J. (1998). Towards an intelligent tutoring system for situation awareness training in complex, dynamic environments. *Intelligent Tutoring Systems, 1452,* 26–35.

Bassin, M. (1988). Teamwork at General Foods: New and improved. *Personnel Journal, 67*(5), 62–70.

Baty, D. L., Wempe, T. E., & Huff, E. M. (1974). A study of aircraft map display location and orientation. *IEEE Transaction on Systems, Man, & Cybernetics, SMC-4,* 560–568.

Beard, D. V., & Walker, J. Q. (1990). Navigational techniques to improve the display of large two-dimensional spaces. *Behavior & Information Technology, 9*(6), 451–466.

Begault, D. R., & Pittman, M. T. (1996). 3-dimensional audio versus head-down traffic alert and collision avoidance system displays. *International Journal of Aviation Psychology, 6*(2), 79–93.

Belkin, N. J., Marchetti, P. G., & Cool, C. (1993). BRAQUE: Design of an interface to support user interaction in information retrieval. *Information Processing & Management, 29*(3), 325–344.

Bell, B. S., & Kozlowski, S. W. J. (2002). A typology of virtual teams: Implications for effective leadership. *Group & Organization Management, 27*(1), 14–49.

Bell, C. A., Stout, N. A., Bender, T. R., Conroy, C. S., Crouse, W. E., & Meyers, J. R. (1990). Fatal occupational injuries in the United States, 1980 through 1985. *JAMA, 263,* 3047–3050.

Belz, S.M., Robinson, G.S., & Casali, J.G., (1999). A new class of auditory warning signals for complex systems: auditory icons. *Human Factors, 41*(4), 608–618.

Bellenkes, A. H., Wickens, C. D., & Kramer, A. F. (1997). Visual scanning and pilot expertise: The role of attentional flexibility and mental model development. *Aviation, Space, & Environmental Medicine, 68*(7), 569–579.

Benbassat, D., & Abramson, C. I. (2002). Landing flare accident reports and pilot perception analysis. *The International Journal of Aviation Psychology, 12*(2), 137–152.

Bendix, T., & Hagberg, M. (1984). Trunk posture and load on the trapezius muscle whilst sitting at sloping desks. *Ergonomics, 27,* 873–882.

Bendix, T., Winkel, T., & Tersen, F. (1985). Comparison of office chairs with fixed forwards and backwards inclining, or tiltable seats. *European Journal of Applied Physiology, 54,* 378–385.

Bennett, K. B., & Flach, J. M. (1992). Graphical displays: Implications for divided attention, focused attention, and problem solving. *Human Factors, 34*(5), 513–533.

Bensel, C. K., & Santee, W. R. (1997). Climate and clothing. In G. Salvendy (ed.), *Handbook of human factors and ergonomics.* New York: Wiley.

Bergus, G. R., Levin, I. P., & Elstein, A. S. (2002). Presenting risks and benefits to patients—The effect of information order on decision making. *Journal of General Internal Medicine, 17*(8), 612–617.

Bernstein, M. (1988). The bookmark and the compass. *ACM SIGOIS Bulletin, 9*(4), 34–45.

Bernstein, D., Clark-Stewart, A., Roy, E., & Wickens, C. D. (1997). *Psychology* (4th ed.). Houghton-Mifflin.

Bessant, J., Levy, P., Ley, C., Smith, S., & Tranfield, D. (1992). Organization design for factory 2000. *The International Journal of Human Factors in Manufacturing, 2*(2), 95–125.

Bettman, J. R., Johnson, E. J., & Payne, J. (1990). A componential analysis of cognitive effort and choice. *Organizational Behavior & Human Performance, 45,* 111–139.

Bhatnager, V., Drury, *C. G., &* Schiro, S. G. (1985). Posture, postural discomfort and performance. *Human Factors, 27,* 189–199.

Bias, R. G., & Mayhew, D. (1994). *Cost-justifying usability.* New York: Academic Press.

Billings, C. E. (1996). *Toward a human-centered approach to automation.* Englewood Cliffs, NJ: Erlbaum

Bink, B. (1962). The physical working capacity in relation to working time and age. *Ergonomics, 5*(1), 25–28.

Bisseret, A. (1981). Application of signal detection theory to decision making in supervisory control. *Ergonomics, 24,* 81–94.

Bisson, R. V. et al. (1992). Digital flight data as a measure of pilot performance associated with continued operations during Desert Storm. *In Nutritional Metabolic Disorders and Lifestyle of Aircrew Members* (pp. 12-54). Neuilly-Surseine; France: NATO.

Bittner, R., Smrcka, P., Vysoky, P., Hana, K., Pousek, L., & Schreib, P. (2000). Detecting of fatigue states of a car driver, *Medical Data Analysis, Proceedings* (vol. 1933, pp. 260–273).

Bliss, J., & Gilson, R. (1998). Emergency signal failure: Implications and recommendations. *Ergonomics, 41*(1), 57–72.

Bjork, R. (ed). (1994). *Learning, remembering, believing: enhancing human performance.* Washington DC: National Academy of Sciences Press.

Bjork, R. A. (1999). Assessing our own competence: Heuristics and illusions. In D. Gopher & A. Koriat (eds.), *Attention and performance XVII* (pp. 435–459). Cambridge, MA: Bradford Book.

Blair, D. C., & Maron, M. E. (1985). An evaluation of retrieval effectiveness for a full-text document-retrieval system. *Communications of the ACM, 28*(3), 289–299.

Blake, R. F., Mouton, J. S., & McCanse, A. A. (1989). *Change by design.* Reading, MA: Addison-Wesley.

Blanchard, B. S., & Fabrycky, W. T. (1990). *Systems engineering and analysis.* Englewood Cliffs, NJ: Prentice Hall.

Boff, K. R., Kaufman, L., & Thomas, J. P. (eds.). (1986). *Handbook of perception and human performance.* New York: Wiley.

Boff, K. R., Monk, D. L., Swierenga, S. J., Brown, C. E., & Cody, W. T. (1991). Computer-aided human factors for systems designers. *Proceedings of the 35th Annual Meeting of the Human Factors Society & Ergonomics Society* (pp. 332–336). Santa Monica, CA: HFES.

Boff, K., & Lincoln, T. (1988). *Engineering data compendium: Human perception and performance.* (4 vols.). Wright-Patterson Air Force Base, OH: Armstrong Aerospace Medical Research Laboratory, AAMRL/NATO.

Bogner, M. S. (ed.). (1994). *Human error in medicine.* Hillsdale, NJ: Erlbaum.

Bolia, R. S., D'Angelo, W. R., & McKinley, R. L., (1999) Aided visual search in three dimensional space. *Human Factors, 4*(4), 664–669 (1999).

Bond, D. (2001, Dec. 17). Crisis at Herndon: 11 airplanes astray. *Aviation Week & Space Technology,* 96–99.

Bonnet, M. H., & Arand, D. L. (1995). We are chronically sleep deprived. *Sleep, 18*(10), 908–911.

Booher, H. R. (1975). Relative comprehensibility of pictorial information and printed words in proceduralized instructions. *Human Factors, 17,* 266–277.

Booher, H. R. (ed.). (1990). *MANPRINT: An approach to systems integration.* New York: Van Nostrand Reinhold.

Booher, H. R. (ed.) (2003) *Handbook of human systems integration.* N.Y.: John Wiley & Sons.

Borg, G. (1985). *An introduction to Borg's RPE-Scale.* Ithaca, NY: Movement Publications.

Borman, W. C., Hanson, M. A., & Hedge, J. W. (1997). Personnel selection. *Annual Review of Psychology., 48,* 299–337.

Bos, J. F. T., Stassen, H. G., & van Lunteren, A. (1995). Aiding the operator in the manual control of a space manipulator. *Control Eng. Practice, 3*(2), 223–230.

Bourne, P. G. (1971). Altered adrenal function in two combat situations in Vietnam. In B. E. Elefheriou & J. P. Scott (eds.), *The physiology of aggression and defeat.* New York: Plenum.

Bowers, C. A., Oser, R. L., Salas, E., & Cannon-Bowers, J. A. (1996). Team performance in automated systems. In R. Parasuraman & M. Mouloua (eds.), *Automation and human performance: Theory and applications* (pp. 243–263). Mahwah, NJ: Erlbaum.

Bowers, C. A., Weaver, J. L., & Morgan, B. B. (1996). Moderating the performance effects of stressors. In J. Driskell & E. Salas (eds.), *Stress and human performance.* Mahwah, NJ: Erlbaum.

Bowers, V. A., & Snyder, H. L. (1990). Concurrent versus retrospective verbal protocol for comparing window usability. *Proceedings of the 34th Annual Meeting f the Human Factors & Ergonomics Society* (pp. 1270–1274). Santa Monica, CA: HFES.

Boyce, P. R. (1997). Illumination. In G. Salvendy (ed.), *Handbook of human factors and ergonomics* (2nd ed.). New York: Wiley.

Bradsher, K. (2002). High and mighty: SUVs—The world's most dangerous vehicles and how they got that way. New York: Public Affairs.

Brehmer, B. (1980). In One Word—Not from Experience. *Acta Psychologica, 45*(1–3), 223–241

Bresley, B. (1995, Apr–Jun). 777 flight deck design. *Airliner,* 1–9.

Bridger, R. (1988). Postural adaptations to a sloping chair and work surface. *Human Factors, 30*(2), 237–247.

Bridger, R. S. (1995). *Introduction to ergonomics.* New York: McGraw-Hill.

Broadbent, D. E. (1972). *Decision and stress.* New York: Academic Press.

Broadbent, D., & Broadbent, M. H. (1980). Priming and the passive/active model of word recognition. In R. Nickerson (ed.), *Attention and performance, VIII.* New York: Academic Press.

Brock, J. F. (1997). Computer-based instruction. In G. Salvendy (ed.), *Handbook of human factors and ergonomics* (2nd ed.) (pp. 578–593). New York: Wiley.

Brogan, D. (ed.) (1993). Visual search 2: Proceedings of the 2nd International Conference on Visual Search. London: Taylor & Francis.

Brookhuis, K. A., & de Waard, D. (1993). The use of psychophysiology to assess driver status. *Ergonomics, 36,* 1099–1110.

Brouha, L. (1967). *Physiology in Industry.* 2nd ed. New York: Pergamon Press.

Brouwer, W. H., Waternik, W., Van Wolffelaar, P. C., & Rothengatter, T. (1991). Divided attention in experienced young and older drivers: Lane tracking and visual analysis in a dynamic driving simulator. *Human Factors, 33*(5), 573–582.

Brown, I. D. (1994). Driver fatigue. *Human Factors, 36*(2), 298–314.

Brown, I. D., Groeger, J. A., & Biehl, B. (1988). Is driver training contributing enough towards road safety. In J. A. Rothergatter, & R. A. de Bruin (eds.), *Road users and traffic safety* (pp. 135–156). Assen/Maastricht, Netherlands: Van Corcum.

Brown, I. D., Tickner, A. H., & Simmonds, D. C. V. (1969). Interference between concurrent tasks of driving and telephoning. *Journal of Psychology, 53*(5), 419–424.

Brown, O., Jr. (1990). Macroergonomics: A review. In K. Noro & O. Brown, Jr. (eds.), *Human factors in organizational design and management III* (pp. 15–20). Amsterdam: North-Holland.

Bryan, W. L. & Harter, N. (1899). Studies on the telegraphic language. The acquisition of a hierarchy of habits. *Psychological Review. 6,* 345–375.

Bullinger, H., Kern, P, & Braun, M. (1997). Controls. In G. Salvendy (ed.), *Handbook of human factors and ergonomics, Chapter 2* (2nd ed.). New York: Wiley.

Burg, A., & Hulbert, S. (1961). Dynamic visual acuity as related to age, sex, and static acuity. *Journal of Applied Psychology, 45,* 111–116.

Burgoon, J. K., Buller, D. B., & Woodall, W. G. (1989). *Nonverbal communication: the unspoken dialogue.* New York: Harper & Row.

Burke, E. (1995a). Pilot selection I: The state-of-play. *Proceedings of the 8th International Symposium on Aviation Psychology* (pp. 1341–1346). Columbus: Ohio State University.

Burke, E. (1995b). Pilot selection II: Where do we go from here? *Proceedings of the 8th International Symposium on Aviation Psychology* (pp. 1347–1353). Columbus: Ohio State University.

Burns, C. M. (2000). Navigation strategies with ecological displays. *International Journal of Human-Computer Studies, 52*(1), 111–129.

Cabrera, E. F., & Raju, N. S. (2001). Utility analysis: Current trends and future directions. *International Journal of Selection and Assessment, 9*(1–2), 92–102.

Casali, T., & Porter, D. (1980). On difficulties in localizing ambulance *sirens. Human Factors, 22,* 7l9–724.

Caldwell, J. A., Caldwell, J. L., Crowley, J. S., and Jones, H. D. (1995). Sustaining helicopter pilot performance with Dexedrine during periods of sleep deprivation. *Aviation, Space and Environmental Medicine. 66,* 930–937.

Campbell, B. J., Stewart, J. R., & Campbell, F. A. (1988). *Changes with death and injury associated with safety belt laws* 1985–1987 (Report HSRC-A138). Chapel Hill: University of North Carolina Highway Safety Research Center.

Campbell, J. L., Carney, C., & Kantowitz, B. H. (1998). *Human factors design guidelines for advanced traveler information systems (ATIS) and commercial vehicle operations (CVO)* (FHWA-RD-98–057). Washington, DC: Federal Highway Administration.

Cano, A. R., & Kleiner, B. M. (1996). Sociotechnical design factors for virtual team performance. *Proceedings of the 40th Annual Meeting Human Factors & Ergonomics Society* (pp. 786–790). Santa Monica, CA: HFES.

Caplan, S. (1990). Using focus group methodology for ergonomic design. *Ergonomics, 33*(5), 527–533.

Card, S. K., English, W. K., & Burr, B. J. (1978). Evaluation of mouse, rate-controlled isometric joystick, step keys, and task keys for text selection on a CRT. *Ergonomics, 21*(8), 601–613.

Card, S., Moran, T. P., & Newell, A. (1983). *The psychology of human-computer interactions.* Hillsdale, NJ: Erlbaum.

Card, S., Moran, T., & Newell, A. (1986). The model human processor. In K. Boff, L. Kaufman, & J. Thomas (eds.), *Handbook of perception and human performance* (vol. 2). New York: Wiley.

Carlow International. (1990). Human factors engineering: Part I. Test procedures; Part II: Human factors engineering data guide for evaluation (HEDGE). Washington, DC: Army Test & Evaluation Command. ADA226480.

Carlson, R. A., Sullivan, M. A., & Schneider, W. (1989). Practice and working memory effects in building procedural skill. *Journal of Experimental Psychology: Learning, Memory & Cognition, 3,* 517–526.

Carroll, J. M. (ed.) (1995). *Scenario-based design: Envisioning work and technology in system development.* New York: Wiley.

Carroll, J. M., Mack, R. L., & Kellogg, W. A. (1988). Interface metaphors and the user interface design. In M. Helander (ed.), *Handbook of human-computer interaction* (pp. 67–85). Amsterdam: North-Holland.

Carroll, J. S. (1994). The organizational context for decision making in high-hazard industries. *Proceedings of the 38th Annual Meeting of the Human Factors & Ergonomics Society* (pp. 922–925). Santa Monica, CA: HFES.

Carroll, J. M., & Carrithers, C. (1984). Blocking learner error states in a training-wheels system. *Human Factors, 26,* 377–389.

Carron, A. V., Bray, S. R., & Eys, M. A. (2002). Team cohesion and team success in sport. *Journal of Sports Sciences, 20*(2), 119–126.

Carskadon, M. A., & Dement, W. C. (1975). Sleep studies on a 90-minute day. *Electroencephalogram Clinical Neurophysiology, 39,* 145–155.

Carswell, C. M. (1992). Reading graphs: Interactions of processing requirements and stimulus structure. In B. Burns (ed.), *Percepts, concepts, and categories* (pp. 605–647). Amsterdam: Elsevier.

Casali, J., Lam, S., & Epps, B. (1987). Rating and ranking methods for hearing protector wearability. *Sound & Vibration, 21*(12), 10–l8.

Casey, S. (1993). Set phasers on stun and other true tales of design, technology, and human error. Santa Barbara, CA: Aegean.

Casey, S. (1998). *Set phasers on stun.* Santa Barbara, CA: Aegean.

Casner, S. M. (1994). Understanding the determinants of problem-solving behavior in a complex environment. *Human Factors, 36,* 580–596.

Cattell, R. B., Eber, H. W., & Tatsuoka, M. (1970). *Handbook for the sixteen personality factor questionnaire (16PF).* Champaign, IL: Institute for Personality & Ability Testing.

Chadwick, D., Knight, B., & Rajalingham, K. (2001). Quality control in spreadsheets: A visual approach using color codings to reduce errors in formulae. *Software Quality Journal, 9*(2), 133–143.

Chaffin, D. B. (1997). Biomechanical aspects of workplace design. In G. Salvendy (ed.), *Handbook of human factors and ergonomics,* (2nd ed.). New York: Wiley.

Chaffin, D. B., Andersson, G. B. J., & Martin, B. J. (1999). *Occupational biomechanics* (3rd ed). New York: Wiley.

Chaffin, D. B., Nelson, C., Ianni., S. D. & Punte, P. A. (2001) *Digital human monitoring for vehicle and workplace design.* Warrendale, PA.: Society of Automotive Engineers.

Chao, B. P. (1993). Managing user interface design using concurrent engineering. *Proceedings of the 37th Annual Meeting of the Human Factors & Ergonomics Society* (pp. 287–290). Santa Monica, CA: HFES.

Chao, C. D., Madhavan, D., & Funk, K. (1996). Studies of cockpit task management errors. *The International Journal of Aviation Psychology, 6*(4), 307–320.

Chapanis, A. (1970). Human factors in systems engineering. In K. B. DeGreene (ed.), *Systems psychology* (pp. 28–38). New York: McGraw-Hill.

Chapanis, A. (1991). To communicate the human factors message, you have to know what the message is and how to communicate it. *Human Factors Society Bulletin, 34*(11), 1–4.

Chapanis, A., Ochsman, R. B., Parrish, R. N., & Weeks, G. D. (1972). Studies in interactive communication: I. The effects of four communication modes on the behavior of teams during cooperative problem-solving. *Human Factors, 14*(6), 487–509.

Charness, N., & Schultetus, R. S. (1999). Knowledge and expertise. In F. T. Durso (ed.), *Handbook of applied cognition* (pp. 57–82). New York: Wiley.

Chase, W., & Chi, M. (1979). *Cognitive skill: implications for spatial skill in large-scale environments* (Technical Report #1). Pittsburgh: University of Pittsburgh Learning & Development Center.

Chen, S. J., & Li, L. (2003). Decomposition of interdependent task group for concurrent engineering. *Computers & Industrial Engineering, 44*(3), 435–459.

Clark, H. H., & Chase, W. G. (1972). On the process of comparing sentences against pictures. *Cognitive Psychology, 3,* 472–517.

Clark, R. C. (1989). Developing technical training: A structured approach for the development of classroom and computer-based instructional materials. Reading, MA: Addison-Wesley.

Clauser, C. E., Tucker, P. E., McConville, J. T., Churchill, E., Laubach, L. L., & Reardon, J. A. (1972). *Anthropometry of Air Force women* (pp. 1–1157). AMRL-TR-70–5. Wright-Patterson Air Force Base, OH: Aerospace Medical Research Labs.

Cohen, J. T., & Graham, J. D. (2003). A revised economic analysis of restrictions on the use of cell phones while driving. *Risk Analysis, 23*(1), 1–14.

Cohen, M. S., Freeman, J. T., & Wolf, S. (1996). Metarecognition in time-stressed decision making: Recognizing, critiquing, and correcting. *Human Factors, 38*(2), 206–219.

Cole, W. G. (1986). Medical cognitive graphs. *Proceedings of the ACM-SIGCHI: Human Factors in Computing Systems* (pp. 91–95). New York: Association for Computing Machinery.

Comperatore, C. A., Liberman, H., Kirby, A. W., Adams, B., & Crowley, J. S. (1996, Oct.). *Melatonin efficacy in aviation missions requiring rapid deployment and night operations* (USAARL 97–03). Ft. Rucker, AL: U. S. Army Aeromedical Res. Lab.

Cook, R. I., & Woods, D. D. (1994). Operating at the sharp end: The complexity of human error. In M. S. Bogner (ed.), *Human error in medicine* (pp. 255–301). Hillsdale, NJ: Erlbaum.

Cook, R. I., & Woods, D. D. (1996). Adapting to new technology in the operating room. *Human Factors, 38*(4), 593–613.

Cook, R., Woods, D., & McDonald, J. (1991). Human performance in anesthesia: A corpus of cases (Technical Report CSEL91. 003). Columbus: Ohio State University, Cognitive Systems Engineering Laboratory.

Coombs, C. H., Dawes, R. M., & Tversky, A. (1970). *Mathematical psychology.* Englewood Cliffs, NJ: Prentice Hall.

Cooper, A. (1999). The inmates are running the asylum: Why high tech products drive us crazy. Indianapolis, IN: SAM.

Cooper, C. L., & Cartwright, S. (2001). A strategic approach to organizational stress management. In P. A. Hancock & P. A. Desmond (eds.), *Stress, workload, and fatigue* (235–248). Mahwah, NJ: Erlbaum.

Cooper, K. H., Pollock, M. L., Martin, R. P., White, S. R., Linnerud, A. C., & Jackson, A. (1975). Physical fitness levels versus selected coronary risk factors. *Journal of the American Medical Association, 236*(2), 166–169.

Cooper, P. J., & Zheng, Y. (2002). Turning gap acceptance decision-making: The impact of driver distraction. *Journal of Safety Research, 33*(3), 321–335.

Cotrambone, R., & Carroll, J. M. (1987). Learning a word processing system with training wheels and guided exploration. *Proceedings of CHI & GI Human Factors in Computing Systems & Graphics Conference* (pp. 169–174). New York: ACM.

Cowan, N. (2001). The magical number 4 in short-term memory: a reconsideration of mental storage capacity. *Behavioral & Brain Sciences, 24,* 87–185.

Craik, F. I. M., & Lockhart, R. S. (1972). Levels of processing: a framework for memory research. *Journal of Verbal Learning & Verbal Behavior, 11,* 671–684.

Cream, B. W., Eggemeier, F. T., & Klein, G. A. (1978). A strategy for the development of training devices. *Human Factors, 20,* 145–158.

Creasy, R. (1980). Problem solving, the FAST way. *Proceedings of Society of Added-Value Engineers Conference* (pp. 173–175). Irving, TX: Society of Added-Value Engineers.

Crocker, M. J. (1997). Noise. In G. Salvendy (ed.), *Handbook of human factors and ergonomics* (Chpt. 24). New York: Wiley.

CTD News. (1995). CTDs taking bigger bite of corporate bottom line. *CTD News 4*(6), 1.

Curtis, K., Kindlin, C., Reich, K., and White, D. (1995). Functional reach in wheelchair users: The effects of trunk and lower extremity stabilization. *Arch Phys Med Rehabil., 76,* 360–367.

Cutting, J. E., & Vishton, P. M. (1995). Perceiving layout and knowing distances: The integration, relative patency, and contextual use of different information about depth. In W. Epstein & S. J. Rogers, eds., *Handbook of perception and cognition: Perception of space and motion* (vol. 5). New York: Academic Press.

Czeisler, C. A., Kennedy, W. A., & Allan, J. S. (1986). Circadian rhythms and performance decrements in the transportation industry. In A. M. Coblentz (ed.), *Proceedings of a Workshop on the Effects of Automation on Operator Performance* (pp. 146–171). Paris: Universite Rene Descartes.

Czeisler, C. A., Moore-Ede, M. C., & Coleman, R. M. (1982, 30 Jul). Rotating shift work schedules that disrupt sleep are improved by applying circadian principles. *Science, 217,* 460–462.

Czeisler, C. A. et al. (1989). Bright light induction of strong resetting of human circadian pacemaker. *Science, 244,* 1328–1333.

Czeisler, C. A., Weitzman, E. D., Moore-Ede, M. C., Zimmerman, J. C., & Knauer, R. S., (1980). Human sleep: Its duration and organization depend on its circadian phase. *Science, 210,* 1264–1267.

Czerwinski, M., Cutrell, E., & Horvitz, E. (2000). Instant messaging: effects of relevance and timing. In S. Turner & P. Turner (eds.), *People & computers XIV: Proceedings of HCI 2000,* vol. 2, 71–76. British Computer Society.

Damos, D. (1991). *Multiple task performance.* Taylor & Francis.

Das, B., & Kozey, J. (1999). Structural anthropometroic measurements for wheel chair mobile adults. *Applied Ergonomics, 30,* 385–390.

Davies, D. R., & Parasuraman, R. (1982). *The psychology of vigilance.* New York: American Elsevier.

Davis, W., Grubbs, J. R., & Nelson, S. M. (1995). *Safety made easy: A checklist approach to OSHA compliance.* Rockville, MD: Government Institutes.

Davis, F. D., & Kottemann, J. E. (1994). User perceptions of decision support effectiveness two production planning experiments. *Decision Sciences, 25*(1), 57–78.

Dawes, R. M., Faust, D., & Meehl, P. E. (1989). Clinical versus statistical judgment. *Science, 243,* 1668–1673.

Dayton, T., McFarland, A., & White, E. (1994). Software development: Keeping users at the center. *Exchange: Information technology at work, 10*(5), 12–17.

De Raedt, R., & Ponjaert-Kristoffersen, I. (2000). Can strategic and tactical compensation reduce crash risk in older drivers? *Age & Ageing, 29*(6), 517–521.

Degani, A., & Wiener, E. L. (1993). Cockpit checklists: Concept, design, and use. *Human Factors, 35*(2), 345–360.

DeJoy, D. M. (1991). A revised model of the warning process derived from value-expectancy theory. *Proceedings of the 35th Annual Meeting of the Human Factors Society* (pp. 1043–1047). Santa Monica, CA: HFES.

Dempsey, P. G. (1999). Utilizing criteria for exposure and compliance assessments of multiple-task manual materials handling jobs. *International Journal of Industrial Ergonomics, 24*(4), 405–416.

Dempsey, P. G., Sorock, G. S., Cotnam, J. P., Ayoub, M. M., Westfall, P. H., Maynard, W., Fathallah, F., and O'Brien, N. (2000). Field evaluation of the revised NIOSH lifting equation. In *Proceedings of the International Ergonomics Association/Human Factors and Ergonomics Society 2000 Congress* (CD-ROM). San Diego, CA. 537–540.

Denton, G. G. (1980). The influence of visual pattern on perceived speed. *Perception, 9,* 393–402.

Department of Defense. (1984). m.l-STD-882B Washington D.C.: U.S. Government Printing Office.

Department of Health and Human Services. (1991). Code of Federal Regulations, Title 45, Part 46: Protection of human subjects. Washington, DC: HHS.

Department of Labor. (1993). *Occupational safety and health standards for general industry.* Washington, DC: U. S. Government Printing Office.

Derrick, W. L. (1988). Dimensions of operator workload. *Human Factors, 30*(1), 95–110.

DeSanctis, G., & Gallupe, R. B. (1987). A foundation for the study of group decision support systems. *Management Science, 33*(5), 589–609.

Desmond, P. A., & Hancock, P. A. (2001). Active and passive fatigue states. In P. A. Hancock & P. A. Desmond (eds.), *Stress, workload, and fatigue* (pp. 455–365). Mahwah, NJ: Erlbaum.

Desurvire, H., & Thomas, T. C. (1993). Enhancing the performance of interface evaluators using non-empirical usability methods. *Proceedings of the 37th Annual Meeting of the Human Factors & Ergonomics Society* (pp. 1132–1136). Santa Monica, CA: HFES.

deSwart, T. (1989). Stress and stress management in the Royal Netherlands Army. *Proceedings of the User's Stress Workshop.* Washington, DC: U.S. Army Health Services Command.

Devenport, J. L., Studebaker, C. A., & Penrod, S. D. (1999). Perspectives on jury decision-making: Cases with pretrial publicity and cases based on eyewitness identifications. In F. T. Durso (ed.), *Handbook of applied cognition* (pp. 819–845). Chichester, England: Wiley.

Dewar, R. (1993, July). Warning: Hazardous road signs ahead. *Ergonomics in Design,* 26–31.

Diaper, D. (1989). *Task analysis for human-computer interaction.* Chichester, UK: Ellis Horwood.

DiBerardinis, L. (ed.) (1998). *Handbook of occupational safety and health.* New York: Wiley-Interscience.

Diehl, A. E. (1991). The effectiveness of training programs for preventing aircrew error. In R. S. Jensen (ed.), *Proceedings of the 6th International Symposium on Aviation Psychology* (pp. 640–655). Columbus, OH: Dept. of Aviation, Ohio State University.

Dinges, D. F., Orne, M T., Whitehouse, W. G., & Orne, E. C. (1987). Temporal placement of a nap for alertness: Contributions of circadian phase and prior wakefulness. *Sleep, 10,* 313–329.

Dingus, T. A., & Hulse, M. C. (1993). Some human factor design issues and recommendations for automobile navigation information systems. *Transportation Research, 1C(2),* 119–131.

Dingus, T. A., Antin, J. F., Hulse, M. C., & Wierwille, W. (1988). Human factors issues associated with in-car navigation system usage. *Proceedings of the 32nd Annual Meeting of the Human Factors Society* (pp. 1448–1453). Santa Monica, CA: HFS.

Dingus, T. A., Hathaway, J. A., & Hunn, B. P. (1991). A most critical warning variable: Two demonstrations of the powerful effects of cost on warning compliance. *Proceedings of the 35th Annual Meeting of the Human Factors Society* (pp. 1034–1038). Santa Monica, CA: HFS.

Dingus, T. A., McGehee, D. V., Manakkal, N., Johns, S. K., Carney, C., & Hankey, J. (1997). Human factors field evaluation of automobile headway maintenance/collision warning devices. *Human Factors, 39,* 216–229.

Dismukes, K. (2001). The challenge of managing interruptions, distractions, and deferred tasks. *Proceedings of the 11th International Symposium on Aviation Psychology.* Columbus, OH: The Ohio State University.

Dix, A., Finlay, T., Abowd, G., & Beale, R. (1993). *Human-computer interaction.* Englewood Cliffs, NJ: Prentice Hall.

Dixon, S. R., & Wickens, C. D. (2003). Control of multiple-UAVs: A workload analysis. *Proceedings of the 12th International Symposium on Aviation Psychology.* Dayton, OH: Wright State University.

Dockery, C. A., & Neuman, T. (1994). Ergonomics in product design solves manufacturing problems: Considering the users' needs at every stage of the product's life. *Proceedings of the 38th Annual Meeting of the Human Factors & Ergonomics Society* (pp. 691–695). Santa Monica, CA: HFES.

Donchin, Y., Gopher, D., Olin, M., et al. (1995). A look into the nature and causes of human errors in the intensive care unit. *Critical Care Medicine, 23,* 294–300.

Dornheim, M. P. (1995). Dramatic incidents highlight mode problems in cockpit. *Aviation Week & Space Technology.* (Jan. 30), 55–59.

Dougherty, E. M. (1990). Human reliability analysis – where shouldst thou turn? *Reliability Engineering & System Safety, 29,* 283–299.

Douglas, S., & Moran, T. P. (1983). Learning text editor semantics by analogy. *Proceedings CHI '83. Human factors in computing systems* (pp. 207–211). New York: Association of Computing Machinery.

Driskell, J. E., & Salas, E. (eds.) (1996). *Stress and human performance.* Mahwah, NJ: Erlbaum.

Driskell, J. E., Johnston, J. H., & Salas, E. (2001). Does stress training generalize to novel settings? *Human Factors, 43*(1), 99–110.

Druckman, D., & Bjork, R. (1994). *Learning, remembering, believing: enhancing human performance.* Washington, DC: National Academic Press.

Drury, C. (1975a). Inspection of sheet metal: Model and data. *Human Factors, 17,* 257–265.

Drury, C. (1975b). Application to Fitts' Law to foot pedal design. *Human Factors, 17,* 368–373.

Drury, C. (1982). Improving inspection performance. In G. Salvendy (ed.), *Handbook of industrial engineering.* New York: Wiley.

Duffy, V. G., & Salvendy, G. (1999). The impact of organizational ergonomics on work effectiveness: with special reference to concurrent engineering in manufacturing industries. *Ergonomics, 42*(4), 614–637.

Dunn, W., & Swierczek, F. (1977). Planned organizational change: Toward grounded theory. *Journal of Applied Behavioral Science, 13*(2), 135–157.

Durding, B. M., Becker, C. A., & Gould, J. D. (1977). Data organization. *Human Factors, 19,* 1–14.

Durlach, N. I., & Mavor, A. S. (eds.) (1995). *Virtual reality: Scientific and technological challenges.* Washington, DC: National Academy Press.

Durnin, T. V. G. A., & Passmore, R. (1967). *Energy, work, and leisure.* London, UK: Heinemann.

Durso, F. (ed.). (1999). *Handbook of applied cognition.* New York: Wiley.

Durso, F., & Gronlund, S. (1999). Situation awareness In F. T. Durso (ed.), *Handbook of applied cognition* (pp. 283–314). New York: Wiley.

Dzindolet, M. T., Pierce, L. G., Beck, H. P., & Dawe, L. A. (2002). The perceived utility of human and automated aids in a visual detection task. *Human Factors, 44*(1), 79–94.

Eastman Kodak Company, Ergonomics Group. (1986). *Ergonomic design for people at work,* vol. 1. New York: Van Nostrand Reinhold.

Eastman Kodak Company, Ergonomics Group. (1986). *Ergonomic design for people at work,* vol. 2, New York: Van Nostrand Reinhold.

Eastman, M. C., & Kamon, E. (1976). Posture and subjective evaluation at flat and slanted desks. *Human Factors, 18*(1), 15–26.

Eberts, R. E., & MacMillan, A. G. (1985). Misperception of small cars. In R. E. Eberts & C. G. Eberts (eds.), *Trends in ergonomics/human factors II* (pp. 33–39). Amsterdam: Elsevier.

Eckbreth, K. A. (1993). The ergonomic evaluation and improvement of a cable forming process: A case study. *Proceedings of the 37th Annual Meeting of the Human Factors & Ergonomics Society* (pp. 822–825). Santa Monica, CA: HFES.

Edholm, O. G. (1967). *The biology of work.* New York: McGraw Hill.

Edland, A., & Svenson, O. (1993). Judgment and decision making under time pressure: Studies and findings. In O. Svenson & A. J. Maule (eds.), *Time pressure and stress in human judgment and decision making* (pp. 27–40). New York: Plenum.

Edwards, W. (1954). The theory of decision making. *Psychological Bulletin, 51,* 380–417.

Edwards, W. (1961). Behavioral decision theory. *Annual Review of Psychology, 12,* 473–498.

Edwards, W. (1987). Decision making. In G. Salvendy (ed.), *Handbook of human factors* (pp. 1061–1104). New York: Wiley.

Edworthy, J., Loxley, S., & Dennis, I. (1991). Improved auditory warning design: Relations between warning sound parameters and perceived urgency. *Human Factors, 33,* 205–231.

Egan, D. E., Remde, J. R., Gomez, L. M., Landauer, T. K., Eberhardt, J., & Lochbaum, C. C. (1989). Formative design-evaluation of SuperBook. *ACJvI Transactions on Information Systems, 7*(1), 30–57.

Einhorn, H. J., & Hogarth, R. M. (1978). Confidence in judgment: Persistence of the illusion of validity. *Psychological Review, 85,* 395–416.

Elkerton, J. (1988). Online aiding for human-computer interfaces. In M. Helander (ed.), *Handbook of human-computer interaction* (pp. 345–364). Amsterdam: North-Holland.

Elkind, J. I., Card, S. K., Hochberg, J., & Huey, B. M. (eds.) (1990). *Human performance models for computer-aided engineering.* San Diego: Academic Press.

Elmes, D. G., Kantowitz, B. H., & Roediger III, H. L. (1995). *Research methods in psychology.* St. Paul, MN: West Publishing.

Elstein, A. S., Schulman, L. S., & Sprafka, S. A. (1978). *Medical problem solving: An analysis of clinical reasoning.* Cambridge, MA: Harvard University Press.

Endsley, M. R. (1995). Toward a theory of situation awareness in dynamic systems. *Human Factors, 37,* 85–104.

Endsley, M. R., & Garland, D. J. (2000). *Situation awareness analysis and measurement.* Mahwah, NJ: Erlbaum.

Endsley, M. R., & Kiris, E. O. (1995). The out-of-the-loop performance problem and level of control in automation. *Human Factors, 37*(2), 381–394.

Environmental Protection Agency. (1974). Information on levels of environmental noise requisite to protect public health and welfare with an adequate margin of safety (EPA 550/9–74–004). Washington, DC.

Epps, B. W. (1987). A comparison of cursor control devices on a graphic editing task. *Proceedings of the31st Annual Meeting of the Human Factors Society* (pp. 442–446). Santa Monica, CA: HFS.

Ericsson, K. A. (ed.) (1996). *The road to excellence.* Mahwah, NJ: Erlbaum.

Eriksson, M., & Papanikolopoulos, N. P. (2001). Driver fatigue: A vision-based approach to automatic diagnosis. *Transportation Research Part C-Emerging Technologies, 9*(6), 399–413.

Evans, L. (1988). Older driver involvement in fatal and severe traffic crashes. *Journal of Gerontology: Social Sciences, 43*(5), 186–193.

Evans, L. (1991) *Traffic safety and the driver.* New York: Van Nostrand.

Evans, L. (1996). A crash course in traffic safety. *1997 Medical & Health Annual.* Chicago: Encyclopedia Britannica.

Eysenck, H. T., & Eysenck, S. B. G. (1964). *Manual of the Eysenck personality inventory.* London: University of London Press.

Fadden, S., Ververs, P. M., & Wickens, C. D. (2001). Pathway HUDS: Are they viable? *Human Factors, 43*(2), 173–193.

Fairclough, S. H., May, A. J., & Carter, C. (1997). The effect of time headway feedback on following behaviour. *Accident Analysis & Prevention, 29*(3), 387–397.

Fanger, P. (1977). *Thermal comfort.* New York: McGraw Hill.

Farfan, H. (1973). *Mechanical disorders of the low back.* Philadelphia: Lea & Febiger.

Farley, R. R. (1955). Some principles of methods and motion study as used in development work. *General Motors Engineering Journal, 2*(6), 20–25.

Farmer, C. M., Lund, A. K., Trempel, R. E., & Brover, E. R. (1997). Fatal crashes of passenger vehicle systems before and after adding antilock braking systems. *Accident Analysis & Prevention. 29*, 745–757.

Farr, M. J., & Psotka, J. (eds.) (1992). *Intelligent instruction by computer: Theory and practice.* New York: Taylor & Francis.

Federal Aviation Administration. (1987). U. S. Federal Aviation Administration Advisory Circular #25–11. *Transport Category Airplane Electronic Display Systems.* Washington, DC: U.S. Department of Transportation.

Fellner, D. J., & Sulzer-Azaroff, B. (1984). Increasing industrial safety practices and conditions through posted feedback. *Journal of Safety Research, 15*(1), 7–21.

Fenchel, R. S., & Estrin, G. (1982). Self-describing systems using integral help. *IEEE Transactions on Systems, Man, & Cybernetics, SMC-12*, 162–167.

Fennema, M. G., & Kleinmuntz, D. N. (1995). Anticipations of effort and accuracy in multiatttribute choice. *Organizational Behavior & Human Decision Processes, 63*, 21–32.

Ferguson, E. S. (1992). *Engineering and the Mind's Eye.* Cambridge, MA: MIT Press.

Ferguson, S. A. (2003). Other high-risk factors for young drivers: How graduated licensing does, doesn't, or could address them. *Journal of Safety Research, 34*(1), 71–77.

Feyer, A-M., & Williamson, A. M. (2001). Broadening our view of effective solutions to commercial driver fatigue. In P. A. Hancock & P. A. Desmond (eds.), *Stress, workload, and fatigue* (pp. 550–565). Mahwah, NJ: Erlbaum.

Finegold, L. S., Harris, S. S., and von Gierke, H. E., (1994) Community annoyance and sleep disturbance. *Noise Control Engineering Journal. 44*(3), 25–30.

Firenzie, R. J. (1978). *Industrial Safety: Management and Technology.* N.Y.: Kendall/Hunt.

Fischer, R. P. (1999). Probing knowledge structures. In D. Gopher and R. Koriat (eds.) *Attention and performance,* Vol. XVII (537–556). Cambridge, MA: MIT Press.

Fischer, E., Haines, R. F., & Price, T. A. (1980). *Cognitive issues in head-up displays* (NASA Technical Paper 1711). Moffett Field, CA: NASA Ames Research Center.

Fischhoff, B. (1975). Hindsight foresight: The effect of outcome knowledge on judgment under uncertainty. *Journal of Experimental Psychology-Human Perception and Performance., 1*, 288–299.

Fischhoff, B. (1982). Debiasing. In D. Kahneman, P. Slovic, & A. Tversky (eds.) (1982). *Judgment under uncertainty: Heuristics and biases.* New York: Cambridge University Press.

Fischhoff, B., & MacGregor, D. (1982). Subjective confidence in forecasts. *Journal of Forecasting, 1,* 155–172.

Fisher, D. L., & Tan, K. C. (1989). Visual displays: The highlighting paradox. *Human Factors, 31*(1), 17–30.

Fisher, D. L., Yungkurth, E. J., & Moss, S. M. (1990). Optimal menu hierarchy design: Syntax and semantics. *Human Factors, 32*(6), 665–683.

Fisher, R. P. (1999). Probing knowledge structures. In D. Gopher & A. Koriat (eds.), *Attention and performance, vol. XVII (537–556).* Cambridge, MA: MIT Press.

Fisher, R. P., & Geiselman, R. E. (1992). Memory-enhancing techniques for investigative interviewing: the cognitive interview. Springfield, IL: Thomas.

Fisk, A. D. (1987). High performance cognitive skill acquisition: perceptual/rule learning. *Proceedings of the 31st Annual Meeting of the Human Factors Society* (pp. 652–656). Santa Monica, CA: HFS.

Fisk, A. D., & Gallini, J. K. (1989). Training consistent components of tasks: Developing an instructional system based on automatic/controlled processing principles. *Human Factors, 31*(4), 453–463.

Fisk, A. D., & Hodge, K. A. (1992). Retention of trained performance in consistent mapping search after extended delay. *Human Factors, 34*(2), 147–164.

Fitts, P. M. (1951). Human engineering for an effective air-navigation and traffic-control system. Columbus, OH: Ohio State University Foundation.

Fitts, P. M. (1954). The information capacity of the human motor system in controlling the amplitude of movement. *Journal of Experimental Psychology,* 47, 381–391.

Fitts, P. M. & Posner, M. I. (1967). *Human Performance.* Belmont, CA: Brooks/Cole.

Fitts, P. M., & Seeger, C. M. (1953). S-R compatibility: Spatial characteristics of stimulus and response codes. *Journal of Experimental Psychology, 46,* 199–210.

Flin, R., Slaven, G., & Stewart, K. (1996). Emergency decision making in the offshore oil and gas industry. *Human Factors, 38,* 262–277.

Fowler, R. H., Fowler, W. A. L., & Wilson, B. A. (1991). Integrating query, thesaurus, and documents through a common visual representation. *Proceedings of CHI '91* (pp. 142–151). New York: Association for Computing Machinery.

Fowler, R. L., Williams, W. E., Fowler, M. G., & Young, D. D. (1968). *An investigation of the relationship between operator performance and operator panel layout for continuous tasks.* Technical Report Number 68–170. Wright Patterson Air Force Base, Ohio.

Fracker, M. L., & Wickens, C. D. (1989). Resources, confusions, and compatibility in dual axis tracking: Display, controls, and dynamics. *Journal of Experimental Psychology: Human Perception & Performance, 15,* 80–96.

Francis, G. (2000). Designing multifunction displays: An optimization approach. *International Journal of Cognitive Ergonomics, 4*(2), 107–124.

Freeman, D. (1996). How to make spreadsheets error-proof. *Journal of Accountancy, 181*(5), 75–77.

Friedman, B. A., & Mann, R. W. (1981). Employee assessment methods assessed. *Personnel, 58*(6), 69–74.

Friedman, R. C., Bigger, J. T., & Kornfield, D. S. (1971). The intern and sleep loss. *New England Journal of Medicine, 285,* 201–203.

Frymoyer, T. W., Pope, M. H., Constanza, M., Rosen, T., Goggin, T., & Wilder, D. (1980). Epidemiological studies of low back pain. *Spine, 5,* 419–423.

Fuld, R. B. (2000). The fiction of function allocation, revisited. *International Journal of Human-Computer Studies, 52*(2), 217–233.

Fuller, R. (1988). Psychological aspects of learning to drive. In T. A. Rothergatter & R. A. de Bruin (eds.), *Road users and traffic safety* (pp. 527–537). Assen/Maastricht, Netherlands: Van Gorcum.

Furnas, G. W., Landuaer, T. K., Gomez, L. M., & Dumais. S. T. (1987). The vocabulary problem in human system communication. *Communications of the ACM, 30*(11), 964–971.

Galitz, W. O. (1985). *Handbook of screen format design.* Wellesley Hills, MA: QED Information Sciences.

Galitz, W. O. (1993). *User-interface screen design.* New York: Wiley.

Garg, A., Herrin, G., & Chaffin, D. (1978). Prediction of metabolic rates form manual materials handing jobs. *American Industrial Hygiene Association Journal, 39*(8), 661–674.

Garland, D. J., Endsley, M. R., Andre, A. D., Hancock, P. A., Selcon, S. J., & Vidulich, M. A. (1996). Assessment and measurement of situation awareness. *Proceedings of the 40th Annual Meeting of the Human Factors & Ergonomics Society* (pp. 1170–1173). Santa Monica, CA: HFES.

Garland, D. J., Wise, J. A. & Hopkin, V. D. (eds.) (1999). *Handbook of aviation human factors.* Mahwah, NJ: Erlbaum.

Gaver, W. W. (1986). Auditory icons: Using sound in computer interfaces. *Human-Computer Interaction, 2,* 167–177.

Gawron, V. J., French, J., & Funke, D. (2001). An overview of fatigue. In P. A. Hancock & P. A. Desmond (eds.), *Stress, workload, and fatigue* (pp. 581–595). Mahwah, NJ: Erlbaum.

Geber, B. (1991). HELP'. The rise of performance support systems. *Training, 28,* 23–29.

Gentner, D., & Stevens, A. L. (1983). *Mental models.* Hillsdale, NJ: Erlbaum.

George, P. S. (1987). Team building without tears. *Personnel Journal, 66*(11), 122–129.

Gerard, M., & Martin, B. (1999). Post-effects of long term hand vibration frequency on visuo-manual performance in a tracking task. *Ergonomics, 42*(2), pp. 314–325.

Gery, G. J. (1989). Training versus performance support: Inadequate training is now insufficient. *Performance Improvement Quarterly, 2*(3), 51–71.

Gibson, J. J. (1979). *The ecological approach to visual perception.* Boston: Houghton Mifflin.

Gibson, J. J., & Crooks, L. E. (1938). A theoretical field-analysis of automobile driving. *American Journal of Psychology, 51,* 453–471.

Gigerenzer, G., & Todd, P. (1999). *Simple heuristics that make us smart.* New York: Oxford University Press.

Gillan, D. J., Wickens, C. D., Hollands, J. G., & Carswell, C. M. (1998). Guidelines for presenting quantitative data in HFES publications. *Human Factors, 40*(1), 28–41.

Gillie, T., & Broadbent, D. E. (1989). What makes interruptions disruptive? A study of length, similarity, and complexity. *Psychological Research, 50,* 243–250.

Gilliland, K., & Schlegal, R. E. (1995, Jan. 3). Readiness to perform testing and the worker. *Ergonomics & Design*, 14–19.

Gilmore, W. E. (1985). *Human engineering guidelines for the evaluation and assessment of video display units* (NUREG-CR-4227). Washington, DC: U.S. Nuclear Regulatory Commission.

Godley, S. (1997). Perceptual countermeasures for speeding: Theory literature review and empirical research. In D. Harris (ed.), *Engineering psychology and cognitive ergonomics*. Brookfield, VT: Ashgate.

Godley, S., Fildes, B. N., & Triggs, T. J. (1997). Perceptual counter measures to speeding. In D. Harris (ed.), *Engineering psychology and cognitive ergonomics*. London: Ashgate.

Goetsch, D. I. (2001). *Occupational safety and health* (4th ed.). Englewood Cliffs, NJ: Prentice Hall.

Goh, J., & Wiegmann, D. A. (2001). Visual flight rules light into instrument meteorological conditions: an empirical investigation of the possible causes. *The International Journal of Aviation Psychology, 11*(4), 359–379.

Goldman, S. R., Petrosino, A. J., & Cognition and Technology Group at Vanderbilt. (1999). Design principles for instruction in content domains: lessons from research on expertise and learning. In F. T. Durso (ed.), *Handbook of applied cognition* (pp. 595–627). New York: Wiley.

Goldstein, I. L. (1986). Training in organizations: needs assessment, development, and evaluation (2nd ed.). Monterey, CA: Brooks/Cole.

Gong, Q., & Salvendy, G. (1994). Design of skill-based adaptive interface: The effects of a gentle push. *Proceedings of the 38th Annual Meeting of the Human Factors & Ergonomics Society* (pp. 295–299). Santa Monica, CA: HFES.

Gong, R., & Kieras, D. (1994). A validation of the GOMS model methodology in the development of a specialized, commercial software application. *CHI '94* (pp. 351–357). New York: Association for Computing Machinery.

Gopher, D. (1982). A selective attention test as a predictor of success in flight training. *Human Factors, 24*, 173–183.

Gopher, D. (1993). The skill of attention control: Acquisition and execution of attention strategies. In D. E. Meyer & S. Kornblum (eds.), *Attention and performance XIV*. Cambridge, MA: MIT Press.

Gopher, D., & Kahneman, D. (1971). Individual differences in attention and the prediction of flight criteria. *Perceptual & Motor Skills, 33*, 1335–1342.

Gopher, D., & Raij, D. (1988). Typing with a two hand chord keyboard-Will the QWERTY become obsolete? *IEEE Trans. in System, Man, & Cybernetics, 18*, 601–609.

Gopher, D., Olin, M., Badhih, Y., Cohen, G., Donchin, Y., Bieski, M., & Cotev, S. (1989). The nature and causes of human errors in a medical intensive care unit. *Proceedings of the 32nd Annual Meeting of the Human Factors Society*. Santa Monica, CA: HFS.

Gopher, D., Weil, M., & Baraket, T. (1994). Transfer of skill from a computer game trainer to flight. *Human Factors, 36*(3), 387–405.

Gordon, S. E. (1988). Focusing on the human factor in future expert systems. In M. C. Majumdar, D. Majumdar, & J. Sackett (eds.), *Artificial intelligence and other innovative computer applications in the nuclear industry* (pp. 345–352). New York: Plenum.

Gordon, S. E. (1994). Systematic training program design: Maximizing effectiveness and minimizing liability. Englewood Cliffs, NJ: Prentice Hall.

Gordon, S. E., & Gill, R. T. (1992). Knowledge acquisition with question probes and conceptual graph structures. In T. Lauer, E. Peacock, & A. Graesser (eds.), *Questions and information systems* (pp. 29–46). Hillsdale, NJ: Erlbaum.

Gordon, S. E., & Gill, R. T. (1997). Cognitive task analysis. In C. Zsambok & G. Klein (eds.), *Naturalistic decision making.* Hillsdale, NJ: Erlbaum.

Gould, T. D., & Lewis, C. (1985). Designing for usability: Key principles and what designers think. *Communications of the ACM, 28*(3), 360–411.

Grabinger, R., Jonassen, D., & Wilson, B. G. (1992). The use of expert systems. In H. D. Stolovitch & E. J. Keeps (eds.), *Handbook of human performance technology* (pp. 365–380). San Francisco: Jossey-Bass.

Grabowski, M. R., & Hendrick, H. (1993). How low can we go?: Validation and verification of a decision support system for safe shipboard manning. *IEEE Transactions on Engineering Management, 40*(1), 41–53.

Grabowski, M., & Roberts, K. H. (1999). Risk mitigation in virtual organizations. *Organization Science, 10*(6), 704–721.

Graeber, R. C. (1988). Aircrew fatigue and circadian rhythmicity. In E. L. Wiener & D. C. Nagel (eds.), *Human factors in aviation* (pp. 305–344). San Diego: Academic Press.

Grandjean, E. (1988). *Fitting the task to the man* (4th ed.). London: Taylor & Francis.

Grandjean, E., Hunting, W., & Pidermann, M. (1983). VDT workstation design: Preferred settings and their effects. *Human Factors, 25,* 161–175.

Gray, W. D. (2000). The nature and processing of errors in interactive behavior. *Cognitive Science, 24*(2), 205–248.

Gray, W. D., & Fu, W. T. (2001). Ignoring perfect knowledge in-the-world for imperfect knowledge in-the-head: Implications of rational analysis for interface design. *CHI Letters, 3*(1).

Gray, W. D., John, B. E., & Atwood, M. E. (1993). Project Ernestine: Validating GOMS for predicting and explaining real-world task performance. *Human Computer Interaction, 8.*(3), 237–309.

Green, A. E. (1983). *Safety systems reliability.* Chichister, U.K.: John Wiley.

Green, D. M., & Swets, J. A. (1988). *Signal detection theory and psychophysics.* New York: Wiley.

Green, M. (2000). "How long does it take to stop?" Methodological analysis of driver perception-response times. *Transportation Human Factors, 2*(3), 195–216.

Green, P. (1995). Automotive techniques. In J. Weimer (ed.), *Research techniques in human engineering* (pp. 165–201). San Diego: Academic Press.

Green, P. A. (1999). *Visual and task demands of driver information systems* (UMTRI 98–16). Ann Arbor: University of Michigan Transportation Research Institute.

Greenbaum, T. L. (1993). *The handbook of focus group research.* New York: Lexington Books.

Greenberg, L., & Chaffin, D. B. (1976). *Workers and their tools.* Midland, MI: Pendell.

Greenberg, S. (1993). The computer user as toolsmith: The use, reuse, and organization of computer-based tools. Cambridge, UK: Cambridge University Press.

Griener T. M., & Gordon, C. C. (1990). *An assessment of long-term changes in anthropometric dimensions: Secular trends of u. s. Army males* (Natick/TR-91/006). Natick, MA: U. S. Army Natick Research, Development & Engineering Center.

Grieve, D., & Pheasant, S. (1982). Biomechanics. In W. T. Singleton (ed.), *The body at work.* Cambridge, England: Cambridge University Press.

Griffin, M. (1997a). *Handbook of human vibration.* New York: Academic Press.

Griffin, M. (1997b). Vibration and motion. In G. Salvendy (ed.), *Handbook of human factors and ergonomics.* New York: Wiley.

Gronlund, S. D., Canning, J. M., Moertl, P. M., Johansson, J., Dougherty, R. P., & Mills, S. H. (2002). An information organization tool for planning in air traffic control. *The International Journal of Aviation Psychology, 12*(4), 377–390.

Guerlain, S. A., Smith, P. J., Obradovich, J. H., Rudman, S., Strohm, P., Smith, J. W., Svirbely, J., & Sachs, L. (1999). Interactive critiquing as a form of decision support: An empirical evaluation. *Human Factors, 41*(1), 72–89.

Gugerty, L. J. (1997). Situation awareness during driving: Explicit and implicit knowledge in dynamic spatial memory. *Journal of Experimental Psychology: Applied, 3,* 42–66.

Gunasekaran, A. (1999). Agile manufacturing: A framework for research and development. *International Journal of Production Economics, 62*(1–2), 87–105.

Hagberg, M. (1981). Muscular endurance and surface electromyogram in isometric and dynamic exercise. *Journal of Applied Physiology: Respiration, Environment, & Exercise Physiology, 51,* 1.

Halasz, F., & Moran, T. P. (1982). Analogy considered harmful. *Human Factors in Computer Systems Proceedings* (pp. 383–386). Washington, DC: National Bureau of Standards.

Hallett, P. E. (1986). Eye movements. In K. R. Boff, L. Kaufman, & J. P. Thomas (eds.), *Handbook of perception and human performance,* vol. 1 (pp. 10–1/10–112). New York: Wiley.

Hamelin, P. (1987). Lorry driver's time habits in work and their involvement in traffic accidents. *Ergonomics 30,* 1323–1333.

Hamil, P., Drizo, T., Johnson, C., Reed, R., & Roche, A. (1976). *NCHS growth charts,* Monthly Vital Statistics Report, Health Examination Survey Data, HRA 76–1120, vol. 25, no. 3. National Center for Health Statistics.

Hammer, J. M. (1999). Human factors of functionality and intelligent avionics. In D. J. Garland, J. A. Wise, & V. D. Hopkin (eds.), *Handbook of aviation human factors* (pp. 549–566). Mahwah, NJ: Erlbaum.

Hammer, M., & Champy, J. (1993). *Reengineering the corporation.* New York: HarperCollins.

Hammer, W (2000). *Occupational safety management and engineering* (5th ed.). Englewood Cliffs, NJ. Prentice Hall.

Hammond, K. R. (1993). Naturalistic decision making from a Brunswikian viewpoint: Its past, present, future. In G. A. Klein & J. Orasanu & R. Calderwood & C. Zsambok (Eds.), *Decision Making in Action: Models and Methods* (pp. 205–227). Norwood, NJ: Ablex Publishing.

Hammond, K. R., Hamm, R. M., Grassia, J., & Pearson, T. (1987). Direct comparison of the efficacy of intuitive and analytical cognition in expert judgment. *IEEE Transactions on Systems, Man, & Cybernetics, SMC-17*(5), 753–770.

Hancock, P. A., & Desmond, P. A., (2001) *Stress, Workload and Fatigue.* Mahwah, N.J.: Lawrence Erlbaum Associates.

Hancock, P. A., & Warm, J. S. (1989). A dynamic model of stress and sustained attention. *Human Factors, 31,* 519–537.

Hart, S., & Staveland, L. (1988). Development of NASA TLX (Task Load Index). In P. Hancock & N. Meshkati (Eds.). *Human Mental Workload.* Amsterdam, N.L.: North Holland.

Hanson, M. A., Hedge, J. W., Logan, K. K., Bruskiewicz, K. T., & Borman, W. C. (1995). Application of the critical incident technique to enhance crew resource management training. *Proceedings of the Eighth International Symposium on Aviation Psychology* (pp. 568–573). Ohio State University.

Hare, A. P. (1992). Groups, teams, and social interaction: theories and applications. New York: Praeger.

Harris, J. E., & Wilkins, A. J. (1982). Remembering to do things: A theoretical framework and illustrative experiment. *Human Learning, 1,* 123–136.

Harris, W. (1977). Fatigue, circadian rhythm, and truck accidents. In R. Mackie (ed.), *Vigilance theory, operational performance, and physiological correlates* (pp. 133–146). New York: Plenum.

Harrison, Y. & Horne J. A.(2000). The impact of sleep deprivation on decision making: a review. *Journal of Experimental Psychology: Applied. 6* (3), pp. 236–358.

Harrison, B. L., & Vicente, K. J. (1996). A case study of transparent user interfaces in a commercial 3D modeling and paint application. *Proceedings of the 40th Annual Meeting of the Human Factors & Ergonomics Society.* Santa Monica, CA: HFES.

Harrison, M. H., Brown, G. A., & Belyavin, A. T. (1982). The "Oxylog": An evaluation. *Ergonomics, 25,* 809.

Hartley, L. R., & Hassani, J. E. (1994). Stress, violations and accidents. *Applied Ergonomics, 25*(4), 221–231.

Haskell, I. D., & Wickens, C. D. (1993). Two- and three-dimensional displays for aviation: A theoretical and empirical comparison. *International Journal of Aviation Psychology, 3*(2), 87–109.

Haslem, D. R. (1982). Sleep loss, recovery sleep, and military performance. *Ergonomics, 25,* 163–178.

Hauser, J. R., & Clausing, D. (1988). The house of quality. *Harvard Business Review,* May–June, 63–73.

Hawkins, F. H. (1993). *Human factors in flight.* Brookfield, VT: Ashgate.

Hawkins, F. & Orlady, H. (1993). *Human factors in flight: 2nd Ed.* Brookfield, VT: Ashgate.

Hecht, T. D., Allen, N. J., Klammer, J. D., & Kelly, E. C. (2002). Group beliefs, ability, and performance: The potency of group potency. *Group Dynamics-Theory Research & Practice, 6*(2), 143–152.

Hedge, A., McCrobie, D., Morimoto, S., Rodriguez, S., & Land, B. (1996). Toward pain-free computing. *Ergonomics in Design, 4*(1), 4–10.

Heenefrund, W. (1985). The fine art of team building. *Association Management, 37*(8), 98–101.

Helander, M. (ed.). (1988). *Handbook of human-computer interaction.* Netherlands: Elsevier.

Helander, M. G. (1987). Design of visual displays. In G. Salvendy (ed.), *Handbook of human factors* (pp. 507–548). New York: Wiley.

Hellier, E., Edworthy, J., Weedon, B., Walters, K., & Adams, A. (2002). The perceived urgency of speech warnings: Semantics versus acoustics. *Human Factors, 44,* 1–17.

Helmreich, R. (1997, May). Managing human error in aviation *Scientific American,* 62–67.

Helmreich, R. L., & Wilhelm, J. A. (1991). Outcomes of Crew Resource Management training. *The International Journal of Aviation Psychology, 1,* 287–300.

Hendrick, H. (1996). The ergonomics of economics is the economics of ergonomics. *Proceedings of the 40th Annual Meeting of the Human Factors & Ergonomics Society* (pp. 1–10). Santa Monica, CA: HFES.

Hendrick, H. W. (1986). Macroergonomics: A conceptual model for integrating human factors with organizational design. In O. Brown, Jr. & H. W. Hendrick (eds.), *Human factors in organizational design and management II* (pp. 467–477). Amsterdam: North-Holland.

Hendrick, H. W. (1995). Humanizing re-engineering for true organizational effectiveness: A macroergonomic approach. *Proceedings of the 39th Annual Meeting of the Human Factors & Ergonomics Society* (pp. 761–765). Santa Monica, CA: HFES.

Hendrickson, J. (1987). Hiring the right stuff. *Personnel Administrator, 32*(11), 70–74.

Hendy, K., Jian Quiao, L., & Milgram, P. (1997). Combining time and intensity effects in assessing operator information processing load. *Human Factors, 39,* 30–47.

Henley, J., & Kumamoto, J. (1981). *Reliability engineering and risk assessment.* New York: Prentice Hall.

Hennessy, R. T. (1990). Practical human performance testing and evaluation. In H. R. Booher (ed.), *MANPRINT: An approach to systems integration* (pp. 433–479). New York: Van Nostrand Reinhold.

Herrin, G. D., Taraiedi, M., & Anderson, C. K. (1986). Prediction of overexertion injuries using biomechanical and psychophysical models. *American Industrial Hygiene Association Journal, 47*(6), 322–330.

Herrmann, D., Brubaker, B., Yoder, C., Sheets, V., & Tio, A. (1999). Devices that remind. In F. T. Durso (ed.), *Handbook of applied cognition* (pp. 377–407). Chichester, England: Wiley.

Hertzberg, H. T. E. (1972). Engineering anthropometry. In H. P. Van Cott & R. G. Kinkade (eds.), *Human engineering guide to equipment design.* Washington, DC: U.S. Government Printing Office.

Hess, S. M., Detweiler, M. C., & Ellis, R. D. (1999). The utility of display space in keeping track of rapidly changing information. *Human Factors, 41,* 257–281.

Hick, W. E. (1952). On the rate of gain of information. *Quarterly Journal of Experimental Psychology, 4,* 11–26.

Hill, G. W. (1982). Group versus individual performance: are N + 1 heads better than one? *Psychological Bulletin, 91,* 517–539.

Hockey, G. R. J. (1986). Changes in operator efficiency as a function of environmental stress, fatigue, and circadian rhythms. In K. R. Boff, L. Kaufman, & J. P. Thomas (eds.), *Handbook of perception and human performance, vol. II* (pp. 44–1/44–49). New York: Wiley.

Hockey, G. R. J., Wastell, D. G., & Sauer, J. (1998). Effects of sleep deprivation and user interface on complex performance: A multilevel analysis of compensatory control. *Human Factors, 40*(2), 233–253.

Hogan, R., Johnson, J., & Briggs, S. (1997). *Handbook of Personality Psychology.* San Diego: Academic Press.

Holcom, M. L., Lehman, W. E. K., & Simpson, D. D. (1993). Employee accidents: Influences of personal characteristics, job characteristics, and substance use in jobs differing in accident potential. *Journal of Safety Research, 24,* 205–221.

Holding, D. H. (1987). Concepts of training. In G. Salvendy (ed.), *Handbook of human factors* (1st ed.) (pp. 939–962). New York: Wiley.

Hollingshead, A. B. (1998). Group and individual training-The impact of practice on performance. *Small Group Research, 29*(2), 254–280.

Holmes, T. H., & Rahe, R. H. (1967). The social readjustment rating scale. *Journal of Psychosomatic Research, 11,* 213–218.

Hopkin, V. D. (1995). *Human factors in air traffic control.* London: Taylor & Francis.

Hopkin, V. D., & Wise, J. A. (1996). Human factors in air traffic system automation. In R. Parasuraman & M. Mouloua (eds.), *Automation and human performance: Theory and applications* (pp. 319–336). Mahwah, NJ: Erlbaum.

Horne, J. A. (1988). *Why we sleep.* Oxford: Oxford University Press.

Horne, J. A., Anderson, N. R., & Wilkinson, R. T. (1983). Effects of sleep deprivation on signal detection measures of vigilance: Implications for sleep function. *Sleep, 6,* 347–358.

Horrey, W. J. & Wickens, C. D. (2002). *Driving and side task performance: The effects of display clutter, separation, and modality* (AHFD-02–13/GM-02–2). Savoy, IL: University of Illinois, Aviation Human Factors Division.

Hosea, T. M., Simon, S. R., Delatizky, T., Wong, M. A., & Hsieh, C. C. (1986). Myoelectric analysis of the paraspinal musculature in relation to automobile driving. *Spine, 11,* 928–936.

Houghton, R. C. (1984). Online help systems: A conspectus. *Communications of the ACM, 27,* 126–133.

Houston, C. (1987). Going higher: The story of man at high altitudes. Boston: Little Brown.

Huey, M. B., & Wickens, C. D. (eds.) (1993). *Workload transition: Implications for individual and team performance.* Washington, DC: National Academy Press.

Humphreys, P., McIvor, R., & Huang, G. (2002). An expert system for evaluating the make or buy decision. *Computers & Industrial Engineering, 42*(2–4), 567–585.

Hunt, R., & Rouse, W. (1981). Problem-solving skills of maintenance trainees in diagnosing faults in simulated power plants. *Human Factors, 23,* 317–328.

Hunter, D. R., & Burke, E. F. (1994). Predicting aircraft pilot training success: a meta-analysis of published research. *The International Journal of Aviation Psychology, 4*, 1–12.

Hunter, D. R., & Burke, E. F. (1995). *Handbook of pilot selection.* Aldershot, UK: Avebury Aviation.

Hunter, J. E., & Hunter, R. F. (1984). Validity and utility of alternative predictors of job performance. *Psychological Bulletin, 96*, 72–98.

Huse, E. F., &Cummings, T. G. (1985). *Organizational development and change* (3rd ed.). St. Paul, MN: West.

Hutchins, E. (1995). *Cognition in the wild.* Cambridge, MA: MIT Press.

Hutchins, E. L., Holland, J. D., & Norman, D. A. (1985). Direct manipulation interfaces. *Human-Computer Interaction, 1*(4), 31 1–338.

Hyman, R. (1953). Stimulus information as a determinant of reaction time. *Journal of Experimental Psychology, 45*, 423–432.

Imada, A. S. (1991). The rationale and tools of participatory ergonomics. In K. Noro & A. S. Imada (eds.), *Participatory ergonomics* (pp. 30–49). London: Taylor & Francis.

Imada, A. S., & Feiglstok, D. M. (1990). An organizational design and management approach for improving safety. In K. Noro & 0. Brown, Jr. (eds.), *Human factors in organizational design and management III* (pp. 479–482). Amsterdam: North-Holland.

Inzana, C. M., Willis, R. P., & Kass, S. J. (1994). The effects of physical distribution of team members on team cohesiveness and performance. *Proceedings of the 38th Annual Meeting of the Human Factors & Ergonomics Society* (p. 953). Santa Monica, CA: HFES.

Irving, S., Polson, P., & Irving, J. E. (1994). A GOMS analysis of the advanced automation cockpit, *Proceedings of the Conference on Human Factors in Computing Systems* (pp. 344–350). Boston, MA.

Jacko, J. A., & Sears, A. (Eds.). (2002). *Handbook for Human-Computer Interaction.* Mahwah: Lawrence Erlbaum & Associates.

Jackson, J. (1994). A multimodal method for assessing and treating airsickness. *International Journal of Aviation Psychology, 4*(1), 85–96.

Jagacinski, R. J., & Flach, J. M. (2003). *Control theory for humans.* Mahwah, NJ: Erlbaum.

Jagacinski, R. & Miller, D. (1978). Describing the human operator's internal model of dynamic system. *Human Factors, 22*, 425–434.

Jager, M., & Luttman, A. (1989). Biomechanical analysis and assessment of lumbar stress during load lifting using a dynamic I9-segment human model. *Ergonomics, 32*, 93–112.

Janis, I. L. (1982). Decision making under stress. In L. Goldberger & S. Breznitz (eds.), *Handbook of stress: Theoretical and clinical aspects* (pp. 69–87). New York: Free Press.

Janis, I. L., & Mann, L. (1977). Decision making: A psychological analysis of conflict, choice, and commitment. New York: Free Press.

Jarvenpaa, S. L., & Leidner, D. E. (1999). Communication and trust in global virtual teams. *Organization Science, 10*(6), 791–815.

Jenness, J. W., Lattanzio, R. J., O'Toole, M., Taylor, N., & Pax, C. (2002). Effects of manual versus voice-activated dialing during simulated driving. *Perceptual & Motor Skills, 94*(2), 363–379.

Jensen, R. S. (1982). Pilot judgment: training and evaluation. *Human Factors, 24,* 61–74.

Johannesen, L. J., Cook, R. I., & Woods, D. D. (1994). Cooperative communications in dynamic fault management. *Proceedings of the 38th Annual Meeting of the Human Factors & Ergonomics Society* (pp. 225–229). Santa Monica, CA: HFES.

Johnson, E. M., Cavanagh, R. C., Spooner, R. L., & Samet, M. G. (1973). Utilization of reliability measurements in Bayesian inference: Models and human performance. *IEEE Transactions on Reliability, 22,* 176–183.

Johnson, P. (1992). Human-computer interaction: Psychology, task analysis and software engineering. London: McGraw-Hill.

Johnson, S. T. (March–April 1993). Work teams: What's ahead in work design and rewards management. *Compensation & Benefits Review,* 35–41.

Johnston, J. A., & Cannon-Bowers, J. A. (1996). Training for stress exposure. In J. E. Driskell & E. Salas (eds.), *Stress and human performance.* Mahwah, NJ: Erlbaum.

Jonassen, D. H. (ed.) (1988). *Instructional design for microcomputer courseware.* Hillsdale, NJ: Erlbaum.

Jones, D. G., & Endsley, M. R. (1996). Sources of situation awareness errors in aviation. *Aviation, Space, & Environmental Medicine, 67,* 507–512.

Jones, P. E., & Roelofsma, P. (2000). The potential for social contextual and group biases in team decision-making: biases, conditions and psychological mechanisms. *Ergonomics, 43*(8), 1129–1152.

Kaczmarer, K., & Bach-T-Rita, P. (1995). Haptic displays. In W. Barfield & T. Furness (eds.), *Virtual environments and advanced interface design.* New York: Oxford University Press.

Kahane, C. J. (1989). *An evaluation of center high mounted stop lamps based on 1987 data* (DOT HS 807 442). Washington, DC: National Highway Traffic Safety Administration.

Kahn T. F., & Monod, H. (1989). Fatigue induced by static work, *Ergonomic 5*(32), 839–846.

Kahneman, D., Ben-Ishai, R., & Lotan, M. (1973). Relation of a test of attention to road accidents. *Journal of Applied Psychology, 58,* 113–115.

Kahneman, D., Slovic, P., & Tversky, A. (eds.) (1982). *Judgment under uncertainty: Heuristics and biases.* New York: Cambridge University Press.

Kahneman, D., & Tversky, A. (1984). Choices, values and frames. *American Psychologist, 39,* 341–350.

Kalsher M. J., Wogalter, M. S., Brewster, B. M., & Spunar, M. E. (1995). Hazard level perceptions of current and proposed warning sign and label panels. *Proceedings of the 39th Annual Meeting of the Human Factors & Ergonomic Society* (pp. 351–355), Santa Monica, CA: HFES.

Kamon, E., & Goldfuss, A. (1978) In-plant evaluation of the muscle strength of workers. *American Industrial Hygiene Association Journal, 39,* 801–807.

Kanawattanachai, P., & Yoo, Y. (2002). Dynamic nature of trust in virtual teams. *Journal of Strategic Information Systems, 11*(3–4), 187–213.

Kantowitz, B. H. (1990). Can cognitive theory guide human factors measurement? *Proceedings of the 34th Annual Meeting of the Human Factors Society* (pp. 1258–1262). Santa Monica, CA: HFS.

Kantowitz, B. H. (1992). Selecting measures for human factors research. *Human Factors, 34*(4), 387–398.

Kantowitz, B. H., & Simsek, O. (2001). Secondary-task measures of driver workload. In P. A. Hancock & P. A. Desmond (eds.), *Stress, workload, and fatigue* (pp. 395–408). Mahwah, NJ: Erlbaum.

Kantowitz, B. H., & Sorkin, R. D. (1987). Allocation of functions. In G. Salvendy (ed.), *Handbook of human factors* (pp. 355–369). New York: Wiley.

Kantowitz, B. H., Hanowski, R. J., & Kantowitz, S. C. (1997). Driver acceptance of unreliable traffic information in familiar and unfamiliar settings. *Human Factors, 39*(2), 164–176.

Kaptein, N. A. (1994). *Benefits of in-car head-up displays* (Technical Report #TNO-TM 1994B-20). Soesterberg, Netherlands: TNO Human Factors Research Institute.

Kaptein, N. A., Theeuwes, J., & Van Der Horst, R. (1996). *Driving simulator validity: Some considerations* (Report 96–13 38). Transportation Research Board 75th Annual Meeting, Washington, DC, Jan. 7–11.

Karat, C. (1990). Cost-benefit analysis of usability engineering techniques. *Proceedings of the 34th Annual Meeting of the Human Factors Society* (pp. 839–843). Santa Monica, CA: HFS.

Karis, D. (1987). Fine motor control with CBR protective gloves. *Proceedings of the 31st Annual Meeting of the Human Factors Society* (pp. l206–1210). Santa Monica, CA: HFS.

Karvonen, K., & Parkkinen, J. (2001). Signs of trust: A semiotic study of trust formation in the web. *1st International Conference on Universal Access in Human-Computer Interaction, 1*, 1076–1080.

Karwowski, W., Genaidy, A., & Asfour, S. (eds.) (1990). *Computer-aided ergonomics.* London: Taylor & Francis.

Karwowski, W., Warnecke, H. J., Hueser, M., & Salvendy, G. (1997). Human factors in manufacturing. In G. Salvendy (ed.), *Handbook of human factors and ergonomics* (2nd ed.). New York: Wiley.

Katzenbach, J. R., & Smith, D. K. (1993). *The wisdom of teams: Creating the high-performance organization.* Boston: Harvard Business School Press.

Keegan, T. T. (1953). Alternations of the lumbar curve related to posture and seating. *Journal of Bone & Joint Surgery, 35A,* 589–603.

Keller, J. J., & Nussbaum, J. (2000). *OSHA compliance manual: Application of key OSHA topics.* New York: Keller.

Kelley, C. R. (1968). *Manual and automatic control.* New York: Wiley.

Kelly, M. L. (1955). A study of industrial inspection by the method of paired comparisons. *Psychological Monographs, 69*(394), 1–16.

Keppel, G. (1992). *Design and analysis: A researcher's handbook* (3rd ed.). Englewood Cliffs, NJ: Prentice Hall.

Kerns, K. (1999). Human factors in air traffic control/flight deck integration: Implications of data-link simulation research. In D. J. Garland, J. A. Wise, & V. D.

Hopkin (Eds.), *Handbook of aviation human factors* (pp. 519–546). Mahwah, NJ: Lawrence Erlbaum.

Kerstholt, J. H., Passenier, P. O., Houttuin, K., & Schuffel, H. (1996). The effect of a priori probability and complexity on decision making in a supervisory control task. *Human Factors, 38*(10), 65–78.

Kiefer, R. & Gellatly, A. W. (1996). *Quantifying the consequences of the "Eyes on the Road" benefit attributed to head-up displays.* Society of Automotive Engineers Publication 960946. Warrendale, PA.: Society for Automotive Engineers.

Kiefer, R. J. (1991). Effect of a head-up versus head-down digital speedometer on visual sampling behavior and speed control performance during daytime automobile driving (SAE Technical Paper Series 910111). Warrendale, PA: Society of Automotive Engineers.

Kiefer, R. J. (1995). Defining the "HUD benefit time window." In *Vision in vehicles VI Conference.* Amsterdam: Elsevier.

Kieras, D. E. (1988a). Towards a practical GOMS model methodology for user's interface design. In M. Helander (ed.), *Handbook of human-computer interaction* (pp. 135–157). Amsterdam: North-Holland.

Kieras, D., & Polson, P. G. (1985). An approach to the formal analysis of user complexity, *International Journal of Man-Machine Studies, 22,* 365–394.

Kies, J. K., Williges, R. C., & Rosson, M. B. (1998). Coordinating computer-supported cooperative work: A review of research issues and strategies. *Journal of the American Society for Information Science, 49*(9), 776–791.

Kim, J., & Moon, J. Y. (1998). Designing towards emotional usability in customer interfaces—trustworthiness of cyber-banking system interfaces. *Interacting With Computers, 10*(1), 1–29.

King, P. M. (1994). Participatory ergonomics: A group dynamics perspective. *Work, 4*(3), 195–200.

Kintsch, W. & Van Dijk, T. A. (1978). Toward a model of text comprehension and reproduction. *Psychological Review, 85,* 363–394.

Kirlik, A. (1993). Modeling strategic behavior in human-automation interaction: Why an "aid" can (and should) go unused. *Human Factors, 35*(2), 221–242.

Kirlik, A., Walker, N., Fisk, A. D., & Nagel, K. (1996). Supporting perception in the service of dynamic decision making. *Human Factors, 38*(2), 288–299.

Kirwan, B., & Ainsworth, L. K. (eds.) (1992). *A guide to task analysis.* London: Taylor & Francis.

Klein, G. (1989). Recognition-primed decisions. *Advances in Man-Machine Systems Research, 5,* 47–92.

Klein, G. (1993). A recognition-primed decision (RPD) model of rapid decision making. In G. Klein, J. Orasanu, R. Calderwood, & C. E. Zsambok (eds.), *Decision making in action: Models and methods* (pp. 138–147). Norwood, NJ: Ablex.

Klein, G. A., & Crandall, B. W. (1995). The role of mental simulation in naturalistic decision making. In P. Hancock, J. Flach, J. Caird, & K. Vicente (eds.), *Local applications of the ecological approach to human machine systems* (vol. 2, pp. 324–358). Hillsdale, NJ: Erlbaum.

Klein, G., & Calderwood, R. (1991). Decision models: Some lessons from the field. *IEEE Transactions on Systems, Man & Cybernetics, 21,* 1018–1026.

Klein, G., Orasanu, J., Calderwood, R., & Zsambok, C. E. (eds.) (1993). Decision making in action: Models and methods. Norwood, NJ: Ablex.

Klein, R. (1991). Age-related eye disease, visual impairment, and driving in the elderly. *Human Factors, 33*(5), 521–526.

Kohn, T. P., Friend, M. A., & Winterberger, C. A. (1996). *Fundamentals of occupational safety and health.* Rockville, MD: Government Institutes.

Konz, S. (1997). Toxology and human comfort. In G. Salvendy (ed.), *Handbook of human factors and ergonomics.* New York: Wiley.

Korteling, J. E. (1994). Effects of aging, skill modification and demand alternation on multiple task performance. *Human Factors, 36,* 27–43.

Korteling, J. E., & van Emmerik, M. L. (1998). Continuous haptic information in target tracking from a moving platform. *Human Factors, 40,* 198–208.

Kosslyn, S. M. (1994). *Elements of graph design.* New York: W. H. Freeman.

Kraemer, K. L., & King, J. (1988). Computer-based systems for cooperative work and group decision making. *Computing Surveys, 20,* 115–146.

Kraemer, K. L., & Pinsonneault, A. (1990). Technology and groups: Assessments of the empirical research. In J. Galegher, R. E. Kraut, & C. Egido (eds.), *Intellectual teamwork: Social and technical foundations of cooperative work* (pp. 375-405). Hillsdale, NJ: Lawrence Erlbaum Associates.

Kramer, A. (1991). Physiological metrics of mental workload: A review of recent progress. In D. Damos (ed.), *Multiple task performance* (pp. 279–328). London: Taylor & Francis.

Kramer, A. F., Coyne, J. T., & Strayer, D. L. (1993). Cognitive function at high altitude. *Human Factors, 35*(2), 329–344.

Kramer, R. M. (1999). Trust and distrust in organizations: emerging perspectives, enduring questions. *Annual Review of Psychology, 50,* 569–598.

Kroemer, K. H. E. (1987). Biomechanics of the human body. In G. Salvendy (ed.), *Handbook of human factors* (pp. 169–181). New York: Wiley.

Kroemer, K. H. E., & Grandjean, E. (1997). *Fitting the task to the human: A textbook of occupational ergonomics* (5th ed.). London: Taylor & Francis.

Kroemer, K. H. E., Kroemer, H. B., & Kroemer-Elbert, K. E. (1994). *Ergonomics: How to design for ease and efficiency.* Englewood Cliffs, NJ: Prentice Hall.

Kroft, P., & Wickens, C.D. (in press, 2003). Displaying multi-domain graphical database information: an evaluation of scanning, clutter, display size, and user interactivity. *Information Design Journal, 11*(1.)

Kryter, K. D. (1972). Speech communications. In H. P. Van Cott & R. G. Kinkade (eds.), *Human engineering guide to system design.* Washington, DC: U. S. Government Printing Office.

Kryter, K. D. (1983). Presbycusis, sociocusis and nosocusis. *Journal of the Acoustical Society of America, 73*(6), 1897–1917.

Kusiak, A. (1999) *Engineering Design: Products, Processes & Systems.* NY.: Academic Press.

Lamble, D., Kauranen, T., Laakso, M., & Summala, H. (1999). Cognitive load and detection thresholds in car following situations: safety implications for using mobile (cellular) telephones while driving. *Accident Analysis & Prevention, 31*(6), 617–623.

Landauer, T. K. (1995). *The trouble with computers: Usefulness, usability, and productivity.* Cambridge, MA: MIT Press.

Landsdown, T. C. (2001). Causes, measures, and effects of driver visual workload. In P. A. Hancock & P. A. Desmond (eds.), *Stress, workload, and fatigue* (pp. 351–369). Mahwah, NJ: Erlbaum.

Langolf, C. D., Chaffin, D. B., & Foulke, S. A. (1976). An investigation of Fitts's law using a wide range of movement amplitudes. *Journal of Motor Behavior, 8,* 113–128.

Langon-Fox, J., Code, S., & Langford-Smith, K. (2000). Team mental models: Techniques, methods and analytic approaches. *Human Factors, 42,* 242–271.

Laughery, K. R., Jr. & Corker, K. (1997). Computer modeling and simulation. In G. Salvendy (ed.), *Handbook of human factors and ergonomics* (2nd ed., pp. 1375–1408). New York: Wiley.

Layton, C., Smith, P. J., & McCoy, C. E. (1994). Design of a cooperative problem-solving system for en-route flight planning: An empirical evaluation. *Human Factors, 36*(4), 94–119.

Lazarus, R. S., & Folkman, S. (1984). *Stress, appraisal and coping.* New York: Springer.

Lee, D. N. (1976). A theory of visual control of braking based on information about time to collision. *Perception, 5,* 437–459.

Lee, E., & MacGregor, J. (1985). Minimizing user search time in menu retrieval systems. *Human Factors, 27*(2), 157–162.

Lee, J. D. (1997). A functional description of ATIS/CVO systems to accommodate driver needs and limits. In Y. I. Noy (ed.), *Ergonomics and safety of intelligent driver interfaces* (pp. 63–84). Mahwah, NJ: Erlbaum.

Lee, J. D. (2001). Emerging challenges in cognitive ergonomics: Managing swarms of self-organizing agent-based automation. *Theoretical Issues in Ergonomics Science, 2*(3), 238–250.

Lee, J. D., & Sanquist, T. F. (1996). Maritime automation. In R. Parasuraman & M. Mouloua (eds.), *Automation and human performance* (pp. 365–384). Mahwah, NJ: Erlbaum.

Lee, J. D., & Sanquist, T. F. (2000). Augmenting the operator function model with cognitive operations: Assessing the cognitive demands of technological innovation in ship navigation. *IEEE Transactions on Systems, Man, & Cybernetics-Part A: Systems & Humans, 30*(3), 273–285.

Lee, J. D., & See, K. A. (2002). Trust in computer technology and the implications for design and evaluation. In C. Miller (ed.), *Etiquette for human-computer work: Technical report FS-02–02* (pp. 20–25). Menlo Park, CA: American Association for Artificial Intelligence.

Lee, J. D., & See, K. A. (in press). Trust in technology: designing for appropriate reliance. *Human Factors.*

Lee, J. D., Caven, B., Haake, S., & Brown, T. L. (2001). Speech-based interaction with in-vehicle computers: The effect of speech-based e-mail on drivers' attention to the roadway. *Human Factors, 43,* 631–640.

Lee, J. D., Gore, B. F., & Campbell, J. L. (1999). Display alternatives for in-vehicle warning and sign information: Message style, location, and modality. *Transportation Human Factors Journal, 1*(4), 347–377.

Lee, J. D., McGehee, D. V., Brown, T. L., & Reyes, M. L. (2002). Collision warning timing, driver distraction, and driver response to imminent rear end collisions in a high-fidelity driving simulator. *Human Factors, 44*(2), 314–334.

Lee, J. D., Morgan, J., Wheeler, W. A., Hulse, M. C. & Dingus, T. A. (1997). *Development of human factors guidelines for advanced traveler information systems* (U. S. Federal Highway Administration Report FHWA-RD-95–201). Washington, DC.

Lee, J., & Moray, N. (1992). Trust, control strategies and allocation of function in human-machine systems. *Ergonomics, 35*(10), 1243–1270.

Lee, J. D., & Morzy, N. (1994). Trust, self-confidence and operator's adaptation to automation. *International Journal of Human Computer Studies, 40,* 153–184.

Lehman, W. E. K., & Simpson, D. D. (1992). Employee substance use and on-the-job behaviors. *Journal of Applied Psychology, 77*(3), 309–321.

Lehner, P. E., & Zirk, D. A. (1987). Cognitive factors in user/expert-system interaction. *Human Factors, 29*(1), 97–109.

Leibowitz, H. (1988). The human senses in flight. In E. Wiener & D. Nagel (eds.), *Human factors in aviation* (pp. 83–110). San Diego: Academic Press.

Leibowitz, H. W., Owens, D. A., & Helmreich, R. L. (1995). Transportation. In R. Nickerson (ed.), *Emerging needs and opportunities for human factors research* (pp. 241–261). Washington, DC: National Academy Press.

Leonard, S. D., Hill, G. W., & Otani, H. (1990). Factors involved in risk perception. *Proceedings of the 34th Annual Meeting of the Human Factors Society* (pp. 1037–1041). Santa Monica, CA: HFS.

Lesgold, A. M., & Curtis, M. E. (1981). Learning to read words efficiently. In A. M. Lesgold & C. A. Perfetti (eds.), *Interactive processes in reading*. Hillsdale, NJ: Erlbaum.

Leveson, N. G. (1995). *Safeware: System safety and computers.* New York: Addison-Wesley.

Levett, L. M., & Kovera, M. B. (2002, Dec.). Psychologists battle over the general acceptance of eyewitness research. *Monitor on Psychology,* 23.

Levin, I. P., Gaeth, G. J., Schreiber, J., & Lauriola, M. (2002). A new look at framing effects: Distribution of effect sizes, individual differences, and independence of types of effects. *Organizational Behavior and Human Decision Processes, 88*(1), 411–429.

Levine, M. (1982). You-are-here maps: Psychological considerations. *Environment & Behavior, 14,* 221–237.

Lewandowsky, S., Mundy, M., & Tan, G. (2000). The dynamics of trust: Comparing humans to automation. *Journal of Experimental Psychology, Applied, 6*(2), 104–123.

Lhose, J. (1993). A cognitive model for perception and understanding. In S. P. Rebertsoll et al. (eds.) Human Factors in Computing Systems. *CHI '91 Conference Proceedings* (pp. 137–144). New York: Association for Computing Machinery.

Liebler, S. N., & Parkman, A. W (1992). Personnel selection. In H. D. Stolovitch & E. J. Keeps (eds.), *Handbook of human performance technology* (pp. 259–276). San Francisco: Jossey-Bass.

Lin, L., Vicente, K. J., & Doyle, D. J. (2001). Patient safety, potential adverse drug events, and medical device design: A human factors engineering approach. *Journal of Biomedical Informatics, 34*(4), 274–284.

Lind, A. R., & McNichol, G. W. (1967). Circulatory responses to sustained handgrip contractions performed during exercise, both rhythmic and static. *Journal of Physiology,* 192, 595–607.

Lindstrom, L., Kadefors, R., & Petersen, I. (1977). An electromyographic index for localized muscle fatigue. *Journal of Applied Physiology: Respiration, Environment, & Exercise Physiology, 43,* 750.

Lintern, G., & Roscoe, S. (1980). Visual cue augmentation in contact flight simulation. In S. Roscoe (Ed.). *Aviation Psychology.* Ames, Iowa: Iowa State University Press.

Lipschutz, L., Roehrs, T., Spielman, A., Zwyghuizen H., Lamphere, J., & Roth, T. (1988). Caffeine's alerting effects in sleepy normals. *Journal of Sleep Research, 17,* 49.

Lipshitz, R., & Bell Shaul, O. (1997). Schemata and mental models in recognition-primed decision making. In C. E. Zsambok & G. Klein (eds.), *Naturalistic decision making* (pp. 293–303). Mahwah, NJ: Erlbaum.

Liu, Y. (in press). The aesthtic and the ethic dimensions of human factors and design. *Theoretical Issues in Ergonomics Science.*

Liu, Y., Fuld, R., & Wickens, C. D. (1993). Monitoring behavior in manual and automated scheduling systems. *International Journal of Man-Machine Studies, 39,* 1015–1029.

Loftus, E. F., Loftus, G. R., & Messo, J. (1987). Some facts about "weapon focus." *Law & Human Behavior, 11,* 55–62.

Loftus, G. R., Dark, V. J., & Williams, D. (1979). Short-term memory factors in ground controller/pilot communication. *Human Factors, 21,* 169–181.

Logan, G. D. (1985). Skill and automaticity: Relations, implications, and future directions. *Canadian Journal of Psychology, 39,* 367–386.

Logie, R. H. (1995). *Visuo-spatial working memory.* Hove, UK: Erlbaum.

Long, L., & Churchill, E. (1965). Anthropometry of USAF basic trainees contrasts of several subgroups. Paper presented to the 1968 meeting of the American Association of Physical Anthropometrists.

Loomis, T. M., & Lederman, S. J. (1986). Tactual perception. In K. Boff, L. Kaufman, & J. Thomas (eds.), *Handbook of human perception and performance.* New York: Wiley.

Luce, R. D., & Raiffa, H. (1957). Games and decisions: Introduction and critical survey. New York: Wiley.

Luehmann, G. (1958). Physiological measurements as a basis of work organization in industry. *Ergonomics, 1,* 328–344.

Lund, A. M. (1994). Navigating on the information highway. *Proceedings of the 38th Annual Meeting of the Human Factors & Ergonomics Society* (pp. 271–274). Santa Monica, CA: HFES.

Lundberg, C., Johansson, K., Ball, K., Bjerre, B., Blomqvist, C., Braekhus, A., Brouwer, W. H., Bylsma, F. W., Carr, D. B., Englund, L., Friedland, R. P., Hakamies-Blomqvist, L., Klemetz, G., Oneill, D., Odenheimer, G. L., Rizzo, M., Schelin, M., Seideman, M., Tallman, K., Viitanen, M., Waller, P. F., & Winblad, B. (1997). Dementia and driving: An attempt at consensus. *Alzheimer Disease & Associated Disorders, 11*(1), 28–37.

Lunenfeld, H., & Alexander, G. (1990). *A user's guide to positive guidance* (3rd ed.). Washington, DC: Federal Highway Administration.

Lusk, S. L., Ronis, D. L., & Kerr, M. T. (1995). Predictors of hearing protection use among workers: Implications for training programs. *Human Factors, 37*(3), 635–640.

Lusted, L. B. (1976). Clinical decision making. In D. Dombal & J. Grevy (eds.), Decision making and medical care. Amsterdam: North-Holland.

Lynch, P. J., & Horton, S. (1999). *Web style guide : Basic design principles for creating web sites*. New Haven, CT: Yale University Press.

Lyng, S. (1990). Edgework: A social psychological analysis of voluntary risk-taking. *American Journal of Sociology, 95*(4), 851–886.

MacGregor, D., Fischhoff, B., & Blackshaw, L. (1987). Search success and expectations with a computer interface. *Information Processing & Management, 23*, 419–432.

Mackinlay, J. D., Rao, R., & Card, S. K. (1995). An organic user interface for searching citation links. *CHI '95* (pp. 67–73). New York: Association for Computing Machinery.

Maes, P. (1994). Agents that reduce work and information overload. *Communications of the ACM, 37*(7), 31–40.

Magers, C. S. (1983). An experimental evaluation of on-line HELP for non-programmers. In *Proceedings of CHI '83: Human Factors in Computing Systems* (pp. 277–281). New York: Association for Computing Machinery.

Malaterre, G. (1990). Error analysis and in-depth accident studies. *Ergonomics, 33*, 1403–1421.

Mane, A. M., Adams, J. A., & Donchin, E. (1989). Adaptive and part-whole training in the acquisition of a complex perceptual-motor skill. *Acta Psychologica, 71*, 179–196.

Manning, M. V. (1996). So you're the safety director: An introduction to loss control and safety management. Rockville, MD. Government Institutes.

Mansdorf, S. Z. (1995). *Complete manual of industrial safety*. Englewood Cliffs, NJ: Prentice Hall.

Mantei, M., & Teorey, T. J. (1988). Cost/benefit for incorporating human factors in the software lifecycle. *Communications of the ACM, 31*(4), 428–439.

Manuele, F. A. (1997). *On the practice of safety* (2nd ed.). New York: Van Nostrand Reinhold.

Marcotte, A. T., Marvin, S., & Lagemann, T. (1995). Ergonomics applied to product and process design achieves immediate, measurable cost savings. *Proceedings of the 39th Annual Meeting of the Human Factors & Ergonomics Society* (pp. 660–663). Santa Monica, CA: HFES.

Mark, M. A., & Greer, J. E. (1995). The VCR tutor: Effective instruction for device operation. *Journal of the Learning Sciences, 4*(2), 209–246.

Marras, W. S., & Kim, J. Y. (1993). Anthropometry of industrial populations. *Ergonomics, 36*(4), 371–378.

Marrelli, A. F. (1993, Nov-Dec). Determining costs, benefits, and results. *Technical & Skills Training,* 8–14.

Martin, G. (1989). The utility of speech input in user-computer interfaces. *International Journal of Man-Machine System Study, 18,* 355–376.

Massaro, D. W., & Cohen, M. M. (1995). Perceiving talking faces. *Current Directions in Psychological Science,* 4, 104–109.

Masson, M. E. J., Hill, W. C., & Conner, J. (1988). Misconceived misconceptions? *Proceedings of CHI '88* (pp. 151–156). New York: Association for Computing Machinery.

Mayer, D. L., Jones, S. F., and Laughery, K. R. (1987). Accident proneness in the industrial setting. *Proceedings of the Human Factors Society 31st Annual Meeting* (pp. 196–199). Santa Monica, CA: Human Factors Society.

Mayhew, D. R., Simpson, H. M., Williamson, S. R., & Ferguson, S. A. (1997). Effectiveness and role of driver education in a graduated licensing system. *Journal of Public Health Policy.* Mayhew, D. T. (1990). Cost-justifying human factors support—A framework. *Proceedings of the 34th Annual Meeting of the Human Factors Society* (pp. 834–838). Santa Monica, CA: HFES.

Mayhew, D. T. (1992). *Principles and guidelines in software user interface design.* Englewood Cliffs, NJ: Prentice Hall.

McAlindon, P. J. (1994). The development and evaluation of the keybowl: A study on an ergonomically designed alphanumeric input device. *Proceedings of the 38th Annual Meeting Human Factors & Ergonomics Society* (pp. 320–324). Santa Monica, CA: HFES.

McCartt, A. T. (2001). Graduated driver licensing systems-Reducing crashes among teenage drivers. *Journal of the American Medical Association, 286*(13), 1631–1632.

McCartt, A. T., Rohrbaugh, J. W., Hammer, M. C., & Fuller, S. Z. (2000). Factors associated with falling asleep at the wheel among long-distance truck drivers. *Accident Analysis & Prevention, 32*(4), 493–504.

McClumpha, A. M., & James, M. (1994). Understanding automated aircraft. In M. Mouloua & R. Parasuraman (eds.), *Human performance in automated systems: Current research and trends* (pp. 183–190). Hillsdale, NJ: Erlbaum.

McDaniel, J. W., & Hofmann, M. A. (1990). Computer-aided ergonomic design tools. In H. R. Booher (ed.), *MANPRINT: An approach to systems integration.* New York: Van Nostrand Reinhold.

McFarlane, D. C., & Latorella, K. A. (2002). The scope and importance of human interruption in human-computer interaction design. *Human-Computer Interaction, 17,* 1–61.

McGraw, K., & Harbison, K. (1997). User-centered requirements. The scenario-based engineering process. Mahwah NJ: Erlbaum.

McGreevy, M. W., & Ellis, S. R. (1986). The effect of perspective geometry on judged direction in spatial information instruments. *Human Factors, 28,* 439–456.

McKenna, F. P. (1988). What role should the concept of risk play in theories of accident involvement? *Ergonomics, 31,* 469–484.

McKinley, R. L., Ericson, M.A., & D'Angelo, W. R. (1994). Three dimensional auditory displays: Development, application and performance. *Aviation Space and Environmental Medicine, 65,* A31-A38.

McKnight, A. J., & Shinar, D. (1992). Brake reaction time to center high-mounted stop lamps on vans and trucks. *Human Factors, 34*(2), 205–213.

McMillan, G., Eggleston, R. G., & Anderson, T. R. (1997). Nonconventional Controls. In G. Salvendy (ed.) *Handbook of human factors and ergonomics.* New York: Wiley.

McNeil, B. J., Pauker, S. G., Cox, H. C., Jr., and Tversky, A. (1982). On the elicitation of preferences for alternative therapies. *New England Journal of Medicine, 306,* 1259–1262.

McRuer, D. (1980). Human dynamics in man-machine systems. *Automatica, 16,* 237–253.

McVey, G. F. (1990). The application of environmental design principles and human factors guidelines to the design of training and instructional facilities: Room size and viewing considerations. *Proceedings of the 34th Annual Meeting of the Human Factors Society* (pp. 552–556). Santa Monica, CA: HFS.

Means, B., Salas, E., Crandall, B., & Jacobs, T. O. (1993). Training decision makers for the real world. In G. Klein, J. Orasallu, R. Calderwood, & C. E. Zsambok (eds.), *Decision making in action: Models and methods* (pp. 306–326). Norwood, NJ: Ablex.

Medin, D. L., & Ross, B. H. (1992). *Cognitive psychology.* Orlando, FL: Harcourt Brace Jovanovich.

Meecham, W. (1983, May 10). Paper delivered at Acoustical Society of America Meeting, Cincinnati, Ohio, as reported in *Science News, 123,* p. 294.

Mehle, T. (1982). Hypothesis generation in an automobile malfunction inference task. *Acta Psychologica, 52,* 87–116.

Meister, D. (1971). *Human factors: Theory and practice.* New York: Wiley.

Meister, D. (1986). *Human factors testing and evaluation.* New York: Elsevier.

Meister, D. (1987). System design, development, and testing. In G. Salvendy (ed.), *Handbook of human factors* (pp. 17–41). New York: Wiley.

Meister, D. (1989). *Conceptual aspects of human factors.* Baltimore: Johns Hopkins University Press.

Meister, D. (2002). Complexity as a dimension of ergonomics design. *Ergonomics in Design, 10*(2), 10–14.

Melzer, J. E., and Moffitt, K. (1997) *Head Mounted Displays.* New York: McGraw-Hill.

Metzger, U., & Parasuraman, R. (2001). The role of the air traffic controller in future air traffic management: an empirical study of active control versus passive monitoring. *Human Factors, 43*(4), 519–528.

Meyer, R. E. (1999). *Instructional Technology.* In F. Durso (Ed.). *Handbook of Applied Cognition* (pp. 551–570). Chichester, U.K.: John Wiley.

Michon, J. A. (1989). Explanatory pitfalls and rule-based driver models. *Accident Analysis & Prevention, 21*(4), 341–353.

Miller, G. A. (1956). The magical number seven plus or minus two: Some limits on our capacity for processing information. *Psychological Review, 63,* 81–97.

Miller, G. A., & Nicely, P. E. (1955). An analysis of some perceptual confusions among some English consonants. *J. Acoust. Soc. Amer. 27*, 338.

Miller, G. A., Heise, G. A., & Lichten, W. (1951). The intelligibility of speech as a function of the text of the test materials. *Journal of Experimental Psychology, 41*, 329–335.

Miller, J. (1996, Apr 4). Fit for duty? *Ergonomics in Design,* 11–17.

Mitchard, H., & Winkes, J. (2002). Experimental comparisons of data entry by automated speech recognition, keyboard, and mouse. *Human Factors, 44*(2), 198–209.

Monk, T. H., & Wagner, J. A. (1989). Social factors can outweigh biological ones in determining night shift safety. *Human Factors, 31*(6), 721–724.

Monty, R. A., & Senders, J. W. (1976). Eye movements and psychological processes. Hillsdale, NJ: Erlbaum.

Moran, M. M. (1996). Construction safety handbook: A practical guide to OSHA compliance and injury prevention. Rockville, MD: Government Institutes.

Moray, N. (1969). *Listening and attention.* Baltimore, MD: Penguin.

Moray, N. (1986). Monitoring behavior and supervising control. In K. R. Boff, L. Kaufman, & J. P. Thomas (eds.), *Handbook of perception and human performance* (Vol. 2, Ch. 40, pp. 41–51). New York: Wiley.

Moray, N. (1997). Human factors in process control. In G. Salvendy (ed.), *The handbook of human factors and ergonomics* (2nd ed.). New York: Wiley.

Moray, N. (2003). Monitoring, complacency, scepticism and eutactic behaviour. *International Journal of Industrial Ergonomics, 31*(3), 175–178.

Moray, N., & Rotenberg, I. (1989). Fault management in process control: eye movements and action. *Ergonomics, 32*(11), 1319–1342.

Morgan, B. B., Glickman, A. S., Woodward, E. A., Blaiwes, A. S., & Salas, E. (1986). *Measurement of team behaviors in a navy environment* (Tech. Report NTSC TR-86–014). Orlando, FL: Naval Training Systems Center.

Moroney, W. F. (1994). Ethical issues related to the use of humans in human factors and ergonomics. *Proceedings of the 38th Annual Meeting of the Human Factors & Ergonomics Society* (pp. 404–407). Santa Monica, CA: HFES.

Morris, N. M., Lucas, D. B., & Bressler, M. S. (1961). Role of the trunk in the stability of the spine. *Journal of Bone & Joint Surgery, 43A,* 327–351.

Mortimer, R. G. (1993). The high mounted brake lamp. A cause without a theory. *Proceedings of the 37th Annual Meeting of the Human Factors & Ergonomics Society* (pp. 955–959). Santa Monica, CA: HFES.

Mortimer, R. G., Goldsteen, K., Armstrong, R. W., & Macrina, D. (1990). Effects of incentives and enforcement on the use of seat belts by drivers. *Journal of Safety Research, 21,* 25–37.

Mosier, J. N., & Smith, S. L. (1986). Application guidelines for designing user interface software. *Behavior & Information Technology,* 5, 39–46.

Mosier, K. L., Skitka, L. J., Heers, S., & Burdick, M. (1998). Automation bias: decision making and performance in high-tech cockpits. *International Journal of Aviation Psychology,* 8, 47–63.

Mourant, R. R., & Rockwell, T. H. (1972). Strategies of visual search by novice and experienced drivers. *Human Factors, 14*(4)), 325–335.

Muckler, F. A. (1987). The human-computer interface: The past 35 years and the next 35 years. In G. Salvendy (ed.), Cognitive engineering in the design of human-computer interaction and expert systems: *Proceedings of the 2nd International Conference on Human-Computer Interaction.* Amsterdam: Elsevier.

Muckler, F. A. (1992). Selecting performance measures: "Objective" versus "subjective" measurement. *Human Factors, 34*(4), 441–455.

Muir, B. (1987). Trust between humans and machines, and the design of decision aids. *International Journal of Man-Machine Studies, 27,* 527–549.

Muir, B. M. (1988). Trust between humans and machines, and the design of decision aids. In E. Hollnagel, G. Mancini, & D. D. Woods (eds.), *Cognitive engineering in complex dynamic worlds* (pp. 71–83). London: Academic Press.

Mumaw, R. T., & Roth, E. M. (1995). Training complex tasks in a functional context. *Proceedings of the 39th Annual Meeting of the Human Factors & Ergonomics Society* (pp. 1253–1257). Santa Monica, CA: HFES.

Mumaw, R. J., Roth, E. M., Vicente, K. J., & Burns, C. M. (2000). There is more to monitoring a nuclear power plant than meets the eye. *Human Factors, 42*(1), 36–55.

Murray, J., & Liu, Y. (1997). Hortatory operations in highway traffic management. *IEEE Transactions on Systems, Man, and Cybernetics—Part A: Systems and Humans, 27*(3), 340–350.

Muthard, E. K., & Wickens, C. D. (2003). Factors that mediate flight plan monitoring and errors in plan revision: Planning under automated and high workload conditions. *Proceedings of the 12th International Symposium on Aviation Psychology.* Dayton, OH: Wright State University.

Mykityshyn, M. G., Kuchar, J. K., & Hansman, R. J. (1994). Experimental study of electronically based instrument approach plates. The *International Journal of Aviation Psychology, 4*(2), 141–166.

Nachemson, A., & Morris, T. M. (1964). In vivo measurements of intradiscal pressure. *Journal of Bone & Joint Surgery, 46A,* 1077.

Nagamachi, M., & Imada, A. S. (1992). A macroergonomic approach for improving safety and work design. *Proceedings of the 36th Annual Meeting of the Human Factors Society* (pp. 859–861). Santa Monica, CA: HFS.

Naitoh, P. (1981). Circadian cycles and restorative power of naps. In L. C. Johnson, D. I. Tepas, W. P. Colquhoun, & M. J. Colligan (eds.), *Biological rhythms, sleep and shift work* (pp. 553–580). New York: Spectrum.

Nanda, R. (1986). Training in team and consensus building. *Management Solutions, 31*(9), 31–36.

Nass, C., Moon, Y., Fogg, B. J., Reeves, B., & Dryer, D. C. (1995). Can computer personalities be human personalities? *International Journal of Human-Computer Studies, 43,* 223–239.

NASA. (1996). *User-interface guidelines* (NASA DSTL-95-033). Greenbelt, MD: National Aeronautics and Space Administration—Goddard Space Flight Center.

National Aeronautics & Space Administration (NASA) (1969).

National Aeronautics & Space Administration (NASA) (1978). Anthropometric source book, vol. 1: Anthropometry for designers; vol. 2: A handbook of anthropometric data; vol. 3: Annotated bibliography (NASA Reference Publication 1024). Houston, TX: NASA.

National Highway Traffic Safety Administration (1989). Interim report on the safety consequences of raising the speed limit on rural interstate highways. Washington, DC.

National Institute for Occupational Safety & Health (NIOSH) (1981). *Work practices guide for the design of manual handling tasks.* NIOSH.

National Research Council (1995). Human factors in the design of tactical displays for the individual soldier: Phase 1 report. Washington, DC: National Academy Press.

National Safety Council (1989). *Accident facts.* Chicago: National Safety Council.

National Safety Council (1990). *Accident facts.* Chicago: National Safety Council.

National Safety Council (1993a). *Accident facts.* Chicago: National Safety Council.

National Safety Council (1993b). *Safeguarding concepts illustrated* (6th ed.). Chicago, IL: National Safety Council.

National Safety Council (1996). *Accident facts.* Chicago: National Safety Council.

National Transportation Safety Board (1986). *China Airlines B-747 Northwest of San Francisco, Cal. 2/09/35* (NTSB Report # AAR-86/03. Washington, DC.

National Transportation Safety Board (1991). Aircraft Accident Report: Runway Collision of US Air Flight 1493, Boeing 737 and Skywest Flight 5569 Fairchild Metroliner, Los Angeles International Airport (PB91–910409–NTSB/AAR-91/08). Washington, DC, Feb. 1.

National Transportation Safety Board (1992). *Aircraft accident report. Runway collision of USAIR FLIGHT 1493 and Skywest Flight 5569.* NTSB/ AAR-91/08. Washington, DC: National Transportation Safety Board.

Nature.(1986). Whole issue on Chernobyl. (vol. 323), p. 36.

Navon, D., & Gopher, D. (1979). On the economy of the human processing system. *Psychological Review, 86,* 254–255.

Neisser, U. (1982). Memory observed: Remembering in natural contexts. In U. Neisser (ed.), *John Dean's memory: A case study* (pp. 139–159). San Francisco: W. H. Freeman.

Neisser, U., Novick, R., & Lazar, R. (1964). Searching for novel targets. *Perceptual & Motor Skills, 19,* 427–432.

Newcomb, L. C., & Jerome, G. C. (1995). A statistical model for predicting success in aviation. *Proceedings of the 8th International Symposium on Aviation Psychology* (pp. 1113–1116). Columbus: Ohio State University, Department of Aviation.

Newell, A., & Rosenbloom, P. S. (1981). Mechanisms of skill acquisition and the law of practice. In T. R. Anderson (ed.), *Cognitive skills and their acquisition* (pp. 1–55). Hillsdale, NJ: Erlbaum.

Newman, R. L. (1995). *Head-up displays: Designing the way ahead.* Brookfield, VT: Avebury.

Nickerson, R. S., & Moray, N. P. (1995). Environmental change. In R. Nickerson (ed.), *Emerging needs and opportunities for human factors research* (pp. 158–176). Washington, DC: National Academy Press.

Nicolle, C. (1995, July). Design issues for older drivers. *Ergonomics in Design, 3,* 14–18.

Nielson, J. (1987). Using scenarios to develop user friendly videotex systems. *Proceedings of NordDATA '87 Joint Scandinavian Computer Conference* (pp. 133–138), Trondheim, Norway, June 1987.

Nielson, J. (1989). Executive summary In J. Nielson (ed.), *Coordinating user interfaces for consistency* (pp. 1–7). Boston: Academic Press.

Nielson, J. (1993). *Usability engineering.* Cambridge, MA: AP Professional.

Nielson, J. (1994a). Enhancing the explanatory power of usability heuristics. *Chi '94 Proceedings* (pp. 152–158). New York: Association for Computing Machinery.

Nielson, J. (1994b). Heuristic evaluation. In J. Nielsen & R. L. Mack (eds.), *Usability inspection methods* (pp. 25–64). New York: Wiley.

Nielson, J. (1994c). As they may work. *Interactions: New Visions of Human-Computer Interaction* (October, pp. 19–24). New York: Association for Computing Machinery.

Nielson, J., & Molich, R. (1990). Heuristic evaluation of user interfaces. *CHI '90 Proceedings* (pp. 249–256). New York: Association for Computing Machinery.

Nielson, J. (1993). *Usability engineering.* Cambridge, MA: Academic Press Professional.

Norman, D. A. (1981). Categorization of action slips. *Psychological Review,* 88, 1–15.

Norman, D. A. (1986). Cognitive engineering. In D. A. Norman & S. W. Draper (eds.), *User centered system design: New perspectives on human-computer interaction* (pp. 31–61). Hillsdale, NJ: Erlbaum.

Norman, D. A. (1988). *The psychology of everyday things.* New York: Harper & Row.

Norman, D. A. (1992). *Turn signals are the facial expressions of automobiles.* Reading, MA: Addison-Wesley.

Norman, D. A. (1998). *The Invisible Computer: Why good products can fail, the PC is so complex, and information appliances the answer.* Cambridge, MA: MIT Press.

Norman, D. A., & Draper, S. W. (eds.) (1986). *User centered system design.* Hillsdale, NJ: Erlbaum.

Norman, D., & Bobrow, D. (1975). On data-limited and resource-limited processing. *Journal of Cognitive Psychology, 7,* 44–60.

Norman, D. A., Ortony, A., & Russell, D. M. (2003). Affect and machine design: Lessons for the development of autonomous machines. *IBM Systems Journal, 42*(1), 38–44.

Noro, K., & Imada, A. S. (1991). *Participatory ergonomics.* London: Taylor & Francis.

Noy, Y. I. (1997). Human factors in modern traffic systems. *Ergonomics, 40*(10), 1016–1024.

NTSB. (1990). Marine accident report-Grounding of the U. S. Tankship Exxon Valdez on Bligh Reef, Prince William Sound, Valdez, Alaska, March 24, 1989 (NTSB/MAR90/04). Washington, DC: NTSB.

NTSB. (1997). Marine accident report-Grounding of the Panamanian Passenger Ship ROYAL MAJESTY on Rose and Crown Shoal near Nantucket, Massachusetts June 10, 1995 (NTSB/MAR97/01). Washington, DC: NTSB.

Nugent, W. A. (1987). A comparative assessment of computer-based media for presenting job task instructions. *Proceedings of the 31st Annual Meeting of the Human Factors Society* (pp. 696–700). Santa Monica, CA: HFS.

O'Brien, R., & Sheldon, w. C. (1941). *Women's measurements for garment and pattern construction*, U. S. Department of Agriculture, Misc. Pub. No. 454. Washington, DC: U.S. Government Printing Office.

O'Donnell, R. D., & Eggemeier, F. T. (1986). Workload assessment methodology. In K. R. Boff, L. Kaufman, & J. Thomas (eds.), *Handbook of perception and human performance: vol. II: Cognitive processes and performance* (Chapter 42). New York: Wiley.

O'Hara, J., & Brown, W. (1994). *Advanced human system interface design review guideline (*NUREG/CR-5908*)*. Washington, DC: U.S. Nuclear Regulatory Commission.

O'Hara, J., Brown, W., Stubler, W., Wachtel, J., & Persensky, J. (1995). *Human system interface design review guideline* (Draft NUREG-0700, Rev 1.). Washington, DC: U.S. Nuclear Regulatory Commission.

O'Hare, D., & Roscoe, S. N. (1990). *Flightdeck performance: The human factor*. Ames, IA: Iowa State University Press.

Occupational Safety & Health Administration (1983). Occupational noise exposure: Hearing conservation amendment. *Federal Register, 48*, 9738–9783.

Ohnemus, K. R., & Biers, D. W. (1993). Retrospective versus concurrent thinking-out loud in usability testing. *Proceedings of the 37th Annual Meeting of the Human Factors & Ergonomics Society* (pp. 1127–1131). Santa Monica, CA: HFES.

Olson, G. M., & Olson, J. S. (2003). Human-computer interaction: Psychological aspects of the human use of computing. *Annual Review of Psychology, 54*, 491–516.

Oman, C. M. (1993). Sensory conflict in motion sickness. In S. R. Ellis (ed.), *Pictorial communications in virtual and real environments* (2nd ed. pp. 362–376). London: Taylor & Francis.

Orasanu, J. (1990). *Shared mental models and crew decision making* (Tech. Report 46). Princeton, NJ: Princeton University, Cognitive Sciences Laboratory.

Orasanu, J. (1993). Decision-making in the cockpit. In E. L. Weiner, B. G. Kanki, & R. L. Helmreich (eds.), *Cockpit resource management* (pp. 137–168). San Diego: Academic Press.

Orasanu, J., & Backer, P. (1996). Stress and military performance. In J. Driskell & E. Salas (eds.), *Stress and human performance*. Mahwah, NJ: Erlbaum.

Orasanu, J., & Connolly, T. (1993). The reinvention of decision making. In G. Klein, J. Orasanu, R. Calderwood, & C. E. Zsambok (eds.), *Decision making in action: Models and methods* (pp. 3–20). Norwood, NJ: Ablex.

Orasanu, J., & Salas, E. (1993). Team decision making in complex environments. In G. Klein, J. Orasanu, & R. Calderwood (eds.), *Decision making in action: Models and methods* (pp. 327–345). Norwood, NJ. Ablex.

Orasanu, J., Martin, L., & Davison, J. (2001). Cognitive and contextual factors in aviation accidents: decision errors. In E. Salas & G. A. Klein (eds.), *Linking expertise and naturalistic decision making* (pp. 209–225).

Orlady, H. W., & Orlady, L. M. (1999). *Human factors in multi-crew flight operations*. Brookfield, VT: Ashgate Publishing LTD.

Osburn, H. G. (1987). Personnel selection. In G. Salvendy (ed.), *Handbook of human factors* (pp. 911–938). New York: Wiley.

Owens, D. P., Antonoff, R. J., & Francis, E. (1994). Biological motion and nighttime pedestrian conspicuity. *Human Factors, 36,* 718–732.

Owsley, C., Ball, K., McGwin, G., Sloane, M. E., Roenker, D. L., White, M. F., & Overley, E. T. (1998). Visual processing impairment and risk of motor vehicle crash among older adults. *Journal of the American Medical Association, 279*(14), 1083–1088.

Palmer, B., Gentner, F., Schopper, A., & Sottile, A. (1996) Review and analysis: Scientific review of air mobility command and crew rest policy and fatigue issues. *Fatigue Issues, 1–2.*

Panko, R. R. (1998). What we know about spreadsheet errors. *Journal of End User Computing, 10*(2), 15–21.

Parasuraman, R. (1986). Vigilance, monitoring and search. In K. Boff, L. Kaufman, & J. Thomas (eds.), *Handbook of perception and performance* (vol. 2, pp. 43/1–43/39). New York: Wiley.

Parasuraman, R. (1987). Human-computer monitoring. *Human Factors, 29,* 695–706.

Parasuraman, R., & Byrne, E. A. (2003). Automation and human performance in aviation. In P. S. Tsang & M. A. Vidulich (eds.), *Principle and practice of aviation psychology* (pp. 311–356). Mahwah, NJ: Erlbaum.

Parasuraman, R., & Riley, V. (1997). Humans and Automation: Use, misuse, disuse, abuse. *Human Factors, 39*(2), 230–253.

Parasuraman, R., Davies, D. R., & Beatty, J. (1984). *Varieties of attention.* New York: Academic Press.

Parasuraman, R., Hancock, P., & Olofinboba, O. (1997). Alarm effectiveness in driver-centered collision warning systems. *Ergonomics, 40,* 390–399.

Parasuraman, R., Molloy, R., & Singh, I. L. (1993). Performance consequences of automation-induced complacency. *International Journal of Aviation Psychology, 3*(1), 1–23.

Parasuraman, R., & Riley, V. (1997). Humans and automation: Use, misuse, and abuse. *Human Factors, 39,* 230—253.

Parasuraman, R., Sheridan, T. B., & Wickens, C. D. (2000). A model for types and levels of human interaction with automation. *IEEE Transactions on Systems, Man, & Cybernetics : Part A: Systems and Humans, 30*(3), 286–297.

Parasuraman, R., Warm, J. S., & Dember, W. N. (1987). Vigilance: Taxonomy and utility. In L. S. Mark, J. S. Warm, & R. L. Huston (eds.), *Ergonomics and human factors* (pp. 11–31). New York: Springer-Verlag.

Park, I., & Hannafin, M. J. (1993). Empirically-based guidelines for the design of interactive multimedia. *ETR&D-Educational Technology Research and Development, 41*(3), 63–85.

Park, K. S. (1997). Human error. In G. Salvendy (ed.), *Handbook of human factors and ergonomics* (2nd ed.). New York: Wiley.

Parkes, A. M. (1993). Voice communications in vehicles. In A. M. Parkes & S. Franzen (eds.), *Driving future vehicles* (pp. 219–228). Washington, DC: Taylor & Francis.

Parkes, A. M., & Coleman, N. (1990). Route guidance systems: A comparison of methods of presenting directional information to the driver. In E. J. Lovesey (ed.), *Contemporary ergonomics 1990* (pp. 480–485). London: Taylor & Francis.

Parks, D. L, & Boucek, G. P,. Jr. (1989). Workload prediction, diagnosis, and continuing challenges. In G. R. McMillan, D. Beevis, E. Salas, M. H. Strub, R. Sutton, & L. Van Breda (eds.), *Applications of human performance models to system design* (pp. 47–64). New York: Plenum.

Passaro, P. D., Cole, H. P., & Wala, A. M. (1994). Flow distribution changes in complex circuits: Implications for mine explosions. *Human Factors, 36*(4), 745–756.

Passmore, R., & Durnin, T. V G. A. (1955). Human energy expenditure, *Physiological Review, 35,* 83–89.

Patel, V. L., Cytryn, K. N., Shortliffe, E. H., & Safran, C. (2000). The collaborative health care team: the role of individual and group expertise. *Teaching & Learning in Medicine, 12*(3), 117–132.

Patten, T. H., Jr. (1981). Organizational development through teambuilding. New York: Wiley.

Patterson, R. D. (1990). Auditory warning sounds in the work environment. *Philosophical Transactions of the Royal Society of London Series B-Biological Sciences, 327*(1241), 485–492.

Pauls, J. (1980). Building evacuation: Research findings and recommendations. In D. Canter (ed.), *Fires and human behavior* (pp. 251–275). New York: Wiley.

Pauls, J. (1994). Vertical evacuation in large buildings: Misses opportunities for research. *Disaster Management, 6*(3), 128–132.

Pauls, J., & Groner, N. (2002). Human factors contributions to building evacuation research and system design: Opportunities and obstacles. *Proceedings of Workshop to identify innovative research needs to foster improved fire safety in the U.S.* Washington, DC: National Academy of Sciences, National Research Council.

Pausch, R. (1991). Virtual reality on five dollars a day. *Computer Human Interaction (CHI) Proceedings* (pp. 265–269). New York: American Society for Computer Machinery.

Payne, J. W. (1982). Contingent decision behavior. *Psychological Bulletin, 92,* 382–402.

Payne, J. W., Bettman, J. R., & Johnson, E. J. (1988). Adaptive strategy selection in decision making. *Journal of Experimental Psychology: Learning, Memory, & Cognition, 14,* 534–552.

Peacock, B., & Karwowski., W. (1993). *Automotive ergonomics.* Washington, DC: Taylor & Francis.

Peacock, B., & Peacock-Goebel, G. (2002). Wrong number: They didn't listen to Miller. *Ergonomics in Design, 10*(2), 4, 22.

Perrow, C. (1984). *Normal accidents.* New York: Basic Books.

Peters, R. H. (1991). Strategies for encouraging self-protective employee behavior. *Journal of Safety Research, 22,* 53–70.

Peters, T. J. (1988). *Thriving on chaos.* New York: Knopf.

Pew, R. W., & Mavor, A (1998*). Modeling human and organizational performance.* Washington, DC: National Academy of Sciences.

Pheasant, S. T. (1986). *Bodyspace.* London: Taylor & Francis.

Pheasant, S. T., & O'Neill, D. (1975). Performance in gripping and turning: A study in hand/handle effectiveness. *Applied Ergonomics, 6,* 205–208.

Phillips, E. H. (2001, Jan. 8). CFIT declines, but threat persists. *Aviation Week & Space Technology,* 28–30.

Picard, R. W. (1997). *Affective computing.* Cambridge, Mass.: MIT Press.

Pisoni, D. B. (1982). Perception of speech: The human listener as a cognitive interface. *Speech Technology, 1,* 10–23.

Pitz, G. F., & Sachs, N. J. (1984). Judgment and decision: Theory and application. *Annual Review of Psychology, 35,* 139–163.

Pollock, E., Chandler, P., & Sweller, J. (2002). Assimilating complex information. *Learning & Instruction 12,* 61–86.

Pomerantz, J. R., & Pristach, E. A. (1989). Emergent features, attention, and perceptual glue in visual form perception. *Journal of Experimental Psychology: Human Perception & Performance, 15,* 635–649.

Pope, M. H., Andersson, G. B. T., Frymoyer, T. W., & Chaffin, D. B. (eds.) (1991). *Occupational low back pain.* St. Louis: Mosby Year Book.

Posch, J. L., & Marcotte, D. R. (1976). Carpal tunnel syndrome, an analysis of 1201 cases. *Orthopedic Review, 5,* 25–35.

Post, D. (1992). Colorometric measurement, calibration, and characterization of self-lumious displays. In H. Widdel & D.L. Post (eds.), *Color in electronic displays* (pp 299–312). NY: Plenum Press.

Poulton, E. C. (1976). Continuous noise interferes with work by masking auditory feedback and inner speech. *Applied Ergonomics, 7,* 79–84.

Povenmire, H. K., & Roscoe, S. N. (1973). Incremental transfer effectiveness of a ground-based general aviation trainer. *Human Factors, 15,* 534–542.

Preczewski, S. C., & Fisher, D. L. (1990). The selection of alphanumeric code sequences. *Proceedings of the 34th Annual Meeting of the Human Factors Society* (pp. 224–228). Santa Monica, CA: HFS.

President's Commission (1986). Report of the President's Commission on the space shuttle Challenger. Wash. D. C.: U.S. Government Printing Agency.

Price, H. E. (1985). The allocation of functions in systems. *Human Factors, 27*(1), 33–45.

Price, H. E. (1990). Conceptual system design and the human role. In H. R. Booher (ed.), *Manprint: An approach to systems integration* (pp. 161–203). New York: Van Nostrand Reinhold.

Prinzel, L. J., Freeman, F. C., Scerbo, M. W., Mikulka, P. J., & Pope, A. T. (2000). A closed-loop system for examining psychophysiological measures for adaptive task allocation. *International Journal of Aviation Psychology, 10*(4), 393–410.

Pritchett, A. (2001). Reviewing the role of cockpit alerting systems. *Human Factors & Aerospace Safety, 1,* 5–38.

Proctor, R. W., & Van Zandt, T. (1994). *Human factors in simple and complex systems.* Needham Heights, MA: Allyn & Bacon.

Proulx, G. (2001). Highrise evacuation: A questionable concept. *Proceedings of the 2nd International Symposium on Human Behavior in Fire,* pp. 221–230.

Raby, M., & Lee, J. D. (2001). Fatigue and workload in the maritime industry. In P. A. Hancock & P. A. Desmond (eds.), *Fatigue and workload* (pp. 566–578). Mahwah, NJ: Erlbaum.

Raby, M., McGehee, D. V., Lee, J. D., & Norse, G. E. (2000). Defining the interface for a snowplow lane tracking device using a systems-based approach. *Proceedings of the IEA2000/HFES2000 Congress* (pp. 369–372). Santa Monica, CA: HFES.

Raby, M., & Wickens, C. D. (1994). Strategic workload management and decision biases in aviation. *International Journal of Aviation Psychology, 4*(3), 211–240.

Rajalin, S., & Summala, H. (1997). What surviving drivers learn from a fatal road accident. *Accident Analysis & Prevention, 29*(3), 277–283.

Ramachandran, V. S. (1988). Perceiving shape from shading. *Scientific American, 259,* 76–83.

Rantanen, E. M., & Gonzalez de Sather, J. C. M. (2003). Human factors evaluation for a new boiler control interface at the University of Illinois at Urbana-Champaign's Abbott Power Plant: An Avi/Psych 258/IE 240 special project, Fall 2002. University of Illinois Technical Report (AHFD-03–07). Savoy, IL: Institute of Aviation, Aviation Human Factors Division.

Rasmussen, J. (1981). Models of mental strategies in process plant diagnosis. In J. Rasmussen & W. B. Rouse (eds.), *Human detection and diagnosis of system failures.* New York: Plenum.

Rasmussen, J. (1983). Skills, rules, knowledge: Signals, signs, and symbols and other distinctions in human performance models. *IEEE Transactions on Systems, Man, & Cybernetics, 13*(3), 257–267.

Rasmussen, J. (1986). Information processing and human-machine interaction: An approach to cognitive engineering. New York: Elsevier.

Rasmussen, J. (1993). Deciding and doing: Decision making in natural contexts. In G. Klein, J. Orasallu, R. Calderwood, & C. E. Zsambok (eds.), *Decision making in action: Models and methods* (pp. 158–171). Norwood, NJ: Ablex.

Rasmussen, J., Pejtersen, A., & Goodstein, L. (1995). *Cognitive engineering: Concepts and applications.* New York: Wiley.

Ray, P. S., Purswell, T. L., & Bowen, D. (1993). Behavioral safety program: Creating a new corporate culture. *International Journal of Industrial Ergonomics 12,* 193–198.

Read, P. P. (1993). Ablaze: The story of the heroes and victims of Chernobyl. New York: Random House.

Reason, J. (1990). *Human error.* New York: Cambridge University Press.

Reason, J. (1997). *Managing the risks of organizational accidents.* Brookfield, VT: Ashgate.

Reason, T. T., & Brand, T. T. (1975). *Motion sickness.* New York: Academic Press.

Recarte, M. A., & Nunes, L. M. (2000). Effects of verbal and spatial-imagery tasks on eye fixations while driving. *Journal of Experimental Psychology: Applied, 6*(1), 31–43.

Redelmeier, D. A., & Tibshirani, R. J. (1997). Association between cellular-telephone calls and motor vehicle collisions. *New England Journal of Medicine, 336*(7), 453–458.

Reder, L. M. (ed.). (1996). *Implicit memory and metacognition.* Mahwah, NJ: Erlbaum.

Reed, P., & Billingsley, P. (1996). Software ergonomics comes of age: The ANSI/HFES-200 standard. *Proceedings of the 40th Annual Meeting of the Human Factors & Ergonomics Society* (pp. 323–327). Santa Monica, CA: HFES.

Reeves, B., & Nass, C. (1996). *The Media Equation: How people treat computers, television, and new media like real people and places.* New York: Cambridge University Press.

Regan, D. M., Kaufman, L., & Lincoln, J. (1986). Motion in depth and visual acceleration. In K. Boff, L. Kaufman, & J. Thomas (eds.), *Handbook of perception and human performance* (pp. 19–1/19–46). New York: Wiley.

Reigeluth, C. M. (1989). Instructional design theories and models: an overview of their current status. Hillsdale, NJ: Erlbaum.

Reilly, R. R., & Chao, G. T. (1982). Validity and fairness of some alternative employee selection procedures. *Personnel Psychology, 35*, 1–62.

Reinfurt, D. W., Campbell, B. J., Stewart, J. R., & Stutts, J. C. (1990). Evaluating the North Carolina safety belt wearing law. *Acid. Anal. & Prev., 22*(3), 197–210.

Reithel, B. J., Nichols, D. L., & Robinson, R. K. (1996). An experimental investigation of the effects of size, format, and errors on spreadsheet reliability perception. *Journal of Computer Information Systems, 36*(3), 54–64.

Rensink, R. A. (2002). Change detection. *Annual Review Psychology, 53*, 245–277.

Rettig, M. (1991). Nobody reads documentation. *Communications of the ACM, 34*(7), 19–24.

Reynolds, L. (1994). Colour for air traffic control displays. *Displays, 15*, 215–225.

Richardson, R. M. M., Telson, R. U., Koch, C. G., & Chrysler, S. T. (1987). Evaluations of conventional, serial, and chord keyboard options for mail encoding. *Proceedings of the 31st Annual Meeting of the Human Factors Society* (pp. 911–915). Santa Monica, CA: HFS.

Robertson, G. G., Card, S. K., & Mackinlay, J. D. (1993). Information visualization using 3D interactive animation. *Communications of the ACM, 36*(4), 57–71.

Roche, A. F., & Davila, G. H. (1972). Late adolescent growth in stature. *Pediatrics, 50*, 874–880.

Roebuck, J. A., Kroemer, K. H. E., & Thomson, w G. (1975). *Engineering anthropometry methods.* New York: Wiley.

Rogers, T. G., & Armstrong, R. (1977). Use of human engineering standards in design. *Human Factors, 19*(1), 15–23.

Rogers, T. G., & Pegden, C. D. (1977). Formatting and organizing of a human engineering standard. *Human Factors, 19*(1), 55–61.

Rohmert, W. (1965). Physiologische Grundlagell der Erholungszeitbestimmung, *Zeitblatt derArbeitswissenschaft,* 19, p. 1. Cited in E. Simonson, 1971, p. 246.

Roland, H. E., & Moriarty, B. (1990). *System safety engineering and management* (2nd ed.). New York: Wiley.

Rolt, L. T. C. (1978). *Red for danger.* London: Pan Books.

Romiszowski, A. J. (1984). Producing instructional systems: Lesson planning for individualized and group learning activities. New York: Nichols Publishing.

Rosa, R. (2001). Examining work schedules for fatigue: Its not just hours of work. In P. A. Hancock and P. A. Desmond (Eds.) *Stress, Workload, and Fatigue.* Mahwah, N.J.: Lawrence Erlbaum Associates.

Rosenthal, R. (1991). Meta-analytic procedures for social research. Newbury Park, Calif.: Sage.

Roscoe, S. N. (1968). Airborne displays for flight and navigation. *Human Factors, 10,* 321–332.

Roscoe, S. N. (2002). Ergavionics: Designing the job of flying an airplane. *International Journal of Aviation Psychology, 12*(4), 331–339.

Roscoe, S. N., Corl, L., & Jensen, R. S. (1981). Flight display dynamics revisited. *Human Factors, 23,* 341–353.

Rosekind, M. R., Graeber, R. C., Dinges, D. F., Connell, L. J., Rountree, M. S., Spinweber, C. L., & Gillen, K. A. (1994). Crew factors in flight operations: IX. effects of preplanned cockpit rest on crew performance and alertness in long-haul operations (NASA Technical Memorandum 103884). Moffett Field, CA: NASA Ames Research Center.

Rosenthal, L. J., & Reynard, W. (1991, Fall). Learning from incidents to avert accidents. *Aviation Safety Journal,* 7–10.

Roske-Hofstrand, R. J., & Paap, K. R. (1986). Cognitive networks as a guide to menu organization: An application in the automated cockpit. *Ergonomics, 29,* 1301–1311.

Rosow, J. M., & Zager, R. (1990). *Training—The competitive edge.* San Francisco: Jossey-Bass.

Ross, H. L. (1988). Deterrence-based policies in Britain, Canada, and Australia. In M. D. Lawrence, J. R. Stortum, & F. E. Zimrig (eds.), *Social control of the drinking driver* (pp. 64–78). Chicago: University of Chicago Press.

Roth, E. M. (1994). Operator performance in cognitive complex simulated emergencies: Implications for computer-based support systems. *Proceedings of the 38th Annual Meeting of the Human Factors & Ergonomics Society* (pp. 200–204). Santa Monica, CA: HFES.

Roth, E. M. (1997). Analysis of decision making in nuclear power plant emergencies: An investigation of aided decision making. In C. E. Zsambok & G. Klein (eds.), *Naturalistic decision making* (pp. 175–182). Mahwah, NJ: Erlbaum.

Roth, E., Bennett, K., & Woods, D. D. (1987). Human interaction with an "intelligent" machine. *International Journal of Man-Machine Studies, 27,* 479–525.

Roth, E. M., & Woods, D. D. (1989). Cognitive task analysis: An approach to knowledge acquisition for intelligent system design. In G. Guida & C. Tasso (eds.), *Topic in expert system design.* The Netherlands: Elsevier.

Rouse, W. B. & Valusek, J. (1993). Evolutionary design of systems to support decision making. In G. Klein, J. Orasanu, R. Calderwood, & C. E. Zsambok (eds.), *Decision making in action: Models and methods* (pp. 270–286). Norwood, NJ: Ablex.

Rouse, W. B. (1988). Adaptive aiding for human/computer control. *Human Factors, 30*(4), 431–443.

Rouse, W. B. (1990). Designing for human error. Concepts for error tolerant systems. In H. R. Booher (ed.), *Manprint: An approach to systems integration* (pp. 237–255). New York: Van Nostrand Reinholt.

Rouse, W. B., & Morris, N. M. (1986). On looking into the black box: Prospects and limits in the search for mental models. *Psychological Bulletin, 100,* 349–363.

Rubin, T. (1994). Handbook of usability testing: How to plan, design and conduct effective tests. New York: Wiley.

Rubinstein, T., & Mason, A. F. (1979, Nov). The accident that shouldn't have happened: An analysis of Three Mile Island. *IEEE Spectrum,* 33–57.

Sagberg, F., Fosser, S., & Saetermo, I. A. F. (1997). An investigation of behavioral adaptation to airbags and antilock brakes among taxi drivers. *Accident Analysis & Prevention, 29*(3), 293–302.

Salas, E., Bowers, C. A., & Rhodenizer, L. (1998). It is not how much you have but how you use it: Toward a rational use of simulation to support aviation training. *International Journal of Aviation Psychology, 8*(3), 197–208.

Salas, E., & Burke, C. S. (2002). Simulation for training is effective when. *Quality & Safety in Health Care, 11*(2), 119–120

Salas, E., Cannon-Bowers, J. A., & Johnston, J. H. (1997). How can you turn a team of experts into an expert team? Emerging training strategies. In C. E. Zsambok & G. Klein (eds.), *Naturalistic decision making* (pp. 359–370). Mahwah, NJ: Erlbaum.

Salas, E., Rozell, D., Mullen, B., & Driskell, J. E. (1999). The effect of team building on performance-an integration. *Small Group Research, 30*(3), 309–329.

Salasoo, A., White, E. A., Dayton, T., Burkhart, B. J., & Root, R. W. (1994). Bellcore's user-centered design approach. In M. E. Wiklund (ed.), *Usability in practice: How companies develop user-friendly products* (pp. 489–515). Boston: AP Professional.

Salvendy, G. (ed.). (1997). *The handbook of human factors and ergonomics* (2nd ed.). New York: Wiley.

Sanders, A. F. (1970). Some aspects of the selective process in the functional visual field. *Ergonomics,* 13, 101–117.

Sanders, M. S. (1977). Anthropometric survey of truck and bus drivers: Anthropometry, control reach and control force. Westlake Village, CA: Canyon Research Group.

Sanders, M. S., & McCormick, E. J. (1993). *Human factors in engineering and design* (7th ed.). New York: McGraw Hill.

Sanderson, P. M. (1989). The human planning and scheduling role in advanced manufacturing systems: An emerging human factors domain. *Human Factors, 31,* 635–666.

Sanderson, P. M., Flach, J. M., Buttigieg, M. A., & Casey, E. J. (1989). Object displays do not always support better integrated task performance. *Human Factors, 31,* 183–98.

Santhanam, R., & Wiedenbeck, S. (1993). Neither novice nor expert: The discretionary user of software. *International Journal of Man-Machine Studies, 38*(2), 201–229.

Salvendy, G. & Carayon, P. (1997). Data collection and evaluation of outcome measures. In G. Salvendy (ed.), *Handbook of human factors and ergonomics.* New York: Wiley.

Sarno, K. J., & Wickens, C. D. (1995). The role of multiple resources in predicting time-sharing efficiency: An evaluation of three workload models in a multiple task setting. *International Journal of Aviation Psychology, 5*(1), 107–130.

Sarter, N. B., & Woods, D. D. (2000). Teamplay with a powerful and independent agent: A full-mission simulation study. *Human Factors, 42*(3), 390–402.

Sarter, N. B., & Schroeder, B. (2001). Supporting decision making and action selection under time pressure and uncertainty: The case of in-flight icing. *Human Factors, 43*(4), 573–583.

Sarter, N. B., Woods, D. D., & Billings, C. E. (1997). Automation surprises. In G. Salvendy (ed.), *Handbook of human factors and ergonomics* (2nd ed.) (pp. 1926–1943). New York: Wiley.

Satish, U., & Streufert, S. (2002). Value of a cognitive simulation in medicine: towards optimizing decision making performance of healthcare personnel. *Quality & Safety in Health Care, 11*(2), 163–167.

Sato, H., Ohashi, J., Iwanaga, K., Yoshitake, R., and Shimada, K. (1984). Endurance time and fatigue in static contractions. *Journal of human ergology, 13,* 147–154.

Scerbo, M. W. (1996). Theoretical perspectives on adaptive automation. In R. Parasuraman & M. Mouloua (eds.), Automation and human performance: Theory and applications (pp. 37–64). Mahwah, NJ: Erlbaum.

Schacherer, C. W. (1993). Toward a general theory of risk perception. Proceedings of the 37th Annual Meeting of the Human Factors & Ergonomics Society (pp. 984–988). Santa Monica, CA: HFES.

Schacter, D. L. (2001). Seven sins of memory: How the mind forgets and remembers. Boston: Houghton Mifflin.

Schank, R. C., & Abelson, R. (1977). Scripts, plans, goals, and understanding. Hillsdale, NJ: Erlbaum.

Schmidt, F. L., & Hunter, J. E. (1981). Employment testing: old theories and new research findings. American Psychologist, 36(10), 1128–1137.

Schmidt, R. A. & Bjork, R. A. (1992). New conceptualizations of practice: Common principles in three paradigms suggest new concepts for training. *Psychological Science, 3*(4), 207–217.

Schmidt, J. K., & Kysor K. P. (1987). Designing airline passenger safety cards. *Proceedings of the 31st annual meeting of the Human Factors Society* (pp. 51–55). Santa Monica, CA: Human Factors Society.

Schneider, W. (1985). Training high-performance skills: Fallacies and guidelines. *Human Factors, 27*(3), 285–300.

Schottelius, B. A., & Schottelius, D. D. (1978). *Textbook of physiology* (18th ed.). St. Louis: Mosby. Really 18th Ed.

Schraagen, J. M. (1997). Discovering requirements for a naval damage control decision support system. In C. E. Zsambok & G. Klein (eds.), *Naturalistic decision making* (pp. 269–283). Mahwah, NJ: Erlbaum.

Schraagen, J. M., Chipman, S. F., & Shalin, V. L. (2000). *Cognitive task analysis.* Mahwah, NJ: Erlbaum.

Schum, D. (1975). The weighing of testimony of judicial proceedings from sources having reduced credibility. *Human Factors, 17,* 172–203.

Schustack, M. W., & Sternberg, R. J. (1981). Evaluation of evidence in causal inference. *Journal of Experimental Psychology, General, 110,* 101–120.

Schwing, R. C., & Kamcrud, D. B. (1988). The distribution of risks: Vehicle occupant fatalities and time of the week. *Risk Analysis, 8,* 127–133.

Seagull, J., & Gopher, D. (1995). Training head movement in visual scanning: An embedded approach to the development of piloting skills with helmet-mounted displays. *Proceedings of the 39th Annual Meeting of the Human Factors & Ergonomics Society.* Santa Monica, CA: HFES.

Seagull, F. J., & Sanderson, P. (2001). Anesthesia alarms in context: An observational study. *Human Factors, 43,* 66–78.

Segal, L. D. (1994). Actions speak louder than words: How pilots use nonverbal information for crew communications. *Proceedings of the Human Factors and Ergonomics Society 38th Annual Meeting* (pp. 21–25). Santa Monica, CA: Human Factors and Ergonomics Society.

Seibel, R. (1964). Data entry through chord, parallel entry devices. *Human Factors, 6,* 189–192.

Seidler, K., & Wickens, C. D. (1992). Distance and organization in multifunction displays. *Human Factors, 34,* 555–569.

Selcon, S. J., Taylor, R. M., & Koritas, E. (1991). Workload or situational awareness?: TLX vs. SART for aerospace systems design evaluation. *Proceedings of the Human Factors Society, 35* (pp. 62–6).

Senders, J. W. (1964). The human operator as a monitor and controller of multidegree of freedom systems. *IEEE Transactions on Human Factors in Electronics, HFE-5,* 2–6.

Sengupta, A., Das, B. (2000). Maximum reach envelope for the seated and standing male and female for industrial workstation design. *Ergonomics, 43*(9), 1390–1404.

Shanteau, J. and Dino, G.A., (1993). Environmental stressor effects on creativity and decision making. In O. Svenson & J. A. Maule (Eds.) *Time pressure and stress in human judgment and decision making.* (pp. 293–308). New Yourk: Plenum Press.

Sharda, R., Barr, S. H., & McDonnell, J. C. (1988). Decision support system effectiveness: a review and an empirical test. *Management Science, 34,* 139–159.

Sheridan, T. (1981). Understanding human error and aiding human diagnostic behavior in nuclear power plants. In J. Rasmussen & W. Rouse (eds.), *Human detection and diagnosis of system failures.* New York: Plenum.

Sheridan, T. (1997). Supervisory control. In G. Salvendy (ed.), *Handbook of human factors.* New York: Wiley.

Sheridan, T. B. (2002). Humans and automation: System design and research issues. New York: Wiley.

Sherman, W. R., & Craig, A. B. (2003). Understanding virtual reality: Interfaces, applications and design. San Francisco: Elsevier.

Sherry, L., & Polson, P. G. (1999). Shared models of flight management system vertical guidance. *International Journal of Aviation Psychology, 9,* 139–154.

Shinar, D. (1995). Field evaluation of an advance brake warning system. *Human Factors, 37*(4), 746–751.

Shinar, D., & Schieber, F. (1991). Visual requirements for safety and mobility of older drivers. *Human Factors, 33*(5), 507–520.

Shneiderman, B. (1992). Designing the user interface: Strategies for effective human-computer interaction (2nd ed.). Reading, MA: Addison-Wesley.

Shope, J. T., & Molnar, L. J. (2003). Graduated driver licensing in the United States: evaluation results from the early programs. *Journal of Safety Research, 34*(1), 63–69.

Shortliffe, E. H. (1976). *Computer-based medical consultations: MYCIN*. New York: Elsevier.

Shults, R. A., Elder, R. W., Sleet, D. A., Nichols, J. L., Alao, M. O., Carande-Kulis, V. G., Zaza, S., Sosin, D. M., & Thompson, R. S. (2001). Reviews of evidence regarding interventions to reduce alcohol-impaired driving. *American Journal of Preventive Medicine, 21*(4), 66–88.

Siegel, A. I. (1983). The miniature job training and evaluation approach: additional findings. *Personnel Psychology, 36,* 41–56.

Sime, J. D. (1993). Crowd psychology and engineering: Designing for people or ball bearings? In R. A. Smith & J. F. Dicjie (eds.), *Engineering for crowd safety* (pp. 119–131), New York: Elsevier Science Publishers.

Simon, F., & Corbett, C. (1996). Road traffic offending, stress, age, and accident history among male and female drivers. *Ergonomics, 39*(5), 757–780.

Simon, H. A. (1957). *Models of man.* New York: Wiley.

Simon, H. A. (1987). Decision making and problem solving. *Interfaces, 17,* 11–31.

Simonson E., & Lind, A. R. (1971). Fatigue in static work. In E. Simonson (ed.), *Physiology of work capacity and fatigue.* Springfield, IL: Charles Thomas Publisher.

Simonson, E. (ed.) (1971). *Physiology of work and capacity and fatigue.* Springfield, IL: Charles Thomas Publisher.

Simpson, C. (1976, May). Effects of linguistic redundancy on pilot's comprehension of synthesized speech. *Proceedings of the 12th Annual Conference on Manual Control* (NASA TM-X-73, 170). Washington, DC: U.S. Government Printing Office.

Skitka, L. J., Mosier, K. L., Burdick, M., & Rosenblatt, B. (2000). Automation bias and errors: Are crews better than individuals? *International Journal of Aviation Psychology, 10*(1), 85–97.

Sklar, A., & Sarter, N. (1999). Good vibrations: Tactile feedback in support of attention allocation and human-automation coordination in event-driven domains. *Human Factors, 41*(4), 543–552.

Slamecka, N. J., & Graf, P. (1978). The generation effect: delineation of a phenomena. *Journal of Experimental Psychology: Human Learning & Memory, 4,* 592–604.

Slappendel, C., Laird, I., Kawachi, I., Marshall, S., & Cryer, C. (1993). Factors affecting work-related injury among forestry workers: A review. *Journal of Safety Research, 24,* 19–32.

Slovic, P., Fischhoff, B., & Lichtenstein, S. (1977). Behavioral decision theory. *Annual Review of Psychology, 28,* 1–39.

Smith, L. J. (1993). The scientific basis of human factors-A behavioral cybernetic perspective. *Proceedings of the 37th Annual Meeting of the Human Factors & Ergonomics Society* (pp. 534–538). Santa Monica, CA: HFES.

Smith, P. J., McCoy, E., & Layton, C. (1997). Brittleness in the design of cooperative problem-solving systems: The effects on user performance. *IEEE Transactions on Systems Man and Cybernetics Part A-Systems and Humans, 27*(3), 360–371.

Smither, R. D. (1994). *The psychology of work and human performance* (2nd ed.). New York: HarperCollins.

Sniezek, J. A., Wilkins, D. C., Wadlington, P. L., & Baumann, M. R. (2002). Training for crisis decision-making: Psychological issues and computer-based solutions. *Journal of Management Information Systems, 18*(4), 147–168.

Snook, S. H., & Ciriello, V. M. (1991). The design of manual handling tasks: Revised tables of maximum acceptable weights and forces. *Ergonomics, 34*, 1197–1213.

Somes Job & Dalziel, R. F. (2001). Defining fatigue as a condition of the organism and distinguishing it from habituation, adaptation, and boredom. In P. A. Hancock & P. A. Desmond (eds.), *Stress, workload, and fatigue* (pp. 466–475). Mahwah, NJ: Erlbaum.

Society of Automotive Engineers (SAE). (2002). *Civilian American and European Surface Anthropometry Resource (CAESAR), CAESAR Project Data Set—North American Edition.* ISBN 0-7680-1105-1.

Sohn, S. Y., & Stepleman, R. (1998). Meta-analysis on total braking time. *Ergonomics, 41*(8), 1129–1140.

Sojourner, R. J., & Antin, J. F. (1990). The effect of a simulated head-up display speed meter on perceptual task performance, *Human Factors, 32*, 329–340.

Sollenberger, R. L., & Milgram, P. (1993). Effects of stereoscopic and rotational displays in a 3D path-tracing task. *Human Factors, 35*(3), 483–499.

Solomon, Z., Mikulincer, M., & Hobfoll, S. E. (1987). Objective versus subjective measurement of stress and social support: Combat-related reactions. *Journal of Consulting & Clinical Psychology, 55*, 577–583.

Sorkin, R. (1988). Why are people turning off our alarms? *Journal of Acoustical Society of America, 84*, 1 107–1 108.

Sorkin, R. D. (1989). Why are people turning off our alarms? *Human Factors Bulletin, 32*(4), 3–4.

Sorkin, R. D., Kantowitz, B. H., & Kantowitz, S. C. (1988). Likelihood alarm displays. *Human Factors, 30*, 445–460.

Spence, C., & Driver, J. (2000). Audiovisual links in attention: implications for interface design. In D. Harris (ed.), *Engineering psychology and cognitive ergonomics.* Hampshire: Ashgate.

Spencer, K. (1988). The psychology of educational technology and instructional media. London & UK: Routledge.

Spool, J. M., & Snyder, C. (1993). *Product usability. Survival techniques.* Tutorial presented for IBM Santa Teresa Laboratory. Andover, MA: User Interface Engineering.

Squires, P. C. (1956). *The shape of the normal work area.* Report No. 275. New London, CT: Navy Department, Bureau of Medicine & Surgery, Medical Research Laboratory.

Srinivasan, R., & Jovanis, P. P (1997). Effect of selected in-vehicle route guidance systems on driver reaction times. *Human Factors, 39*, 200–215.

Stager, P., & Angus, R. (1978). Locating crash sites in simulated air-to-ground visual search. *Human Factors, 20*, 453–466.

Stager, P., Hameluck, D., & Jubis, R. (1989). Underlying factors in air traffic control incidents. *Proceedings of the 33rd Annual Meeting of the Human Factors Society.* Santa Monica, CA: HFS.

Stanton, N. (1994). *Human factors in alarm design.* London: Taylor & Francis.

Stanton, N. A., & Marsden, P. (1996). From fly-by-wire to drive-by-wire: Safety implications of automation in vehicles. *Safety Science.* 24, 35–49.

Status Report (1994, Dec 17). *All the 16-year-olds didn't make it home. 29. #13.* Arlington, VA: Insurance Institute for Highway Safety.

Status Report (1995, Mar 18). *Airbags save lives. 30. #3.* Arlington, VA: Insurance Institute for Highway Safety.

Steeb, R., & Johnson, S. C. (1981). A computer-based interactive system for group decision-making. *IEEE Transactions, 11*, 544–552.

Stern, J. A., Boyer, D., & Schroeder, D. (1994). Blink rate: A possible measure of fatigue. *Human Factors, 36*(2), 285–297.

St. John, M, Cowen, M. B., Smallman, H.S., and Oonk, H.M. (2001). The use of 2D and 3D displays for shape-understanding versus relative positioning tasks. *Human Factors. 43*, (1), pp. 79–98.

Stokes, A. F., & Kite, K. (1994). *Flight stress: Stress, fatigue and performance in aviation.* Brookfield, VT: Ashgate Aviation.

Stokes, A. F., Wickens, C. D., & Kite, K. (1990). *Display technology: Human factors concepts.* Warrendale, PA: Society of Automotive Engineers.

Stout, R. J. (1995). Planning effects on communication strategies: A shared mental model perspective. *Proceedings of the 39th Annual Meeting of the Human Factors & Ergonomics Society* (pp. 1278–1282). Santa Monica, CA: HFES.

Strauch, B. (1997). Automation and decision making: lessons learned from the Cali accident. *Proceedings of the 41st Annual Meeting of the Human Factors & Ergonomics Society* (pp. 195–199). Santa Monica, CA: HFES.

Strawbridge, T. (1986). The influence of position, highlighting, and imbedding on warning effectiveness. *Proceedings of the 30th Annual Meeting of the Human Factors Society* (pp. 716–720). Santa Monica, CA: HFS.

Strayer, D. L., Drews, F. A., & Johnston, W. A. (2003). Cell phone-induced failures of visual attention during simulated driving. *Journal of Experimental Psychology: Applied, 9*(1), 23–32.

Strayer, D. L. & Johnston, W. A. (2001). Driven to distraction: Dual-task studies of simulated driving and conversing on cellular telephone. *Psychological Science, 12*(6), 462–466.

Street, W. (1974). Brainstorming by individuals, coacting and interacting groups. *Journal of Applied Psychology, 59*, 433–436.

Streeter, L. A., Vitello, D., & Wonsiewicz, S. A. (1985). How to tell people where to go: Comparing navigational aids. *International Journal on Man-Machine Studies, 22*, 549–562.

Stubler, W. F., & O'Hara, J. M. (1995). Group-view displays for enhancing crew performance. *Proceedings of the 39th Annual Meeting of the Human Factors & Ergonomics Society* (pp. 1199–1203). Santa Monica, CA: HFES.

Su, Y. L., & Lin, D. Y. M. (1998). The impact of expert-system-based training on calibration of decision confidence in emergency management. *Computers in Human Behavior, 14*(1), 181–194.

Sulc, S. (1996). Speech characteristics in the course of coping with in-flight emergencies. In *Situation Awareness: Limitations and Enhancements In the Aviation Environment* (NATO AGARD CP-575). Neuilly-Sur-Seine, France: AGARD.

Sumby, W., & Pollack, I. (1954). Visual contribution to speech intelligibility in noise. *Journal of the Acoustical Society of America, 26,* 212–215.

Summala, H. (1981). Driver/vehicle steering response latencies. *Human Factors, 23,* 683–692.

Summala, H. (1988). Zero-risk theory of driver behaviour. *Ergonomics, 31,* 491–506.

Summala, H. (2000). Brake reaction times and driver behavior analysis. *Transportation Human Factors, 2*(3), 217–226.

Summala, H., & Mikkola, T. (1994). Fatal accidents among car and truck drivers: Effects of fatigue, age, and alcohol consumption. *Human Factors, 36*(2), 315–326.

Sundstrom, E., & Altman, I. (1989). Physical environments and work-group effectiveness. *Research in Organizational Behavior, 11,* 175–209.

Sundstrom, E., De Meuse, K. P., & Futrell, D. (1990). Work teams: Applications and effectiveness. *American Psychologist, 45,* 120–133.

Svenson, O. (1981). Are we less risky and more skillful than our fellow drivers? *Acta Psychologica, 47,* 143–148.

Svenson, O., & Maule, A. J. (eds.) (1993). Time pressure and stress in human judgment and decision making. New York: Plenum.

Swain, A. D., and Guttmann, H. E. (1983). *Handbook of Human Reliability Analysis with Emphasis on Nuclear Power Plant Applications.* Sandia National Laboratories, NUBERG/CR-1278, US Nuclear Regulatory Commission, Washington, DC.

Sweller, J. (1994). Cognitive load theory, learning difficulty and instructional design. *Learning & Instruction 4,* 295–312.

Sweller, J. (1999). Instructional designs in technical areas. Melbourne: ACER.

Sweller, J., & Chandler, P. (1994). Why some material is difficult to learn. *Cognition & Instruction 12*(3), 185–233.

Swennsen, R. G., Hessel, S. J., & Herman, P. G. (1977). Omissions in radiology. Faulty search or stringent reporting criteria? *Radiology, 123,* 563–567.

Swets, J. A. (1996). Signal detection theory and ROC analysis in psychology and diagnostics. Mahwah, NJ: Erlbaum.

Swezey, R. W. (1987). Design of job aids and procedural writing. In G. Salvendy (ed.), *Handbook of human factors* (pp. 1039–1057). New York: Wiley.

Swezey, R. W., & Llaneras, R. E., (1997). Models in training and instruction. In G. Salvendy (ed.), *Handbook of human factors and ergonomics* (2nd ed.) (pp. 514–577). New York: Wiley.

Taggart, R. W. (1989). Results of the drug testing program at Southern Pacific Railroad. In S. W. Gust & T. M. Walsh (eds.), *Drugs in the workplace: Research and evaluation data* (NIDA Research Monograph No. 91). Washington, DC: U.S. Government Printing Office.

Talebzadeh, H., Mandutianu, S., & Winner, C. F. (1995). Countrywide loan underwriting expert system. *AI Magazine, 16*(1), 51–64.

Tang, J. C. (1991). Findings from observational studies of collaborative work. In S. Greenberg (ed.), *Computer-supported cooperative work and groupware* (pp. 11–28). San Diego: Academic Press.

Tattersall, A. J., & Hockey, G. R. J. (1995). Level of operator control and changes in heart rate variability during simulated flight maintenance. *Human Factors, 37*(4), 682–698.

Taylor, W., Pearson, J., Mair, A., & Burns, W. (1965). Study of noise and hearing in jute weavers. *Journal of the Acoustical Society of America, 38,* 113–120.

Teague, R. C., & Allen, J. A. (1997). The reduction of uncertainty and troubleshooting performance. *Human Factors, 39*(2), 254–267.

Tenner, E. (1996). Why things bite back: Technology and the revenge of unanticipated consequences. New York: Knopf.

Theeuwes, J. (1994). Visual attention and driving behavior. In C. Santos (ed.), *Human factors in road traffic* (pp. 103–123). Lisbon, Portugal: Escher.

Theeuwes, J., & Godthelp, H. (1995). Self-explaining roads. *Safety Science, 19,* 217–225.

Theeuwes, J., & Hagenzieker, M. P. (1993). Visual search of traffic scenes: On the effect of location expectations. In A. G. Gale et al. (eds.), *Vision in vehicles-IV* (pp. 149–158). Amsterdam: Elsevier.

Theeuwes, J, Alferdinck, J.W.A., & Perel, M. (2002). Relation between glare and driving performance. *Human Factors, 44*(1). 79–94.

Tichauer, E. R. (1978). *The biomechanical basis of ergonomics.* New York: Wiley.

Tindall-Ford, S., Chandler, P., & Sweller, J. (1997). When two sensory modes are better than one. *Journal of Experimental Psychology: Applied, 3,* 257–287.

Tipping, J., Freeman, R. F., & Rachlis, A. R. (1995). Using faculty and student perceptions of group-dynamics to develop recommendations for PBL training. *Academic Medicine, 70*(11), 1050–1052.

Todd, P., & Benbasat, I. (2000). Inducing compensatory information processing through decision aids that facilitate effort reduction: An experimental assessment. *Journal of Behavioral Decision Making, 13*(1), 91–106.

Tolcott, M. A., Marvin, F. F., & Bresnick, T. A. (1989). *The confirmation bias in military situation assessment.* Reston, VA: Decision Science Consortium.

Tränkle, U., Gelau, C., & Metker, T. (1990). Risk perception and age-specific accidents of young drivers. *Accident Analysis & Prevention, 22*(2), 119–125.

Treisman, A. (1986). Properties, parts, and objects. In K. R. Boff, L. Kaufman, & J. P. Thomas (eds.), *Handbook of perception and human performance.* New York: Wiley.

Triggs, T., & Harris, W. G. (1982). *Reaction time of drivers to road stimuli* (Human Factors Report HFR-12). Clayton, Australia: Monash University.

Trotter, M., & Gleser, G. (1951). The effect of aging upon stature. *American Journal of Physical Anthropology, 9*(31), 1–324.

Tsang, P. S., & Viduich, M. A. (eds.) (2003). *Principles and practice of aviation psychology.* Mahwah, NJ: Erlbaum.

Tsang, P., & Wilson, G. (1997). Mental Workload. In G. Salvendy (Ed.). *Handbook of Human Factors.* NY: John Wiley.

Tseng, S., & Fogg, B. J. (1999). Credibility and computing technology. *Communications of the ACM, 42*(5), 39–44.

Tufte, E. R. (1983). *The visual display of quantitative information.* Cheshire, CT: Graphics Press.

Tufte, E. R. (1990). *Envisioning information.* Cheshire, CT. Graphics Press.

Tulga, M. K., & Sheridan, T. B. (1980). Dynamic decisions and workload in multitask supervisory control. *IEEE Transactions on Systems, Man, & Cybernetics, SMC-10*, 217–232.

Tversky, A. (1977). Features of similarity. *Psychological Review, 84*, 327–352.

Tversky, A., & Kahneman, D. (1981). The framing of decisions and the psychology of choice, *Science, 211*, 453–458.

Tversky, A., & Kahneman, D. (1974). Judgment under uncertainty: Heuristics and biases. *Science, 185*, 1124–1131.

Tversky, B., & Franklin, N. (1990). Searching imagined environments. *Journal of Experimental Psychology: General, 119*, 63–76.

U.S. Department of Defense (1989). *Human engineering design criteria for military systems, equipment, and facilities* (M1L-STD-1472D). Washington, DC: Department of Defense.

Ulrich, L., & Trumbo, D. (1965). The selection interview since 1949. *Psychological Bulletin, 63*, 100–116.

Urban, J. M., Weaver, J. L., Bowers, C. A., & Rhodenizer, L. (1996). Effects of workload and structure on team process and performance: Implications for complex team decision making. *Human Factors, 38*(2), 300–310.

Uttal, W. R., Baruch, T., & Allen, L. (1994). Psychophysical foundations of a model of amplified night vision in target detection tasks. *Human Factors, 36*, 488–502.

Van Cott, H. P., & Kinkade, R. G. (eds.) (1972). *Human engineering guide to equipment design.* Washington, DC: U.S. Government Printing Office.

Van Der Horst, R. (1988). Driver decision making at traffic signals. In *Traffic accident analysis and roadway visibility* (pp. 93–97). Washington, DC: National Research Council.

Van Houten, R., & Nau, P. A. (1983). Feedback interventions and driving speed: A parametric and comparative analysis. *Journal of Applied Behavior Analysis, 16*, 253–281.

Vanderheiden, G. C. (1997). Design for people with functional limitations resulting from disability, aging, or circumstances. In G. Salvendy (ed.) *Handbook of human factors and ergonomics* (2nd ed.).(Chpt 60). New York: Wiley.

Varhelyi, A. (2002). Speed management via in-car devices: Effects, implications, perspectives. *Transportation, 29*(3), 237–252.

Vas, H. W. (1973). Physical workload in different body postures, while working near to or below ground level. *Ergonomics, 16*, 817–828.

Vazquez-Abad, J., & Winer, L. R. (1992). Emerging trends in instructional interventions. In H. D. Stolovitch & E. J. Keeps (eds.), *Handbook of human performance technology* (pp. 672–687). San Francisco: Jossey-Bass.

Vicente, K. (1999). Cognitive work analysis: Towards safe, productive and healthy computer-based work. Mahwah NJ: Lawrence Erlbaum Associates.

Vicente, K. J. (2002). Ecological interface design: Progress and challenges. *Human Factors, 44*(1), 62–78.

Vicente, K. J., & Williges, R. C. (1988). Accommodating individual differences in searching a hierarchical file system. *International Journal of Man-Machine Studies, 29*, 647–668.

Vicente, K., & Rasmussen, J. (1992). *Ecological interface design.* IEEE *Transactions on systems, man & cybernetics, 22*(4), 589–606.

Violanti, J. M. (1998). Cellular phones and fatal traffic collisions. *Accident Analysis & Prevention, 30*(4), 519–524.

Violanti, J. M., & Marshall, J. R. (1996). Cellular phones and traffic accidents: An epidemiological approach. *Accident Analysis & Prevention, 28*(2), 265–270.

Volpe, C. E. (1993). Training for team coordination and decision making effectiveness: theory, practice, and research directions. *Proceedings of the 37th Annual Meeting of the Human Factors & Ergonomics Society* (pp. 1226–1227). Santa Monica, CA: HFES.

Volpe, C. E., Cannon-Bowers, J. A., Salas, E., & Spector, P. E. (1996). The impact of cross-training on team functioning: an empirical investigation. *Human Factors, 38*(1), 87–100.

von Breda, L. (1999). *Anticipating behavior in supervisory vehicle control.* Delft University, Netherlands: Delft University Press.

Wagemann, L. (1998). Analysis of the initial representations of the human-automation interactions (HAI). *Travail Humain, 61*(2), 129–151.

Wagenaar, W. A., & Groeneweg, J. (1987). Accidents at sea: Multiple causes and impossible consequences. *International Journal of Man-Machine Studies, 27,* 587–598.

Walker, G. H., Stanton, N. A., & Young, M. S. (2001). Where is computing driving cars? *International Journal of Human-Computer Interaction, 13*(2), 203–229.

Waller, P. F. (1991). The older driver. *Human Factors, 33*(5), 499–506.

Ward, N. J., & Parkes, A. (1994). Head-up displays and their automotive application: An overview of human-factors issues affecting safety. *Accident Analysis & Prevention, 26*(6), 703–717.

Warm, J. S. (1984). *Sustained attention in human performance.* Chichester, UK: Wiley.

Warm, J. S., & Parasuraman, R. (eds.) (1987). Vigilance: Basic and applied. *Human Factors, 29,* 623–740.

Warm, J. S., Dember, W. N., & Hancock, P. A. (1996). Vigilance and workload in automated systems. In R. Parasuraman & M. Mouloua (eds.), *Automation and human performance: Theory and applications* (pp. 183–199). Mahwah, NJ: Erlbaum.

Wasielewski, P. (1984). Speed as a measure of driver risk: Observed speeds versus driver and vehicle characteristics. *Accident Analysis & Prevention, 16,* 89–103.

Wasserman, D. E. (1987). Motion and vibration. In G. Salvendy (ed.), *Handbook of human factors* (pp. 650–669). New York: Wiley.

Wasserman, S., & Faust, K. (1994). *Social Network Analysis.* NY: Cambridge University Press.

Waters, T. R., Putz-Anderson, V, Garg, A., & Fine, L. (1993). Revised NIOSH equation for the design and evaluation of manual lifting tasks, *Ergonomics, 36*(7), 749–776.

Weimer, J. (ed.) (1995). *Research techniques in human engineering.* Englewood Cliffs, NJ: Prentice Hall.

Weintraub, D. J., & Ensing, M. J. (1992). *Human factors issues in head-up display design: The book of HUD* (SOAR CSERIAC State of the Art Report 92–2). Dayton, OH: Crew System Ergonomics Information Analysis Center, Wright-Patterson AFB.

Weiss, E. H. (1991). *How to write usable user documentation* (2nd ed.). Phoenix: Oryx Press.

Wells, G. L. (1993). What do we know about eyewitness identification? *American Psychologist, 48,* 553–571.

Wells, G. L., & Seelau, E. P. (1995). Eyewitness identification: psychological research and legal policy on lineups. *Psychology, Public Policy & Law, 1,* 765–91.

Wells, G. L., Lindsay, R. C., & Ferguson, T. I. (1979). Accuracy, confidence, and juror perceptions in eyewitness testimony. *Journal of Applied Psychology, 64,* 440–448.

West, J. B. (1985). *Everest, the testing place.* New York: McGraw-Hill.

Wetzel, C. D., Radke, P. H., & Stern, H. W. (1994). *Instructional effectiveness of video media.* Multi media instruction. Mahwah NJ: Lawrence Erlbaum.

Wexley, K. M., & Latham, G. P. (1991). *Developing and training human resources in organizations* (2nd ed.). New York: HarperCollins.

White, C. C. (1990). A survey on the integration of decision analysis and expert systems for decision support. *IEEE Transactions on Systems, Man, & Cybernetics, 20*(2), 358–364.

White, R. M., & Churchill, E. (1971). *The body size of soldiers, U.S. Army anthropometry—1966* (pp. 1–329). Tech. Report 72–51–CE. Natick, MA: U.S. Army Natick Labs.

Whiteside, Bennet, & Holtzblat (1988).

Wickelgren, W. A. (1964). Size of rehearsal group in short-term memory. *Journal of Experimental Psychology, 68,* 413–419.

Wickens, C. D. (1984). Processing resources in attention. In R. Parasuraman & R. Davies (eds.), *Varieties of attention* (pp. 63–101). New York: Academic Press.

Wickens, C. D. (1986) The effects of control dynamics on performance. In K. R. Boff, L. Kaufman, & J. P. Thomas (eds.), *Handbook of perception and performance,* vol. II (pp. 39–1/39–60). New York. Wiley.

Wickens, C. D. (1989). Attention and skilled performance. In D. Holding (ed.), *Human skills* (2nd ed.) (pp. 71–105). New York: Wiley.

Wickens, C. D. (1992a). *Engineering psychology and human performance* (2nd ed.). New York: HarperCollins.

Wickens, C. D. (1992b). The human factors of graphs at Human Factors Society annual meetings. *Human Factors Bulletin, 35*(7), 1–3.

Wickens, C. D. (1995). Aerospace techniques. In J. Wiemer (ed.), *Research techniques in human engineering* (pp. 1 12–142). Englewood Cliffs, NJ: Prentice Hall.

Wickens, C. D. (1996). Designing for stress. In J. E. Driskell & E. Salas (eds.), *Stress and human performance.* Mahwah, NJ: Erlbaum.

Wickens, C. D. (1997). Attentional issues in head-up displays. In D. Harris (ed.), *Engineering psychology and cognitive ergonomics: Transportation systems,* vol. 1 (pp. 3–21). Aldershot, UK: Ashgate Publishing.

Wickens, C. D. (1998, Oct.). Commonsense statistics. *Ergonomics & Design,* 18–22.

Wickens, C. D. (1999). Frame of reference for navigation. In D. Gopher & A. Koriat (eds.), *Attention and performance,* vol. 16. Orlando, FL: Academic Press.

Wickens, C. D. (2000). The tradeoff of design for routine and unexpected performance: Implications of situation awareness. In D. J. Garland & M. R. Endsley (eds.), *Situation awareness analysis and measurement.* Mahwah, NJ: Erlbaum.

Wickens, C. D. (2000a). The when and how of using 2-D and 3-D displays for operational tasks. *Proceedings of the IEA2000/HFES2000 Congress* (pp. 3–403/3–406). Santa Monica, CA: HFES.

Wickens, C. D. (2000b). Human factors in vector map design: The importance of task-display dependence. *Journal of Navigation, 53*(1), 54–67.

Wickens, C. D. (2002a). Multiple resources and performance prediction. *Theoretical Issues in Ergonomic Science, 3*(2), 159–177.

Wickens, C. D. (2002b). Situation awareness and workload in aviation. *Current Directions in Psychological Science, 11*(4), 128–133.

Wickens, C. D. (2002c). Aviation psychology. In L. Backman & C. von Hofsten (eds.), *Psychology at the turn of the millennium, vol. 1: Cognitive biological, and health perspectives.* East Sussex, Great Britain: Psychology Press.

Wickens, C. D. (2003a). Aviation displays. In P. S. Tsang & M. A. Vidulich (eds.), *Principles and practice of aviation psychology* (pp. 147–199). Mahwah, NJ: Erlbaum.

Wickens, C. D. (2003b). Pilot actions and tasks: Selection, execution, and control. In P. Tsang & M. Vidulich (eds.), *Principles and practice of aviation psychology* (pp. 239–263). Mahwah, NJ: Erlbaum.

Wickens, C. D., & Baker, P. (1995). Cognitive issues in virtual reality. In W. Barfield & T. A. Furness III (eds.), *Virtual environments and advanced interface design* (pp. 515–541). New York: Oxford University Press.

Wickens, C. D., & Carswell, C. M. (1995). The proximity compatibility principle: Its psychological foundation and its relevance to display design. *Human Factors, 37*(3), 473–494.

Wickens, C. D., & Carswell, C. M. (1997). Information processing. In G. Salvendy (ed.), *Handbook of human factors and ergonomics* (2nd ed.). New York: Wiley.

Wickens, C. D., Goh, J., Helleberg, J., Horrey, W., & Talleur, D. A. (2003, in press). Attentional models of multi-task pilot performance using advanced display technology. *Human Factors.*

Wickens, C. D., & Hollands, J. (2000). *Engineering psychology and human performance* (3rd ed.). Upper Saddle River, NJ: Prentice Hall.

Wickens, C. D., & Kessel, C. (1980). The processing resource demands of failure detection in dynamic systems. *Journal of Experimental Psychology: Human Perception & Performance, 6,* 564–577.

Wickens, C. D., & Liu, Y. (1988). Codes and modalities in multiple resources: A success and a qualification. *Human Factors, 30,* 599–616.

Wickens, C. D., & Long, J. (1995). Object- vs. space-based models of visual attention: Implications for the design of head-up displays. *Journal of Experimental Psychology: Applied, 1*(3), 179–194.

Wickens, C. D., & Prevett, T. (1995). Exploring the dimensions of egocentricity in aircraft navigation displays. *Journal of Experimental Psychology, Applied, 1*(2), 110–135.

Wickens, C. D., & Seidler, K. S. (1995). Information access, representation and utilization. In R. Nickerson (ed.), *Emerging needs and opportunities for human factors research.* Washington, DC: National Academy of Sciences.

Wickens, C. D., & Seidler, K. S. (1997). Information access in a dual task context. *Journal of Experimental Psychology, Applied, 3,* 1–20.

Wickens, C. D., & Seppelt, B. (2002). *Interference with driving or in-vehicle task information: The effects of auditory versus visual delivery* (AHFD-02–18/GM-02–03).). Savoy: University of Illinois, Aviation Human Factors Division.

Wickens, C. D., & Xu, X. (2003). How does automation reliability influence workload? Proceedings, 1st Annual Robotics Consortium. U. S. Army Research Laboratory Collaborative Technology Alliance Program.

Wickens, C. D., Gempler, K., & Morphew, M. E. (2000). Workload and reliability of predictor displays in aircraft traffic avoidance. *Transportation Human Factors Journal, 2*(2), 99–126.

Wickens, C. D., Helleberg, J., & Xu, X. (2002). Pilot maneuver choice and workload in free flight. *Human Factors, 44*(2), 171–188.

Wickens, C. D., & Hollands, J. (2000). *Engineering psychology and human performance* (3rd ed.). Upper Saddle River, NJ: Prentice Hall.

Wickens, C. D., Liang, C-C, Prevett, T., & Olmos, O. (1996). Electronic maps for terminal area navigation: Effects of frame of reference on dimensionality. *International Journal of Aviation Psychology, 6*(3), 241–271.

Wickens, C. D., Mavor, A. S., Parasuraman, R., & McGee, J. P. (eds.). (1998). *The future of air traffic control: human operators and automation.* Washington, DC: National Academy Press.

Wickens, C. D., Mavor, A., & McGee, J. (eds.) (1997). *Flight to the future: Human factors in air traffic control.* Washington, DC: National Academy of Sciences.

Wickens, C. D., Merwin, D. H., & Lin, E. (1994). Implications of graphics enhancements for the visualization of scientific data: Dimensional integrality, stereopsis, motion, and mesh. *Human Factors, 36*(1), 44–61.

Wickens, C. D., Sandry, D., & Vidulich, M. (1983). Compatibility and resource competition between modalities of input, central processing, and output: Testing a model of complex task performance. *Human Factors, 25,* 227–248.

Wickens, C. D., Stokes, A. F., Barnett, B., & Hyman, F. (1991). The effects of stress on pilot judgment in a MIDIS simulator. In O. Svenson & J. Maule (eds.), *Time pressure and stress in human judgment and decision making.* Cambridge, UK: Cambridge University Press.

Wickens, C. D., Thomas, L. C., & Young, R. (2000). Frames of reference for display of battlefield terrain and enemy information: Task-display dependencies and viewpoint interaction use. *Human Factors, 42*(4), 660–675.

Wickens, C. D., Todd, S., & Seidler, K. S. (1989). *Three-dimensional displays: Perception, implementation, and applications* (CSERIAC SOAR 89–001). Crew System Ergonomics Information Analysis Center, Wright-Patterson AFB, OH.

Wickens, C. D., Ververs, P., & Fadden, S. (2003). Head-up displays. In D. Harris (ed.), *Human factors for commercial flight deck.* Ashgate.

Wickens, C. D., Vidulich, M., & Sandry-Garza, D. (1984). Principles of S-C-R compatibility with spatial and verbal tasks: The role of display-control location and voice-interactive display-control interfacing. *Human Factors, 26,* 533–543.

Wickens, C. D., Vincow, M. A., Schopper, A. W., & Lincoln, S. E. (1997). *Computational models of human performance in the design and layout of controls and displays.* Wright-Patterson AFB, OH: CSERIAC.

Wickens, C. D., Zenyuh, J., Culp, V., & Marshak, W. (1985). Voice and manual control in dual task situations. *Proceedings of the 29th Annual Meeting of the Human Factors Society.* Santa Monica, CA: HFS.

Wickens, T. D. (2002). *Elementary Signal Detection Theory.* New York, Oxford.

Wiegmann, D. A., & Shappell, S. A (2001). Human Error analysis of commercial aviation accidents. *Aviation, Space, & Environmental Medicine. 72*(11), 1006–1016.

Wiegmann, D. A., and Shappell, S.A. (2003). *A human error approach to aviation accident analysis.* Burlington Vt.: Ashgate.

Wiener, E. L. (1977). Controlled flight into terrain accidents: System-induced errors. *Human Factors, 19,* 171–181.

Wiener, E. L. (1988). Cockpit automation. In E. L. Wiener & D. C. Nagel (eds.), *Human factors in aviation* (pp. 433–461). San Diego: Academic Press.

Wiener, E. L., & Curry, R. E. (1980). Flight deck automation: Promises and problems. *Ergonomics, 23*(10), 995–1011.

Wiener, E. L., Kanki, B. G., & Helmreich, R. L. (eds.) (1993). *Cockpit resource management.* San Diego: Academic Press.

Wierwille, W. W. (1995). Development of an initial model relating driver in-vehicle visual demands to accident rate, *Proceedings of the 3rd Annual Mid-Atlantic Human Factors Conference* (pp. 1–7). Blacksburg: Virginia Polytechnic and State University.

Wierwille, W. W., & Casali, J. G. (1983). A validated rating scale for global mental workload measurement applications. *Proceedings of the 27th Annual Meeting of the Human Factors Society* (pp. 129–133). Santa Monica, CA: HFS.

Wierwille, W. W., & Tijerina, L. (1996). An analysis of driving accident narratives as a means of determining problems caused by in-vehicle visual allocation and visual workload. In A. G. Gale, I. D. Brown, C. M. Haslegrave, & S. P. Taylor (eds.), *Vision in vehicles* (vol. 79–86). Amsterdam, Netherlands: Elsevier.

Wightman, D. C., & Lintern, G. (1985). Part-task training for tracking and manual control. *Human Factors, 27*(3), 267–283.

Wiklund, M. E. (ed.) (1994). Usability in practice: How companies develop user-friendly products. Boston: AP Professional.

Wikman, A. S., Nieminen, T., & Summala, H. (1998). Driving experience and time-sharing during in-car tasks on roads of different width. *Ergonomics, 41*(3), 358–372.

Wilde, G. J. S. (1988). Risk homeostasis theory and traffic accidents: Propositions, deductions and discussion of dissension in recent reactions. *Ergonomics, 31*(4), 441–468.

Wilkinson, R. T. (1992). How fast should night shift rotate? *Ergonomics, 35,* 1425–1446.

Williams, A. F. (2003). Teenage drivers: Patterns of risk. *Journal of Safety Research, 34*(1), 5–15.

Williams, A. F., & O'Neill, B. (1974). On-the-road driving records of licensed race drivers. *Accident Analysis & Prevention, 6,* 263–270.

Williams, H. L., Gieseking, C. F., & Lubin, A. (1966). Some effects of sleep loss on memory. *Perceptual Motor Skills, 23,* 1287–1293.

Williges, R. C. (1995). Review of experimental design. In J. Weimer (ed.), *Research techniques in human engineering.* Englewood Cliffs, NJ: Prentice Hall.

Williges, R. C., Johnston, W. A., & Briggs, G. E. (1966). Role of verbal communication in teamwork. *Journal of Applied Psychology, 50,* 473–478.

Williges, R. C., Williges, B. H., & Elkerton, J. (1987). Software interface design. In G. Salvendy (ed.), *Handbook of human factors* (pp. 1416–1449). New York: Wiley.

Williges, R., (1992). (Ed.) *The education and training of human factors specialists.* Washington, DC: National Academy Press.

Wilson, J. R., & Corlett, E. N. (1991). *Evaluation of human work.* London: Taylor & Francis.

Wilson, J. R., & Rutherford, A. (1989). Mental models: Theory and application in human factors. *Human Factors, 31*(6), 617–634.

Wine, J. (1971). Test anxiety and direction of attention. *Psychological Bulletin, 76,* 92–104.

Winkel, T., & Jorgensen, K. (1986). Evaluation of foot swelling and lower-limb temperatures in relation to leg activity during long-term seated office work. *Ergonomics, 29*(2), 313–328.

Wixon, D., Holtzblatt, K., & Knox, S. (1990, April). Contextual design: An emergent view of system design. *CHI '90 Proceedings,* 329–336.

Wogalter, M. S., Desaulniers, D. R., & Brelsford, T. W. (1987). Consumer products: How are the hazards perceived. *Proceedings of the 31st Annual Meeting of the Human Factors Society* (pp. 615–619). Santa Monica, CA: HFS.

Wogalter, M S., Kalsher, M. J., & Racicot, B. M. (1993). Behavioral compliance with warnings: Effects of voice, context, and location. *Safety Science, 16,* 113–120.

Wogalter, M. S., Tarrard, S. W., & Simpson, S. N. (1992). Effects of warning signal words on consumer-product hazard perceptions. *Proceedings of the 36th Annual Meeting of the Human Factors Society* (pp. 935–939). Santa Monica, CA: HFS.

Wogalter, M., Allison, S., & McKenna, N. (1989). Effects of cost and social influence on warning compliance. *Human Factors, 31*(2), 133–140.

Wohl, J. (1983). Cognitive capability versus system complexity in electronic maintenance. *IEEE Transactions on Systems, Man, & Cybernetics, 13,* 624–626.

Wolff, J. S., & Wogalter, M. S. (1998). Comprehension of pictorial symbols: Effects of context and test method. *Human Factors, 40,* 173–186.

Wolff, T. (1892). Das Gesetz der Transformation der Knochen. Berlin: Hirschwald.

Wood, J. M., & Troutbeck, R. (1994). Effect of visual impairment on driving. *Human Factors, 36*(3), 476–487.

Wood, N. L., & Cowan, N. (1995). The cocktail party phenomenon revisited: attention and memory in the classic selective listening procedure of Cherry (1953).

Journal of Experimental Psychology: Learning, Memory, & Cognition, 21, 255–260.

Woods, D. D. (1984). Visual momentum: A concept to improve the cognitive coupling of a person and computer. *International journal of Man-Machine Studies, 21,* 229–244.

Woods, D. D. (1995). The alarm problem and directed attention in dynamic fault management. *Ergonomics, 38*(11), 2371–2394.

Woods, D., & Cook, R. (1999). Perspectives on human error: Hindsight biases and local rationality. In F. Durso (Ed.) Handbook of applied cognition. NY: John Wiley.

Woods, D., Patterson, E., Corban, J., & Watts, J. (1996). Bridging the gap between user-centered intentions and actual design practice, *Proceedings of the Human Factors and Ergonomics Society 40th Annual Meeting* (Vol. 2, pp. 967–971). Santa Monica, CA: Human Factors and Ergonomics Society.

Woods, D. D., & Roth, E. (1988). Aiding human performance: II. From cognitive analysis to support systems. *Le Travail Humain, 51,* 139–172.

Woods, D. D., Roth, E. M., Stubler, W. F., & Mumaw, R. J. (1990). Navigating through large display networks in dynamic control applications. *Proceedings of the 34th Annual Meeting of the Human Factors Society* (pp. 396–399). Santa Monica, CA: HFS.

Woods, D., Wise, J., & Hanes, L. (1981). An evaluation of nuclear power plant safety parameter display systems. *Proceedings of the 25th Annual Meeting of the Human Factors Society* (pp. 110–114). Santa Monica, CA: HFS.

Wozny, L. A. (1989). The application of metaphor, analogy, and conceptual models in computer systems. *Interacting with Computers, 1*(3), 273–283.

Wright, D. B., & Davies, G. M. (1999). Eyewitness testimony. In F. T. Durso (ed.), *Handbook of applied cognition* (pp. 789–818). Chichester, England: Wiley, Ltd.

Wright, D. B., & McDaid, A. T. (1996). Comparing system and estimator variables using data from real line-ups. *Applied Cognitive Psychology, 10,* 75–84.

Wright, P. (1974). The harassed decision maker: Time pressures, distractions, and the use of evidence. *Journal of Applied Psychology, 59,* 555–561.

Wright, P. (1977). Presenting technical information: a survey of research findings. *Instructional Science, 6,* 93–134.

Wright, P., Lickorish, A., & Milroy, R. (2000). Route choices, anticipated forgetting, and interface design for on-line reference documents. *Journal of Experimental Psychology: Applied, 6*(2), 158–167.

Wroblewski, L. & Rantanen, E. M. (2001). Design guidelines for web-based applications. *Proceedings of the 45th Annual Meeting of the Human Factors and Ergonomics Society* (pp. 1191–1195) Santa Monica, CA: HFES.

Wyszecki, C. (1986). Color appearance. In K. Boff, L. Kaufman, & J. Thomas (eds.), *Handbook of perception and human performance,* vol. I. New York: Wiley.

Xiao, Y., Hunter, W. A., MacKenzie, C. F., & Jefferies, N. J. (1996). Task complexity in emergency medical care and its implications for team coordination. *Human Factors, 38*(4), 636–645.

Xiao, Y., Mackenzie, C. F., & the LOTAS Group (1995). Decision making in dynamic environments: Fixation errors and their causes. *Proceedings of the 39th Annual*

Meeting of the Human Factors & Ergonomics Society (pp. 469–473). Santa Monica, CA: HFES.

Yantis, S. (1993). Stimulus-driven attentional capture. *Current Directions in Psychological Science, 2,* 156–161.

Yeh, M., & Wickens, C. D. (2001a). Attentional filtering in the design of electronic map displays: A comparison of color-coding, intensity coding, and decluttering techniques. *Human Factors, 43*(4), 543–562.

Yeh, M., & Wickens, C. D. (2001b). Display signaling in augmented reality: The effects of cue reliability and image realism on attention allocation and trust calibration. *Human Factors, 43*(3), 355–365.

Yeh, M., Merlo, J., Wickens, C. D., & Brandenburg, D. L. (in press, 2003). Head-up vs. head-down: The costs of imprecision, unreliability, and visual clutter on cue effectiveness for display signaling. *Human Factors.*

Yeh, M., Wickens, C. D., & Seagull, F. J. (1999). Target cueing in visual search: The effects of conformality and display location on the allocation of visual attention. *Human Factors, 41*(4), 524–542.

Yeh, Y. Y., & Wickens, C. D. (1988). Dissociation of performance and subjective measures of workload. *Human Factors, 30,* 111–120.

Yerkes, R. M., & Dodson, J. D. (1908). The relation of strength of stimulus to rapidity of habit formation. *Journal of Comparative Neurological Psychology, 18,* 459–482.

Yetton, P., & Bottger, P. (1983). The relationships among group size, member ability social decision schemes, and performance. *Organizational Behavior & Human Performance, 32,* 145–159.

Yokohori, E. (1972). *Anthropometry of JASDF personnel and its implications for human engineering.* Tokyo: Aeromedical Laboratory, Japanese Air Self Defense Force, Tachikawa Air Force Base.

Yoon, W. C., & Hammer, J. M. (1988). Deep-reasoning fault diagnosis: An aid and a model. *IEEE Transactions on Systems, Man, & Cybernetics, 18*(4), 659–675.

Yost, W. A. (1994). *Fundamentals of hearing* (3rd ed.). San Diego: Academic Press.

Young, L. R. (2003) Spatial Orientation. In P. S. Tsang & M. A. Vidulich (Eds). *Principles and Practice of Aviation Psychology.* Mahwah, N.J.: Lawrence Erlbaum Associates.

Young, M., & Stanton, N. (1997). Automotive automation: Effects, problems and implications for driver mental workload. In D. Harris (ed.), *Engineering psychology and cognitive ergonomics.* Aldershot, UK: Ashgate Publishing Ltd.

Young, S. L., & Laughery, K. R. (1994). Components of perceived risk: A reconciliation of previous findings. *Proceedings of the 38th Annual Meeting of the Human Factors & Ergonomics Society* (pp. 888–892). Santa Monica, CA: HFES.

Young, S. L., Wogalter, M. S., & Brelsford, J. W. (1992). Relative contribution of likelihood and severity of injury to risk perceptions. *Proceedings of the 36th Annual Meeting of the Human Factors & Ergonomics Society* (pp. 1014–1018). Santa Monica, CA: HFES.

Zachary, W. W. (1988). Decision support systems: Designing to extend the cognitive limits. In M. Helander (ed.), *Handbook of human-computer interaction* (pp. 997–1030). Amsterdam: North-Holland.

Zakay, D., & Wooler, S. (1984). Time pressure, training, and decision effectiveness. *Ergonomics, 27,* 273–284.

Zheng, J., Bos, N., Olson, J. S., & Olson, G. M. (2001). Trust without touch: Jumpstart trust with social chat. *CHI Extended Abstracts* (pp. 293–294). New York: Association for Computing Machinery.

Zsambok, C. E., & Klein, G. (1997). *Naturalistic decision making.* Mahwah, NJ: Erlbaum.

Zuboff, S. (1988). In the age of smart machines: The future of work technology and power. New York: Basic Books.

Author Index

Subject Index